Hydraulic Engineering

Hydraulic Engineering

Edited by
Leslie York

Willford Press
www.willfordpress.com

Published by Willford Press,
118-35 Queens Blvd., Suite 400,
Forest Hills, NY 11375, USA

ISBN: 978-1-68285-419-8

Cataloging-in-Publication Data

Hydraulic engineering / edited by Leslie York.
p. cm.
Includes bibliographical references and index.
ISBN 978-1-68285-419-8
1. Hydraulic engineering. 2. Hydraulics. I. York, Leslie.
TC145 .H93 2018
627--dc23

For information on all Willford Press publications
visit our website at www.willfordpress.com

Willford Press

Printed in China.

Contents

Preface

This book aims to highlight the current researches and provides a platform to further the scope of innovations in this area. This book is a product of the combined efforts of many researchers and scientists, after going through thorough studies and analysis from different parts of the world. The objective of this book is to provide the readers with the latest information of the field.

Hydraulic engineering is an important sub-discipline of civil engineering and it studies the flow of fluids, mainly of water and sewage. The subject helps in developing designs for various hydraulic structures such as dams, canals, etc. This book contains some path-breaking studies in the field of hydraulic engineering. For all readers who are interested in hydraulic engineering, the case studies included in this book will serve as an excellent guide to develop a comprehensive understanding. Through this book, we attempt to further enlighten the readers about the new concepts in this field.

I would like to express my sincere thanks to the authors for their dedicated efforts in the completion of this book. I acknowledge the efforts of the publisher for providing constant support. Lastly, I would like to thank my family for their support in all academic endeavors.

Editor

1

Drought-Tolerance of Wheat Improved by Rhizosphere Bacteria from Harsh Environments: Enhanced Biomass Production and Reduced Emissions of Stress Volatiles

Salme Timmusk[1]*, Islam A. Abd El-Daim[1], Lucian Copolovici[2], Triin Tanilas[2], Astrid Kännaste[2], Lawrence Behers[1], Eviatar Nevo[3], Gulaim Seisenbaeva[4], Elna Stenström[1], Ülo Niinemets[2,5]

1 Dept. of Forest Mycology and Plant Pathology, Uppsala BioCenter, SLU, Uppsala, Sweden, **2** Dept. of Plant Physiology, Estonian University of Life Sciences, Tartu, Estonia, **3** Institute of Evolution, The University of Haifa, Haifa, Israel, **4** Dept. of Chemistry and Biotechnology, Uppsala BioCenter, SLU, Uppsala, Sweden, **5** Estonian Academy of Sciences, Tallinn, Estonia

Abstract

Water is the key resource limiting world agricultural production. Although an impressive number of research reports have been published on plant drought tolerance enhancement via genetic modifications during the last few years, progress has been slower than expected. We suggest a feasible alternative strategy by application of rhizospheric bacteria coevolved with plant roots in harsh environments over millions of years, and harboring adaptive traits improving plant fitness under biotic and abiotic stresses. We show the effect of bacterial priming on wheat drought stress tolerance enhancement, resulting in up to 78% greater plant biomass and five-fold higher survivorship under severe drought. We monitored emissions of seven stress-related volatiles from bacterially-primed drought-stressed wheat seedlings, and demonstrated that three of these volatiles are likely promising candidates for a rapid non-invasive technique to assess crop drought stress and its mitigation in early phases of stress development. We conclude that gauging stress by elicited volatiles provides an effectual platform for rapid screening of potent bacterial strains and that priming with isolates of rhizospheric bacteria from harsh environments is a promising, novel way to improve plant water use efficiency. These new advancements importantly contribute towards solving food security issues in changing climates.

Editor: Girdhar K. Pandey, University of Delhi South Campus, India

Funding: Financial support for the study was provided by the European Commission through European Regional Fund (the Centre of Excellence in Environmental Adaptation), and the European Research Council (advanced grant 322603, SIP-VOL+), European Union FP7:247669, the Swedish Research Council for Environment, Agricultural Sciences and Spatial Planning 2009-243, Carl Tryggers Stiftelse for vetenskaplig forskning CTS09:385, Stiftelsen Oscar och Lili Lamms Minne DO2009-0052, and by the Estonian Ministry of Science and Education (institutional grant IUT-8-3). The funders had no role in study design, data collection and analysis, decision to publish, or preparation of the manuscript.

Competing Interests: The authors have declared that no competing interests exist.

* E-mail: salme.timmusk@slu.se

Introduction

The world population is predicted to increase beyond 8 billion by 2030 implying major challenges for agricultural sector to secure food availability [1]. A key challenge for plant growth is global water shortage, limiting crop yields already today in more than 70% of arable lands, and the drought limitations further gain in importance in the near future as agricultural activities expand to less fertile areas to satisfy growing demands for food [2,3]. Accordingly, understanding and improving plant survival and growth under restricted water availability is of central significance in contemporary plant science.

A variety of strategies has been used to improve the drought tolerance of crops, including traditional selection methods and genetic engineering [4]. Drought stress tolerance is a complex trait, and although a large number of genes involved in plant drought stress responses has been identified, large gaps still remain [4–6]. Finding of suitable phenotypes is further complicated by exposure of plants to multiple environmental stressors in the field either simultaneously or sequentially. Although a certain overlap may occur in plant physiological responses to different stressors, plant responses to each stress alone or in combination with other stresses can be different [7]. Furthermore, plant response to drought stress is also developmentally regulated [8–10].

Given the strongly varied environments with complex stress environment, ontogenetic variations, large number of different crops with huge variety of cultivars, and the plethora of genes expression of which need to be altered or novel genes engineered into plants, it is currently unclear whether the engineering technology will develop fast enough to cope with rapidly increasing food demands in near future [11]. As a relatively simple and low-cost alternative strategy, the use of free-living plant growth-promoting bacteria (PGPB) has been highlighted as a promising broad-spectrum means to improve plant growth [11–14]. The PGPB ability to increase plant growth and provide protection against various pathogens has been frequently reported and applied since 50ies in agricultural systems. The very first report on PGPB-induced drought stress tolerance was published by Timmusk and Wagner only in 1999 [14,15]. From the huge variety of potential microorganisms, gram-positive rhizosphere bacteria are most likely to be commercially applied in wide range of areas of limited water resources due to the ease of their handling and endospore-forming ability that facilitate efficient colonization under environmental stress conditions [11,14,16–19].

Although the exact mechanisms of plant drought stress tolerance enhancement by rhizosphere bacteria remain largely speculative, possible explanations include: (1) production of hormones like abscisic acid, gibberellic acid, cytokinins, and auxin; (2) production of essential enzymes, 1-aminocyclopropane-1-carboxylate (ACC) deaminase to reduce the level of ethylene in the root of developing plants; (3) induced systemic resistance by bacterially-produced compounds; (4) formation of bacterial biofilm i.e. extracellular matrix [12,13,20,21] [17]. In particular, an extracellular matrix formed by bacterial biofilm can provide almost infinite range of macromolecules beneficial for plant development and growth. Biofilms contain sugars and oligo- and polysaccharides that can play various roles in bacteria-plant interactions, e.g., in improving water availability in root medium. The water retention capacity of some polysaccharides can exceed several-fold their mass [17]. In fact, it has been demonstrated that even a small polysaccharide alginate content in the biofilm facilitates maintenance of hydrated microenvironment [22].

Here we studied the importance of habitat of rhizospheric bacteria in plant drought stress tolerance enhancement. The bacteria included in this study were isolated from harsh environments such as South Facing Slope (SFS) in 'Evolution Canyon' (EC), Israel, Mt. Lemmon, Arizona, USA and Tina plain, Giza, Egypt as described by us earlier [23]. We showed that the bacterial capacities to solubilize phosphorus, form biofilms, tolerate high salt concentrations and the bacterial activity of ACC deaminase were significantly higher in stressed environments [23]. The bacteria have coevolved with their hosts over long periods of time, and thus, they are expected to control to a large extent plant adaptation to their belowground environment. Furthermore, bacteria intimately associated with host plant roots have likely contributed to plant adaptation to harsh environmental conditions at evolutionary time-scales as well [23].

Higher plants respond to water shortage by multiple physiological and morphological modifications [19,24–26]. Characteristic drought responses include stomatal closure, osmotic adjustment by increased content of cellular osmotica, changes in foliage structure and total whole plant leaf area and root system characteristics, collectively improving water conservation, water use efficiency (rate of production per water use) and water extraction from drying soil [24–28]. Despite adaptive modifications, reactive oxygen species (ROS) and accumulation of free radicals associated with damage to membranes and buildup of lipid peroxides can occur in severely water-stressed plants [29–31]. While changes in ROS formation after reaching the threshold water stress level may initially activate host defense mechanisms, their continued production under drought stress can be the ultimate cause of damage in water-stressed plants, indicating that the balance between formation and detoxification of activated oxygen species is critical for cell survival. Induced changes in different ROS-scavenging enzymes and enhanced photosynthetic efficiency have been reported in PGPB-inoculated plants, helping the plants to cope with drought stress [32,33]. Some of key ROS-scavenging enzymes include glutathione-ascorbate cycle enzymes glutathione reductase (GR) (EC 1.8.1.7) and monodehydroascorbate reductase (MDHAR) (EC 1.6.5.4) [34]. Superoxide dismutase (SOD) (EC 1.15.1.1) and catalase (CAT) (EC 1.11.1.6) act as main enzymatic scavengers for superoxide (O_2^-) and hydrogen peroxide (H_2O_2) [33].

In plant responses to a multitude of stresses, induction of volatiles plays a key role [35,36]. The stress-induced volatiles serve as signals for development of priming and systemic responses within the same and in neighboring plants [37–39], and the emission rates of volatiles are generally strongly associated with the severity of stress [40–44]. Thus, monitoring stress-elicited emissions can provide key insight into immediate stress responses, priming and acclimation. In fact, plant volatiles constitute as yet unexplored, potentially hugely valuable resource for studies on plant stress responses and for rapid phenotyping of plant genotypes according to stress resistance.

The primary goal of this work was to systematically compare bacterial isolates from stressed and mild environments in their capacity to enhance drought stress tolerance of wheat (*Triticum aestivum*). We suggest that bacteria coevolved with plant roots under stress over millennia have a superior potential to enhance drought tolerance of wheat. We further propose a rapid non-invasive method for assessing differences in plant stress tolerance using plant volatiles. To our knowledge, importance of PGPB habitat and involvement of plant volatiles in bacterially-driven drought tolerance enhancement have not been reported before.

Experimental Procedures

Bacteria and plant treatment

The bacterial strains from North-Israeli 'Evolution Canyon' both from more stressed SFS and less stressed NFS sites (SFS and NFS strains) were isolated as described by us earlier [23]Firmicute *Bacillus thuringiensis* AZP2 was isolated from ponderosa pine (*Pinus ponderosa*) roots grown on gneiss rock at Mt Lemmon, AZ, USA (32.38568° N, 110.69486° W elevation 2150 m) as described by Timmusk et al. [23]. Firmicute *Paenibacillus polymyxa* B was isolated from salty rice (*Oryza sativa*) rhizosphere at Tina plain, Giza, Egypt (31.044° N, 32.6661° E, elevation 13 m) as described earlier [23]. Briefly, the plant roots were carefully shaken and washed in sterile distilled water to remove all loosely attached rock powder and to collect bacteria intimately adhered to plant roots. Plants were placed in sterile plastic bags, transferred to the laboratory, and then stored at +4°C until they were processed in the next day. Plant rhizosphere material (1 g) was homogenized by FastPrep Instrument (BIO 101 Systems), and hence, the rhizosphere macerate contains inner root bacteria. Plant rhizosphere material was suspended in sterile PBS medium (137 mM NaCl, 2.7 mM KCl, 10 mM Na_2HPO_4, 2 mM KH_2PO_4, pH = 7.4).

The content of endospore-forming bacteria was determined after heat treatment of the soil or plant material suspension at 80°C for 30 min. Tryptic soy agar (TSA) plates were inoculated with 100 μL of heat-treated bacterial suspensions, corresponding to 10^{-3}–10^{-5} g soil or plant rhizosphere material per plate. All agar media (pH = 7) contained 15 g agar and 50 mg cycloheximide, to reduce fungal growth. The inoculated Petri dishes were incubated for several weeks at room temperature (~21°C), and at 30°C, 37°C, or 40°C in boxes together with a beaker of water to keep the agar medium moist.

Screening for bacterial metabolic properties and plant drought tolerance enhancement. Bacterial metabolic properties were screened as described by us earlier [23]. Briefly, a small amount of material from 60 bacterial colonies was picked from the TSA plates and transferred to a tryptic soy broth (TSB) medium for a biofilm production assay. The remainder of each colony was streaked on to Dworkin and Foster (DF) plates for the 1-aminocyclopropane-1-carboxylate (ACC) deaminase assay [45]; National Botanical Research Institute's phosphate growth medium (NBRIP) [46] for the P solubilization assay; and high salt medium for halophilicity assay. The TSA plates were used to ensure that some of each colony remained. The endospore-forming ACC deaminase-containing, P-solubilizing and salt tolerant bacteria were retained and studied for their capacity to enhance wheat (*Triticum aestivum*) drought stress tolerance (survival). The colonies that were able to enhance wheat seedling survival were 16S rDNA sequenced from the TSA plate.

Three wheat (*Triticum aestivum*) cultivars (spring wheat cv. Sids1, drought-sensitive winter wheat cv. Stava and drought-tolerant winter wheat cv. Olivin) were used in the present study. Seeds were surface sterilized using 5% chlorine solution. Bacteria were grown in tryptone soy broth (TSB) medium at 28°C overnight. Culture density was determined by colony forming unit analysis (CFU). Inoculation was performed by soaking grains in solutions containing 10^7 bacteria ml^{-1} for 4 hours at 28°C. For the control treatment, another set of grains was soaked in sterile TSB media. Seeds were sown in plastic pots filled with 450 g sand or sand mixed with 10% greenhouse soil. Both inoculated and non-inoculated treatments were replicated twelve times and each treatment had three plants per pot. The pots were incubated in controlled environment in a MLR-351H (Phanasonic, IL, USA) growth chamber with 24/16°C (day/night) temperature, and 16 h photoperiod at a quantum flux density of 250 μmol m^{-2} s^{-1}. The soil moisture was adjusted at 75% of soil water-holding capacity (12.5% of soil dry mass). Soil moisture content was kept constant during the first 10 days of seedling growth.

In 10 days after seed germination, drought stress was induced by stopping watering. Plants grown in sand were stressed for 10 days and plants grown in sand mixed with 10% greenhouse soil were stressed for 14 days. Soil volumetric water content was evaluated using 5TE soil moisture sensors (Decagon Devices, Inc, Pullman, WA, USA). We used 10-days-old wheat plants in our experimental settings as it is well-known that drought tolerance is developmentally regulated and increases with plant age [8–10]. plant drought genomics Accordingly, using young plants provides most informative insight into the potential of bacterial priming in increasing plant survival at the most critical stage of plant ontogeny.

AZP2 identification and quantification

Randomly selected plants were used to confirm that bacteriofilm on AZP2-inoculated plants consisted of AZP2 after 10 days of growth. The roots of selected plants were carefully shaken and washed in sterile distilled water to remove loosely attached soil and to collect bacteria intimately linked to the plant root. Thereafter, the roots were homogenized and the content of endospore-forming bacteria was determined after heat treatment of the soil or plant material suspension at 80°C for 30 min. Tryptic soy agar (TSA) plates were inoculated with 100 mL of these suspensions, corresponding to 10^{-3}–10^{-5} g soil or plant rhizosphere material per plate. DNA was isolated from 1-day-old cultures on agar plates. Single colonies were resuspended to obtain a bacterial density of about 10^5 colony forming units per ml suspension. A 0.3 ml aliquot of the bacterial culture was suspended in 4.7 ml of buffer (10 mM Tris-HCl, pH = 7.6, 50 mM KCl, 0.1% Tween 20). For lysing, the suspension was heated and immediately cooled on ice. The mixture was centrifuged at 6000 g for 5 min and the supernatant was used for PCR analysis. Aliquots of 10 mM of primers 1492R (5′-GGTTACCTTGTTACGACTT-3′) and 27F (5′-AGAGTTTGATCCTGGCTCAG-3′) and 1 ml of template were used. The reaction was performed in 10 ml. The reactions conditions were 95°C for 2 min followed by 30 cycles of denaturation at 95°C for 15 s, annealing at 55°C for 20 s, primer extension at 72°C for 1 min, followed by the final extension at 72°C for 5 min. For sequencing, the PCR products were purified with QIAquickTM Gel Extraction kit (QIAGEN, Hilden, Germany).

Plant growth analysis

Plant survival was calculated daily after stress application using 32 stressed plants that were randomly selected and divided into two groups with 16 plants each. After stress application, the plants were watered and allowed to recover for 4 days. The recovered plants were counted as survived plants. Eight days after application of drought stress, the survived and recovered plants were harvested, washed and blotted dry between filter paper. Plant roots were counted and their length was estimated with Root Reader3D Imaging and Analysis system [47]. Root-adhering soil was evaluated in twelve plants per treatment. Roots with adhering soil (RAS) were carefully separated from bulk sand and sand soil mix by shaking. Shoot, soil and root dry mass (RT) were recorded after drying the samples at 105°C to a constant mass, and RAS/RT ratio was calculated. Water use efficiency (WUE) was calculated as the ratio of total plant dry mass to total water use during the experiment.

Root hair length and density were evaluated using twelve plants. Plants were carefully separated from soil by shaking. After separation of loosely attached soil, plant roots were washed in distilled water and left to drain in Petri dishes containing 5 ml of water. The other set of plant was homogenised and used for AZP2 identification and quantification. The dried root system characteristics (root hair density and length) were evaluated using Zeiss LSM 710 microscope.

Protein extraction and antioxidant enzyme activity measurements

Leaf samples for enzyme activity determination were taken after 8 days from drought-treated and well-watered plants. Plant tissue was mixed with 10 ml extraction buffer as described by Knöerzer et al. [48]. The mixture was centrifuged at 14,000 rpm (Eppendorf, 5415C) for 10 min at 5°C, and the supernatant was used to determine protein content and activity of key antioxidant enzymes. Monodehydroascorbate reductase (MDAR) activity was determined following the decrease in light absorbance at 340 nm due to NADH oxidation as described by Hossain et al. [49]. Glutathione reductase (GR) activity was determined by increase in absorbance at 412 nm according to Smith et al. [50]. Superoxide dismutase (SOD) activity was determined by reduction in light absorbance at 490 nm using an Oxiselect SOD activity assay kit (Cell Biolabs, San Diego, CA, U.S.A.) according to manufacturer's instructions. Catalase (CAT) activity was measured by reduction in light absorbance at 520 nm, using an OxiselectTM CAT activity assay kit (Cell Biolabs). For CAT and SOD, enzyme activities were determined per gram of fresh mass (FM). Ultimately, the enzyme activities were normalized with respect to the activities in well-watered plants. The corresponding activities were 20 nmol (mg $FM)^{-1}$ min^{-1} for MDHAR, 25 nmol (mg $FM)^{-1}$ min^{-1} for GR, 0.45 μmol (g $FM)^{-1}$ min^{-1} for SOD and 53 μmol (mg $FM)^{-1}$ for CAT.

Scanning Electron Microscopy

Environmental scanning electron microscopy (ESEM) micrographs of the samples were obtained with a Hitachi TM-1000-μDex variable pressure scanning electron microscope. Samples were deposited on a carbon tape and coated by gold using Sputter Coater 108 auto (Cressington). Twelve root systems grown in sand soil under water stress were studied for AZP2 biofilm formation.

Foliage gas exchange measurements

Steady-state net assimilation and transpiration rates and stomatal conductance were recorded immediately after stress application (day 0) and in 2, 5, 8 and 10 days from start of stress application using a Walz GFS-3000 portable gas exchange system equipped with a LED-array/PAM-fluorimeter 3055-FL (H. Walz GmbH, Effeltrich, Germany) [51,52]. Leaf temperature was set at 25°C, incident quantum flux density at 1000 μmol m^{-2} s^{-1}, chamber CO_2 concentration at 390 μmol mol^{-1} and air humidity at 60%. Multiple leaves were inserted side by side to fill the leaf chamber area. After insertion of leaves, leaves were stabilized under measurement conditions until steady-state gas exchange rates were established, typically 20–30 min before recording the values. After completion of gas-exchange measurements, the leaf chamber fluorimeter was removed and a photograph of the chamber window was taken for enclosed leaf area estimation.

VOCs sampling and analysis

Volatiles were trapped by sampling 4 L of chamber air from the Walz GFS-3000 cuvette outlet onto a multibed stainless steel cartridge, and analyzed by a Shimadzu 2010Plus GC-MS (Shimadzu Corporation, Kyoto, Japan) as in our previous studies [40,51,52]. Ethylene emission rate was measured at days 0, 2, 5, 8 and 10 days after stress application using Picarro G1106 real-time ethylene analyzer (Picarro, Inc, CA, USA). The ethylene analyzer was linked to GFS-3000 gas-exchange system through a bypass loop.

Data confirmation and validation

Experiments were repeated three times to confirm reproducibility of plant phenotypes. Replicated data were analysed by three-way ANOVA (stress x strain x stress exposure time), and all treatment effects were considered significant at a conservative level of significance of $P \leq 0.01$.

Results

Screening of two sets of rhizosphere bacteria originating from harsh and moderately stressed environments for their drought tolerance enhancement ability

Twelve bacterial isolates from the rhizosphere of wild progenitors of cereals at South Facing Slope (SFS, more stressful site) and twelve isolates from North Facing Slope (NFS, less stressful site) in North-Israeli 'Evolution Canyon' isolated and characterized by us earlier were assessed for their capacity to enhance wheat (*Triticum aestivum*, cultivars Stava, Olivin, Sids 1) drought stress tolerance [23]. Briefly, twelve SFS strains were able to grow on ACC as a sole N source, in 2 M NaCl medium and had a high P solubilization index (3–4) and a high biofilm formation value (OD_{590}>1). Twelve NFS strains were not able to grow on ACC as a sole N source, neither in 2 M NaCl medium and had a low P solubilization index ≤1.5) and moderate biofilm formation potential ($OD_{590} \leq 0.5$) [23]. Two to four subpopulations of *Bacillus cereus, B. megaterium, B. pumilus* from SFS and NSF were used for comparison. Only one *Paenibacillus polymyxa* subpopulation was isolated from NSF and was screened along with its four subpopulations from SFS. On average (±SD 16±2% of *P. polymyxa* 15±3%, *B. cereus* 16±2% *B. megaterium* and 10±3%of *B. pumilus* treated plants survived the 10-day drought treatment followed by a four-day recovery period after re-watering. None of the wheat seedlings treated with the same bacterial species representative to NFS survived the drought stress. These results confirmed our hypothesis that harsh environments constitute the source for highly potent rhizosphere bacteria. Therefore, more strains from differently stressed environments were isolated. Two additional isolates from ponderosa pine (*Pinus ponderosa*) rhizosphere at Mt Lemmon AZ, USA, and rice (*Oryza sativa*) rhizosphere at Tina plain, Giza, Egypt were used in this study.

In comparison to other isolates, *Bacillus thuringiensis* AZP2 isolated from ponderosa pine ecosystem grown under particularly severe nutrient limitation and drought, heat and UV stress exhibited the best potential to enhance plant drought stress tolerance and was chosen for further studies (Table 1). Both the strains *Bacillus thuringiensis* AZP2 and *P. polymyxa* B were able to grow on ACC as the sole N source, 2 M NaCl medium and had high P solubilization index (3.1±0.2). Their biofilm formation value at OD_{590} was 1.0±0.2. On average (±SE), 43±3% of *B. thuringiensis* AZP2- and 23±2% of *P. polymyxa* B-treated wheat seedlings survived the 10-day drought stress. Two additional strains similarly enhanced wheat survival (Fig. 1A, Fig. S2), yet AZP2 priming potential was slightly higher. Hence *B. thuringiensis* AZP2 was the very best strain and this study focuses on AZP2 priming. In addition, *Paenibacillus polymyxa* B strain was used to evaluate the robustness of our method to assess the success of bacterial priming by monitoring alterations in volatile emission and photosynthetic activity.

Effect of AZP2 on wheat traits and phenotype

The AZP2-inoculated plants were exposed to severe drought stress for 10 days by stopping watering. Cessation of irrigation resulted in strong reductions in soil volumetric water content as evaluated by soil moisture sensors (Fig. S1). During the first two days of stress exposure, no visible signs of drought were recorded. However, after the third day, plant survivorship decreased significantly, and this trend was reinforced with increasing the stress period. Less than 20% of non-inoculated plants could survive for seven days of stress, and none could survive for eight days of drought (Fig. 1A).

In contrast, AZP2-inoculated plants exhibited a delayed initial response to drought stress. After six days of drought stress, 80% of *B. thuringiensis* AZP2-primed plants survived and more than 20% of *B. thuringiensis* AZP2-primed plants survived after nine days of stress (Fig. 1A). All AZP2-primed plants were severely wilted after 10-day water stress (Fig. 1A). However, after 4 days of recovery treatment (re-watered to soil field capacity, i.e. 12.5% of soil dry mass), 43% of primed seedlings were able to recover, while none of the non-primed seedlings recovered (Table 1). Root and shoot dry mass of both control and drought-stressed plants were significantly increased in the case of AZP2 priming (Fig. 1B). Taken together, the bacterial priming resulted in 100% seedling germination, 78% increase in plant dry mass and five times greater average plant survival grown under severe drought and nutritional stress.

Slightly improved soil nutrition further improved the beneficial effects of AZP2 priming. Growing wheat in sand mixed with 10% greenhouse soil and drought-stressed for 14 days resulted in major phenotypic differences (Fig. 2A–C). The non-primed plants were severely damaged after such acute water stress, but *B. thuringiensis* AZP2 priming strongly reduced the drought-induced damage, and increased plant survival and final dry mass (Fig. 2A–C). While about 43% of AZP2 plants survived in sand after 10 days of drought stress followed by 4 days recovery after re-watering, 89% survived in sand with greenhouse soil after 14 days of drought stress followed by 4 days recovery (Table 1 and Fig. 2A–C).

Detailed analysis of AZP2-primed wheat seedlings revealed that AZP2 drought tolerance improvement was always associated with longer and denser root hair (Table 2). AZP2 effect on root hair formation was observed under normal watering regime, and this trait was gradually more expressed with increasing the length of the drought stress treatment (Table 2). Changed root hairiness was the major morphological difference in comparison to

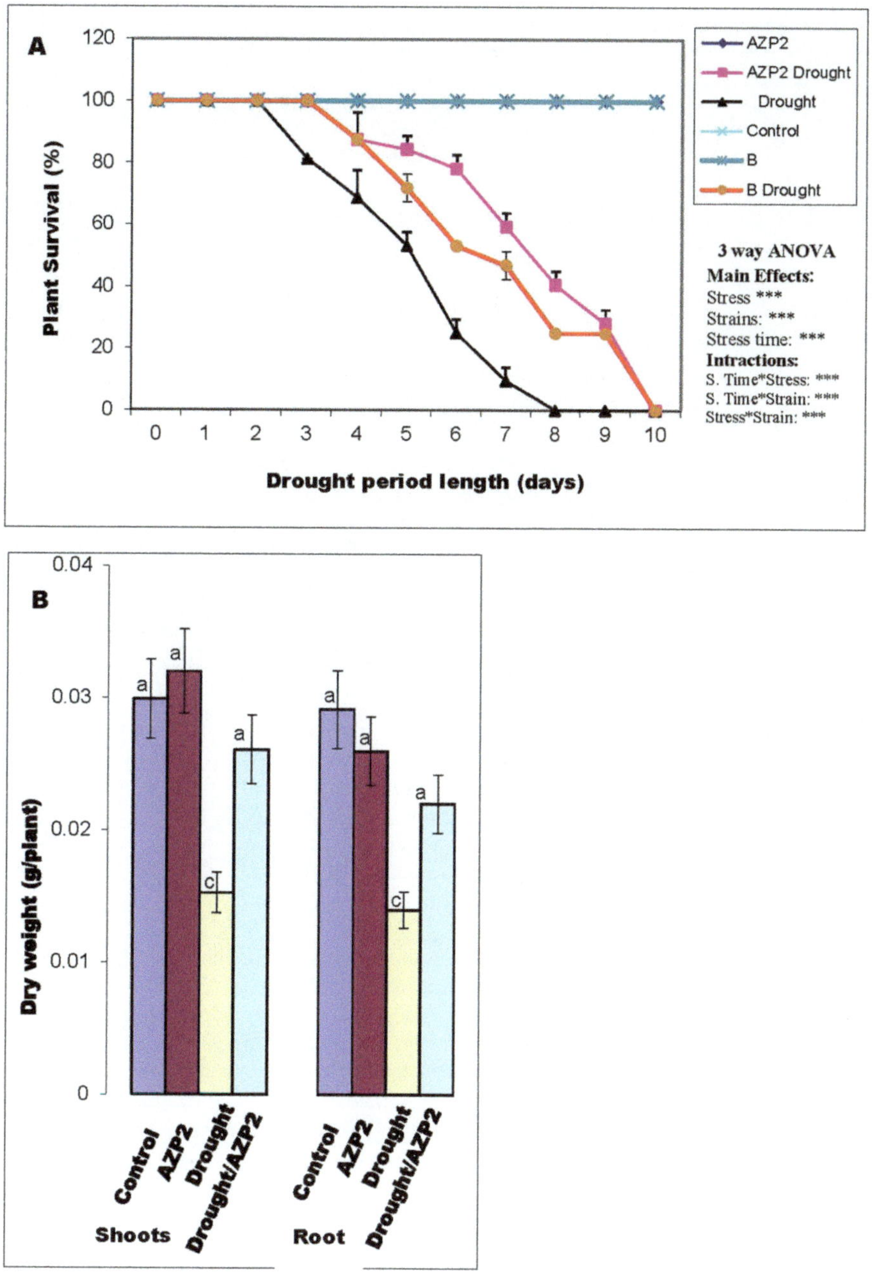

Figure 1. Enhancement of wheat (*Triticum aestivum*) drought tolerance by *Bacillus thuringiensis* AZP2 and *Paenibacillus polymyxa* B in sand soil. Panel A demonstrates the effect of AZP2 and B priming on seedlings survival after a severe 10-day drought stress episode. Panel B shows the effect of AZP2 priming on whole plant dry mass after 8 days growth without watering. The statistical analysis in (A) is based on a three-way ANOVA (stress, strains (i.e. AZP2 and B) and stress exposure time). ANOVA was conducted on two plant groups with 16 replicates in each group. *** indicate highly significant effects for the tested factor at $P \leq 0.01$. In B, eight independent experiments were performed, and treatments labelled with the same letter are not significantly different at $P \leq 0.01$.

non-inoculated plants observed during first five days of drought stress both in sand as well as in sand supplemented with 10% greenhouse soil. Furthermore, AZP2 reduced root length under watering, but significantly increased density and length of lateral roots under drought stress in sand soil (Table 2). Compared to non-primed controls, two to three times more lateral roots were observed in AZP2-primed plants (Table 2).

Bacterial biofilm formation on plant root surface was estimated as the amount of soil attached to roots. Two to three times more soil was attached to AZP2-primed roots under drought stress and up to two times more under normal watering (Table 2). Electron microscopic imaging of the AZP2-primed wheat seedlings grown under drought stress confirmed the bacterial biofilm formation on root hairs (Fig. 3). Overall, in AZP2-primed plants, water use efficiency increased by 63% in comparison to non-primed control plants.

Plant photosynthetic activity is improved by the drought tolerance enhancing bacteria

Foliage net assimilation rate and stomatal conductance were monitored through the drought stress treatment and in

Table 1. Effect of rhizosphere bacterial priming on wheat (*Triticum aestivum L. cv. Stava*) survival under drought stress.

Bacterial strain[1]	Origin	Average (±SD) plant survival (%)[2]
Control		0
Bacillus thuringiensis AZP2	Mt. Lemmon, AZ, USA	43±3
Paenibacillus polymyxa B	Tina Plain, Giza, Egypt	23±2
P. polymyxa A26,27, 58,64	South Facing Slope, 'Evolution Canyon'SFS,EC) Israel	16±2
B.cereus A4, 7, 8	SFS, EC	15±3
B. megaterium A2, 3, 5	SFS, EC	16±2
B. pumilus A1, 3	SFS, EC	10±3
P. polymyxa E1	North Facing Slope, 'Evolution Canyon', (NFS,EC) Israel	0
B.cereus E1, 3, 4, 5	NFS, EC	0
B. megaterium E2, 4, 5, 6	NFS, EC	0
B. pumilus E3, 5, 6	NFS, EC	0

[1]Patent pending.
[2]Plant survival in sand soil after 10-day drought stress followed by 4 days recovery after re-watering.

well-watered control treatment in both primed and non-primed plants. Under drought-stress, a steady decline in net assimilation rate was recorded in all stressed wheat plants (Fig. 4A). Net assimilation rate was almost totally inhibited in non-primed wheat plants in 8 days since withholding water. This result was qualitatively the same in plants inoculated with *P. polymyxa* E1 and in plants inoculated with *B. cereus* E1 from NFS. (Fig. 4A). However, *B. thuringiensis* AZP2-primed plants exhibited much higher net assimilation rate compared to the non-inoculated controls, independently of whether the plants were stressed or not. The only exception was the 10-day drought-stressed treatment in primed plants where the net assimilation rate was almost completely inhibited. No significant difference in net assimilation rates among *B. thuringiensis* AZP2- and *P. polymyxa* B-primed plants was observed. A regression analysis demonstrated a very strong positive correlation between net assimilation rate and plant survivorship through the drought-stress period ($r^2 = 0.95$, $P < 0.001$; Fig. S2A).

Stress- and priming-driven modifications in stomatal conductance (g_s) mirrored those in net assimilation rate. Non-primed control treatment showed relatively stable g_s between 150 and 200 mmol m^{-2} s^{-1} throughout the course of the experiment (Fig. 4B). In contrast, drought-stressed non-primed plants showed a steady reduction in g_s during the drought treatment. *Bacillus thuringiensis* AZP2-primed plants had significantly higher g_s whether drought-stressed or not (Fig. 4B). Changes in stomatal conductance during the drought cycle were positively correlated with plant survivorship ($r^2 = 0.91$, $P < 0.001$; Fig. S2B).

AZP2-colonization upregulates the ascorbate-glutathione cycle, CAT and SOD

Due to the importance of antioxidant enzymes in ROS scavenging, the activities of monodehydroascorbate reductase (MDHAR), glutathione reductase (GR), catalase (CAT) and superoxide dismutase (SOD) were studied after 8 days in drought-stressed and well-watered plants. Enzyme activities were normalized to those of well-watered seedlings. The relative activity of MDHAR was affected increased by water stress and AZP2 colonization (Table 2). Activity of GR was slightly enhanced by drought, and strongly enhanced by AZP2. Both SOD and CAT activities were increased by AZP2 under drought stress (Table 2).

Benzaldehyde, geranyl acetone and β-pinene show distinct patterns in response to bacterially induced plant drought stress tolerance

The GC-MS analysis showed that seven terpenoid and benzenoid compounds were emitted from wheat leaves including α-pinene, limonene, *para*-cymene, α-phellandrene and camphene. Among the compounds, benzaldehyde, β-pinene and geranyl acetone were most responsive to drought stress and exhibited greatest differences among the treatments, while some other compounds such as ethylene emission rate were similar among the treatments (data not shown).

Benzaldehyde emissions increased with increasing the drought stress period. The emission reached a maximum value when non-primed wheat plants were grown without water for 8 days (Fig. 5A). On the other hand, *B. thuringiensis* AZP2-primed stressed plants showed modest benzaldehyde emission compared to the non-primed stressed plants. Both *B. thuringiensis* AZP2- and *P. polymyxa B*-primed stressed plants showed similar benzaldehyde emission patterns and the benzaldehyde emission rate was negatively correlated with plant survival under stress conditions ($r^2 = 0.96$, Fig. S3A and Fig. 5, three-way ANOVA).

Low-level β-pinene emissions were detected from well-watered wheat plants independent of whether the plants were primed or not. However, two-fold higher β-pinene emissions were detected already in two days after drought stress, and the emissions further stabilized to a somewhat lower steady level for the rest of the study. *B. thuringiensis* AZP2-primed stressed plants kept their β-pinene emissions at a low level similar to the non-stressed plants (Fig. 5B).

Drought stress induced a steady increase in geranyl acetone emission (Fig. 5C). *Bacillus thuringiensis* AZP2-primed stressed plants maintained a significantly lower geranyl acetone emission compared with their non-primed counterparts. Strong negative correlations were observed between geranyl acetone emissions and plant survivorship ($r^2 = 0.97$, $P < 0.001$) and net assimilation rate ($r^2 = 0.994$, $P < 0.001$) (Figs.S3B and C).

Among the compounds, benzaldehyde, β-pinene and geranyl acetone showed significant differences in response to drought stress in wheat plants primed with different strains. Compared with *P. polymyxa B*-primed plants, in plants primed with *Bacillus thuringiensis* AZP2 these compounds were elicited to a lower degree under the most severe drought stress of 6–8 days since withholding irrigation

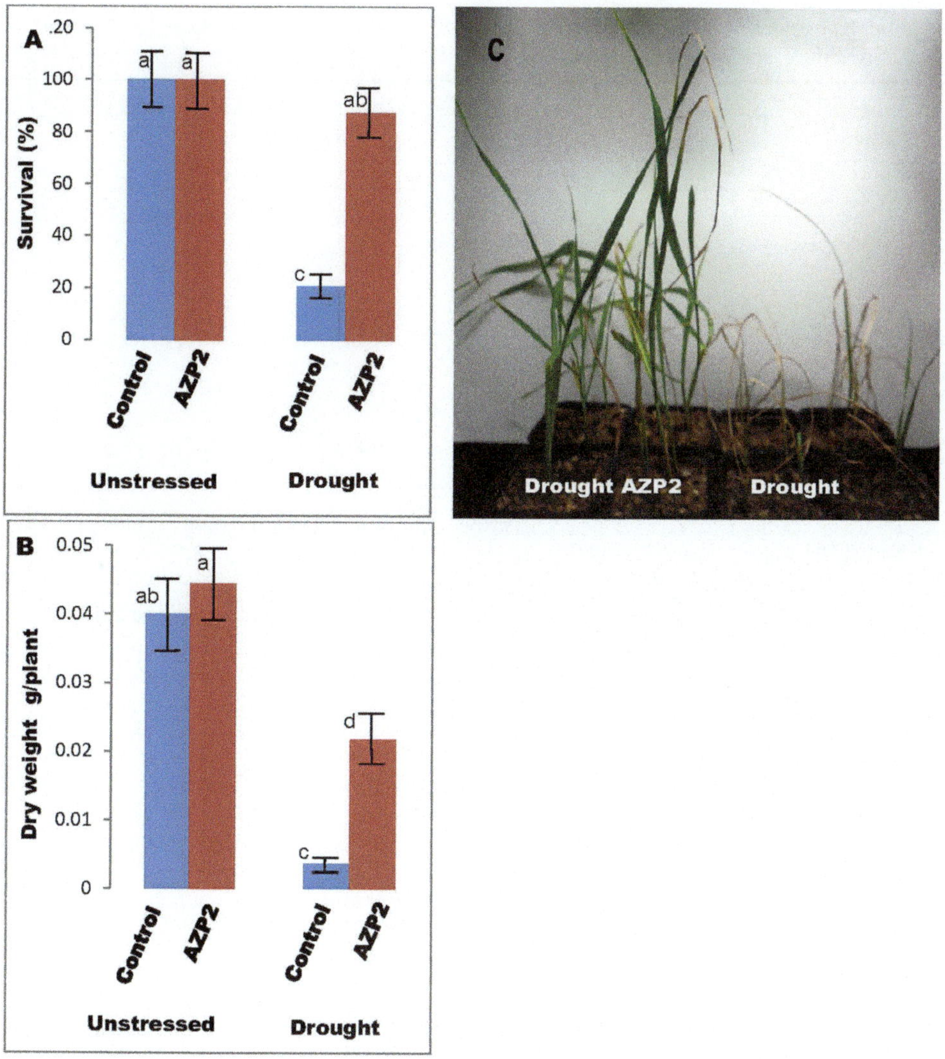

Figure 2. Increase of wheat drought stress tolerance by *Bacillus thuringiensis* AZP2 in sand mixed with 10% greenhouse soil. Effect of AZP2 priming on wheat survival (A, C) and dry mass (B) after 14 days of drought stress. Eight independent experiments were performed, and treatments with the same letter are not significantly different at $P \leq 0.01$.

water (Fig. 5 and SIFig.3). In wheat inoculated with *P. polymyxa* E1 and *B. cereus* E1 from NFS no differences in the degree of elicitation of volatile emissions compared with controls were observed (data not shown).

Discussion

Rhizospheric bacteria from harsh environments induce drought stress tolerance in modern wheat

Achieving high yields of crop plants under global irrigation water shortage is the main challenge for agriculture. Microorganisms could play a significant role in providing viable solutions, but we first must understand and exploit their unique properties and develop methods for monitoring their performance [11,14,16–19,53].

The first report on enhancement of plant drought stress tolerance by rhizosphere bacteria has been published in 1999, and several *Paenibacillus* sp. and *Bacillus sp.* and some other gram-positive bacterial isolates were shown to be effective in improving plant drought stress tolerance (8). In our study, we observed five times greater survival and 78% higher biomass in inoculated plants under drought stress (Table 1), confirming the potential of bacterial priming in enhancing plant performance under drought. To the best of our knowledge, such a strong improvement of drought tolerance by bacterial inoculation has not been reported before.

Until now, there has been no studies investigating the importance of the habitat of free-living plant growth-promoting bacteria (PGPB) on their capacity to affect plant drought tolerance. *Bacillus thuringiensis* AZP2 strain was isolated as a ponderosa pine root endophyte from highly stressed environment in Arizona, USA. The tree seedlings had been growing on gneiss rock under UV, heat and drought stress and severe nutritional deprivation. In our study, 43% of *B. thuringiensis* AZP2 treated wheat plants survived the severe 10-day drought stress exposure followed by 4 days recovery after re-watering (Table 1). The second best isolate, *Paenibacillus polymyxa B* was isolated from Tina plain, Giza, Egypt, where rice plants had been growing under severe salt stress over prolonged period. In our study, 23% of *P. polymyxa B* treated wheat plants survived the 10-day drought stress.

Table 2. Effect of priming by *Bacillus thuringiensis* AZP2 on wheat *(Triticum aestivum L. cv. Stava)* on average (±SD) growth characteristics, water use efficiency and antioxidant enzyme activities.

	Well-watered		**Drought-stressed**	
	Control	**AZP2-primed**	**Control**	**AZP2-primed**
Average plant survival improvement (%)				**500**
Plant dry mass increase (%)		9		78
Germination rate (%)	72	**100**	50	**100**
Lateral root count[1]	266 ± 76^{ab}*	235 ± 21^{ab}	181 ± 16^{c}	$\mathbf{192.5\pm3.5^{bc}}$
Total lateral root length (cm)	263 ± 11^{a}	183 ± 23^{b}	89 ± 11^{d}	$\mathbf{142\pm31^{bc}}$
Longest root length (cm)	28.4 ± 1.4^{ab}	27.6 ± 2.2^{ab}	20.7 ± 1.2^{c}	23.2 ± 1.2^{bc}
Soil attached to root[2] ($g\ g^{-1}$ dry root)	$\mathbf{62\pm19^{a}}$	$\mathbf{92\pm30^{ab}}$	$\mathbf{10\pm5^{c}}$	$\mathbf{25\pm11^{bc}}$
Average root hair length[3] (mm)	$\mathbf{0.74\pm0.21^{a}}$	$\mathbf{1.5\pm0.5^{b}}$	$\mathbf{0.84\pm0.21^{a}}$	$\mathbf{1.91\pm0.22^{b}}$
Root hair density (number per mm^3)	$\mathbf{24.1\pm3.1^{a}}$	$\mathbf{30.1\pm2.0^{ab}}$	$\mathbf{24.1\pm2.0^{a}}$	$\mathbf{32.0\pm1.1^{b}}$
Number of fresh roots per plant	nt	nt	$\mathbf{3.0\pm1.1^{a}}$	$\mathbf{10.1\pm2.0^{b}}$
Water use efficiency[4] ($g\ g^{-1}$)	0.1053 ± 0.0031^{d}	0.1087 ± 0.0021^{d}	$\mathbf{0.0802\pm0.0032^{b}}$	$\mathbf{0.135\pm0.012^{c}}$
Relative Enzyme Activities:[5]				
MDHAR	1 ± 0.1^{a}	1.4 ± 0.2^{a}	$\mathbf{1.8\pm0.1^{b}}$	$\mathbf{2.6\pm0.2^{c}}$
GR	1 ± 0.2^{a}	1.3 ± 0.1^{a}	$\mathbf{1.4\pm0.3^{a}}$	$\mathbf{1.8\pm0.2^{c}}$
SOD	1 ± 0.2^{a}	1.3 ± 0.1^{ab}	$\mathbf{0.7\pm0.1^{ab}}$	$\mathbf{1.9\pm0.2^{c}}$
CAT	1 ± 0.1^{a}	1 ± 0.2^{a}	$\mathbf{1.5\pm0.1^{bc}}$	$\mathbf{2\pm0.2^{c}}$

[1]Analysis of plant root was conducted by Root Reader3D Imaging and Analysis System and manually [7].
[2]Twelve plants per treatment were sampled. Roots with adhering soil (RAS) were carefully separated from bulk soil by shaking. Soil and root dry mass (RT) was recorded after drying the samples at 105°C, and RAS/RT ratio was calculated.
[3]Twelve plants were carefully separated from soil by shaking followed by washing the roots in distilled water and left to drain in Petri dishes with water to maintain humidity. Root system characteristics were evaluated by Zeiss LSM 710 microscope.
[4]Water use efficiency is defined as the ratio of total plant dry mass per total water used.
[5]MDHAR - Monodehydroascorbate reductase, GR- Glutathione reductase, SOD- Superoxide dismutase, CAT-Catalase.
See Materials and Methods for enzyme extraction and activity measurements.
*Means followed by the same letter are not significantly different at $p\leq0.01$. See Experimental procedures.

Our results further revealed that 12 bacterial subpopulations from harsh SFS environments in 'Evolution Canyon' in Israel had much greater potential to alter drought resistance of wheat than 12 subpopulations from more moderate NFS environments (Table 1). 16% of *P. polymyxa*, 15% of *B. cereus*, 16% of *B. megaterium* and 10% of *B. pumilus* treated plants from SFS survived the drought stress (Table 1). In contrast, the isolates of the same species from moderate NFS did not significantly affect wheat drought tolerance (Table 1) [23].

All drought tolerance-enhancing bacteria tested were biofilm-forming, ACC deaminase-containing and had a high P-solubilizing activity. These metabolic qualities have likely contributed to their survival in harsh environments, and are likely responsible for their capacity to increase plant drought tolerance [23]. We argue that rhizospheric bacteria have coevolved with their host plant roots under stressful conditions, leading to fine-tuned regulatory gene expression networks at multiple levels to respond with high flexibility to diverse stresses in the belowground environment. Taken together, our results suggest that the bacteria from harsh environments are more likely to contribute plant stress tolerance enhancement.

Enhancement of plant drought tolerance by rhizosphere bacteria depends on dynamic interactions among bacteria, roots, soil, and water in the rhizosphere

Despite several mechanisms suggested, the mode of how rhizosphere bacteria enhance plant drought stress tolerance is largely unknown. In particular, we need to ask how *B. thuringiensis* AZP2 priming enhances wheat drought tolerance. *Bacillus thuringiensis* AZP2 genome has been sequenced using Ion Torrent PGMTM system (unpublished data of Salme Timmusk). Comparative genomic analysis of the sequence reveals gene clusters for alginate, ACC deaminase, and auxin (IAA) production and regulation. Any of the traits alone, and in combination could have been responsible for the bacterial drought tolerance enhancement [22,54,55].

Our results indicate that dynamic interactions among bacteria, roots, soil, and water in the rhizosphere determine what mechanism is responsible for the bacterial-induced wheat drought tolerance enhancement at specific stages of seedling growth Table 2 and [56]. Plant seedling successful establishment is required for its survival, and hence, AZP2-induced high germination rate is of prime importance at this stage of development. In fact, AZP2 inoculation resulted in 100% germination of wheat seeds (Table 2).

Following germination, the capacity of roots to extract moisture and nutrients from the soil become the key traits determining plant survival. Improved nutrient and water extraction capacity can be achieved by various ways. Our results indicate that AZP2 inoculation resulted in two to three times longer root hairs, and longer and denser lateral roots (Table 2). These effects were even more pronounced under longer drought stress periods [56]. Root hair length and density are critical when it comes to water and nutrient acquisition from surrounding environment. Although root hair formation can be massively enhanced, this increase should not

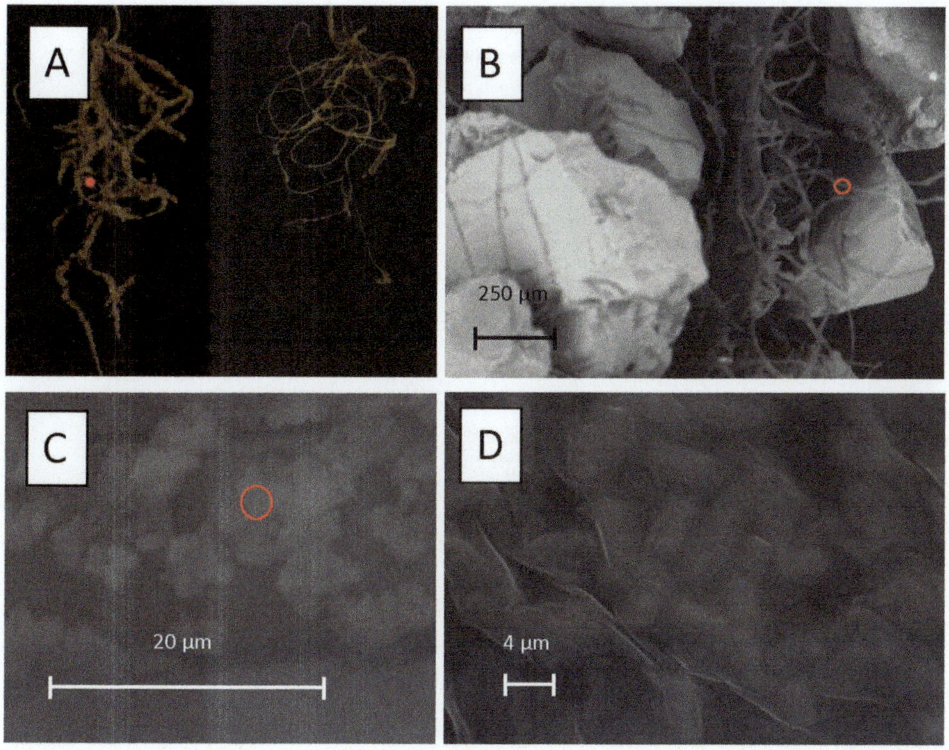

Figure 3. Formation of sand soil mulch and biofilm on root hairs of winter wheat (cv. Stava) by *Bacillus thuringiensis* AZP2. Scanning electron micrographs were made of AZP2-primed wheat root systems after 10-day drought stress and show sand mulch (A, B) and bacterial biofilm formation on root hair (C, D). Significantly more soil mulch is attached to the AZP2 treated plant (A, right) in comparison to control (A, left). Red circles indicate the areas magnified.

necessarily show up as an increase in total root dry mass. Hence our results also draw attention to a general misinterpretation of the importance of 'root mass' as a criterion in showing the effect of inoculation. Based on our results, root hair microscopic observation are highly recommended in parallel of studying total root mass.

In our experiments, especially in the sand, plants were exposed to both nutrient limitation and drought stress. Under any abiotic stress, there is a significant decrease in photosynthesis and, consequently, a reduction in the amount of metabolites and energy. It is important for plants to use this reduced amount of resources to maximize their growth and reproductive potential. Therefore, under particularly severe nutrient starvation in sand soil, increased proliferation of lateral roots was observed (Table 2). This contributes to enhanced exploitation of the topsoil where bioavailable nutrients are typically located in the field relative to the subsoil which is depleted of nutrients. On the other hand, under drought stress, lateral root growth inhibition and enhancement of vertical root growth are usually adaptive responses, because water availability is usually higher in deeper soil layers. Thus, restriction of lateral root growth in the top-layer of sand soil supplemented with more fertile greenhouse soil can be an adaptive modification contributing to greater drought tolerance of wheat. Far more genes than generally believed (more than 300) have been found to be involved in lateral root development [57], and clearly more experimental work is needed to gain insight into regulation of lateral root development as driven by soil nutrient and water availabilities.

Another important root trait in plant protection against drought stress is the creation of bacterial biofilm with attached soil mulch. AZP2-induced denser and longer root hair framework forms an excellent matrix for the bacterially-excreted biofilm comprised of cells and extracellular matrix producing a thick sticky layer around root hair (Fig. 3). Hence, induction of long and dense root hair should be considered as an important drought stress tolerance enhancement strategy. The dense biofilm matrix also limits diffusion of biologically active compounds secreted by bacteria and these are therefore concentrated on the root surface, facilitating plant uptake. In addition, biofilm formation on root hair substantially improves root-to-soil contact, enhancing plant nutrient acquisition from soil and suggesting that biofilm formation importantly contributes to improving plant nutrition as well (Table 2).

Alginate, a hygroscopic bacterial polysaccharide, can play an important role in determining the biofilm capacity to enhance water status of seedlings [58]. Bacterial alginate water-holding capacity is very high and it loses water slowly [58–60], thereby keeping root cells hydrated long enough to allow for cellular metabolic adjustments necessary to enhance drought stress tolerance. The drought tolerance enhancement of alginate might be due to its hygroscopic properties, but can also result from its role in biofilm architecture that contributes to reduced evaporation loss [58]. We have shown that small amounts of alginate are produced by AZP2 under controlled conditions (manuscript in preparation), and thus, the alginate mechanism in AZP2-induced drought tolerance is plausible.

Another important trait for plant successful establishment under stressful conditions is regulation of root development. After 10 days of drought stress, untreated plant root systems were composed of dry thin roots with very few fresh vital roots

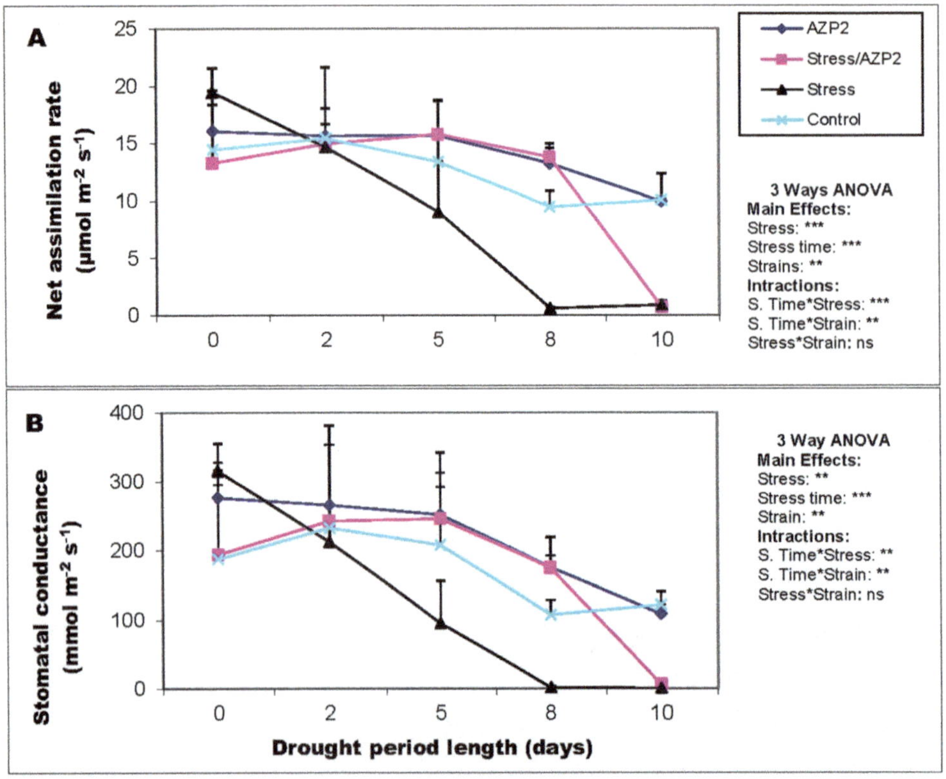

Figure 4. Net assimilation rate (A) and stomatal conductance (B) of *Bacillus thuringiensis* AZP2-primed wheat seedlings under drought stress. The data are shown for plants grown for 0, 2, 5, 8 and 10 days without water. The error bars indicate +SE for three biological replicates. Statistical analysis is based on three-way ANOVA with stress, strains (*Bacillus thuringiensis* AZP2 vs. *P. polymyxa* B) and stress exposure time as factors. ***, ** and ns, indicate highly significant, significant or non-significant effects for the tested factor at $P<0.05$.

(Table 2). Three times more fresh lateral roots were counted on AZP2-primed wheat seedlings. This is in agreement with our earlier suggestion that the biofilm of rhizosphere bacteria together with soil mulch has protective role against drought stress [17,61]. Yet the results also indicate production of biologically active compounds for root maturity regulation in the bacterial biofilm.

Taken together, AZP2 inoculation increases lateral root density and length, and the inoculation enhanced even more strongly root hair density and length (59% and 200% respectively). We conclude that both lateral root and root hair increase in drought-stressed AZP2-inoculated wheat have likely resulted from bacterial production of IAA and ACC deaminase, and perhaps from additional chemical modifications [56]. Any bacterially-secreted compound required for plant drought stress tolerance enhancement is kept in place inside the bacterially-formed and root hair-supported alginate-containing biofilm. The compounds are thus likely to be efficiently taken up by the plant and result in elicitation of gene expression networks leading to more stress-adapted wheat phenotypes (Table 2, Fig. 3) [56].

Gauging volatiles provides an effectual platform for rapid screening of potent bacterial strains

Our results show that ROS scavenging enzyme activities increase in response to bacterial priming, especially under drought stress (Table 2). Similar alterations in antioxidative capacity by stress-alleviating microorganisms have been reported earlier [32,33]. In principle, ROS detoxification capacity could be used as a parameter in screening for drought stress-alleviating potential of microorganisms. Any of rapid and effective screening methods should detect plant stress in early phases before the stress becomes that severe that it can be detected by visual symptoms. Such an early monitoring gives the possibility to cope with the stress situation in its early stage before irreversible damage occurs and the yield of crop plants is seriously reduced. Yet, the major shortcoming of many stress detection methods, including ROS detoxification capacity, is that they are invasive [7] for a review, i.e. to measure plant stress status, the plant has to be entirely or partly harvested and processed. We have aimed to contributing towards development of a non-invasive strategy to screen plant stress and its possible alleviation by potent bacterial strains.

Different volatile organic compounds (VOC) are commonly emitted from plants leaves and these emissions are known to increase substantially under stress situations [35,41–43,52,62]. Some of these VOCs are by-products of stress-elicitation of various physiological processes, while others are used as sophisticated signals by which plants communicate stress within and among the plants and that trigger stress tolerance in receiver organs and plants situated farther from the stress source [63–65]. Although volatile emissions result in significant carbon losses and can lead to plant growth penalties in non-stressed situations [66–68], many of the stress-released volatiles play important physiological and ecological roles under stress.

We have previously demonstrated that abiotic stress can lead to elicitation of monoterpene emissions from wheat, and the emissions were quantitatively correlated with the severity of stress [42,43]. Our results clearly showed that the elevated emission of VOCs was always negatively correlated with plant growth and fitness under drought stress conditions. It has been demonstrated

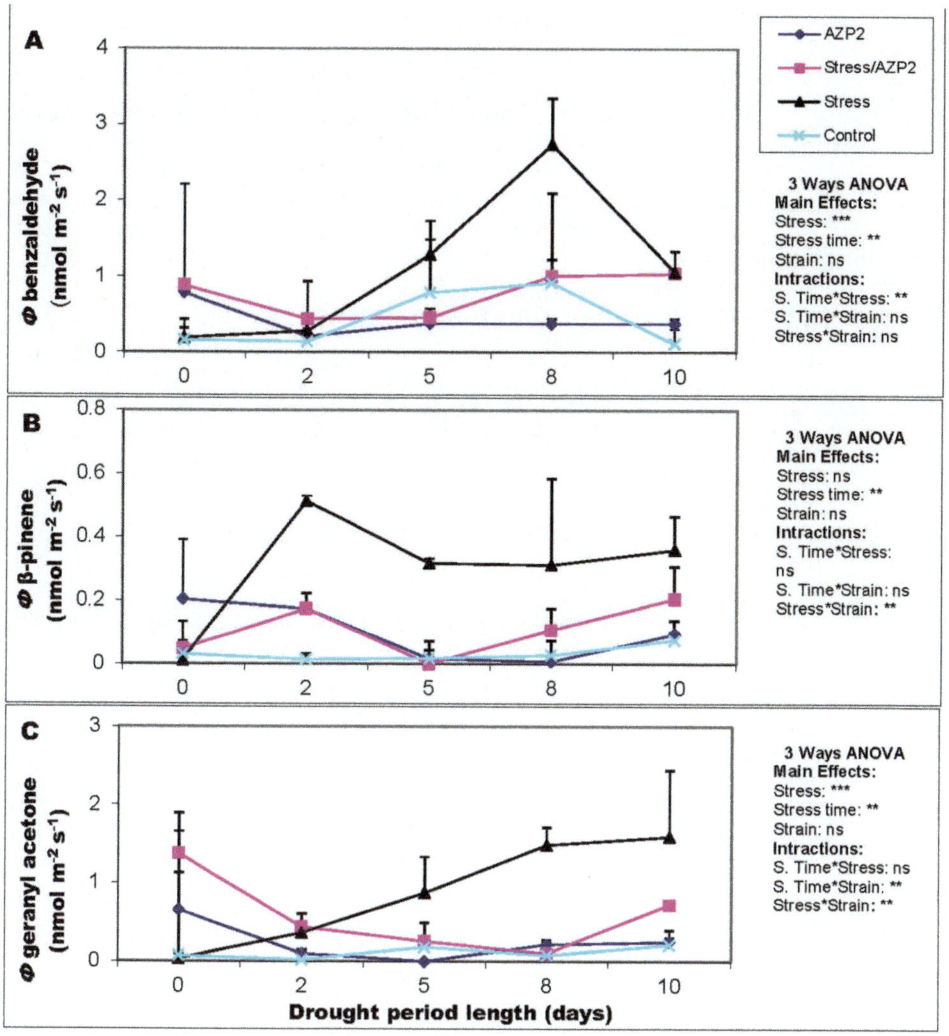

Figure 5. Temporal variations in the emission rates of some benzenoids and terpenoids emitted by wheat plants. Benzaldehyde (A), β-pinene (B) and geranyl acetone (C) emission rates from leaves of drought-stressed (0, 2, 5, 8 and 10 days without water) wheat plants after priming with *Bacillus thuringiensis* AZP2 are demonstrated. The error bars indicate +SE for three biological replicates. Statistical analysis and levels of significance as in Fig. 4.

that plants may lose up to 10%, (exceptionally up to 50%), of the carbon fixed by photosynthesis as cost for VOCs emission under stressful conditions [69]. In addition to typical stress volatile hormones such as ethylene, stresses result in upregulation of several secondary metabolic pathways including terpenoid and shikimic acid pathway [70], and the observation of enhanced emission of certain terpenoids and benzenoids in stress is in agreement with upregulation of these key secondary metabolic pathways.

Overall, the emission rates of induced VOCs are quantitatively associated with the severity of the stress [41,71,72]. Here we further observed that the emission rates of key stress-elicited volatiles increased with increasing the severity of drought stress (Fig. S3). The observation that VOC emissions were significantly reduced and were correlated with higher photosynthesis and plant survival under drought stress in plants primed by rhizosphere bacteria suggests that bacterial inoculation improved plant stress tolerance. Reduced emissions of stress-induced VOCs further imply lower cost for VOC emission, potentially contributing to greater productivity under stress. Volatile detectors capable of detecting very small amounts of volatiles in outdoor conditions are already commercially available, although the cost is still quite high e.g. [73,74–76]. Hence, monitoring of drought stress development and its potential alleviation by rhizosphere bacteria by VOC in realistic biological settings in field might be feasible in the near future. Such measurements could be conducted in parallel with quantitative PCR screening of bacterial strains involved, and quantification of bacterial bioactive compound production.

Conclusions

Certainly, more bacterial strains should be tested in the future to gain insight into the sensitivity of the suggested method. In particular, in discriminating among different bacterial strains at different stages of plant development. Yet, our results demonstrate that plant inoculation with bacteria from harsh environments resulted in significantly higher survival of drought-stressed plants, and in greater photosynthesis and biomass production. This was reflected in modifications in volatile profiles and total emission rates, consistent with the overall hypothesis that measurements of volatiles can provide a promising novel methodology to gauge stress resistance. The results collectively point out that inoculation with bacteria from harsh environments could be the way for rapid enhancement of plant stress tolerance. We suggest that monitoring of the elicitation of VOC emissions is a promising method to characterize the efficiency of different bacterial strains in priming for drought stress resistance.

Supporting Information

Figure S1 Volumetric water content of the rhizosphere of *Bacillus thuringiensis* AZP2- and *P. polymyxa* B-primed and non-primed wheat seedlings under watering and after 10 days of drought exposure.

Figure S2 Correlations of plant survival with net assimilation rate (A), and with stomatal conductance (B) in wheat seedlings drought-stressed for 0, 2, 5, 8 and 10 days. The correlations are significant at $P<0.01$.

Figure S3 Plant survival in relation to emission rates of benzaldehyde (A), and geranyl acetone (B) and geranyl acetone emission rate in relation net assimilation rate (C) on wheat seedlings drought-stressed for 0, 2, 5, 8 and 10 days. The relationships are significant at $P<0.01$.

Author Contributions

Conceived and designed the experiments: ST ÜN LB IED LC GS. Performed the experiments: ST IED LC TT GS. Analyzed the data: ST ÜN IED LB LC AK GS ES. Contributed reagents/materials/analysis tools: ST ÜN LB EN GS. Wrote the paper: ST ÜN LB IED LC.

References

1. Smol JP (2012) Climate Change: A planet in flux. Nature 483: S12–15.
2. Foley JA, Ramankutty N, Brauman KA, Cassidy ES, Gerber JS, et al. (2011) Solutions for a cultivated planet. Nature 478: 337–342.
3. Flexas J, Niinemets U, Galle A, Barbour MM, Centritto M, et al. (2013) Diffusional conductances to CO as a target for increasing photosynthesis and photosynthetic water-use efficiency. Photosynth Res.117: 1–3
4. Fleury D, Jefferies S, Kuchel H, Langridge P (2010) Genetic and genomic tools to improve drought tolerance in wheat. Journal of Experimental Botany 61: 3211–3222.
5. Price AH, Cairns JE, Horton P, Jones HG, Griffiths H (2002) Linking drought-resistance mechanisms to drought avoidance in upland rice using a QTL approach: progress and new opportunities to integrate stomatal and mesophyll responses. Journal of Experimental Botany 53: 989–1004.
6. Wang W, Vinocur B, Altman A (2003) Plant response to drought, salinity and extreme temperatures: towards genetic engineering for stress tolerance. Planta 218: 1–14.
7. Niinemets Ü (2010) Responses of forest trees to single and multiple environmental stresses from seedlings to mature plants: past stress history, stress interactions, tolerance and acclimation. Forest Ecology and Management 260: 1623–1639.
8. van den Boogaard R, de Boer M, Veneklaas EJ, Lambers H (1996) Relative growth rate, biomass allocation pattern and water use efficiency of three wheat cultivars during early ontogeny as dependent on water availability. Physiologia Plantarum 98: 493–504.
9. Cavender-Bares J, Bazzaz FA (2000) Changes in drought response strategies with ontogeny in *Quercus rubra*: implications for scaling from seedlings to mature trees. Oecologia 124: 8–18.
10. Winkel T, Payne W, Renno JF (2001) Ontogeny modifies the effects of water stress on stomatal control, leaf area duration and biomass partitioning of Pennisetum glaucum. The New Phytologist 149: 71–82.
11. Timmusk S, Behers L (2012) Rhizobacterial application for sustainable water management on the areas of limited water resources. Irrigation and Drainage Systems Engineering 12: doi: 10.4172/2168-9768.1000e4111.
12. Yang J, Kloepper JW, Ryu CM (2009) Rhizosphere bacteria help plants tolerate abiotic stress. Trends Plant Sci 14: 1–4.
13. Kim YC, Glick B, Bashan Y, Ryu CM (2013) Enhancement of plant drought tolerance by microbes. In: Aroca R, editor. Plant responses to drought stress. Berlin: Springer Verlag.
14. Timmusk S, Wagner EG (1999) The plant-growth-promoting rhizobacterium *Paenibacillus polymyxa* induces changes in *Arabidopsis thaliana* gene expression: a possible connection between biotic and abiotic stress responses. Mol Plant Microbe Interact 12: 951–959.
15. Mayak S, Tirosh T, Glick BR (2004) Plant growth-promoting bacteria that confer resistance to water stress in tomatoes and peppers. Plant Sci 166: 525–530.
16. Timmusk S (2003) Mechanism of action of the plant growth promoting bacterium *Paenibacillus polymyxa* [PhD thesis]. Uppsala: Uppsala University. 1–54 p.
17. Timmusk S, Nevo E (2011) Plant root associated biofilms. In: Maheshwari DK, editor. Bacteria in agrobiology (vol 3): Plant nutrient management. Berlin: Springer Verlag. pp. 285–300.
18. Timmusk S, Behers L (2012) Enhancing drought tolerance in crops. Available: http://dx.doi.org/10.4172/2168-9768.1000e111 May 2012
19. Timmusk S, Timmusk K, Behers L (2013) Rhizobacterial plant drought stress tolerance enhancement. J Food Security V1: p 10–16.
20. Dimkpa C, Weinand T, Asch F (2009) Plant-rhizobacteria interactions alleviate abiotic stress conditions. Plant Cell Environ 32: 1682–1694.
21. Conrath U, Beckers GJ, Flors V, Garcia-Agustin P, Jakab G, et al. (2006) Priming: getting ready for battle. Mol Plant Microbe Interact 19: 1062–1071.
22. Chang WS, van de Mortel M, Nielsen L, Nino de Guzman G, Li X, et al. (2007) Alginate production by Pseudomonas putida creates a hydrated microenvironment and contributes to biofilm architecture and stress tolerance under water-limiting conditions. J Bacteriol 189: 8290–8299.
23. Timmusk S, Paalme V, Pavlicek T, Bergquist J, Vangala A, et al. (2011) Bacterial distribution in the rhizosphere of wild barley under contrasting microclimates. PLoS One 6: e17968.
24. Niinemets Ü, Keenan TF (in press) Photosynthetic responses to stress in Mediterranean evergreens: mechanisms and models. Environmental and Experimental.
25. Flexas J, Medrano H (2002) Drought-inhibition of photosynthesis in C_3 plants: stomatal and non-stomatal limitations revisited. Annals of Botany 89: 183–189.
26. Chaves MM, Flexas J, Pinheiro C (2009) Photosynthesis under drought and salt stress: regulation mechanisms from whole plant to cell. Annals of Botany 103: 551–560.
27. Chaves MM, Pereira JS, Maroco J, Rodrigues ML, Ricardo CPP, et al. (2002) How plants cope with water stress in the field: photosynthesis and growth. Annals of Botany 89: 907–916.
28. Beis A, Patakas A (2010) Differences in stomatal responses and root to shoot signalling between two grapevine varieties subjected to drought. Functional Plant Biology 37: 139–146.
29. Ionenko IF, Anisimov AV (2001) Effect of water deficit and membrane destruction on water diffusion in the tissues of maize seedlings. Biologia Plantarum 44: 247–252.
30. García-Mata C, Lamattina L (2001) Nitric oxide induces stomatal closure and enhances the adaptive plant responses against drought stress. Plant Physiology 126: 1196–1204.
31. Munns R (2002) Comparative physiology of salt and water stress. Plant, Cell and Environment 25: 239–250.
32. Wang CJ, Yang W, Wang C, Gu C, Niu DD, et al. (2012) Induction of drought tolerance in cucumber plants by a consortium of three plant growth-promoting rhizobacterium strains. PLoS One 7: e52565.
33. Mastouri F, Bjorkman T, Harman GE (2012) Trichoderma harzianum enhances antioxidant defense of tomato seedlings and resistance to water deficit. Mol Plant Microbe Interact 25: 1264–1271.
34. Noctor G, Foyer CH (1998) Ascorbate and glutathione: keeping active oxygen under control. Annual Review of Plant Physiology and Plant Molecular Biology 49: 249–279.
35. Loreto F, Schnitzler JP (2010) Abiotic stresses and induced BVOCs. Trends in Plant Science 15: 154–166.
36. Holopainen JK, Gershenzon J (2010) Multiple stress factors and the emission of plant VOCs. Trends in Plant Science 15: 176–184.
37. Heil M, Silva Bueno JC (2007) Within-plant signaling by volatiles leads to induction and priming of an indirect plant defense in nature. Proceedings of the National Academy of Sciences of the United States of America 104: 5467–5472.
38. Niinemets Ü (2010) Mild versus severe stress and BVOCs: thresholds, priming and consequences. Trends in Plant Science 15: 145–153.
39. Choudhary DK, Johri BN, Prakash A (2008) Volatiles as priming agents that initiate plant growth and defence responses Current Science 94: 595–604.
40. Copolovici L, Kannaste A, Remmel T, Vislap V, Niinemets U (2010) Volatile emissions from Alnus glutionosa induced by herbivory are quantitatively related to the extent of damage. J Chem Ecol 37: 18–28.
41. Copolovici L, Kännaste A, Pazouki L, Niinemets Ü (2012) Emissions of green leaf volatiles and terpenoids from *Solanum lycopersicum* are quantitatively related to the severity of cold and heat shock treatments. Journal of Plant Physiology 169: 664–672.
42. Opriş O, Copaciu F, Soran ML, Ristoiu D, Niinemets Ü, et al. (2013) Influence of nine antibiotics on key secondary metabolites and physiological characteristics

in *Triticum aestivum*: leaf volatiles as a promising new tool to assess toxicity. Ecotoxicology and Environmental Safety 87: 70–79.
43. Copaciu F, Opriş O, Coman V, Ristoiu D, Niinemets Ü, et al. (2013) Diffuse water pollution by anthraquinone and azo dyes in environment importantly alters foliage volatiles, carotenoids and physiology in wheat (*Triticum aestivum*). Water, Air, and Soil Pollution 224: 1478.
44. Brilli F, Ruuskanen TM, Schnitzhofer R, Müller M, Breitenlechner M, et al. (2011) Detection of plant volatiles after leaf wounding and darkening by proton transfer reaction "time-of-flight" mass spectrometry (PTR-TOF). PloS ONE 6: e20419.
45. Penrose DM, Glick BR (2003) Methods for isolating and characterizing ACC deaminase-containing plant growth-promoting rhizobacteria. Physiol Plant 118: 10–15.
46. Nautiyal CS (1999) An efficient microbiological growth medium for screening phosphate solubilizing microorganisms. FEMS Microbiol Lett 170: 265–270.
47. Clark RT, MacCurdy RB, Jung JK, Shaff JE, McCouch SR, et al. (2011) Three-dimensional root phenotyping with a novel imaging and software platform. Plant Physiol 156: 455–465.
48. Knöerzer O, Durner J, Boeger P (1996) Alterations in the antioxidative system of suspension-cultured soybean cells (*Glycine max*) induced by oxidative stress. Physiol Plant 97: 388–396.
49. Hossain M, Nakano Y, Asada K (1984) Mono dehydro ascorbate reductase EC-1.6.5.4 in spinach chloroplasts and its participation in regeneration of ascorbate for scavenging hydrogen per oxide. Plant Cell Physiol 25: 385–396.
50. Smith I, Vierheller T, Thorne C (1989) Properties and functions of glutathione reductase in plants. Physiol Plant 77: 449–456.
51. Copolovici L, Niinemets U (2010) Flooding induced emissions of volatile signalling compounds in three tree species with differing waterlogging tolerance. Plant Cell Environ 33: 1582–1594.
52. Toome M, Randjarv P, Copolovici L, Niinemets U, Heinsoo K, et al. (2010) Leaf rust induced volatile organic compounds signalling in willow during the infection. Planta 232: 235–243.
53. Kim SB, Timmusk S (2013) A simplified method for *Paenibacillus polymyxa* gene knockout and insertional screening. PLoSONE. 101371/journalpone0068092
54. Belimov AA, Dodd IC, Hontzeas N, Theobald JC, Safronova VI, et al. (2009) Rhizosphere bacteria containing 1-aminocyclopropane-1-carboxylate deaminase increase yield of plants grown in drying soil via both local and systemic hormone signalling. New Phytol 181: 413–423.
55. Ivanchenko MG, Muday GK, Dubrovsky JG (2008) Ethylene-auxin interactions regulate lateral root initiation and emergence in Arabidopsis thaliana. Plant J 55: 335–347.
56. Timmusk S, Abd El-Daim I, Rudloff E (in review) Genome wide analysis of *Bacillus thuringiensis* AZP2 role in plant adaptation to environmental stress. manuscript for Plos One.
57. Xiong L, Wang RG, Mao G, Koczan JM (2006) Identification of drought tolerance determinants by genetic analysis of root response to drought stress and abscisic Acid. Plant Physiol 142: 1065–1074.
58. Halverson LJ (2009) Role of alginate in bacterial biofilms. In: Rehm BHA, editor. Alginates: biology and applications. Dordrecht: Springer.pp. 136–141.
59. Hay ID, Rehman ZU, Moradali MF, Wang Y, Rehm BHA (2013) Microbial alginate production, modification and its applications. Microbial Biotechnology 6: 637–650.
60. Donati I, Paoletti S (2009) Material properties of alginates. In: Rehm BHA, editor. Alginates: biology and applications. Dordrecht: Springer. pp. 1–54.
61. Timmusk S, Grantcharova N, Wagner EG (2005) *Paenibacillus polymyxa* invades plant roots and forms biofilms. Appl Environ Microbiol 71: 7292–7300.
62. Copolovici L, Kännaste A, Niinemets Ü (2009) Gas chromatography-mass spectrometry method for determination of monoterpene and sesquiterpene emissions from stressed plants. Studia Universitatis Babes-Bolyai, Chemia 54: 329–339.
63. Pare PW, Tumlinson JH (1999) Plant volatiles as a defense against insect herbivores. Plant Physiol 121: 325–332.
64. Chen F, Tholl D, D'Auria JC, Farooq A, Pichersky E, et al. (2003) Biosynthesis and emission of terpenoid volatiles from Arabidopsis flowers. Plant Cell 15: 481–494.
65. Gouinguene SP, Turlings TCJ (2002) The effects of abiotic factors on induced volatile emissions in corn plants. Plant Physiology 129: 1296–1307.
66. Baldwin IT (2011) Plant Volatiles. Curr Biol 20: 392–397.
67. Holopainen JK (2004) Multiple functions of inducible plant volatiles. Trends Plant Sci 9: 529–533.
68. Niinemets U (2004) Cost of production and physiology of emission of volatile leaf isoprenoids. Adv Plant Physiol 7: 233–268.
69. Sharkey TD, Loreto F (1993) Water stress, temperature, and light effects on the capacity for isoprene emission and photosynthesis of kudzu leaves. Oecologia 95: 328–333.
70. Iriti M, Faoro F (2009) Chemical diversity and defence metabolism: how plants cope with pathogens and ozone pollution. Int J Mol Sci 10: 3371–3399.
71. Copolovici L, Kännaste A, Remmel T, Vislap V, Niinemets Ü (2011) Volatile emissions from *Alnus glutinosa* induced by herbivory are quantitatively related to the extent of damage. Journal of Chemical Ecology 37: 18–28.
72. Beauchamp J, Wisthaler A, Hansel A, Kleist E, Miebach M, et al. (2005) Ozone induced emissions of biogenic VOC from tobacco: relations between ozone uptake and emission of LOX products. Plant, Cell and Environment 28: 1334–1343.
73. Jordan A, Haidacher S, Hanel G, Hartungen E, Märk L, et al. (2009) A high resolution and high sensitivity proton-transfer-reaction time-offlight mass spectrometer (PTR-TOF-MS). International Journal of Mass Spectrometry 286: 122–128.
74. Lindinger W, Hansel A, Jordan A (1998) Proton-transfer-reaction mass spectrometry (PTR-MS): on-line monitoring of volatile organic compounds at pptv levels. Chemical Society Reviews 27: 347–354.
75. Lindinger W, Hansel A, Jordan A (1998) On-line monitoring of volatile organic compounds at pptv levels by means of proton-transfer-reaction mass spectrometry (PTR-MS): medical applications, food control and environmental research. International Journal of Mass Spectrometry and Ion Processes 173: 191–241.
76. Niinemets Ü, Kuhn U, Harley PC, Staudt M, Arneth A, et al. (2011) Estimations of isoprenoid emission capacity from enclosure studies: measurements, data processing, quality and standardized measurement protocols. Biogeosciences 8: 2209–2246.

2

Final Opportunity to Rehabilitate an Urban River as a Water Source for Mexico City

Marisa Mazari-Hiriart[1]*, Gustavo Pérez-Ortiz[1,2], María Teresa Orta-Ledesma[3], Felipe Armas-Vargas[4], Marco A. Tapia[2,5], Rosa Solano-Ortiz[5], Miguel A. Silva[6], Isaura Yañez-Noguez[3], Yolanda López-Vidal[7], Carlos Díaz-Ávalos[8]

1 Laboratorio Nacional de Ciencias de la Sostenibilidad, Departamento de Ecología de la Biodiversidad, Instituto de Ecología, Universidad Nacional Autónoma de México, México, D.F., México, **2** Departamento de Biología, Facultad de Ciencias, Universidad Nacional Autónoma de México, México, D.F., México, **3** Coordinación de Ingeniería Ambiental, Instituto de Ingeniería, Universidad Nacional Autónoma de México, México, D.F., México, **4** Posgrado en Ciencias de la Tierra, Instituto de Geología, Universidad Nacional Autónoma de México, México, D.F., México, **5** Posgrado en Ciencias Biológicas, Facultad de Ciencias-Instituto de Ecología, Universidad Nacional Autónoma de México, México, D.F., México, **6** Posgrado de Ciencias Bioquímicas, Departamento de Ingeniería Celular y Biocatálisis, Instituto de Biotecnología, Universidad Nacional Autónoma de México, Cuernavaca, Morelos, México, **7** Programa de Inmunología Molecular Microbiana, Departamento de Microbiología y Parasitología, Facultad de Medicina, Universidad Nacional Autónoma de México, México, D.F., México, **8** Departamento de Probabilidad y Estadística, Instituto de Investigaciones en Matemáticas Aplicadas y en Sistemas, Universidad Nacional Autónoma de México, México, D.F., México

Abstract

The aim of this study was to evaluate the amount and quality of water in the Magdalena-Eslava river system and to propose alternatives for sustainable water use. The system is the last urban river in the vicinity of Mexico City that supplies surface water to the urban area. Historical flow data were analyzed (1973–2010), along with the physicochemical and bacteriological attributes, documenting the evolution of these variables over the course of five years (2008–2012) in both dry and rainy seasons. The analyses show that the flow regime has been significantly altered. The physicochemical variables show significant differences between the natural area, where the river originates, and the urban area, where the river receives untreated wastewater. Nutrient and conductivity concentrations in the river were equivalent to domestic wastewater. Fecal pollution indicators and various pathogens were present in elevated densities, demonstrating a threat to the population living near the river. Estimates of the value of the water lost as a result of mixing clean and contaminated water are presented. This urban river should be rehabilitated as a sustainability practice, and if possible, these efforts should be replicated in other areas. Because of the public health issues and in view of the population exposure where the river flows through the city, the river should be improved aesthetically and should be treated to allow its ecosystem services to recover. This river represents an iconic case for Mexico City because it connects the natural and urban areas in a socio-ecological system that can potentially provide clean water for human consumption. Contaminated water could be treated and reused for irrigation in one of the green areas of the city. Wastewater treatment plants and the operation of the existing purification plants are urgent priorities that could lead to better, more sustainable water use practices in Mexico City.

Editor: Judi Hewitt, University of Waikato (National Institute of Water and Atmospheric Research), New Zealand

Funding: The authors would like to acknowledge the financial support of Gobierno del Distrito Federal, Secretaría de Medio Ambiente in conducting the project Plan Maestro de Manejo Integral y Aprovechamiento Sustentable de la Cuenca del Río Magdalena del Distrito Federal, through the Programa Universitario de Estudios sobre la Ciudad, Universidad Nacional Autónoma de México (2008–2009), and the project Sistema de Indicadores para el Rescate de los Ríos Magdalena y Eslava, Distrito Federal through the Programa Universitario de Medio Ambiente, UNAM (2009–2010). As well as the support of the Dirección General de Asuntos del Personal Académico through the Programa de Apoyo a Proyectos de Investigación e Innovación Tecnológica, PAPIIT IN219809. The funders had no role in study design, data collection and analysis, decision to publish, or preparation of the manuscript.

Competing Interests: The authors have declared that no competing interests exist.

* Email: mazari@unam.mx

Introduction

In megacities with populations exceeding ten million inhabitants, water is a scarce resource that is in high demand. The exploration and application of new, more sustainable water use and reuse approaches are urgently needed. Water managers must change the perspectives and policies that assume that water can only be used once by considering the benefits obtained from the natural environment surrounding the urban areas.

In emerging economies, cities grow before sufficient hydraulic infrastructure has been developed to handle the water supply and wastewater disposal requirements [1]. This lack of infrastructure significantly affects the surface and groundwater systems, making waterways vulnerable to pollution and leading to deleterious effects on water sources and human health. There is a need for more adequate approaches regarding processes that occur at the basin level, in addition to directly considering hydrological ecosystem services, such as water provision and water quality.

During the evolution of what is now the Mexico City Metropolitan Area (MCMA), lakes and rivers have been transformed into drains or converted into sewers in the vicinity of modern avenues, and freeways transit over piped waterways [2]. The hydraulic system of the Basin of Mexico, where the MCMA and its 22 million inhabitants are located, has been irreversibly altered. Of the 45 rivers in this region, most are piped to avoid flooding and unsanitary conditions.

In the Basin of Mexico, the current water demand is 77 m^3/s for first use, of which 71% is groundwater extracted through wells, 2% is derived from springs and surface water, 21% is from the Cutzamala surface water system, and 6% is from the Lerma groundwater system [3,4]. The subject of this research, the Magdalena-Eslava river system, represents the most important surface water source for human use in Mexico City.

Rivers flow from their origins in high areas to low-lying areas, exposing these water bodies to the integrated effects of human activities along their course and throughout the basin landscape. Anthropogenic impacts have impaired the ability of many river ecosystems to provide the goods and services upon which society depends. Watercourses may be degraded and rendered unable to support healthy aquatic communities such that natural processes are hampered and social value is lost.

Balancing the interactions between natural and constructed systems in urban areas is crucial for the future supply of water for large human settlements. Interest in river rehabilitation and restoration exists at the global level. River restoration is defined as an improvement in the integrity of a river through human intervention. The aim of river restoration is either to recover the hydrologic and ecological processes or to achieve rehabilitation, which means recovering a system that has not returned to its original state or is as healthy as it would have been if it had been fully restored [5].

The Magdalena-Eslava river system is one of the last non-piped rivers remaining in the MCMA. Given its contribution to the water supply in its upper reaches and the potential to recover some of the environmental services provided by the system, there is significant interest in the possibility of improving the water quality in the lower parts of its course. The aim of this study was to evaluate the Magdalena-Eslava river system in terms of the amount and quality of water, based on physicochemical and microbiological attributes that it offers the community. Accordingly, the study was performed based on historical and current data. As one of the last urban rivers, this study provides an opportunity to pose alternatives for more efficient water use.

Study area description

The study area, which is located in Mexico City, comprises the Magdalena and Eslava rivers. The river system covers two climatic regions. The lower area, which reaches to 3,050 masl, presents a temperate sub-humid climate with a summer rainy season and average temperatures that oscillate between 12 and 18°C. The upper basin has a semi-cold climate and a rainy season during the summer in areas higher than 3,050 masl, with an annual mean temperature between 5 and 12°C [6]. The rainy season ranges from May to October, with a monthly precipitation in excess of 250 mm in July [7]. The annual precipitation in most of the area is between 1,200 and 1,500 mm [8].

The Magdalena River originates in *Sierra de Las Cruces* from springs and runoff from the *La Palma*, *El Gavilán* and *El Muñeco* mountains in a location called *Cieneguillas*. The Magdalena River Basin is actually a sub-basin of the Basin of Mexico (hereafter called the Magdalena River Basin) and belongs to what is legally known as the *Zona Protectora Forestal Cañada de Contreras*, which was declared in 1932. The protected zone includes a fringe of 12 km from the river's origin, with a 500 m buffer on each side of the river, as declared in 1947 [9,10].

The Magdalena river basin extends from 2,570 to 3,870 masl [11], covering an area of 29.80 km^2. The Eslava River, which is the basin's largest tributary, flows through an area of 24.05 km^2, with an altitude descending to 2,510 masl at the urban limit.

The Magdalena River is 28.2 km long and flows 14.8 km through a natural forested area known as the Conservation Soil, with an additional 13.4 km crossing through the urban area. In the urban section, 8.8 km of river flows as an open canal, and 4.6 km is piped [12]. The Eslava River is 13.7 km long, with 10.96 km in the Conservation Soil and 2.74 km in urban areas [12]. The basins of the Magdalena and Eslava rivers are shown in Figure 1.

The Magdalena Basin, known as *Cañada de Contreras* in *Sierra de las Cruces*, is the only drainage of significant magnitude [13] within the Basin of Mexico and is primarily classified as temperate sub-humid. Despite its location in a peri-urban area, the Magdalena Basin serves as an important refuge for biodiversity and is home to 1,175 species, of which 212 are considered useful and 39 are listed in the at-risk category [14].

Three vegetation communities are present in the area, according to Ávila-Akerberg (2009) [8] and Nava (2003) [15]. The dominant vegetation types are *Abies religiosa* forest (46%) from 2,750 to 3,500 masl, *Pinus hartwegii* forest (29%) from 3,420 to 3,800 masl, *Quercus* (8.3%) and mixed forest (1.3%) from 2,600 to 3,370 masl, grassland (7.2%), and cloud forest (0.2%). Each type of forest produces different rates of runoff, as calculated by Jujnovsky et al. (2010) [16]. The *Abies* area produces 10,944,800 m^3 of water per year, the *Pinus* forest produces 6,878,000 m^3 of water per year, and the mixed oak forest generates 3,217,500 m^3 of water per year.

The natural vegetation cover has been altered as a result of human activities, resulting in an increase in erosion and the reduction of water and soil humidity retention. Therefore, water flows rapidly during flood events in the lower parts of the basin [13]. The forested area southwest of Mexico City represents the most important surface water source and groundwater recharge area. This area supplies water and regulates both flood events and the microclimate in the MCMA. Urban expansion represents a threat to the Conservation Soil area, which currently provides ecosystem services, such as fresh water.

At the request of the City, the *Gobierno del Distrito Federal* through its Environmental Ministry (*Secretaría del Medio Ambiente*), the two main universities located in Mexico City, the National Autonomous University of Mexico (UNAM) and the Metropolitan Autonomous University (UAM), generated a Master Plan to Rescue the Magdalena and Eslava Rivers [12]. This plan was developed by an interdisciplinary team that included engineers, architects, ecologists, biologists, architects, geographers, and sociologists, who proposed to rehabilitate the river as an urban area that could be used as a community recreation area while recovering the water quality and the watercourse running through the city [12]. A second project developed a system of indicators to measure the improvements resulting from the actions taken as a result of the Master Plan [17], which was not completely implemented.

The topic of river ecosystem recovery and rehabilitation has grown in relevance, receiving increasing attention in the last two decades. From an urban perspective, rivers are crucial for water supply, flood and drought regulation, storm-water management, water quality improvement, and erosion control; additionally, rivers serve as urban wildlife refuges. Overall, rivers can serve a range of purposes from supplying water to offering locations for social and recreational activities; hence, rivers often account for a significant fraction of the few recreational green areas in large cities [18,19].

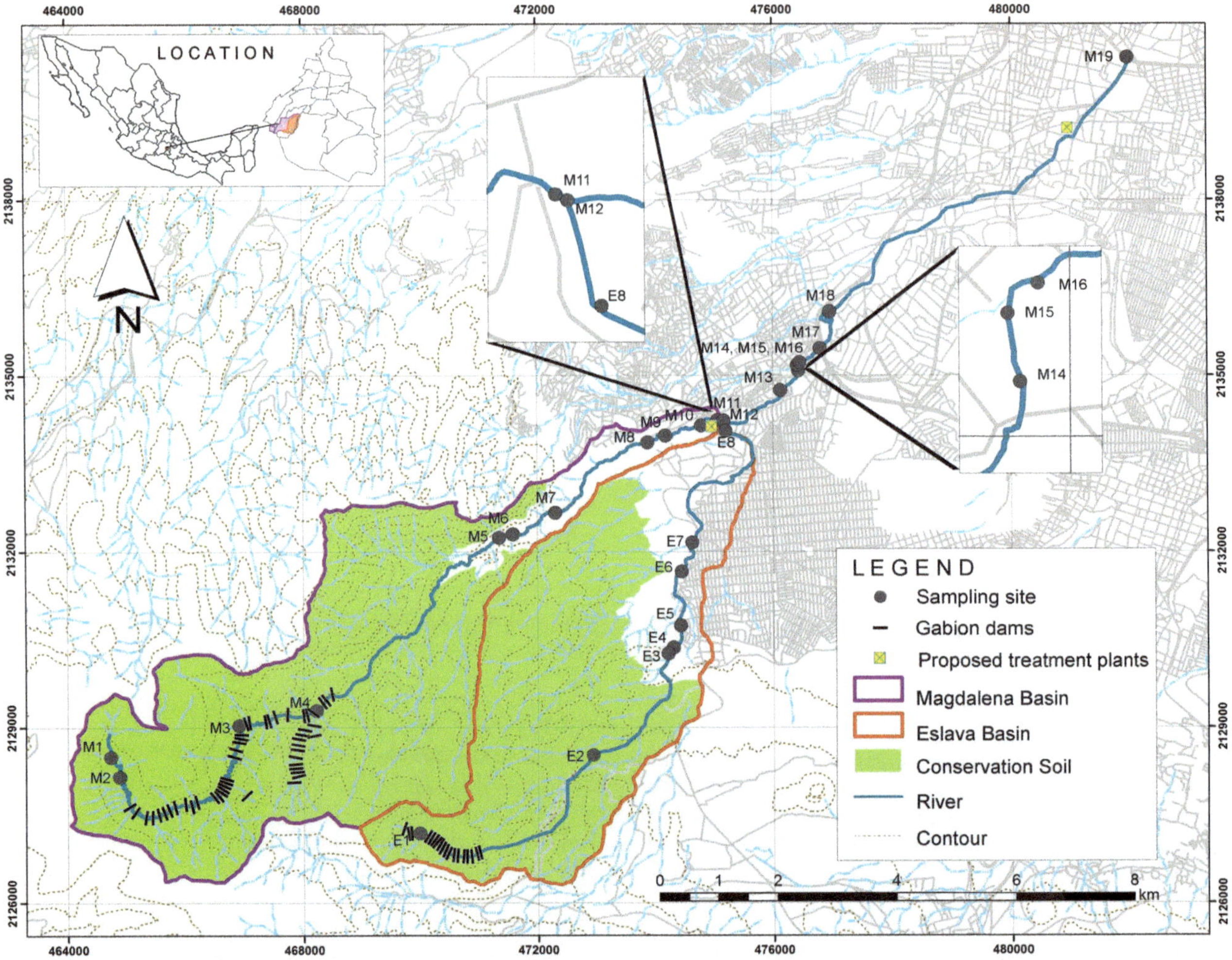

Figure 1. Study area, including the Magdalena and Eslava rivers within the Basin of Mexico and the Mexico City Metropolitan Area.

Methods

To determine the water flow, historical data were collected and analyzed (1973–2010); this approached allowed us to discern important variations due to seasonal or management effect in the river system. Based on current water quality measurements (2008–2012), we describe the actual condition of the river based on physicochemical and microbiological variables. The methods described in this section were applied.

Water flow

A flow analysis was conducted based on historical data recorded at the Santa Teresa hydrometric station (26440) to understand the water quantity and variability of flow at the only existing station of the National Water Commission. This hydrometric station is located after the junction of the Magdalena and Eslava rivers. As mentioned above, the Eslava is the main tributary of the Magdalena River, which also receives inflows from additional small creeks and natural run-off, combined with the discharge of wastewater from the surrounding urban area.

Daily information was obtained from the official flow surveys recorded at the Santa Teresa hydrometric station (26440) and was published in the National Data Bank of Surface Water [20] of the National Water Commission (*Comisión Nacional del Agua*). The data were analyzed for consistency and to detect and correct deviations from the historic series based on the double mass curve method [21,22]. The series of daily data spanning 38 years were then plotted, and two periods for the flow regime were identified.

We calculated the average monthly flow for each regime, i.e., the natural and the regulated areas, so we could compare the regulating effect of water infrastructure on the river. Finally, we plotted the average, minimum and maximum annual flows to observe the river's evolution over time.

Experimental design

After a general survey of the area, sampling sites for water quality evaluation were selected at places with little human or animal influence in the natural area and at others with inputs of urban runoff and wastewater in the urban area. Based on an initial diagnosis performed in 2008 and 2009 [12,17,23], the sampling locations were evaluated for redundancy in terms of the information provided. We computed a correlation matrix for the 19 sampling locations. After 2009, the sampling protocols included ten sampling sites. If two locations had a Pearson's correlation above 0.6 and were neighbors of each other, one was eliminated. This procedure allowed us to reduce the number of sampling locations from 19 to eight stations. We then added two sampling locations a few meters upriver of the intake of the water treatment plants that are located along the river. These ten stations were sampled from 2010 to 2012. The sampling covered the dry and rainy seasons, as shown in Table 1. The selected sampling sites, identified as representative of the Magdalena and Eslava rivers, are depicted in Figure 1. Fieldwork permission was given by the *Secretaría del Medio Ambiente* (SMA) and by the *Comisión de Recursos Naturales* (CORENA), the local authority that belongs to the *Gobierno del Distrito Federal* (GDF).

The behaviors of the two rivers were plotted independently, where the Magdalena River (main course) is presented with a five-year data set (2008–2012), while its tributary, the Eslava River, is shown with a two-year data set (rainy season 2010, 2011–2012) in Table 1.

Physicochemical analyses

As part of a long-term water quality monitoring program in the Magdalena Eslava river system, the attributes of temperature (°C), pH, electric conductivity (µS/cm), turbidity (NTU), total dissolved solids (TDS, mg/L), dissolved oxygen (DO, mg/L), DO saturation percentage, total phosphorus (total P, mg/L), orthophosphate (P-PO_4, mg/L), total nitrogen (total N, mg/L), ammonia (N-NH_3, mg/L), and nitrates (N-NO_3, mg/L) were measured.

The physicochemical variables of temperature, pH, electric conductivity, turbidity, total dissolved solids and dissolved oxygen were measured *in situ* over the course of five years. All measurements were made using a Multiparameter Water Quality Sonde YSI 6600-M (Yellow Springs, OH, USA). Five-hundred-milliliter water samples were taken and analyzed in triplicate for nutrients using a spectrophotometer HACH Model DR2400 (Loveland, CO, USA) following the HACH manual [24]. Total P was measured following the method of PhosVer3 by acid digestion USEPA (0.06 to 3.50 mg/L P) and the molybdovanadate method (1 to 100 mg/L PO_4^{3-}). Orthophosphate was measured using the amino acid method (0.23 to 30.00 mg/L PO_4^{3-}), and ascorbic acid was measured using PhosVer3 USEPA (0.02 to 2.50 mg/L PO_4^{3-}). Total N was measured using the persulfate digestion method (10 to 50 mg/L N) and (0.5 to 25.0 mg/L N). Ammonia was measured following the silicate method (0.02 to 2.50 mg/L NH_3-N), and nitrates were measured using the cadmium reduction method USEPA (0.01 to 0.50 mg/L NO_3-N) [24].

Microbiological analyses

Samples for microbiological analyses were taken at the same time and location as those collected for the physicochemical analyses.

Bacteria counts. One-liter samples for bacteriological analyses were collected from selected sites using polypropylene sterile flasks. Samples taken in triplicate were transported and stored at 4°C according to the American Public Health Association standard procedures [25,26]. Microbiological samples were processed within 24 h of collection. All samples were analyzed in triplicate, and the results were reported in colony-forming units (CFUs) [27,28].

Two bacteriological variables, including fecal coliform (FC, UFC/100 mL) and fecal enterococci (FE, UFC/100 mL), were measured constantly during the period from 2008–2012. Total coliforms (TC, UFC/100 mL) were determined only for the period 2008–2009. Positive samples were processed to identify the pathogenic and opportunistic bacteria.

Water samples were analyzed following the standard membrane filtration procedures to enumerate three bacterial types, namely, FC, FE, and TC. Membrane filters (0.45 µm cellulose acetate, Millipore MF type HA) were placed on M-FC agar for FC, KF Streptococcus agar for streptococci/enterococci, and m-Endo agar MF for total coliforms. Incubation was performed with a WTB Binder brand incubator at 35°C for 24 h for enterococci and total coliforms and at 44.5°C for 24 h for fecal coliforms, according to APHA [25,26]. The results are reported as colony forming units (CFUs).

Bacteria identification. Gram-stain and biochemical tests were performed to identify bacteria using a semi-automatic DADE MicroScan, AutoSCAN-4 (DADE International. West Sacramento, CA) [26]. Positive samples from water filtration isolates were selected based on morphology. Five colonies of each specific morphology were selected and identified by the MicroScan Auto SCAN-4 DADE (West Sacramento, CA).

Additional microbiological variables were analyzed during the period 2008–2009 and during the rainy season of 2010. We evaluated these variables because of their importance and relation to gastrointestinal diseases as a result of their prevalence in water in Mexico. These microorganisms included coliphage, enterovirus (EV), adenovirus F human serotypes 40 and 41 (AdV F), *Cryptosporidium parvum* oocysts and *Giardia lamblia* cysts.

Table 1. The sampling design for the Magdalena-Eslava river system.

Variable	Dry season	Rainy season
Magdalena River		
Physicochemical	2008, 2009, 2010, 2012	2008, 2009, 2010, 2011, 2012
Indicator bacteria	2008, 2009, 2010, 2012	2008, 2009, 2010, 2011, 2012
Pathogens	2008, 2009	2008, 2009, 2010
Coliphages	2008, 2009	2008, 2009
Virus	2008, 2009	2008, 2009
Protozoa	2008, 2009	2008, 2009
Eslava River		
Physicochemical	2011, 2012	2010, 2011, 2012
Indicator bacteria	2011, 2012	2010, 2011, 2012
Pathogens		2010

Coliphage. Coliphages were quantified in duplicate following the double-layer agar method based on the International Standard Water quality-Detection and enumeration of bacteriophages (ISO 10705-1, 1995) [29]. The plaques were quantified after the incubation period, and the results were reported in plaque forming units per milliliter (PFU/mL) [30].

The method detection limit was determined following the double-layer agar technique. Serial dilutions were performed using an MS2 phage with a known titer of 10^8 PFU/mL and *E. coli* K12- Hfr 3000 (ATCC 23631) as the host bacteria. The detection limit for this method was determined to be between 10 and 1 PFU [31].

The coliphage quantification with the *E. coli* Hs (pFamp) as host bacteria was performed in duplicate using the adaptation USEPA "Method 1601: Male-specific (F+) and Somatic Coliphage in Water by Two-step Enrichment Procedure)" [29,30].

Virus determination. For the detection of viruses and protozoa, 10-L water samples collected from the natural sites were concentrated by ultrafiltration following the method of Polaczyk et al. (2008) [32].

Enterovirus (EV) quantification was performed in duplicate on the water samples using RNA extraction techniques based on the RNA viral QIAamp (QIAGEN) mini kit. To quantify the genome number of EV, the basic protocol reported by Monpoeho et al. (2000) [33] was followed.

To obtain the AdV F genome number, the protocol by Xagoraraki et al. (2007) [34] was followed.

Protozoa quantification. *Cryptosporidium parvum* and *Giardia lamblia* were detected and quantified by immunofluorescence microscopy.

One milliliter of each sample (direct for wastewater or concentrated for clean samples) was analyzed in single samples using an indirect immunofluorescence technique in liquid phase according to Method 1623 [35], as modified by Rangel-Martínez [36] and Tapia-Palacios [37]. To count the (oo)cysts in the samples, an Axiostar Plus (Carl Zeiss, Gottingen, Germany) fluorescence microscope was used. Total scans of each slide were performed at 20X and 40X and, in special cases, at 100X for confirmation.

Statistical Analyses

The data gathered during the study were analyzed in a spatial and temporal context. Because the sampling locations were located along river courses, we assumed that the spatial coordinates were one-dimensional. In this coordinate system, we identified each site by a site code and by the distance from the first sampling location in the uppermost part of each river. We fitted generalized linear models (GLMs) with the adequate link function for each attribute [38]. For example, for each variable whose histogram clearly skewed to the left, we fitted GLMs with Gamma error and a logarithmic link function. These analyses were performed to explore the association between the physicochemical and biological variables, using site, season and year as covariates. Such analysis allowed us to detect significant effects of the covariates in each of the attributes considered in the analysis. We also created interaction plots to graphically depict the spatial behavior of the values of the variables along the downhill path of the Magdalena and Eslava rivers and at their confluence. Separating the courses in this way allowed us to compare the Magdalena and Eslava rivers before their merging point and to detect whether differences were present in the water quality variables relative to downriver points.

Results and Discussion

Water flow

Despite their relatively small size, the Magdalena River and its main tributary, the Eslava River, provide water to part of the southern area of Mexico City and are the main surface water source for the city, representing more than half of the surface water used. Several studies have been conducted to estimate the amount of water provided by the Magdalena River. In the 1950s, when water sources for the city were investigated, the Magdalena River registers showed a mean flow of 0.33 m^3/s or 10,400,000 m^3/yr [39]. The mean water volume for the Magdalena River was reported in 1975 as 6,970,000 m^3, equivalent to 0.22 m^3/s, and that for the Eslava River was 3,578,000 m^3, equivalent to 0.11 m^3/s [40]. These volumes total 0.33 m^3/s, with a total volume for potential use of 10,548,000 m^3/yr.

A first view of the hydrologic ecosystem services in terms of water quantity and quality with runoff estimates per environmental unit gives an annual production for 2002–2003 of 21,538,250 m^3, equivalent to 0.67 m^3/s, which is similar to the 0.58 m^3/s reported at the hydrometric station in 1999. From 1999–2001 [41], the average flow was 0.70 m^3/s, of which 81% corresponded to baseflow and 19% to surface runoff, providing water for approximately 118,237 inhabitants. In a complementary study, Jujnovsky et al. (2012) [42] assessed the water supply from the Magdalena River Watershed, determining the water balance based on the SWAT model, and obtained a water provision of 18,400,000 m^3/yr. The baseflow contribution accounted for 85%, while the surface runoff was approximately 15%, providing drinking water for approximately 78,500 inhabitants and being capable of supplying 153,203 potential beneficiaries.

Natural and regulated flow regimes. Our study shows the mean annual flow in the hydrometric station to be 0.78 m^3/s, with contributions of 0.45 m^3/s (57%) from the Magdalena River and 0.34 m^3/s (43%) from the Eslava River. The daily regimen is shown in Figure 2, based on historical data from 1973 to 2010, clearly depicting two periods. The first period is from 1973 to 1989, which represents the natural flow regime (NFR), and the second period is from 1990 to 2010, which represents a regulated flow regimen (RFR). This graph shows a decreasing flow and less seasonal variability in the maximum and minimum under the RFR.

The natural flow regime achieved rates of 0.90 m^3/s per year, surpassing the 2 m^3/s during 16 of the 17 years identified as NFR. Starting in 1990, the maximum flow rarely reached 2 m^3/s. The minimum regulated flow of the second period is three times higher than the natural minimum flow, varying from 0.13 to 0.37 m^3/s, that is, the minimum flow increased by 176%. In contrast, the maxima were reduced by 69% (Figure 2). Regulation structures, such as gabion dams, increase the minimum flow due to retention and water storage. At present, more than 60 regulation structures have been registered [12], as shown in Figure 1. These structures were constructed for different objectives, such as decreasing the flow velocity, creating micro wetlands and diminishing sediment volumes, among others [12]. These hydraulic structures are not recommended because they function as frontiers that break the biological continuity of lotic systems [17].

The hydrometric information is depicted in the curves of average monthly flow (Figure 3). The NFR includes the dry season (January to May), with a mean measurement of 0.29 m^3/s and a lower flow in April of 0.26 m^3/s. The rainy season shows an increase in flow from June to August and a decline in December (0.66, 1.72 and 0.46 m^3/s, respectively). At the end of the 1980s,

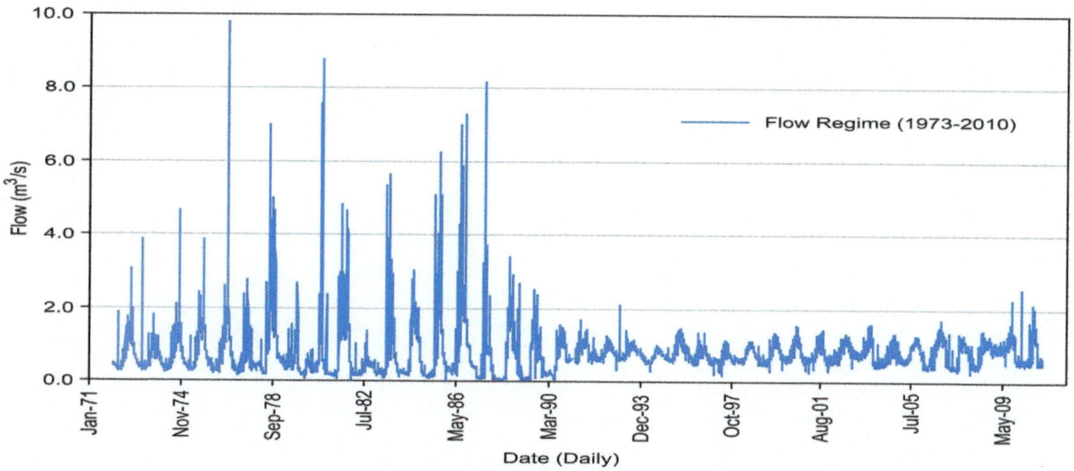

Figure 2. Daily volumetric flow regime (1973–2010) measured at the hydrometric station Santa Teresa (26440).

the basins of the Magdalena and Eslava rivers experienced flow variations, as observed in Figure 3. In the dry season, the RFR curve increased 128% compared with the natural regime (NFR). The RFR value for April was 0.60 m^3/s, compared with 0.26 m^3/s for the NFR in the same month. In the rainy season, the RFR curve diminished with respect to the NFR by 30% from July to October.

Figure 3 shows the changes experienced by the river before and after the hydraulic regulation in the periods mentioned previously, which is defined for four average monthly flow regimes. The inertial behavior of the minimum natural flow regime (mNFR), shows that the dry period (January-May) minimum flow was 0.06 m^3/s, except in March-April, when the minimum flow was 0.04 m^3/s. Starting in the rainy period, the minimum flow responded to the natural increase from June-August (a maximum flow of 0.57 m^3/s) until declining over the period of September to December from 0.51 to 0.09 m^3/s.

The minimum regulated flow regime (mRFR) differs from the natural condition (mNFR curve), with a greater effect observed during the dry season. In March and April, the minimum natural flows increased to 490% (from 0.04 to 0.24 m^3/s) and then increased 202% in January, February and May. From July to October, the mRFR curve increased 64% from its original condition (1973 to 1989); in November and December, the flows increased from 0.20 and 0.09 to 0.57 and 0.47 m^3/s, respectively. This increase can be attributed to the hydraulic structures and to external factors, such as wastewater discharge from irregular settlements in the lower drainage basins and surface runoff during the rainy season. The hydrological histogram (Figure 3) shows the discrepancy in hydrological conditions: under low-flow conditions, the mRFR curve corresponds to the NFR conditions. Therefore, the minimum regulated behavior mimics a natural condition.

Figure 4 depicts the annual flow over a 38-year period. In the first period (1973–1989), the maximum flow reached 10 m^3/s, with a mean annual maximum of 4.94 m^3/s. In the 1990s, the development of hydraulic infrastructure in the river caused a laminar flow effect. During the second period, from 1990 to 2010, the maximum flow decreased 69% (relative to 2.53 m^3/s as the maximum), with a mean annual maximum flow of 1.52 m^3/s.

The behavior of the mean annual flows (Figure 4) also reflects the effect of regulation. Although the mean flow in both periods is very similar, namely, 0.77 and 0.79 m^3/s, its volume has not changed significantly. Over time, extraction from the river has been balanced by wastewater inputs into the river, which has led

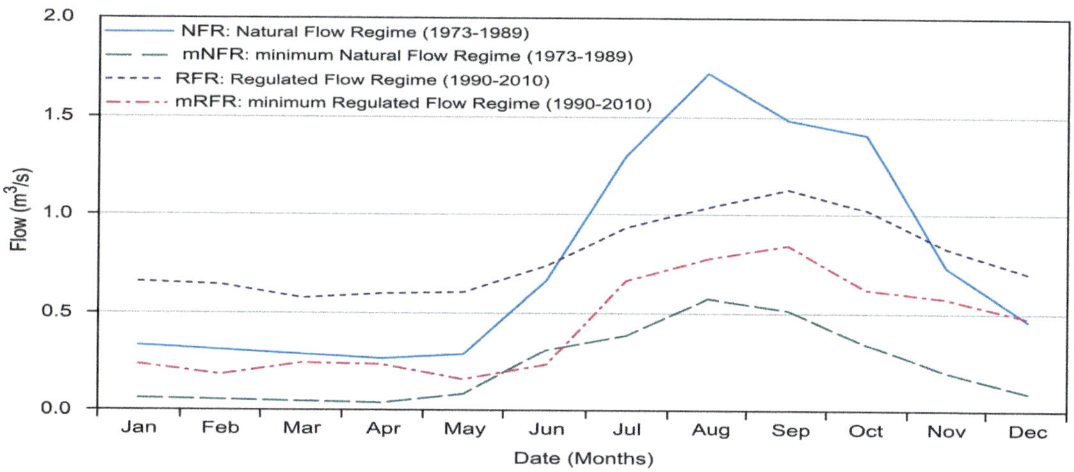

Figure 3. Interannual variation of the NFRs, RFRs, mNFRs and mRFRs measured at hydrometric station 26440.

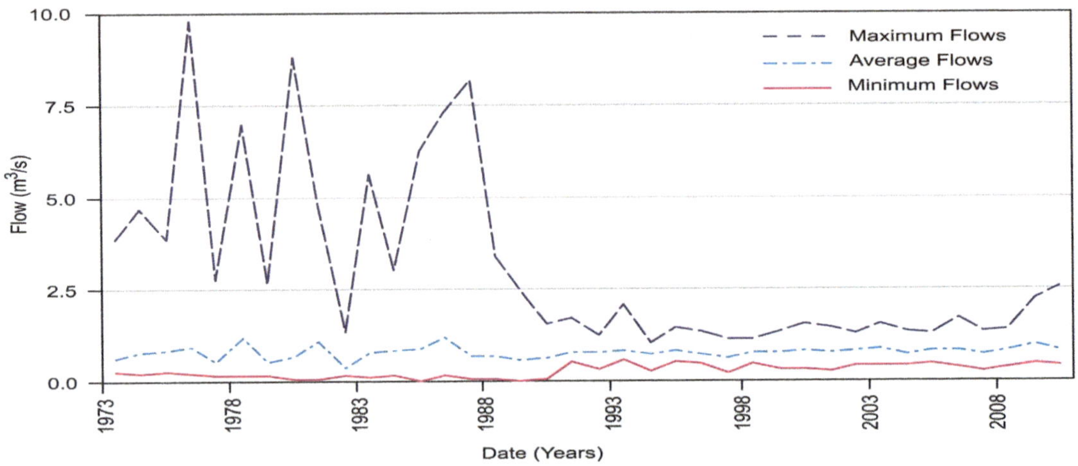

Figure 4. Annual volumetric flow regime at the outlet of the Magdalena and Eslava river basins. Average, maximum and minimum flows.

to an increase of 0.02 m^3/s. The exchange of clean water for untreated wastewater has significantly degraded the environmental condition of the river.

There are only two direct discharge points into the river over the course of the distance from its origin to the treatment plant; in contrast, after the treatment plant, 58 inputs are known [12]. Irregular settlements are problematic because they often lack a sewer system, thereby allowing wastewater to flow directly into the rock or soil and hence toward the lower discharge area where the river is located.

In Figure 5, a historical annual comparative analysis illustrates the natural flow from 1973 to 1989 in two flow-duration curves; during this time, the annual mean flow at the lower point of the Magdalena river was at least 0.54 m^3/s more than 90% of the time, which represents the probable flow that the river used to supply along the year. High flows with flows of approximately 1.14 m^3/s are present during a very short period (10% of the times). In the 1990s, the development of hydraulic infrastructure in the river caused a laminar flow effect; therefore, in the curve for the regulated period from 1990–2010, the mean flow at the river is at least 0.71 m^3/s for 90% of the time, which is an increase of 31% in comparison with the previous natural minimum; consequently, a 25% decrease in high flows is observed. It is also possible to observe changes to the slope of the natural flow curve (reference) with respect to the regulated flow curve, specifically, changing from 0.73 to 0.24%.

The behavior of the natural curve of collected frequencies (1973 to 1989) depicted in Figure 5 is representative of mountain rivers; therefore, based on the flow duration curve observed for the regulated period (1990 to 2010), a new behavior is observed. That is, the river is behaving if it were a floodplain river, where changes are less apparent. Generally, this response is due to changes to the flow of the Magdalena River as a result of anthropogenic activities, with an associated increase in the concentration of several water quality variables.

Figure 6 shows the distribution of water flux for the rainy and dry seasons for the years 2008–2010, starting with the dry season from January-April 2008, followed by the rainy season of 2008, the dry season from November 2008–April 2009, and so on. The figure demonstrates that the mean value is relatively constant for dry and rainy seasons, while the variability changed during those same years; hence, no pattern can be observed for the changes in variability. This is related to the high variability of the amount of rain falling in the upper parts of the basin, particularly during the rainy season of 2010. The low variability for the dry season from November–December 2010 is due in part to the small amount of data for this time lapse.

Water quality

Physicochemical and microbiological analyses. The spatial and temporal analyses for the physicochemical variables and nutrients are presented in Figures 7 and 8. These variables are those most representative of river water quality, such as electrical conductivity, dissolved oxygen, ammonia, total N, orthophosphates and total P, as well as fecal coliforms and fecal enterococci. These variables are considered representative because they have the highest absolute value in the factor loadings of the first three principal components. The factor loadings for these attributes account for more than 80% of the total variability of the Magdalena River system [43].

Although the number of samples at each site was not the same, it only affects the variance of the mean values of the measured water properties. This effect has been considered and corrected in all of our analyses by dividing the mean value estimates by $\sigma/\sqrt{n}$.

The GLM models fitted to the water quality variables [38] indicate statistically significant differences for all of the variables, except dissolved oxygen, in the upper and lower sections of both rivers. The point of change was the location at which human activities and settlements began on either or both river banks. Water quality variables sampled upstream of these urban areas fell within acceptable limits for use as a water supply. The values gradually fell to lower quality as the sampling sites became increasingly affected by the wastewater generated in the irregularly inhabited areas. The spatial trends along the river gradient for the eight most-relevant water quality attributes are shown as box plots in Figure 7 for the Magdalena River and in Figure 8 for the Eslava River.

The plot shows a change point at site M9, which is located at the border between the natural and urban areas: here, increasing concentrations of all values can be observed, except that of dissolved oxygen, which decreases. The highest concentrations were detected at site M15, where part of the water is piped to the city's drainage system. These values then decrease slightly at sites M17 and M18, where other effluents with residual water from the southern area of the city join the waterway.

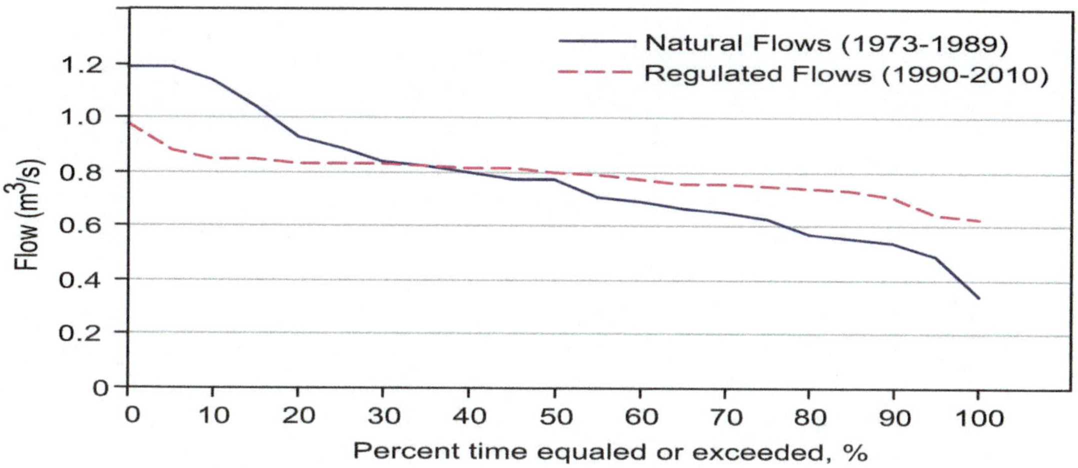

Figure 5. Duration curve of annual natural and regulated flows, Magdalena river basin, 1973–2010.

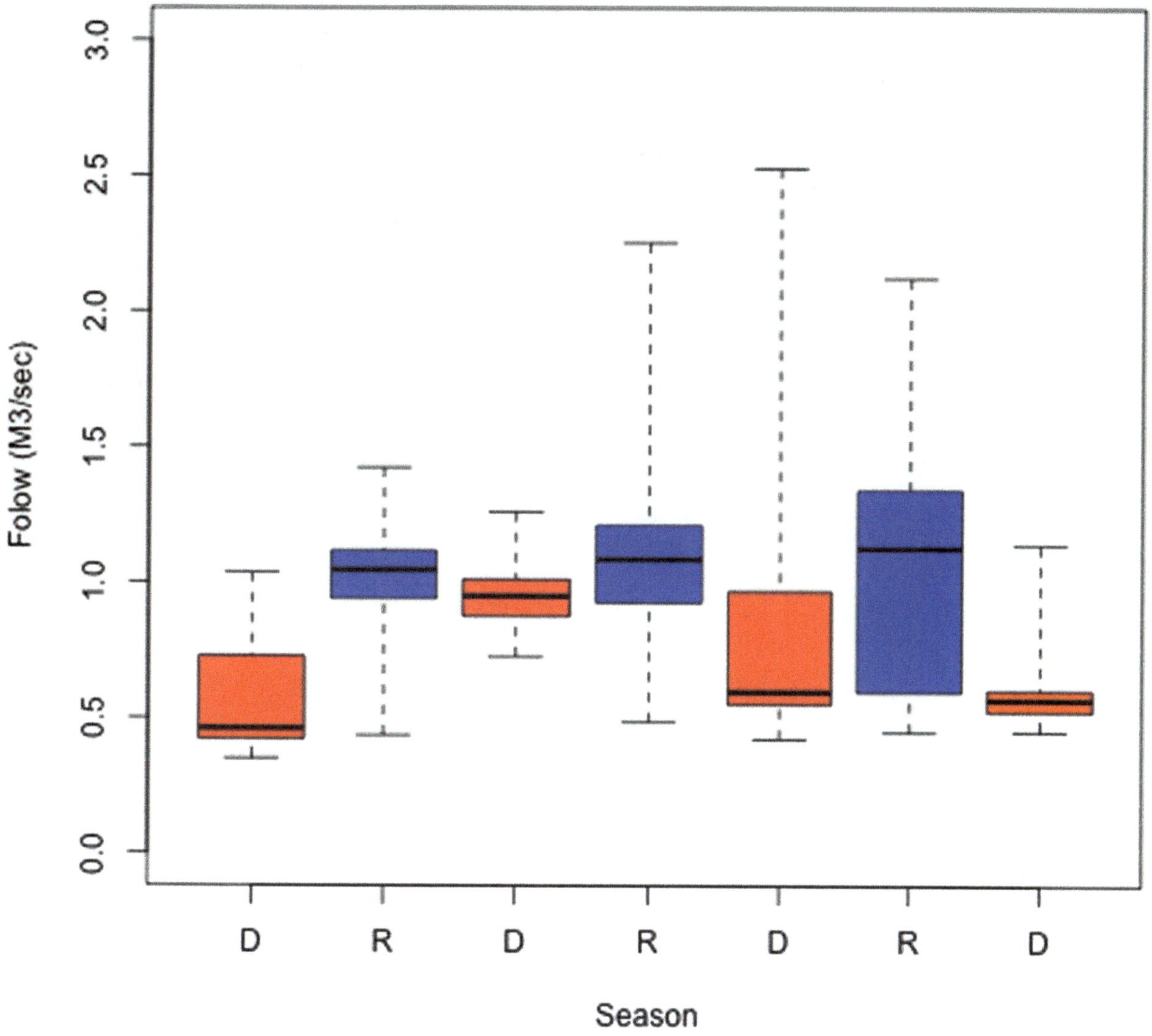

Figure 6. Water flux distribution from 2008–2010 grouped by season.

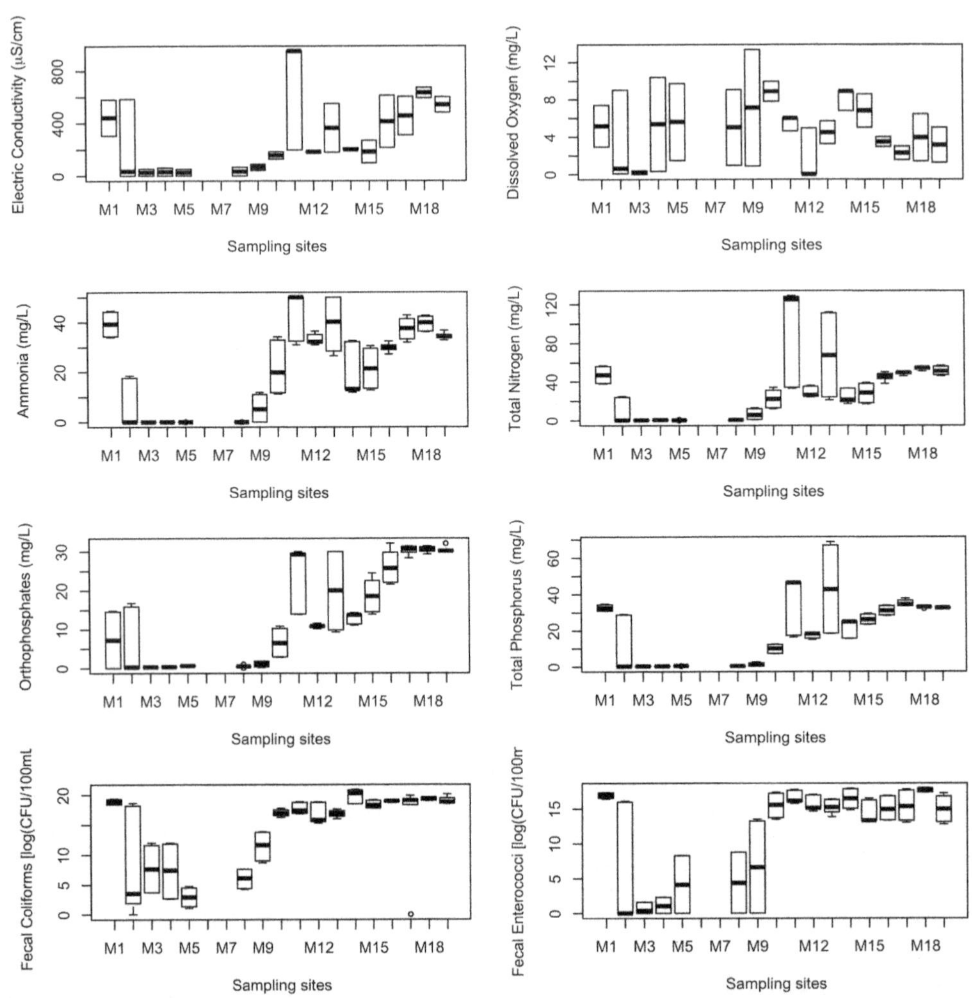

Figure 7. Spatial trend for the eight main water quality variables in the Magdalena River, based on 19 sampling stations.

In the Eslava River, all variables except dissolved oxygen are significantly different between the natural and urban areas, a change that occurs somewhere between E5 and E6. Further, all variables exhibit stable values until the confluence point. The Eslava River has lower water quality than the Magdalena River does at the confluence point, particularly in terms of electrical conductivity, the value of which is six times higher in the Eslava River; additionally, the value of total N is three times higher, while that of total P is twice the concentration of that in the Eslava River. The fecal enterococci values were three times as large on a log scale, which means that this measurement is 1,000 times greater. Please note this difference in the values of the axes on the graphs.

According to Mexican standards, several attributes surpass the NOM-001-SEMARNAT-1996 [44] limits for water discharged to Mexican water bodies. The total N exceeds these levels from M15 to M18 and is double the limit at E8 in the Eslava River. At E7, the values are slightly above the limit.

Because both rivers are used as surface water sources for human use and consumption, the LFDMA [45] limit is 1,000 FC per 100 mL. Taking this into account, the fecal coliforms surpass the limit and require treatment before distribution through the southern parts of the city. The standards are met at the Potabilization Plant La Magdalena, but there are many water intakes near settled populations in an irregular land ownership situation, which means that these populations are exposed to high levels of bacteria; this is an important risk factor that should be taken into further account.

Bacteria identification. The 164 bacterial isolates and identifications obtained from 516 water samples taken from the Magdalena River in the period 2008 to 2010 are presented in Table 2. These belong to nine different families and 18 genera [26,46]. The most abundant genera were *Enterococcus*, *Escherichia coli*, *Enterobacter* and *Klebsiella*. From the Eslava River, 27 bacterial isolates and identifications were obtained from 120 water samples taken during 2012. These belong to four different families and seven genera, as reported in Table 3.

Therefore, a variety of bacteria belonging to a total of 20 different genera were identified in the river system crossing the southern area of Mexico City. The most prevalent bacteria were from the Enterococcaceae family, *Enterococcus*, and from the Enterobacteriaceae family, *Escherichia*, *Enterobacter* and *Klebsiella*. Although these results serve as an initial approach to the identification of bacteria, it is important to note that all of these bacteria are common to the human gut and can be opportunistic pathogens that may cause illness to immune-compromised people who access this water when untreated.

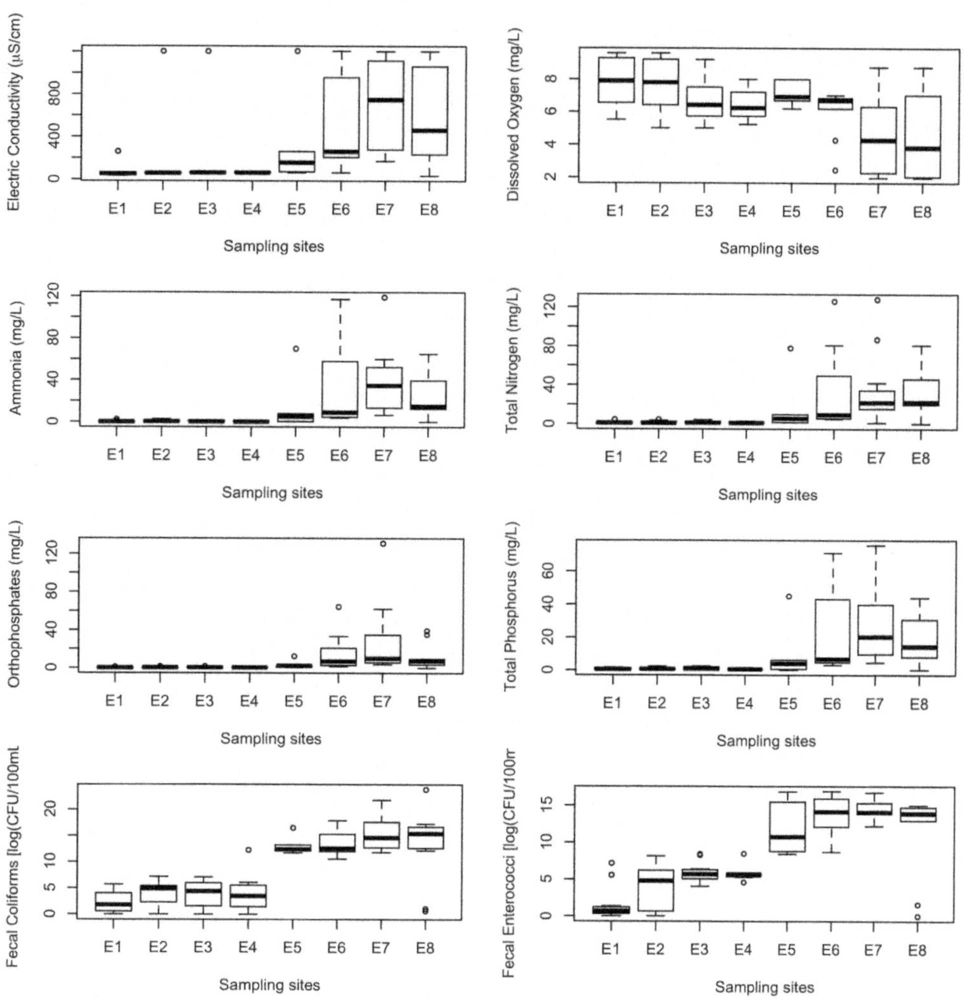

Figure 8. Spatial trend for the eight main water quality variables in the Eslava River, based on eight sampling stations.

Coliphage, virus and parasite analyses. The presence of coliphage was measured by bacterial hosts in samples from the urban zone; this finding is consistent with the inputs of raw and untreated wastewater to the Magdalena River (Table 4). As host bacteria, *E. coli* K12-Hfr registered the higher sensitivity, although larger counts were obtained for *E. coli* Hs (pFamp). It is important to note that there are two infection mechanisms of interest, either by adherence to the cell wall or infection via pilli, as in the coliphage F-specific [47,48,49,50]. Therefore, the host is more susceptible to infection and shows higher counts.

In the natural area sites, no coliphage presence was detected, which suggests that perturbation does not have a detectable impact using this method. Generally, the presence of a phage indicates fecal contamination by a virus; however, this level of analysis is not conclusive because it cannot establish the contaminants' origin, namely, whether human or animal. In urban areas, coliphages are present from the transition zone between the natural and urban zones, where the highest density was detected at site M19; hence, the number of phages increases with human perturbation and influence.

Both adenovirus (AdV F) and enterovirus (EV) exclusively affect humans; their presence indicates human fecal contamination [51,52]. Once these viral particles have been excreted into the environment, their number cannot increase because the presence of the cell host is required for their replication [53]. Therefore, the detection of AdV and EV reflects the fact that in the human population, the virus is circulating and replicating. This constitutes a threat that must be considered regarding the river because these viruses can cause infections to susceptible individuals, yielding viral gastroenteritis.

AdV F replicates quickly in the gastrointestinal track, so the amount of excreted viral particles is on average 1×10^{10}/g in feces [53]. Such infections mainly occur in children under five years of age because adults who were infected as young children often acquire immunity [52]. Nevertheless, if adults are exposed to viral particles, they can be asymptomatically reinfected, with a lower amount of viral particles excreted in their feces. This can favor the dispersion of AdV F because these individuals will not be cautious in their behaviors because they are asymptomatic.

People can present infections and acute gastroenteritis as a result of AdV F [54]. It has been estimated that this virus is the cause of 9 to 12% of acute gastrointestinal cases in children under five years of age in Mexico; however, based on the lack of follow up regarding this type of pathogen and the lack of updated epidemiological information (few studies have been conducted since 1985), this may be an underestimation [55]. Due to the relevance of this virus as a gastroenteritis-causing agent, it is considered the second-most important virus after rotavirus

Table 2. The bacteria identified in the Magdalena River water samples in 2008, 2009 and 2010.

	Sites																			
Isolated microorganisms	**M1**	**M2**	**M3**	**M4**	**M5**	**M6**	**M7**	**M8**	**M9**	**M10**	**M11**	**M12**	**M13**	**M14**	**M15**	**M16**	**M17**	**M18**	**M19**	
n	**24**	**24**	**6**	**24**	**42**	**42**	**18**	**18**	**42**	**42**	**6**	**6**	**42**	**24**	**42**	**24**	**42**	**24**	**24**	**Total**
Moraxellaceae																				
Acinetobacter spp.																		1		1
Aeromonadaceae																				
Aeromonas sp.				1										1		1	1	1		5
Burkholderiaceae																				
Burkholderia sp.								1												1
Alcaligenaceae																				
Bordetella sp.								2												2
Enterobacteriaceae																				
Citrobacter spp.													1				1			2
Enterobacter spp.	1	1			1			3	1		2	1	3	2		1		1	2	19
Escherichia sp.	2	1			2			1	2	1	2	3	2	1	1	3	3	3		27
Hafnia sp.	1																			1
Klebsiella spp.								1	1		1	2	1		1	2	2	2		13
Proteus sp.					1															1
Providencia sp.			1	1																2
Salmonella sp.								1												1
Serratia spp.									2											2
Yersinia sp.						1		1												2
Leuconostocaceae																				
Leuconostoc sp.					1															1
Pseudomonadaceae																				
Pseudomonas sp.			2	3	1				1		1				1					9
Vibrionaceae																				
Vibrio sp.									1			2	1	2	1	1				8
Enterococcaceae																				
Enterococcus spp.	4	2		5	5	1	1	6	5	2	6	3	2	6	2	6	6	5		67
Total	8	4	3	10	11	2	1	16	13	3	12	11	10	12	6	14	13	13	2	164

Table 3. The bacteria identified in the Eslava River water samples in 2010.

	Sites								
Isolated microorganisms	**E1**	**E2**	**E3**	**E4**	**E5**	**E6**	**E7**	**E8**	**Total**
n	**15**	**15**	**15**	**15**	**15**	**15**	**15**	**15**	
Burkholderiaceae									
Ralstonia spp.				1					1
Bacillaceae									
Bacillus spp.	1								1
Enterobacteriaceae									
Escherichia spp.		1	1	1	1	1	1	2	8
Enterobacter spp.				1					1
Klebsiella spp.						1	1	2	4
Yersinia spp.								1	1
Enterococcaceae									
Enterococcus spp.	1	1	1	1	1	1	1	4	11
Total	2	2	2	4	2	3	3	9	27

Table 4. The geometric mean, standard deviation in log scale, maximum and minimum of Coliphage, Enterovirus, Adenovirus, *Cryptosporidium parvum* oocysts and *Giardia lamblia* cysts in the Magdalena River water from 2008 to 2009.

		Sites							
		M1	**M5**	**M8**	**M11**	**M12**	**M17**	**M18**	**M19**
Coliphage with host	Geometric mean	<1	<1		39	52	57	22	114
K12	Standard deviation	<1	<1		17	18	58	51	14
PFU/mL	Min-Max	<1	<1		29–65	27–65	9–140	7–110	97–127
	n	4	4		4	4	6	6	6
Coliphage with host	Geometric mean	<1	<1	<1	4	7	3	3	33
Hs	Standard deviation	<1	<1	<1	10	7	10	3	26
PFU/mL	Min-Max	<1	<1	<1	<1–18	<1–17	<1–20	<1–7	18–80
	n	4	4	2	4	4	6	6	6
Enterovirus Gen/mL	Geometric mean	<1	<1	<1	<1	<1	<1	<1	<1
	Standard deviation	<1	<1	<1	<1	<1	<1	<1	<1
	Min-Max	<1	<1	<1	<1	<1	<1	<1	<1
	n	3	5	6	4	4	6	4	4
Adenovirus Gen/mL	Geometric mean	10	<1	<1	12974	7299	98294	7305	97084
	Standard deviation	577	<1	<1	46297	28951	253772	10157	35432
	Min-Max	<1–1000	<1	<1	1130–930464	982–63008	24930–557800	1618–25186	71486–144840
	n	3	5	4		4	6	4	4
Oocysts	Geometric mean	<1	32	<1	6784	21	575	4244	<1
Cryptosporidium	Standard deviation	<1	707	<1	14849	5196	9504	2121	<1
parvum/L*	Min-Max	<1	0–1000	<1	2000–230002	<1–9000	<1–19000	3000–6000	<1
	n	2	2	1		3	3	2	2
Cysts *Giardia*	Geometric Mean	1000	71	<1	18	18	30	89	<1
lamblia/L*	Standard deviation	<1	3536	<1	2828	3464	16166	5657	<1
	Min-Max	1000–1000	<1–5000	<1	<1–4000	<1–6000	<1–28000	<1–8000	<1
	n	2	2	1	2	2	3	2	2

Cryptosporidium parvum and Giardia lamblia detection limit 1 (oo)cyst/mL [37]; Adenovirus (AdV) detection limit 10 genome copies [34]; Enterovirus (EV) detection limit 10 genome copies [74]; FRNA specific bacteriophage detection limit 1 PFU/mL [30], somatic bacteriophage detection limit 1 PFU/mL [30,34].

worldwide [56]; therefore, it is important to monitor the prevalence of this virus both in the human population and in residual water as an indicator of what is infecting a given population.

According to our results (Table 4), AdV F is present at both the natural and urban sites, indicating human fecal contamination in both locations. The presence of AdV F at the river origin (M1) could represent an isolated contamination event because it was only detected at a low density in the samples from this site. The presence of human fecal contamination coincides with the irregular settlements that discharge their wastewater directly to the river. Because the system once flowed continuously, the viral contamination issue is likely a recent phenomenon. Wastewater discharge is constant; unsurprisingly, the higher amounts of AdV F viral particles were detected in the urban zone due to wastewater discharge without proper treatment [54,57].

It is important to note that the number of AdV F particles can represent an important mechanism of exposure to the population, primarily for children under five, who may be more likely to come in direct contact with the water in urban areas; accordingly, this exposure can lead to moderate to severe gastroenteritis [53]. For the exposure dose, a point estimate was reported by Crabtree *et al.* [58], who considered AdV 4 as a surrogate because no AdV 40/41 dose-response has been published. Based on this finding, a 2 infectious viral particle can be considered the probability for infection based on an average dose of 1.66 [59].

Considering that the detection limit of the protocol applied for AdV 40/41 quantification was ten genome copies, as determined by serial dilutions from the positive control, it would not be possible to identify a hazard related to AdV 40/41. Nevertheless, the fact that the genome detection rate is commonly larger than the number of viral particles [60] clearly demonstrates that the detection capacity of 10 genome copies is adequate for the identification of adenovirus as a hazard.

Cryptosporidium parvum and *Giardia lamblia* are the most common intestinal protozoa in the environment, commonly infecting both humans and domestic and wild animals [61]. These protozoa are among the main disease-causing agents of diarrhea in the world [62]. Both species form resistance structures that allow them to survive outside the host; they can thus be transmitted through water [63]. Additionally, both species have a low infective dose (ID_{50}), specifically, 132 *Cryptosporidium* spp. oocysts [64] and 25–100 *Giardia lamblia* cysts [65]. Thus, a potential host consuming a low number of these microorganisms can readily develop an infection.

Based on these facts, the detection of *C. parvum* and *G. lamblia* in 50% and 44% of the Magdalena River samples, respectively (Table 4), means that the population and animals that use the water for diverse activities are exposed to parasites, which eventually could lead to an outbreak or epidemic. Children between 0 and 14 years of age and immunodeficient persons are susceptible.

For the Cuitzmala river, a rural area in Jalisco, Mexico, Tapia-Palacios (2012) [37] reported a range of 0–87 oocysts/L *C. parvum* and 0–2,655 cysts/L of *G. lamblia*, which in contrast with the results of this analysis for the urban area, though both species were detected in the two areas. The range in the natural zone was lower for both species (0–1,000 oocysts of *C. parvum* and 1,000–5,000 cysts/L of G. lamblia), which is in contrast with the urban areas that exhibited higher ranges (0–23,000 oocysts/L of *C. parvum, and* 0–28,000 cysts/L of *G. lamblia*). This can be explained by the input of wastewater from the urban area [12]. Untreated water could be increasing the number of oocysts present in the zone, leading to a potential population exposure risk because the population densities range from 1–2 log units more than the infective dose.

Because these rivers represent the only surface water source from within the Basin for Mexico City, particular concern should be held regarding the presence of certain microorganisms, such as the protozoa *Giardia* and *Cryptosporidium* and adenovirus, which are relatively resistant to the chlorination process in comparison to other microorganisms. Additionally, the presence of opportunistic and/or potentially pathogenic bacteria in a surface water source that is chlorinated and distributed to the population suggests that there should be additional funds to consider preventive measures, consistent with the practices carried out in other countries concerning water intended for human use.

Physical removal is critical to the control of *Cryptosporidium* because the organism is highly resistant to standard disinfection practices [66,67]. As mentioned by Betancourt and Rose (2004) [68], the effectiveness of conventional disinfection by chlorination and alternative disinfection procedures, such as chlorine dioxide, ozonation and ultraviolet irradiation (UV), toward inactivating *Cryptosporidium* has been the focus of current research. Recent USEPA criteria regarding disinfection processes for water containing this pathogen suggest disinfection with ultraviolet light or ozone and/or filtration [69]. The goals for systems that treat surface water must be disinfection and filtration to remove or inactivate 99.9% of *Giardia lamblia*, 99% of *Cryptosporidium* oocysts and 99.99% of viruses [69]. In Mexico and other developing countries, modern treatment processes should be applied, given the presence of these microorganisms in the surface waters.

One of the first steps when dealing with rivers is understanding their flow and any variations that occur throughout the year and across different years. In Mexico, monitoring practices were initiated but have since been abandoned for many rivers throughout the country. A sustainable approach to water management, especially in light of climate change, would be to monitor and report environmental and social conditions to provide a guide toward a more sustainable water management strategy.

The regimes of the Magdalena and Eslava rivers belong to homonymous basins that have been modified, and both exhibit two distinct historical periods. The periods are defined by the construction of excess hydraulic infrastructure works, such as gabion dams, 50 on the Magdalena and 18 on the Eslava (Figure 1), and marginal drains. The management of the flows in the Magdalena and Eslava rivers has caused significant fluctuations in flow magnitude (69%), has increased the frequency of maximum ordinary flows and has caused a significant increase (176%) in the minimum flow magnitude. This NFR alteration will certainly have consequences for the river communities in the natural area and for system rehabilitation.

These regime changes may contribute to the observed water quality degradation and contamination of the systems because the loss of velocity along a slope and other discontinuities in the river affect the inherent auto-depuration processes. As recorded in our fieldwork, the observed higher concentrations of the physico-chemical variables indicate the presence of higher levels of contaminants in the lower section of both basins, especially as the urban zone is entered.

Wastewater discharges also contribute to the degradation of water quality. Hence, even if discharge does not contribute significantly to flow volume, its effect on water quality is magnified due to the decrease in flow velocity. Further, wastewater discharge increases the concentrations of various water quality attributes.

In some rivers, contaminants are diluted through their tributaries. However, in this case, the Eslava River contributes its own contaminant load, which for some variables is orders of magnitude higher than the Magdalena's water, thereby hampering the auto-depuration process.

It is necessary to treat the residual wastewater dumped into the river and recover the fluvial processes and function of the rivers as ecosystems. This approach would provide several advantages. For example, there would be more water produced in the higher basin, which could be disinfected and distributed to the population. Additionally, a decrease in the amount of water that flows to the drainage system and that must be pumped out of the Basin of Mexico would be realized, as well as additional amounts of water once the wastewater has been treated and reused for irrigation near the river.

In this study, we demonstrate a practical and representative approach to achieving a detailed water quality program that could allow for the detection of change points. Additionally, we identify the problems at some locations and various methods for improving the respective environmental conditions. In this particular case, it is also of important that locations for the required wastewater treatment plants be suggested, as well as the type of plant and the installed capacity.

The water to be treated is considered domestic wastewater based on its Biochemical Oxygen Demand (DBO_5) and Chemical Oxygen Demand (COD), according to Metcalf and Eddy (1996) [70].

It is important to note the need for treatment plants to improve water quality. The wastewater discharge along the river has high concentrations of nitrogen and phosphorus forms, total dissolved solids, high conductivity, and a BOD_5 of approximately 130 mg/L to 240 mg/L (results not shown). These factors, combined with the high microorganism counts, argue in favor of the installation of several wastewater treatment plants (WWTPs) [71]. A general concept for viable plants and processing methods was presented by Orta de Velasquez (2008) [23], although treatability tests would be required, along with a specific design for each plant with an associated project feasibility study.

The wastewater treatment plants considered for this river, in all cases, included tertiary treatment. This treatment is implemented by biological secondary treatment, and the advanced treatment is implemented by physico-chemical treatment. The first treatment of activated sludge helps remove nutrients (N and P) and both soluble and colloidal biodegradable organic matter. Tertiary filtration is used to treat wastewater intended for reuse to remove any suspended matter that was not removed in the previous operations. Microbial standards can be met using both filtration (physical removal) and disinfection (inactivation). The removal and inactivation capabilities vary depending on the filtration technology or disinfectant applied. The first wastewater treatment plant constructed in the area of the Magdalena River was finished at the end of 2013, not clear if it is in operation at the moment. With the goal of removing both nutrients and pathogens, advanced treatment, which includes activated sludge, tertiary filtration and ultraviolet disinfection, was implemented. In this case, an ultraviolet light disinfection process was applied, which can effectively inactivate most viruses (111 mW*s/cm^2 for 6-log inactivation), *Giardia* (121 mW*s/cm^2 for 2-log inactivation) and *Cryptosporidium* (1,900 mW*s/cm^2 for 2-log inactivation) [72]. With this plant being operational, the river must have a basic flow level along its path to preserve some system functions.

At least two WWTP are proposed. The first could be placed near sites M11 and E8, which could treat low-quality water flowing from both rivers. A second WWTP could be placed before site M19, where the treated water could be used for irrigation in one of the few large green areas of the southern part of the city, known as *Viveros de Coyoacán*. The potential locations for these two plants are presented in Figure 1.

From the perspective of the ecosystem services provided by this lotic system, the amount of water provided by the Magdalena-Eslava river system accounts for the ecological flow in the natural area, after which the water goes through a potabilization process and is distributed for human use and consumption in the southern area of the city. The amount of water that is contaminated when clean water from the natural areas of both rivers combines with untreated wastewater is not accounted for, but it is necessary to consider the financial resources that are being wasted on these relatively small rivers.

The current average flow of the Magdalena river is 0.45 m^3/s, based on measured flow, and the cost per m^3 with the water subsidy is $8.17 (equivalent to 0.628 USD at an exchange rate of US $1 = $12.99 Mexican peso) Mexican pesos; without subsidy, the cost is $25.97 (equivalent to US $1.99) (SACM, Official Receipt. *Derechos por el suministro de agua. Magdalena Contreras. Manzana Baja*. September, 2013) [73]. Therefore, the amount of water lost accounts for 14,191,200 m^3 per year, which means a loss of $115,942,104 Mexican pesos (equivalent to US $8,925,489) at the subsidized cost, or $368,545,464 Mexican pesos (equivalent to US $28,371,475). In this example, we have only considered the Magdalena River, which has better water quality and more abundant flow. An equivalent of this at Mexico City prices for bottled water at $5.66 per liter ($5,660 per m^3) is an estimated $79,470,720,000 Mexican pesos (equivalent to US $6,117,838,337).

As part of any megacity water management strategy, there is a need to consider the amount of water lost and the financial capital that those losses represent. There is an urgent need to rehabilitate and recover the flow of the river and its water quality, which is an approach that could be replicated in more than 40 other rivers around Mexico City. The ecosystem service of clean water is being lost in the megacity, where water must be pumped at a very high cost from either groundwater systems or distant basins. These rivers serve as one example of similar situations that have arisen for other rivers in the Basin of Mexico, where water could be managed in a more sustainable way in the vicinity of this megacity. Doing so would yield enormous environmental and economic benefits in the long term.

From the perspective of providing water, water quality represents a fundamental issue. Better use of the existing water sources can be achieved in the MCMA. The rehabilitation of the Magdalena-Eslava river system could provide an example for other local and remote locations to follow.

Health issues are at stake, so action must be taken to ensure water quality in this particular system. The contaminated water flowing through the urban area should be improved to avoid future deleterious effects to the population.

Considering the treatment of wastewater that flows through the urban area, this water could also be used for irrigation in the green areas in the southern area of Mexico City.

Conclusions

The flow regimes of the Magdalena and Eslava rivers have been modified by the increase of hydraulic infrastructure constructed along the river, resulting in decreased maximum flows and increased minimum flows. These changes in flow regimes are directly related to the quality of the river water, due to the loss of velocity in the flow and the loss of continuity of flow.

The water quality of this system deteriorates as the river flows from a natural forested area to a semi-urban area, passing irregular settlements without basic water and drainage services before flowing then into an urbanized area.

The water quality is worst in the Eslava River at the confluence point, M12, after which either a dilution effect or an artifact of the complex drainage system arises, leading to an apparent improvement in the wastewater quality downstream. Organic matter, such as that found in the discharge, affects the oxygen balance and is usually accompanied by microbiological contamination.

The domestic and agricultural wastes contribute phosphates and ammonia nitrogen. Agriculture also contributes to the dissolved solid levels measured in the river.

The Magdalena Eslava river system could be used more efficiently. The river receives domestic discharge, and a portion of the flow passes through the *Delegación Magdalena Contreras*, where no industry is present.

The construction of at least two wastewater treatment plants is proposed, with the goal of restoring the river's water quality and maintaining the flow in the system. The suggested placement of these treatment plants is based on the low flow that occurs during the dry season and the sewer system conditions. By considering the landscape design to be a linear park, the treated wastewater flow was defined to rely on water flow and high water quality for recreational use. Additionally, the potential to reuse this water for green area irrigation and car washing exists. Treating the wastewater in this area of Mexico City would be relatively simple in comparison to treating the wastewater in other industrial areas of this metropolitan area.

Human activities, which have increased significantly in recent decades, have impacted the environment. The scale of urbanization, industrial operation and agricultural production has reached a point at which these activities impact the quality of the hydraulic resources. This is a global phenomenon to which Mexico City is not an exception.

Wastewater in the cities of the developing world transports a combination of liquid and solid residues from regular and irregular residential areas, public institutions, and industrial and commercial establishments. These residues eventually reach pluvial, surface and groundwater. Due to the diversity of compounds and microorganisms in wastewater, such as those observed in this study, it is not possible to provide details for all compounds but only to give an idea of the water technology required for improving water quality, with a view toward reuse and sustainability. Undoubtedly, there exist other compounds that were not measured as part of this evaluation, such as endocrine disruptors or organic compounds; these compounds must be evaluated, especially in the urban section of the lotic system, to develop a complete picture.

The amounts of several types of microorganisms in the urban zone, each with a different resistance to water and wastewater treatment, must be considered from the public health perspective because these microorganisms represent an important exposure factor for the local population.

Mexico City has lost the majority of its rivers due to mismanagement. The case of the Magdalena and Eslava rivers represents an opportunity to rehabilitate a major river system and could serve as a model for other waterways.

Acknowledgments

We thank Lucía Almeida Leñero for her enthusiastic collaboration throughout the years in the study area and the support received for this work. We thank Erick Hjort Colunga, Rosa Isabel Amieva, Ana Cecilia Espinosa, and María del Jazmín Aguilar for their invaluable assistance in the field and with laboratory work. We thank Alma de los Ríos for assistance in creating the figures.

Author Contributions

Conceived and designed the experiments: MMH MTOL CDA FAV. Performed the experiments: GPO MAT RSO MAS IYN. Analyzed the data: CDA MMH GPO TLV MTOL FAV. Contributed reagents/materials/analysis tools: MMH MTOL YLV CDA FAV. Wrote the paper: MMH GPO CDA FAV MATOL YLV MAT RSO. Performed the field work: GPO MAT MMH. Lab work: GPO MAT RSO YLV MAS.

References

1. Foster S, Lawrence A, Morris B (1998) Groundwater in urban development. Assessing management needs and formulating policy strategies. The World Bank. Washington, D.C.
2. Legorreta J (2009) Ríos, lagos y manantiales del Valle de México. Universidad Autónoma Metropolitana-Gobierno del Distrito Federal. Artes Impresas Eón, S.A. de C.V. México, D.F. 365 p.
3. Jiménez B (2009) Risks and integrated management of the urban water cycle in megacities of the developing world: Mexico City. In: Water and Urban Development Paradigms. Towards an integration of engineering, design and management approaches. Feyen J, Shannon K, Neville M (Eds.). CRC Press. Taylor & Francis Group. London. pp. 387–396.
4. Domínguez-Mora R (2011) Diagnóstico resumido de los problemas de abastecimiento de agua y de inundaciones en el Valle de México. Revista Digital Universitaria 12(2): 1067–6079.
5. Bradshaw AD (2002) Introduction and Philosophy. In: Perrow MR, Davy AJ. Handbook of Ecological Restoration. Principles of Restoration. Cambridge University Press. Cambridge, U.K.
6. García E (1988) Modificaciones al sistema de clasificación climática de Köppen. Instituto de Geografía, UNAM, México, D.F. 33 p.
7. Álvarez-Román KE (2000) Geografía de la Educación Ambiental: algunas propuestas de trabajo en el Bosque de Los Dinamos, Área de Conservación Ecológica de la Delegación Magdalena Contreras. Tesis Licenciatura en Geografía. Facultad de Filosofía y Letras, UNAM. México. 127 p.
8. Ávila-Akerberg V (2009) Forest quality in the southwest of Mexico City. Assessment towards ecological restoration of ecosystem services. Ph.D. Dissertation. Institute für Landespflege. Universität Freiburg. Germany. 138 p.
9. Secretaría de Medio Ambiente y Recursos Naturales, Instituto Nacional de Ecología (SEMARNAT-INE) (2003a) Acuerdo que declara Zona Protectora Forestal los bosques de La Cañada de Contreras, D.F. In: Áreas naturales protegidas de México con decretos federales. México, D.F. 302 p.
10. Secretaría de Medio Ambiente y Recursos Naturales, Instituto Nacional de Ecología (SEMARNAT-INE) (2003b) Declaración de Zona Protectora Forestal del Río Magdalena In: Áreas Naturales Protegidas de México con Decretos Federales. México, D.F. 317 p.
11. Ávila-Akerberg V (2004) Autenticidad de los bosques en la cuenca alta del río Magdalena. Diagnóstico hacia la restauración ecológica. Tesis Maestría. Facultad de Ciencias, UNAM. México. 112 p.
12. Programa Universitario de Estudios sobre la Ciudad-Gobierno del Distrito Federal (PUEC-GDF) (2008) Plan Maestro de Manejo Integral y Aprovechamiento Sustentable de la Cuenca del río Magdalena, Distrito Federal. Diagnóstico Integral. 90 p.: http://www.sma.df.gob.mx/sma/links/dowload/biblioteca/diagnostico_integral.pdf Accessed 2013 Aug 20.
13. Rzedowski GC de, Rzedowski J (2001) Flora fanerogámica del Valle de México. 2ª ed. Instituto de Ecología, A.C. y Comisión Nacional para el Conocimiento y Uso de la Biodiversidad. Pátzcuaro, Michoacán. 1406 p.
14. Cantoral-Uriza EA, Almeida-Leñero L, Cifuentes-Blanco L, Paniagua-León L, Luis-Martínez MA, et al. (2009) La diversidad de una cuenca en la Ciudad de México. Ciencias 94: 28–33.
15. Nava M (2003) Los bosques de la cuenca alta del río Magdalena, D.F., México. Un estudio de vegetación y fitodiversidad. Tesis Licenciatura Biología. Facultad de Ciencias. UNAM. México. 64 p.
16. Jujnovsky J, Almeida-Leñero L, Bojorge-García M, Monges YL, Cantoral-Uriza E,et al. (2010) Hydrologic ecosystem services: water quality and quantity in the Magdalena River, Mexico City. Hidrobiológica 20(2): 113–126.
17. Programa Universitario de Medio Ambiente-Gobierno Distrito Federal (PUMA-GDF) (2009) Sistema de Indicadores para el rescate de los ríos Magdalena y Eslava. México, D.F. 134 p. Available: http://www.sma.df.gob.mx/sma/links/download/archivos/sistema_indicadores.pdf. Accessed 2013 Aug 20.
18. Riley AL (1998) Restoring Streams in Cities. A Guide for Planners, Policymakers, and Citizens. Island Press. Washington, D.C.
19. González Reynoso AE, Hernández Muñoz L, Perló Cohen M, Zamora Saenz I (2010) Rescate de ríos urbanos. Propuestas conceptuales y metodológicas para la restauración de ríos. Universidad Nacional Autónoma de México. Programa de Estudios sobre la Ciudad. Ed. Color, S.A. de C.V. México, D.F. 109 p.

20. Banco Nacional de Datos de Aguas Superficiales (BANDAS) (2006) *National Data Bank Surface Waters*. Comisión Nacional del Agua. Available: <ftp://ftp.conagua.gob.mx/Bandas/Bases_Datos_Bandas> Accessed 2013 March 18.
21. Searcy JK, Hardison CH (1960) Double-mass curves. U.S. Geological Survey Water-Supply Paper 1541-B. USGS, Washington, D.C.
22. Yuefeng Y, Tijiu C, Xiaohua W, Mingfang Z, Cunyong J (2011) Effect of forest recovery on summer streamflow in small forested watersheds, Northeastern China. Hydrol Process 26 (8): 1208–1214. DOI: 10.1002/hyp.8204.
23. Orta de Velásquez MT, Yáñez Noguez I, Mazari Hiriart M, Ríos Soriano B, Velázquez Pedroza K, et al. (2008) Restauración de un río urbano. Aspectos ambientales e hidrológicos. XXXI Congreso Interamericano de Ingeniería Sanitaria y Ambiental. Santiago de Chile, Chile. 12–15.
24. HACH (2002) Water analysis Handbook. 4th ed. HACH Co. CA. 1260 p.
25. American Public Health Association (APHA), American Water Works Association, Water Environmental Federation (2005) Standard Methods for Examination of Water and Wastewater. 21a ed. United Book Press. Washington. D.C.
26. Murray PR, Baron EJ, Pfaller MA, Tenover FC, Yolken RH (Eds.) (1995) Manual of Clinical Microbiology. American Society for Microbiology. Washington D. C. 1482 p.
27. Gerba CP (2000) Indicator microorganisms. In: Maier RM, Pepper IM, Gerba CP (Eds.) Environmental Microbiology. Academic Press. San Diego, California. pp. 491–503.
28. Toranzos GA, McFeters GA, Borrego JJ, Savill M (2007) Detection of microorganisms in environmental freshwaters and drinking water. In: Manual of Environmental Microbiology. Edited by Hurst CJ, Crawfortd RL, Garland JL, Lipson DA, Mills AL, et al. Washington D. C. ASM Press. pp. 249–264.
29. ISO 10705-1 (1995), Water quality -Detection and enumeration of bacteriophages-part 1: Enumeration of F-specific RNA bacteriophages. Geneva, Switzerland: International Organization for Standardization.
30. Solano-Ortiz R (2011) Avances metodológicos para la detección del virus de la hepatitis A y bacteriófagos en agua. Tesis Licenciatura, Biología. Facultad de Ciencias, UNAM. México. 68 p.
31. Fonseca-Salazar MA (2010) Bacteriófagos como indicadores de la calidad del agua en Ciudad Universitaria. Tesis Licenciatura, Biología. Facultad de Ciencias, UNAM. México, D.F. 62 p.
32. Polaczyk LA, Narayanan J, Cromeans LT, Hahn D, Roberts MJ, et al. (2008) Ultrafiltration-based techniques for rapid and simultaneous concentration of multiple microbe classes from 100-L tap water samples. J Microbiol Methods 73: 92–99.
33. Monpoeho S, Dehee A, Mignotte B, Schwartzbrod L, Marechal V, et al. (2000) Quantification of enterovirus RNA in sludge samples using single tube real-time RT-PCR. BioTechniques 29: 88–93.
34. Xagoraraki I, Kuo DH, Wong K, Wong M, Rose JB (2007) Occurrence of human adenoviruses at two recreational beaches of the Great Lakes. Appl Environ Microbiol 73: 7874–7881.
35. US Environmental Protection Agency (USEPA) (2005) Method 1623: *Cryptosporidium* and *Giardia* in Water by Filtration/IMS/FA, Cincinnati, OH, December 2005. 68 p.
36. Rangel-Martínez C (2010) Identificación de *Blastocytis hominis* en muestras de agua obtenidas de Ciudad Universitaria en la Ciudad de México. Tesis Licenciatura, Biología. Universidad Simón Bolívar. México D.F. 62 p.
37. Tapia Palacios MA (2012) Detección de *Cryptosporidium parvum* y *Giardia lamblia* en agua del río Cuitzmala, Jalisco. Tesis Licenciatura, Biología. Facultad de Ciencias, UNAM. México. 64 p.
38. McCullagh P, Nedler JA (1989) Generalized Linear Models. Monographs on Statistics and Probability, Vol. 37, Chapman and Hall, New York.
39. Hiriart F, Marsal RJ, Cruickshank G, Key F (1952) Contribución de la Comisión Federal de Electricidad a la Solución del Problema de Abastecimiento de Agua a la Ciudad de México. México. 132 p.
40. Departamento del Distrito Federal (DDF) (1975) Memoria de las Obras del Sistema de Drenaje Profundo del Distrito Federal. Tomo I. DDF-Talleres Gráficos de la Nación. México.
41. González-Martínez MT (2008) Modelación hidrológica como base para el pago por servicios ambientales en la microcuenca del río Magdalena, Distrito Federal. Tesis Maestría. Universidad Autónoma de Querétaro. México. 137 p.
42. Jujnovsky J, González-Martínez TM, Cantoral-Uriza EA, Almeida-Leñero L (2012) Assessment of Water Supply as an Ecosystem Service in a Rural-Urban Watershed in Southwestern Mexico City. Environ Manage 49 (3): 690–702. DOI 10.1007/s00267-011-9804-3.
43. Krzanowski WJ (1988) Principles of multivariate analysis: A user's perspective. Oxford University. Clarendon Press. New York. 563 p.
44. Diario Oficial de la Federación (DOF) (1997) NOM-001-SEMARNAT-1996. Que establece los límites máximos permisibles de contaminantes en las descargas de aguas residuales en aguas y bienes nacionales. 6 enero 1997.
45. Ley Federal de Derechos en Materia de Agua (LFDMA) (2012) Ley Federal en Materia de Agua. México.
46. Bergey's (2012) Bergey's Manual of Systematic Bacteriology (2001–2012), 2nd ed. Springer Verlag. New York.
47. Long CS, El-Khoury SS, Oudejans GS, Sobsey DM, Vinjé J (2005) Assessment of sources and diversity of male-specific coliphages for soured tracking. Environ Eng Sci 22(3): 367–377.
48. Schaper M, Jofre J (2000) Comparison of methods for detecting genotypes of F-specific RNA bacteriophages and fingerprinting the origin of feacal pollution in water samples. J Virol Methods 89: 1–10.
49. Havelaar HA, Hogeboom MW (1984) A method for the enumeration of male-specific bacteriophages in sewage. J Appl Bacteriol 56: 439–447.
50. Mooijman AK, Bahar M, Muniesa M, Havelaar HA (2002) Optimisation of ISO 10705-1 on enumeration of F-specific bacteriophages. Journal of Virological Methods 103: 129–136.
51. Muir P, Van Loon AM (1997) Enterovirus infections of the central nervous system. Intervirology 40: 153–66.
52. Wadell G (1984) Molecular Epidemiology of Human Adenoviruses. Curr Top Microbiol Immunol 110: 191–220.
53. Maier RM, Pepper IL, Gerba CP (2009) Environmental Microbiology. China. Academic Press. 598 p.
54. Hierholzer JC (1992) Adenoviruses in the immunocompromised host. Clin Microbiol Rev 5: 262–274.
55. Luna MM, Leboreiro JI, Zapata IB, Rendón-Macías ME (2013) Los adenovirus como causa de gastroenteritis aguda en los niños. Revista Mexicana de Pediatría 80: 98–104.
56. Enriquez CE, Hurst CJ, Gerba CP (1995) Survival of the enteric adenoviruses 40 and 41 in tap, sea, and waste water. Water Res 29: 2548–2553.
57. Ciruela P, Izquierdo C, Oviedo M, Domínguez A (2004) Rotavirus y adenovirus 40–41 como causantes de gastroenteritis infantil. Vacunas 5: 16–23.
58. Crabtree PA, Gerba CP, Rose JB, Hass CN (1997) Waterborne adenovirus a risk assessment. Water Science and Technology 35 (11–12): 1–6.
59. Kundu A, McBride G, Wuertz S (2013) Adenovirus-associated health risk for recreational activities in a multi-use coastal watershed based on site-specific quantitative microbial risk assessment. Water Res 47: 6309–6325.
60. El-Senousy WM, Osman GA, Melegy AA (2014) Survival of Adenovirus, Rotavirus, Hepatitis A Virus, Pathogenic Bacteria and Bacterial Indicators in Ground Water. World Applied Sciences Journal 29 (3): 337–348. DOI: 10.5829/idosi.wasj.2014.29.03.13849
61. Xiao L, Fayer R (2008) Molecular characterization of species and genotypes of Cryptosporidium and Giardia and assessment of zoonotic transmission. Int J Parasitol 38: 1239–1255.
62. Reynolds KA, Mena KD, Gerba CP (2008) Risk of waterborne illness via drinking water in the United States. Rev Environ Contam Toxicol 192: 117–158.
63. Plutzer J, Karanis P, Domokos K, Törökné A, Márialigeti K (2008) Detection and characterization of Giardia and Cryptosporidium in Hungarian raw, surface and sewage water samples by IFT, PCR and sequence analysis of the SSUrRNA and GDH genes. Int J Hyg Environ Health 211: 524–533.
64. Okhuysen PC, Chappell CL, Crabb JH, Sterling CR, Dupont HL (1999) Virulence of three distinct Cryptosporidium parvum isolates for healthy adults. J Infect Dis, 180: 1275–81.
65. Rendtorff RC (1979). The experimental transmission of Giardia lamblia among volunteer subjects. In:W. Jakubowski and J.C. Hoff (eds) Waterborne Transmission of Giardiasis. Report No. EPA-600/9-79-001, pp. 64-8. US Environmental Protection Agency. Environmental Research Centre, Cincinnati.
66. US Environmental Protection Agency (USEPA) (2001a) *Cryptosporidium*: Drinking Water Health Advisory. Office of Science and Technology EPA-822-R-01-009.Environmental Protection Office of Water: 31 p.
67. US Environmental Protection Agency (USEPA) (2001b) Drinking Water Criteria Document Addendum: *Cryptosporidium*. United States Environmental Protection Agency, Office of Water, Washington, DC.
68. Betancourt WQ, Rose JB (2004) Drinking water treatment processes for removal of *Cryptosporidium* and *Giardia*. Vet Parasitol 126 (1–2): 219–234.
69. US Environmental Protection Agency (USEPA) (2013) Basic Information about Pathogens and Indicators in Drinking Water. December 13, 2013. http://water.epa.gov/drink/contaminants/basicinformation/pathogens.cfm Accessed 2014 Jun 21.
70. Metcalf & Eddy Inc (1996) Ingeniería de Aguas Residuales Tratamiento, Vertido y Reutilización 1: 53–55, 145–46. México.
71. Velázquez PK (2009) Inventario de descargas y propuestas de tratamiento para el saneamiento del Río Magdalena. Tesis Ingeniería Química. FES Zaragoza. Universidad Nacional Autónoma de México.
72. US Environmental Protection Agency (USEPA 1999) *Alternative Disinfectants and Oxidants. Guidance Manual*. Ch. 8. Ultraviolet Radiation. Environmental Protection Office of Water EPA 815-R-99-014. USA.
73. Sistema de Aguas de la Ciudad de México (SACM) (2013) Official Receipt. Derechos por el suministro de agua. Magdalena Contreras. Manzana Baja. September, 2013.
74. Silva-Magaña MA (2013) Uso de indicadores virales para la evaluación de la calidad de agua potable y de reúso en Ciudad Universitaria, México. Tesis Licenciatura. Biología. Facultad de Ciencias, Universidad Nacional Autónoma de México. México: 70 pp.

3

Analysis of the *Citrullus colocynthis* Transcriptome during Water Deficit Stress

Zhuoyu Wang[1], Hongtao Hu[2], Leslie R. Goertzen[2], J. Scott McElroy[3], Fenny Dane[1]*

1 Department of Horticulture, Auburn University, Alabama, United States of America, **2** Department of Biological Sciences, Auburn University, Alabama, United States of America, **3** Department of Crop, Soil and Environmental Sciences, Auburn University, Alabama, United States of America

Abstract

Citrullus colocynthis is a very drought tolerant species, closely related to watermelon (*C. lanatus* var. *lanatus*), an economically important cucurbit crop. Drought is a threat to plant growth and development, and the discovery of drought inducible genes with various functions is of great importance. We used high throughput mRNA Illumina sequencing technology and bioinformatic strategies to analyze the *C. colocynthis* leaf transcriptome under drought treatment. Leaf samples at four different time points (0, 24, 36, or 48 hours of withholding water) were used for RNA extraction and Illumina sequencing. qRT-PCR of several drought responsive genes was performed to confirm the accuracy of RNA sequencing. Leaf transcriptome analysis provided the first glimpse of the drought responsive transcriptome of this unique cucurbit species. A total of 5038 full-length cDNAs were detected, with 2545 genes showing significant changes during drought stress. Principle component analysis indicated that drought was the major contributing factor regulating transcriptome changes. Up regulation of many transcription factors, stress signaling factors, detoxification genes, and genes involved in phytohormone signaling and citrulline metabolism occurred under the water deficit conditions. The *C. colocynthis* transcriptome data highlight the activation of a large set of drought related genes in this species, thus providing a valuable resource for future functional analysis of candidate genes in defense of drought stress.

Editor: Girdhar K. Pandey, University of Delhi South Campus, India

Funding: Funding or sources of support for the study were the Alabama Agricultural Experiment Station and the State of Alabama. The funders had no role in study design, data collection and analysis, decision to publish or preparation of the manuscript.

Competing Interests: The authors have declared that no competing interests exist.

* Email: danefen@auburn.edu

Introduction

Water is essential for plant growth in modern agriculture [1]. Drought delays the development of crops, and strongly affects morphology, as well as physiological processes like transpiration, photosynthesis, respiration and translocation of assimilates [2]. Drought avoidance can be achieved through morphological changes in plants, such as decreased stomatal conductance, reduced leaf area, and extensive root systems [3]. Drought tolerance is achieved by physiological and molecular mechanisms, including osmotic adjustment, antioxidant and scavenger compounds [4]. Both strategies involve the induction of specific genes and proteins, such as dehydrins (dehydration-induced proteins), key enzymes for osmolyte biosynthesis, and detoxification enzymes [5,6].

Plant species have developed diverse strategies to adapt and thrive in all kinds of climates and terrains to deal with extreme changes in the environment. These strategies are supported by rich and complex metabolic networks that enable the plant to synthesize a wide range of compounds. Plant responses to abiotic stresses involve interactions and crosstalk between many molecular pathways. High throughput screening techniques such as transcriptome sequencing have been used to study the adaptability of plants to drought [7]. This led to the discovery of many drought related genes. For example, PIP aquaporins were found to fine-tune the environment in response to declining water availability [8]. However, few natural allelic variants have been cloned for drought related traits, so QTL, RNA sequencing, and other trait isolation methods are needed to improve methodology for exploring complex multivariate data [9,10].

The cucurbit family is a large family with several economically important species, such as watermelon (*Citrullus lanatus*), melon (*Cucumis melo*), cucumber (*Cucumis sativus*) and several *Cucurbita* species with edible fruits [11]. *Citrullus colocynthis* (L.) Schrad ($2n = 2x = 22$), the bitter apple, closely related to domesticated watermelon (*Citrullus lanatus* var. *lanatus*), is a very drought-tolerant perennial herbaceous species in the Cucurbitaceae family [12]. It can survive arid environments by maintaining its water content under severe stress conditions. *C. colocynthis* is an important medicinal plant and a source of valuable oil [12]. Its seeds were found in several early Egyptian, Libyan and Near Eastern sites from about 4000 BC. This species grows in sandy areas throughout northern Africa, southwestern Asia and the Mediterranean region [11,12]. The species has been used as a model to elucidate the function of genes implicated in the stress response ultimately leading to enhancement of stress tolerance in cucurbit crops through genetic manipulation. Si et al. [13] found dynamic gene expression changes in *C. colocynthis* root tissues

Table 1. Paired end read statistics before and after trimming of the *Citrullus colocynthis* transcriptome at 4 time points.

Time point	Day 1		Day 2		Day 3		Day 4	
	Before	After	Before	After	Before	After	Before	After
Reads	21,194,778	20,339,776	23,566,720	22,626,577	23,435,694	22,531,826	27,388,920	26,288,785
	21,379,596	20,537,526	23,774,222	22,846,047	23,626,206	22,738,550	27,737,584	26,649,390
Average	21,287,187	20,438,651	23,670,471	22,736,312	23,530,950	22,635,188	27,563,252	26,469,088
% Trimmed	96.15		96.11		96.19		96.03	

using cDNA amplified fragment length polymorphism (cDNA-AFLP) technique.

Several research groups have used next generation sequencing technologies to study gene expression profiles in species of the cucurbit family. Guo et al [14] used 454 sequencing technology to study the comprehensive profile for watermelon fruit flesh tissues, Grassi et al [15] studied carotenoid pathway regulators in ripening watermelon fruit. The draft genome of watermelon (*C. lanatus*, $2n = 2x = 22$, ~425 Mb) was analyzed by Guo et al [16] using three different watermelon subspecies. Comparative genomic analysis provided an evolutionary scenario for the origin of the 11 watermelon chromosomes derived from a 7-chromosome paleohexaploid eudicot ancestor. The genome sequence of cucumber (*Cucumis sativus*, $2n = 2x = 14$) has been completed, and the genome of melon (*Cucumis melo*, $2n = 2x = 24$) is being sequenced under the Spanish Genomics Initiative (MELONOMICS) [17,18]. Liu et al [19] used sequencing techniques to identify conserved and novel miRNA in watermelon, while Wincker [20] used comparative analysis of genomes between watermelon and sweet orange to detect the traits related to their domestication.

Here, high-throughput sequencing of the leaf transcriptome from *C. colocynthis* provides a glimpse at drought related genes in this uniquely drought tolerant cucurbit species. The lack of extensive genomic and functional genomic resources in *Citrullus* species has hampered research and breeding of the cultivated and economically important *C. lanatus*. This study should facilitate the identification of valuable multiple genes, needed for complex interactions of plant species with the environment.

Materials and Methods

Plant Materials and RNA Extraction

C. colocynthis seedlings were grown in Sunshine Mix #8 under a 16 h light/8 h dark photoperiod at 26°C day, 22°C night temperature. Seedlings with 2–3 true leaves (2–3 week old) in 50 ml containers were exposed to drought by withholding water for 0 (day 1, D1), 24 (day 2, D2), 36 (day 3, D3) or 48 hours (day 4, D4). True leaf samples were collected each day at noon, flash frozen in liquid nitrogen and stored at −80°C. RNA was subsequently extracted using the TRIzol method [21].

Preparation of cDNA Library and Sequencing

Illumina sequencing was performed at the HudsonAlpha Institute of Biotechnology (Huntsville, AL) following manufacturer's instructions. RNA-Seq reads were first processed to remove rRNA sequence contamination. First strand cDNA was synthesized with reverse transcriptase and random primers using the small fragment RNAs as template. Second strand cDNA was then synthesized followed by phosphorylation by T4 DNA polymerase. The cDNA fragments were 3′ adenylated and ligated to the Illumina's paired-end adapters. Enrichment of cDNA templates was conducted following fifteen cycles of PCR amplification. In total, over 20*4Mp were sequenced for mapping assembly and differential expression analysis. Raw sequence data are available for download at NCBI Sequence Read Archive (SAMN02769576, SAMN02769577, SAMN02769578, and SAMN02769579).

Assembly

Raw sequencing data were filtered Z (0.05) using the CLC Genomics workbench (CLC Bio, Aarhus, Denmark). Paired–end sequences from the four samples were used to construct the *de novo C. colocynthis* transcriptome assembly with default parameters. Assembly reads were also assembled against the watermelon

Table 2. Summary details of sequences produced after assembly of *Citrullus colocynthis* reads.

	Length/Number of reads
N50	**1870 bp**
Average length	**1350 bp**
Minimum length	**201 bp**
Maximum length	**1956 bp**
Total number of contigs	**20581**
Full length cDNAs	**5038**

(*C. lanatus*) genome sequences, which were downloaded from the Cucurbit Database (http://www.icugi.org/cgi-bin/ICuGI/index.cgi). Reads were filtered and assembled using the CLC genomics workbench. The parameters used were as follows: 2 points of mismatch cost, 2 points of insertion cost, 2 points of deletion cost, 0.5 as length fraction, 0.95 as similarity fraction. After the *de novo* assembly and watermelon mapping assembly, we used Trinity to assemble all the contigs with the default parameters.

Identification of Full Length cDNAs

Two methods were used to identify full length cDNAs. First, BLASTX searching (E value: $1e^{-10}$) was used to detect the matched cDNAs in SwissProt database; second, ESTScan 2.0 was used to identify the translated sequences. Sequences with either start codon (ATG) and stop codon (TAG/TGA/TAA), or sequences with start codon (ATG) and homologue to a known protein with ≥80% similarity, were chosen as full length cDNAs.

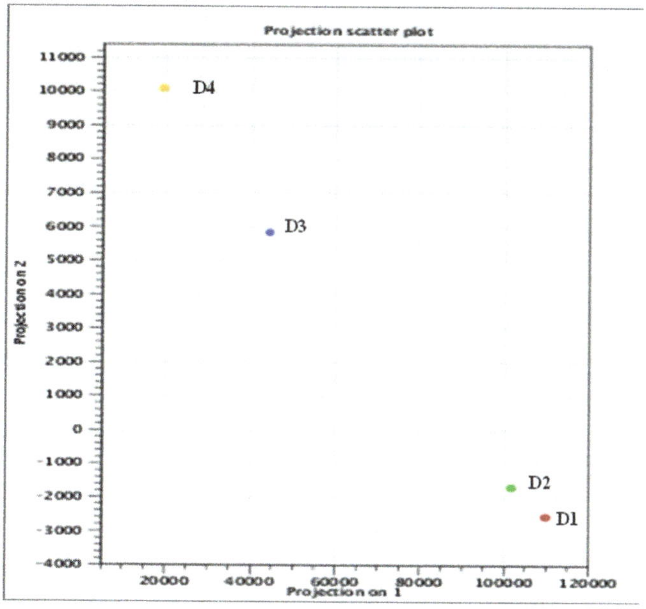

Figure 1. Principal component analysis of the leaf transcriptome of *Citrullus colocynthis* at 0 (D1), 24 (D2), 36 (D3) and 48 (D4) hours of withholding water. PCA analysis was conducted using the CLC workbench. The X-axis indicates principal component 1, the Y-axis principal component 2. Data from D1 and D2 are more closely related than data from D3 and D4.

Expression Analysis Using Transcriptome Reference

Pair-end sequencing reads of the four libraries were filtered using CLC Genomics workbench (0.05) before mapping to the references sequences from assembled cDNAs. First, read counts of each unigene were converted to reads per million (RPM).

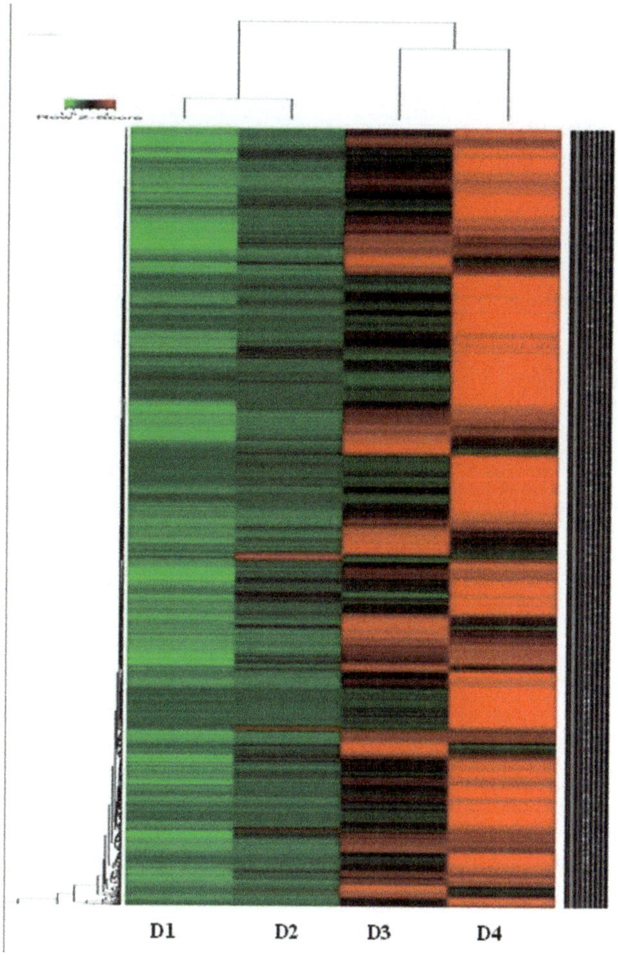

Figure 2. Heat map of the leaf transcriptome of *Citrullus colocynthis* depicting changes in transcript patterns of 2545 gene transcript clusters under water deficit treatment. Heatmap was produced using R. Each red or green block represents normalized expression value of a gene detected at 0 (D1), 24 (D2), 36 (D3) and 48 (D4) hours of withholding water. Red indicates higher gene expression values across treatment, while green indicates lower expression values across treatment.

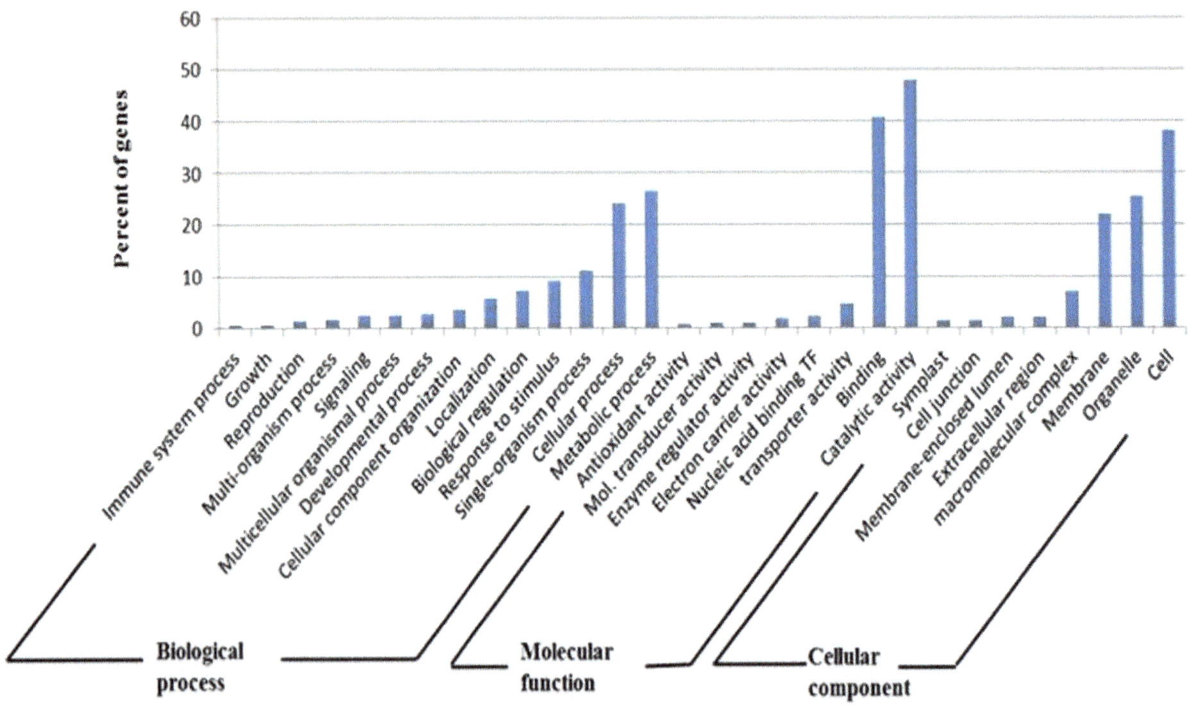

Figure 3. Gene Ontology (GO) of significantly changed *Citrullus colocynthis* genes under drought stress. Based on high-score BLASTX matches in the NR (non-redundant) plant protein database, *C. colocynthis* genes were classified into three main GO categories. The y-axis indicates the percentage of a specific category of genes in each main category.

Secondly, statistical analysis using Kal's test [22] provided in CLC Genomics workbench ($P<0.05$ and fold change ≥ 1.5) was conducted. These transcripts were annotated against the reference sequences.

Gene Ontology Analysis

The functional annotation software BLAST2GO (http://www.blast2go.com/b2ghome) was used to conduct gene ontology (GO) analysis of *C. colocynthis* genes in this study. The databases used were SwissProt and NCBI. BLAST E-value was set at $1.0e^{-3}$. The major GO analysis was determined by BLAST, mapping, and annotation. Results are presented as a bar chart showing the percent of genes belonging to each group.

qRT-PCR Analysis

For cDNA synthesis, 500 ng of the total RNA for each sample (the same RNA was used for RNA-seq analysis) was used in reverse transcription with ProtoScript First Strand cDNA synthesis kit from New England BioLabs (Ipswich, MA). qRT-PCR was performed with SYBR-Green Supermix from Bio-Rad (Hercules, CA) in an Eppendorf Mastercycler ep realplex (Hauppauge, NY). Table S1 contains gene specific primer sequences. Each reaction contained 10 µl of SYBR-Green supermix, 1 µl of cDNA template, 1 µl forward primer (4 µm), 1 µl reverse primer (4 µm), and 7 µl ddH_2O. The qRT-PCR program consisted of one cycle at 95°C for 15 sec, followed 40 cycles of 15 sec at 95°C, 15 sec at 55°C, and 30 sec at 72°C. The relative expression data was compared with actin [13] from *C. colocynthis*. Quantification

Table 3. Validation of the RNA-Seq expression profiles of selected *C. colocynthis* genes by qRT-PCR.

Gene ID	Hit ID	Annotations	Log2FC*-RNA seq	Log2 FC-qRT-PCR
Comp14675	Cla000300	Heat shock protein	7.60	8.37
Comp13927	Cla001351	Cold shock protein	3.41	4.8
Comp2553	Cla002576	MA3 domain-containing protein	2.15	5.33
Comp372	Cla002814	Bax inhibitor-1	1.90	3.56
Comp8117	Cla012143	Drought stress related gene	2.87	4.31
Comp10586	Cla013687	MYB	4.09	5.86
Comp19751	Cla022362	WRKY	6.27	6.00
Comp6823	Cla022169	MADS	2.07	3.67

Transcripts identified as D4 significantly changed genes were compared to transcripts at D1. The table shows the log2 fold change calculated from D4 expression vs D1 expression for RNAseq and qRT-PCR analysis. FC = fold change. Gene ID is gene named in *C. colocynthis*, and Hit ID is blastx of gene with watermelon ID number.

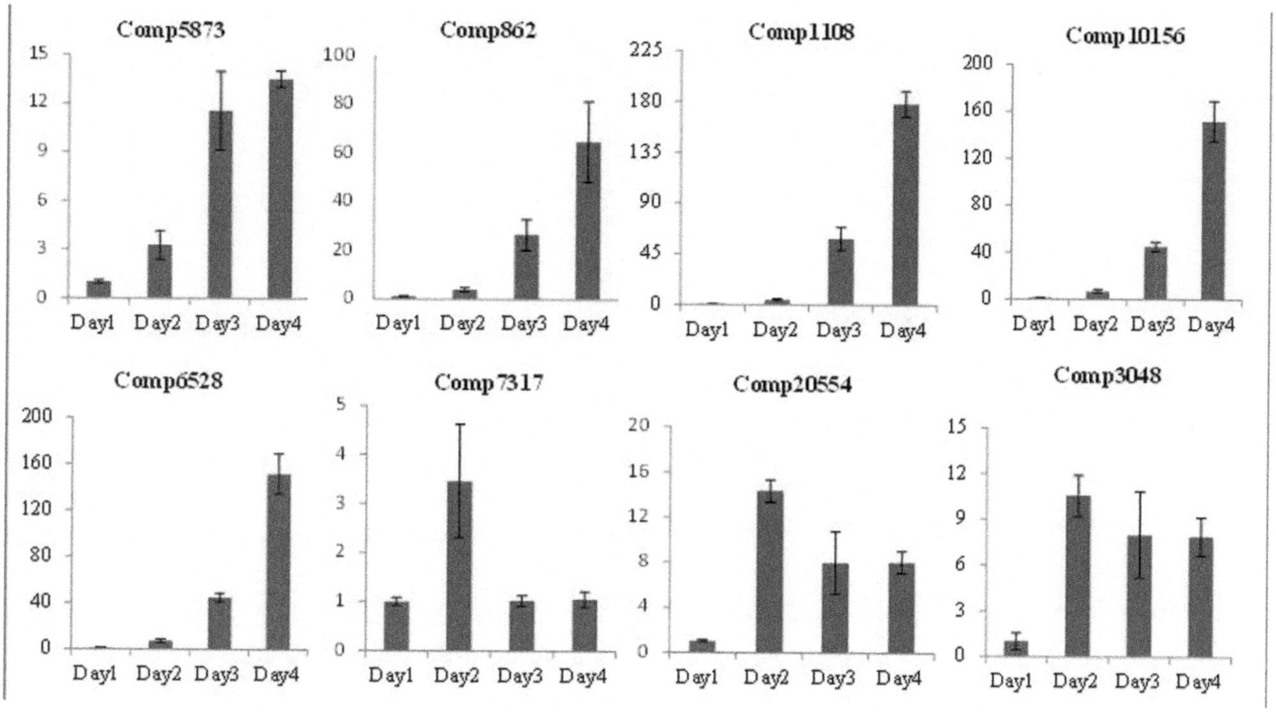

Figure 4. qRT-PCR analysis of 8 selected *Citrullus colocynthis* genes under drought treatment. The y-axis indicates relative expression compared with Day 1 expression of each gene. The x-axis shows days of water-withholding time points.

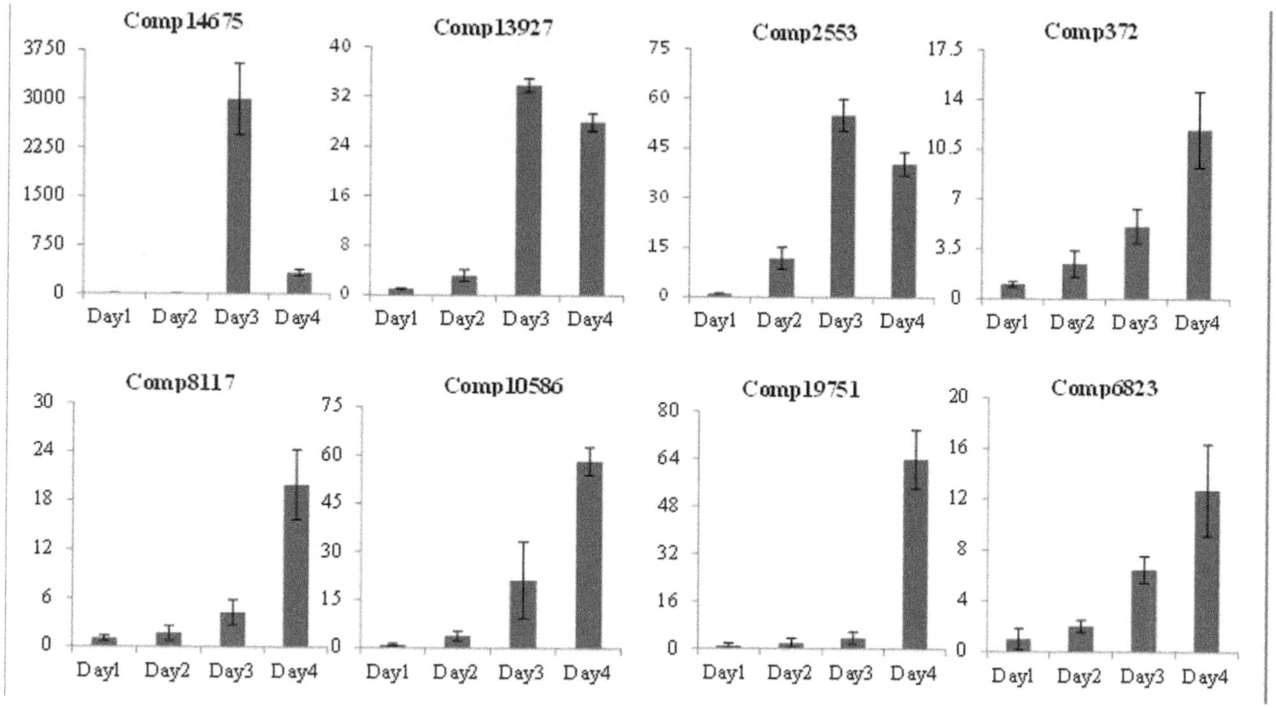

Figure 5. qRT-PCR analysis of 8 additional *Citrullus colocynthis* genes under drought treatment. The y-axis indicates relative expression compared with Day 1 expression of each gene. The x-axis shows days of water-withholding time points.

Table 4. List and number of several drought stress signaling pathway members detected in the *Citrullus colocynthis* transcriptome.

Ion Signaling	**NO.**
Calcium Binding Protein	10
Protein Kinase Pathways For Osmotic Signaling	
MAP kinase	16
Osmotic Stress-Activated Phospholipid Signaling	
PLC-like Phosphodiesterase	9
DAG	8
PIP2-like Aquaporin	5
Detoxification Signaling	
Heavy Metal Transport/Detoxification Protein	32
Transcription Regulators	
Regulatory Proteins	
MYB	131
MYC	43
NAC	42
Leucine-rich Repeat Proteins	150
Functional Proteins	
Heat Shock Protein 70	70
Heat Shock Protein 22	1
grpE Like Protein	4
RBOHD (respiratory burst oxidase)	2
VIP2 (VIRE2-INTERACTING PROTEIN2)	1
ABATransporter-like Protein	100
Synaptobrevin-related protein	5
Toc34-1 (Translocon outer envelope of chloroplast)	1
Beta-amylase	9
Puruvate Kinase	11
TIP1 (TIP GROWTH DEFECTIVE 1)	1
RD22	2

of the relative transcript levels was performed using the comparative C_T method. The induction ratio (IR) was calculated as recommended by the manufacturer and corresponds to $2^{-\Delta\Delta CT}$, where $\Delta\Delta CT = (C_{T,\ target\ gene}, -C_{T,\ actin})_{treatment} - (C_{T,\ target} - C_{T,\ actin})_{control}$. All experiments were replicated three times.

Results and Discussion

Transcriptome Assembly Results

The transcriptome of *C. colocynthis* leaves following 4 days of drought stress was assembled and assessed following paired-end (2*50 bp) Illumina sequencing. The Illumina platform yielded an average of 24 million high-quality reads per sample (Table 1). All sample reads were used to construct a *de novo* assembly. A reference assembly was constructed using the completely sequenced watermelon genome. A total of 20,581 contigs were generated (Table 2). The contigs had an average length of 1350 bp and N50 of 1870 bp. BLASTX was used against SwissProt and ESTscan of translated protein sequences to detect full length cDNAs. A total of 5,038 full length cDNAs were detected in our sequencing assembly (Table 2).

Principle component analysis (PCA) was implemented in the CLC workbench. The results (Figure 1) illustrate differential gene expression patterns in *C. colocynthis* seedlings following four days of withholding water. Gene expression patterns on D1 and D2 were nearly similar, while results from D3 and D4 showed substantial differences as compared to D1 and D2, and gene expression patterns detected on D4 were very different from all other time points as a result of drought stress.

Differentially Expressed Genes Involved in Response to Drought Stress in *C. colocynthis*

The paired-end reads were mapped to the reference genomes after filtration. Read counts of each unigene were converted to reads per million (RPM). The read number of each cDNA was divided by the total number of reads per day (1, 2, 3, or 4) from the data set, and multiplied by 10^6. Statistical analysis was conducted using Kal's test ($p<0.05$ and fold change ≥ 1.5). Genes showing non-significant and significant changes in read counts are shown in Table S2. Each sample was compared to the day 1 sample reads for analysis of their significance level of gene expression. The read results indicated that 59 genes showed significant changes at D2, D3, and D4 as compared to D1; 13 genes showed significant differential expression at D2 and D3 as compared to D1; 30 genes

Table 5. Overview of *Citrullus colocynthis* genes involved in phytohormone signaling.

Function	Ethylene Pathway Major Members	Number
Ethylene-insensitive protein	EIN	5
Ethylene receptors	ETR	3
Constitutive triple response proteins	CTR	1
EIN3-like (EIL) transcription factors	EIL	1
Ethylene response factors	ERF	118
Function	**Auxin Pathway Major Members**	**Number**
Receptors/F-box proteins	TIR	1
Ubiquitin ligase component	SCF	5
Target proteins	Aux/IAA	30
Auxin response factors	ARF	30
Auxin transport protein	PIN	6
Function	**JA Pathway Major Members**	**Number**
Receptor/F-box proteins	COI1	1
Target proteins	JAZ	1
JAZ interacted proteins	NINJA	4
Activator transcription factors	R2R3-MYB transcription factor	3
Activator transcription factors	MYC2,3,4	9
Function	**SA Pathway Major Members**	**Number**
Regulatory proteins	NPR1	2
SA mainly induced genes	WRKY	46
SA mainly induced genes	TGA2,3,5,6	6
Function	**ABA Pathway Major Members**	**Number**
ABA receptors	PYR/PYL/RCAR	19
PYR/PYL/RCAR interacted proteins	PP2C	3
Serine/threonine-protein kinase	SnRK2	1
SnRK2 target	ABI5	1
Function	**GA Pathway Major Members**	**Number**
GA receptors	GID1A/B/C/-like	11
E3 ubiquitin ligase	SLY1/SNZ	2
DELLA proteins	GA1, RGA, RGL1, 2, 3	4

showed significant differential expression early, at D2, as compared to D1; 897 genes showed changes at D3 and D4 as compared to D1; 341 genes were only differentially expressed at D3; 1191 genes were regulated late, at D4 only, under drought stress. In conclusion, the *C. colocynthis* gene expression patterns showed dramatic changes with 2545 genes showing significant changes, mostly occurring late under drought conditions (D3 and D4 of withholding water).

The heat map depicted in Figure 2 corresponded to the principal component analysis. D3 and D4 transcripts were clustered together, and D1 and D2 transcripts were clustered. Also significant changes were seen at D3 and D4 suggesting that the transcriptional response of many genes was up-regulated during drought. Strong effects were especially observed on D3 and D4.

Gene Ontology (GO) Classification

To functionally categorize significantly changed genes in *C. colocynthis* under drought treatment, gene ontology analysis by BLAST2GO was performed. *C. colocynthis* unigenes were categorized in three main GO categorizes: biological process (2672), molecular function (1368) and cellular component (1053). These GO terms were further divided into several sub-categories (Figure 3). In the biological process category, single organism process genes accounted for more than 20% of the biological process genes. In the molecular function category, more than 40% of genes were associated with a catalytic activity. In the cellular category, more than 35% of the genes were associated with the cellular component.

Validation of Illumina Expression Patterns by qRT-PCR Analysis

To confirm the reliability of the Illumina sequencing read analysis, 8 candidate genes were selected and their expression was compared at D4 and D1 using qRT-PCR. The expression patterns resulting from qRT-PCR showed general agreement with those from the Illumina sequencing analysis (Table 3). Discrepancies with respect to ratio of fold changes between sequencing and qRT-PCR analysis can be attributed to the essentially different algorithm and sensitivity of the two techniques [23]. In the deep-sequencing method the absolute expression rather than relative expression as in qRT-PCR analysis is used. Transcrip-

Table 6. Photosynthesis-related and chlorophyll-related gene expression profiles.

Gene ID	Hit ID	Annotations	D1RPM	D2RPM	D3RPM	D4RPM
Comp808	Cla002545	Photosystem I psaA	1180	1066.9	513.6 ↓	418.8 ↓
Comp2604	Cla021635	Photosystem I subunit II	309	319	114 ↓	22.8 ↓
Comp9446	Cla002576	Photosystem I subunit III	470	428 ↓	97 ↓	39.1 ↓
Comp4444	Cla012670	Photosystem I subunit IV	274	219 ↓	70.4 ↓	8.3 ↓
Comp4514	Cla007871	Photosystem I subunit IV A	104.9	98.1 ↓	41.2 ↓	9.9 ↓
Comp2126	Cla004483	Photosystem I subunit V	292.9	281.8 ↓	95.6 ↓	23.5 ↓
Comp4727	Cla007940	Photosystem I subunit XI	257.2	229.5 ↓	61.4 ↓	9.6 ↓
Comp7482	Cla009814	Photosystem I subunit X	225.3	194.1 ↓	60.3 ↓	12.3 ↓
Comp5307	Cla011174	Photosystem I subunit X	199	176 ↓	44 ↓	4.6 ↓
Comp1999	Cla005420	Photosystem II polypeptide	225	194.1 ↓	60.3 ↓	12.3 ↓
Comp2011	Cla008554	Photosystem II 5 kDa protein	138.9	118.7 ↓	45.4 ↓	2.5 ↓
comp7744	Cla013942	Photosystem II Protein	495.8	515.8	331.4 ↓	116.7 ↓
Comp8614	Cla014815	Photosystem II reaction center W protein	310.2	265.5	117.3 ↓	28 ↓
Comp2016	Cla022723	Photosystem II core complex proteins psbY	78	76.3 ↓	24.7 ↓	7.7 ↓
Comp12757	Cla011748	Chlorophyll a-b binding protein 13	290.5	246.2 ↓	3.1 ↓	0.3 ↓
Comp6677	Cla013483	Chlorophyll a-b binding protein 3C	130.6	89.5 ↓	7.9 ↓	3.8 ↓
Comp1476	Cla013826	Chlorophyll a-b binding protein	671.4	583.9 ↓	42.9 ↓	8.8 ↓
Comp7589	Cla015680	Chlorophyll a-b binding protein 37	737.2	489.3 ↓	16.6 ↓	10.5 ↓
Comp2286	Cla017325	Chlorophyll a-b binding protein 3C	577.4	230.8 ↓	4.1 ↓	1 ↓
Comp544	Cla017983	Chlorophyll a-b binding protein 6	235.2	219.1 ↓	72.3 ↓	48.2 ↓
Comp3797	Cla018117	Chlorophyll a-b binding protein 6	632.9	502.3 ↓	154.8 ↓	38.3 ↓
Comp15099	Cla019595	Chlorophyll a-b binding protein 21	228.8	126.3 ↓	0.1 ↓	1
Comp1922	Cla001764	Chlorophyll a-b binding protein 8	609.1	470 ↓	241.6 ↓	58.6 ↓
Comp1584	Cla012368	Chlorophyll a-b binding protein 8	1082.2	884.8 ↓	111.1 ↓	281.1 ↓
Comp940	Cla009752	Chlorophyll a-b binding protein 21	652.6	514.8 ↓	43.3 ↓	3.6 ↓
Comp13569	Cla009753	Chlorophyll a-b binding protein 21	827.5	676.3 ↓	68.1 ↓	5.8 ↓
Comp4958	Cla022963	Chlorophyll a-b binding protein 7	538.2	404.7 ↓	110.1 ↓	18.8 ↓
Comp4043	Cla001790	Oxygen-evolving enhancer protein 1 of photosystem II	1088.6	795.9 ↓	334.6 ↓	82.8 ↓
Comp3077	Cla005429	Oxygen-evolving enhancer protein 2, chloroplastic	604.2	531.1 ↓	395.2 ↓	122.6 ↓
Comp5901	Cla019423	Oxygen-evolving enhancer protein 3	545.6	414.8 ↓	86.7 ↓	17 ↓

D1RPM-D4RMP are gene expression reads per millions for each day. ↓ indicates the down-regulation of each gene at each time point.

tional qRT-PCR analyses of 16 genes during the drought treatments are shown in Figures 4 and 5.

Gene Comp5873 is homeobox-leucine zipper protein, with significant upregulation during all days of drought in *C. colocynthis*. It is known that the expression of several homeobox-leucine zipper proteins is correlated to stress. Athb-12, a homeobox-leucine zipper domain protein from *Arabidopsis*, is functionally involved in salt tolerance in yeast [24]. Hahb-4, a homeobox-leucine zipper gene is potentially involved in water stress in sunflower [25]. Comp862 belongs to the glutathione S-transferase (GST) family, which contains heterogeneous, multi-functional dimeric proteins. This gene is highly up-regulated (60x) during drought in *C. colocynthis*. It is thought that GSTs are involved in cellular detoxification [26]. Comp1108 is a member of the NAC (NAM/ATAF1, 2/CUC2) gene family and NACs are known to be involved in numerous biological processes, including drought stress [27].

Comp10156 is a member of the GID1 (GIBBERELLIN INSENSITIVE DWARF1) family in *C. colocynthis* with high expression under drought conditions. GID1 is a soluble GA receptor in rice [28]. GA-GID1 complex interacts with DELLA proteins, which are negative regulators of GA signaling pathway. RD22, a known dehydration responsive gene in *Arabidopsis*, which is mediated by abscisic acid (ABA), may have physiological and molecular significance for processes underlying memory functions of plants in response to ABA and light pulses [29,30]. One RD22-like protein from soybean can alleviate salinity and osmotic stress [31]. RD22-like genes (Comp 6528) with significant up-regulation (>160x) under drought in *C. colocynthis* were confirmed by qRT-PCR, especially after days of withholding water.

NDR1 (NON-RACE-SPECIFIC DISEASE RESISTANCE1) plays an important role in coordinating broader cellular processes in response to stress and bacterial pathogen infection [32]. The chemical β-aminobutyric acid, which is known to induce resistance

Table 7. Expression profiles of some citrulline metabolic genes during drought stress.

Gene ID	Hit ID	Annotations	D1RPM	D2RPM	D3RPM	D4RPM
Comp1556	Cla006970	Carbamoyl-phosphate synthetase	64.6	70.9 ↑	159 ↑	175.8 ↑
Comp1556	Cla005591	Carbamoyl-phosphate synthetase	196.1	162.5	846.5 ↑	1542 ↑
Comp3843	Cla014787	Carbamoyl-phosphate synthetase	15	20.5 ↑	49.8 ↑	63.3 ↑
Comp11009	Cla016474	Proline dehydrogenase	2.2	3.9 ↑	11.6 ↑	31.4 ↑
Comp01186	Cla023055	Arinosuccinase	5.6	6.2	3.7	5.3
Comp5521	Cla020781	Orinithine carbamoyltransfrase	25.2	19.1 ↓	52.5 ↑	32.5
Comp6244	Cla015337	Acetylornithine aminotransferase	39	45.3 ↑	127 ↑	147.6 ↑
Comp6587	Cla008748	Glutamine amidotransferase	25.6	19.7	12.4	13.8
Comp15261	Cla017928	Glutamate 5-kinase	4.3	5.7	1.7	1.6
Comp3961	Cla019569	Orinithine-oxo-acid transaminase	19.3	30.5	28.1	17.1
Comp7707	Cla023055	Argininosuccinate lyase	5.6	6.2	3.7	5.3
Comp3468	Cla003592	Argininosuccinate lyase	22.9	26.8	35.4	25.6
Comp776_seq2	Cla002611	Arginosuccinate synthase	12.5	25.4	6	7.7
Comp776_seq1	Cla002609	Arginosuccinate synthase	18.2	18.9	2.4 ↓	2 ↓
Comp1556	Cla022915	Carbamoyl-phosphate synthetase	115.4	103.2	53.4 ↓	42 ↓

D1RPM-D4RMP indicate gene expression reads per millions for each day. ↑ and ↓ indicate the significant up-regulation and down-regulation of each gene at each time point.

in plants, primed the expression of many genes, and NDR1/NHL10 was one of them [33]. Comp7317 gene, which is a member of NDR1/NH10 showed significant changes at D2 only during the early stage of water deficit stress.

GRAS (for GA Insensitive, REPRESSOR of gal-3 (RGA), SCARECROW (SCR)) transcription factors, have a major function in plant development and environmental adaption. These TFs are particularly implicated in the modulation of plant tolerance to stressors as cold, drought, salinity by crosstalks via GA to ABA-dependent and ABA-independent pathways [34]. For example, SCL7 confers salt and drought tolerance in *Arabidopsis* [35]. SCL14 is involved in the detoxification of xenobiotics and possibly endogenous harmful metabolites [36]. Comp20554, which belongs to the GRAS transcription factor family, showed significant up-regulation especially on D2.

The expression profile of Comp3048, which codes for a methyltransferase, maintained high levels during drought stress, especially on D2. It was found that myo-inositol-O-methyltransferase (Imt1) responded to low temperature stress in transgenic *Arabidopsis* [37]. Trithorax-like H3K4 methyltransferase from barley is drought inducible [38]. The methylation of myo-inositol catalyzed by myo-inositol methyltransferase (IMT) occurs when plants are under abiotic stress. Over-expression of IMT resulted in improved tolerance to dehydration and salt stress treatment in *Arabidopsis* [39].

Heat-shock proteins (HSPs) are environmentally induced proteins that enable plants to cope with heat and other environmental stresses. For example, *Trichoderma harzianum* Hsp70 transgenic *Arabidopsis* is abiotic stress tolerant [40]. Similarly HSP22 was found to be highly upregulated in *C. colocynthis* roots during drought conditions [13]. Overexpression of GmHsp90S can decrease damage of abiotic stresses in *Arabidopsis* [41]. Comp14675 belongs to the HSP family, and showed up-regulation at later stages of D3 and D4 (Figure 5).

Plant cold shock proteins (CSP) are very conserved among various plant genera [42]. The first CSP identified was WCSP1 from winter wheat, which did accumulate in response to low-temperature stress [43]. Similarly in *C. colocynthis* CSP (Comp13927) was up-regulated during later stages of water deficit.

Comp2553, one MA3 domain-containing protein gene, showed marked changes on D3 and D4 under drought conditions (Figure 5). Loss-of-function of ECIP1 (one MA3 domain-containing protein) resulted in enhanced ethylene response but altered salt response [44] in *Arabidopsis*.

Overexpression of plant BL-1 in *Arabidopsis* resulted in the attenuation of cell death induced by biotic stresses (pathogens) and abiotic stresses such as heat, cold, drought, salt and chemical-induced oxidative stresses [45]. BL-1 might function to control the level of the "pro-survival and pro-death" signals under multiple stress conditions in plants [46]. For example, cucumber BAX inhibitor-1 is a conserved cell death suppressor induced by cold stress and a negative regulator of programmed cell death (PCD) [47]. Comp372 encodes one BL-1 gene, which was up-regulated by drought at later stages (D3 and D4).

Comp8117, which is a homologue of a *Populus* EST (CU233481.1), is a drought stress related gene, up-regulated at D4. Comp10586, which encodes a member of the MYB transcription factor family, is up-regulated during the late stage of drought stress. Some MYB members have been shown to regulate plant responses to biotic and abiotic stress conditions [48]. For example, AtMYB96 acts through the ABA signaling pathway to induce pathogen resistance by promoting salicylic acid biosynthesis, and thus regulating stomata movement, drought tolerance and disease resistance in *Arabidopsis* [49,50]. MYB88 might function directly or indirectly, as positive regulator of stress-responsive genes [51]. TaMYB30-B from wheat did improve drought stress tolerance in transgenic *Arabidopsis* [52]. Another family of transcription factors, the WRKY family, named for the WRKY domain of about 60 amino acids, contains a highly conserved WRKYGQK heptapeptide at its N-terminus and a zinc-finger-like motif at its C-terminus [53]. WRKY transcription factors are involved in multiple aspects of plant growth, development and stress [54]. Several TaWRKY in wheat with roles in the abiotic stress response acted in an ABA-dependent manner [55]. Here, gene Comp19751encodes a WRKY gene, which showed remarkable up-regulation at D4.

MADS-box family members function in reproductive development and stress [56]. For example, OsMADS25 and OsMADS27 transcripts accumulate in response to osmotic stress [57]. Comp6823, which encodes for a MADS-box gene in *C. colocynthis*, is up-regulated markedly at the last stage of drought (D4; Figure 5).

Analysis of the Drought Stress Signaling Transcriptome in *C. colocynthis*

Drought stress signal transduction consists of several pathways including ionic and osmotic homeostasis signaling, detoxification response and growth regulation pathways [58]. Genes detected in *C. colocynthis* leaves during water deficit are listed in Table 4.

In the ion signaling pathway, calcium binding proteins are well known for their involvement in both biotic and abiotic stress response pathways [59]. The calcium ion (Ca^{2+}) as a secondary messenger in plants is sensed by calmodulins (CaMs)/CaM-like protein (CMLs), the caldineurin B-like proteins (CBLs) and Ca2+-dependent protein kinases (CDPKs). CaM binds to CaM-binding proteins (CBPs), which function in different pathways under biotic and biotic stresses [60]. A total of 10 Ca^{2+} binding proteins were detected in the *C. colocynthis* transcriptome.

Protein phosphorylation is a central theme in the cell's response to stress. The MAP kinase cascade in transcript levels consist of a number of protein kinases, such as two-component histidine kinase, MAPKKK, MAPKK, MAPK etc. [61]. Here we detected 16 MAP kinases in the *C. colocynthis* transcriptome.

Membrane phospholipids can activate several types of phospholipases that cleave certain phospholipids to generate lipid messengers (eg. PA, DAG, IP3), which further regulate stress tolerance through modulation of stress-responsive gene expression [62]. Several members in this pathway, such as phospholipase C (PLC), diacylglycerol (DAG) and phosphotidylinositol 4,5-bisphosphate (PIP2)-like aquaporin were detected (Table 4).

Detoxification signaling can ameliorate the damage in plants under stresses [63], as noted in many other plant species. A total of 32 detoxification proteins were detected in the *C. colocynthis* transcriptome.

Molecular mechanisms regulating gene expression in response to drought stress have been studied by analyzing the functional transcription factors in ABA-dependent and ABA-independent pathways [6]. Several regulatory proteins in ABA-dependent or – independent pathway were detected, among which NAC, MYB/C and leucine-rich repeat proteins (LRR). These regulatory proteins can further modulate many responsive transcription factors. Several functional proteins such as heat shock protein (HSP) 70, HSP22, grpE like protein, respiratory burst oxidase D (RBOHD), VIRE2-Interacting protein2 (VIP2), ABA transporter-like protein, synaptobrevin-related protein, translocon outer envelope of chloroplast (Toc34-1), beta-amylase, TIP1 (TIP GROWTH DEFECTIVE 1), and RD22 were detected.

Analysis of Phytohormone Signaling Mediators in *C. colocynthis*

Phytohormones play important roles in regulating plant responses under biotic and abiotic stress. Elaborate phytohormone signaling networks mediate the adaptability of plants to different environmental conditions [64]. Many phytohormones such as ABA, salicylic acid (SA), jasmonic acid (JA), auxin, ethylene and gibberellic acid (GA) are being studied for their role in abiotic stress responses [33,65,66]. It is known that downstream signaling proteins for auxin, GA, JA, and ABA are subjected to ubiquitin-dependent degradation [67]. Putative phytohormone signaling genes detected in *C. colocynthis* during the drought response are listed in Table 5.

Ethylene signaling pathway components were ordered into a hypothetical linear pathway based on both genetic (epistasis) analysis and biochemical interactions [68]. Almost all of the ethylene signaling homologous members (118 ethylene response factors or ERFs) were detected in the *C. colocynthis* transcriptome. APETALA2/ethylene responsive factor (AP2/ERF) transcription factors are well-known for mediating stress responses and development in plants [69].

The auxin response factor (ARF) family contains transcription factors that bind to auxin-responsive elements (AREs) in the promoters of primary auxin-responsive genes. Aux/IAAs are early auxin-response proteins that bind ARFs, therefore inhibiting ARE-mediated gene transcription. Aux/IAAs are involved in ubiquitin-mediated degradation, which is catalyzed by SCF E3 ubiquitin ligase. TIR1 can stimulate Aux/IAA proteolysis by binding auxin to this protein [70]. All of the major auxin signaling related transcription factors found in the *C. colocynthis* transcriptome are shown in Table 5. Similar to the auxin pathway, a novel family of transcriptional regulators, the jasmonate ZIM-domain (JAZ) proteins play a target part as Aux/IAA. In addition, COI1 plays a similar role as TIR1, while MYC and R2R3-MYB transcription factors work as ARFs [71]. Several JA signaling pathway related genes exist in the *C. colocynthis* transcriptome.

In the SA pathway, NPR1 (non-specific disease resistance 1) is a key regulator in SA-dependent defense signaling [72]. Similarly WRKY and TGA play major roles as transcriptional regulators in the SA pathway. We detected 2 NPR1 related proteins, 46 WRKYs and 6 TGAs in *C. colocynthis*, which might function in the *C. colocynthis* SA pathway.

Drought triggers the production of the phytohormone ABA, which in turn causes stomatal closure and induces expression of stress-related genes. The soluble PYR/PYL/RCAR receptors function at the apex of a negative regulatory pathway to directly regulate PP2C phosphatases, which in turn directly regulate SnRK2 kinases [73]. Several of the core transcription factors in the ABA pathway are listed in Table 5.

We also detected 11 GA receptor GIBBERELLIN INSENSITIVE DWARF1 (GID1), 4 DELLA growth inhibitors (DELLAs) and 2 F-box proteins (SLY1) and SNEEZY (SNZ), which play important roles in GA signaling pathways [74,75].

Cellular Metabolism under Drought Stress in *C. colocynthis*

Global gene expression analyses have shown substantial down-regulation of many photosynthetic genes under drought not only in *Arabidopsis* [76], but also several other species such as indica rice [77,78]. Similarly, many photosystem I and II, chlorophyll a, b binding protein, and oxygen evolving enhancer protein genes [79], showed down-regulation during water deficit stress in *C. colocynthis* (Table 6).

Citrulline, a non-protein amino acid intermediate in the arginine biosynthetic pathway, has been found to accumulate in leaves of drought tolerant watermelon under water deficit conditions [80,81,82]. Thus several factors related to the response of this model drought tolerant species to stress have been identified (Table 7, Figure S1). Citrulline metabolic genes (carbamoyl-phosphate synthetase, acetyl glutamate synthase, acetylornithine aminotransferase, aminoglutamate decarboxylase, acetylornithine deacetylase, and glutamate dehydrogenase) were found to be significantly up-regulated during drought.

One of the major research goals is to understand the molecular mechanisms underlying drought tolerance in plants. It is clear that drought triggers a wide variety of responses in *C. colocynthis*. Down regulation of many photosynthetic genes was observed especially at the later stages of drought. Up regulation of many transcription factors, stress signaling factors, detoxification genes, and genes involved in phytohormone signaling occurred throughout the water deficit experiment. Systematic approaches using genomic analyses should lead to the discovery of additional stress factors and provide us with a better understanding of the stress tolerance mechanism of this drought tolerant plant species.

Supporting Information

Figure S1 Citrulline metabolism pathway in *C. colocynthis*.

Table S1 qRT-PCR primer sequence information.

Table S2 Non-significant and significantly changed *C. colocynthis* genes expressed as read counts during drought.

Author Contributions

Conceived and designed the experiments: ZW FD. Performed the experiments: ZW FD. Analyzed the data: ZW FD HH LG. Contributed reagents/materials/analysis tools: FD LG SM. Contributed to the writing of the manuscript: ZW FD.

References

1. Guo Y, Tan J (2013) A biophotonic sensing method for plant drought stress. Sensor Actuat B-Chem 188: 519–524.
2. Do PT, Degenkolbe T, Erban A, Heyer AG, Kopka J, et al. (2013) Dissecting rice polyamine metabolism under controlled long-term drought stress. PLOS ONE 8: e60325.
3. Levitt J (1980) Responses of plants to environmental stress: chilling, freezing and high temperature stresses, 2nd Ed New York: Academic Press.
4. Bartels D, Sunkars R (2005) Drought and salt tolerance in plants. Cr Rev Plant Sci 24: 23–58.
5. Reddy AR, Chaitanya KV, Vivekanandan M (2004) Drought induced responses of photosynthesis and antioxidant metabolism in higher plants. Plant Physiol 161: 1189–1202.
6. Yamaguchi-Shinozaki K, Shinozaki K (2006) Transcriptional regulatory networks in cellular response and tolerance to dehydration and cold stresses. Annu Rev Plant Biol 57: 781–803.
7. Zhao J, Gao Y, Zhang Z, Chen T, Guo W, et al. (2013) A receptor-like kinase gene (GbRLK) from *Gossypium barbadense* enhances salinity and drought-stress tolerance in *Arabidopsis*. BMC Plant Biol 13: 110–125.
8. Sŭrbanovski N, Sargent DJ, Else MA, Simpson DW, Zhang H, et al. (2013) Expression of *Fragaria vesca* PIP aquaporins in response to drought stress: PIP down-regulation correlates with the decline in substrate moisture content. PLOS ONE 8: e74945.
9. Caccamo M, Grotewold E (2013) Turning over a new leaf in plant genomics. Genome Biol 14: 403.
10. Juenger TE (2013) Natural variation and genetic constraints on drought tolerance. Curr Opin Plant Biol 16: 274–281.
11. Jeffrey C (2008) A review of Cucurbitaceae. Bot J Linn Soc 81: 233–24.
12. Dane F, Liu J, Zhang C (2006) Phylogeography of the bitter apple, *Citrullus colocynthis*. Genet Resour Crop Ev 54: 327–336.
13. Si Y, Zhang C, Meng S, Dane F (2009) Gene expression changes in response to drought stress in *Citrullus colocynthis*. Plant Cell Rep 28: 997–1009.
14. Guo S, Liu J, Zheng Y, Huang M, Zhang H, et al. (2011) Characterization of the transcriptome dynamics during watermelon fruit development: sequencing, assembly, annotation and gene expression profiles. BMC Genomics 12: 454.
15. Grassi S, Piro G, Lee JM, Zheng Y, Fei Z, et al. (2013) Comparative genomics reveals candidate carotenoid pathway regulators of ripening watermelon fruit. BMC Genomics 14: 781.
16. Guo S, Zhang J, Sun H, Salse J, William JL, et al. (2013) The draft genome of watermelon (*Citrullus lanatus*) and resequencing of 20 diverse accessions. Nat Genet 45: 51–58.
17. Gonzalez VM, Garcia-Mas J, Arus P, Puigdomenech P (2010) Generation of a BAC-based physical map of the melon genome. BMC Genomics 11: 339.
18. Huang S, Li R, Zhang Z, Li L, Gu X, et al. (2009) The genome of the cucumber, *Cucumis sativus* L. Nat Genet 41: 1275–1281.
19. Liu N, Yang J, Gao S, Xu Y, Zhang M (2013) Genome-wide identification and comparative analysis of conserved and novel microRNAs in grafted watermelon by high-throughput sequencing. PLOS ONE 8: e57359.
20. Wincker P (2013) Genomics and fruit crop selection. Nature Genetics 45: 9–10.
21. Chomczynski P, Sacchi N (1987) Signal-step method of RNA isolation by acid guanidinium thiocyanate–phenol-chloroform extraction. Anal Biochem 162: 156–159.
22. Kal AJ, van Zonneveld AJ, Benes V, van den Berg M, Koerkamp MG, et al. (1999) Dynamics of gene expression revealed by comparison of serial analysis of gene expression transcript profiles from yeast grown on two different carbon sources. Mol Biol Cell 10: 1859–1872.
23. Shi T, Gao Z, Wang L, Zhang Z, Zhuang W, et al. (2012) Identification of differentially expressed genes associated with pistil abortion in Japanese apricot by genome-wide transcriptional analysis. Plos ONE 7: e47810.
24. Shin D, Koo YD, Lee J, Lee H, Baek D, et al. (2004) Athb-12, a homeobox-leucine zipper domain protein from *Arabidopsis thaliana*, increases salt tolerance in yeast by regulating sodium exclusion. Biochem Biophys Res 323: 534–540.
25. Gago GM, Almoguera C, Jordano J, Gonzalez DH, Chan RL (2002) Hahb-4, a homeobox-leucine zipper gene potentially involved in abscisic acid-dependent responses to water stress in sunflower. Plant Cell Environ 25: 633–640.
26. Park H, Choung Y (2010) Evaluation of the biodegradation feasibility of antibiotics by three bacteria involving glutathione S-transferases. Can J Civ Eng 37: 814–819.
27. Shaminuzzaman M, Vodkin L (2013) Genome-wide identification of binding sites for NAC and YABBY transcription factors and co-regulated genes during soybean seedling development by ChIP-Seq and RNA-Seq. BMC Genomics 14: 477.
28. Ueguchi-Tanaka M, Ashikari M, Nakajima M, Itoh M, Katoh E, et al. (2005) GIBBERELLIN INSENSITIVE DWARFS encodes a soluble receptor for gibberellin. Nature 437: 693–69.
29. Goh CH, Nam HG, Park YS (2003) Stress memory in plants: a negative regulation of stomatal response and transient induction of rd22 gene to light in abscisic acid-entrained *Arabidopsis* plants. Plant J 36: 240–255.
30. Xu YH, Liu R, Yan L, Liu ZQ, Jiang SC, et al. (2012) Light-harvesting chlorophyll a/b-binding proteins are required for stomatal response to abscisic acid in *Arabidopsis*. J Exp Bot 63: 1095–1106.
31. Wang H, Zhou L, Fu Y, Cheung M, Wong F, et al. (2012) Expression of an apoplast-localized BURP-domain protein from soybean (GmRD22) enhances tolerance towards abiotic stress. Plant Cell Environ 35: 1932–1947.
32. Knepper C, Savory EA, Brad D (2011) *Arabidopsis* NDR1 is an integrin-like protein with a role in fluid loss and plasma membrane-cell wall adhesion. Plant Physiol 156: 286–300.
33. Chen P, Parshant S, Laurent Z (2011) Priming of the *Arabidopsis* pattern-triggered immunity response upon infection by necrotrophic *Pectobacterium carotovorum* bacteria. Mol Plant Pathol 14: 58–70.
34. Golldack D, Li C, Mohan H, Probst N (2013) Gibberelins and abscisic acid signal crosstalk: living and developing under unfavorable conditions. Plant Cell Rep 32: 1007–1016.
35. Ma H, Liang D, Shuai P, Xia X, Yin W (2010) The salt- and drought-inducible poplar GRAS protein SCL7 confers salt and drought tolerance in *Arabidopsis thaliana*. J Exp Bot 61: 4011–4019.
36. Fode B, Siemsen T, Thurow C, Weigel R, Gatz C (2008) The *Arabidopsis* GRAS protein SCL14 interacts with Class II TGA transcription factors and is essential for the activation of stress-inducible promoters. Plant Cell 20: 3122–3135.
37. Zhu B, Peng RH, Xiong AS, Xu J, Fu XY, et al. (2012) Transformation with a gene for myo-inositol O-methyltransferase enhances the cold tolerance of *Arabidopsis thaliana*. Biol Plantarum 56: 135–139.
38. Papaefthimiou D, Tsataris AS (2012) Characterization of a drought inducible trithorax-like H3K4 methytransferase from barley. Biol Plantarum 56: 683–692.
39. Ahn C, Park U, Park PB (2011) Increased salt and drought tolerance by D-ononitol production in transgenic *Arabidopsis thaliana*. Biochem Biophys Res 415: 669–674.
40. Montero-Barrientos M, Hermosa R, Cardoza RE, Gutiérrez S, Nicolás C, et al. (2010) Transgenic expression of the *Trichoderma harzianum* hsp70 gene increases *Arabidopsis* resistance to heat and other abiotic stresses. Plant Physiol 167: 659–665.
41. Xu J, Xue C, Xue D, Zhao J, Gai J, et al. (2013) Overexpression of GmHsp90s, a heat shock protein (Hsp90) gene family form soybean, decreases damage of abiotic stresses in *Arabidopsis thaliana*. PLOS ONE 8: e69810.
42. Karlson D, Imai R (2003) Conservative of the cold shock domain protein family in plants. Plant Physiol 131: 12–15.
43. Karlson D, Nakaminami K, Toyomasu T, Imai R (2002) A cold-regulated nucleic acid-binding protein of winter wheat shares a domain with bacterial cold shock proteins. J Biol Chem 277: 35248–35256.
44. Lei G, Shen M, Li Z, Zhang B, Duan K, et al. (2011) EIN2 regulates salt stress response and interacts with a MA3 domain-containing protein ECIP1 in *Arabidopsis*. Plant Cell Environ 34: 1678–1692.

45. Ishikawa T, Watanabe N, Nagano N, Kawai-Yamada M, Lam E (2011). Bax inhibitor-1: a highly conserved endoplasmic reticulum-resident cell death suppressor. Cell Death Differ 18: 1271–1278.
46. Watanabe N, Lam E (2009) Bax Inhibitor-1, a conserved cell death suppressor, is a key molecular switch downstream from a variety of biotic and abiotic stress signals in plants. Int. J. Mol. Sci. 10: 3149–3167.
47. Chen XH, Yu H, Deng HJ, Chen JX, MI HB, et al. (2013) Cucumber BAX inhibitor-1, a conserved cell death suppressor and a negative programmed cell death regulator under cold stress. Biol Plantarum 57: 684–690.
48. Cao ZH, Zhang SZ, Wang RK, Zhang RF, Hao YJ (2013) Genome wide analysis of the apple MYB transcription factor family allows the identification of MdoMYB121 gene conferring abiotic stress tolerance in plants. PLOS ONE 8: e69955.
49. Reyes JL, Chua NH (2007) ABA induction of miR159 controls transcript levels of two MYB factors during *Arabidopsis* seed germination. Plant J 49: 592–606.
50. Seo PJ, Xiang F, Qiao M, Park JY, Lee YN, et al. (2009) The MYB96 transcription factor mediates abscisic acid signaling during drought stress response in *Arabidopsis*. Plant Physiol 151: 275–289.
51. Xie Z, Li D, Wang L, Sack FD, Grotewold E (2010) Role of the stomatal development regulators FLP/MYB88 in abiotic stress response. Plant J 64: 731–739.
52. Zhang L, Zhao G, Xia C, Jia J, Liu X, et al. (2012) A wheat R2R3-MYB gene, TaMYB30-B, improves drought stress tolerance in transgenic *Arabidopsis*. J Exp Bot 63: 5873–5885.
53. Rushton PJ, Somssich IE, Ringler P, Shen QJ (2010) WRKY transcription factors. Trends Plant Sci 15: 247–258.
54. Niu CF, Wei W, Zhou QY, Tian AG, Hao YJ, et al. (2012) Wheat WRKY genes TaWRKY2 and TaWRKY19 regulate abiotic stress tolerance in transgenic *Arabidopsis* plants. Plant Cell Environ 35: 1156–1170.
55. Zhu X, Liu S, Meng C, Qin L, Kong L, et al. (2013) WRKY transcription factors in wheat and their induction by biotic and abiotic stress. Plant Mol Biol Rep 31: 1053–1067.
56. Arora R, Agarwal P, Ray S, Singh AK, Singh VP, et al. (2007) MADS-box gene family in rice: genome-wide identification, organization and expression profiling during reproductive development and stress. BMC Genomics 8: 242.
57. Puig J, Meynard D, Khong GN, Pauluzzi G, Guiderdoni E, et al. (2013) Analysis of the expression of the AGL17-like clade of MADS-box transcription factors in rice. Gene Expr Patterns 13: 160–170.
58. Zhu JK (2002) Salt and drought stress signal transduction in plants. Annu Rev Plant Biol 53: 247–273.
59. Wan D, Li R, Zou B, Zhang X, Cong J, et al. (2012) Calmodulin-binding protein CBP60g is a positive regulator of both disease resistance and drought tolerance in *Arabidopsis*. Plant Cell Rep 31: 1269–1281.
60. Ranty B, Aldon D, Galaud JP (2006) Plant calmodulins and calmodulin-related proteins: multifaceted relays to decode calcium signals. Plant Signal Behav 1: 96–104.
61. Morris PC (2001) MAP kinase signal transduction pathways in plants. New Phytol 151: 67–89.
62. Singh A, Kanwar P, Pandey A, Tyagi AK, Sopory SK, et al. (2013) Comprehensive genomic analysis and expression profiling of phospholipase C gene family during abiotic stresses and development in rice. PLOS ONE 8:e62494.
63. Triantaphylidès C, Havaux M (2009) Singlet oxygen in plants: production, detoxification and signaling. Cell 14: 1360–1385.
64. Kohli A, Sreenivasulu N, Lakshmanan P, Kuman PP (2013) The phytohormone crosstalk paradigm takes center stage in understanding how plants respond to abiotic stresses. Plant Cell Rep 32: 945–957.
65. Peleg Z, Blumwald E (2011) Hormone balance and abiotic stress tolerance in crop plants. Curr Opin Plant Biol 14: 290–295.
66. Santner A, Calderon-Villalobos LIA, Estelle M (2009) Plant hormones are versatile chemical regulators of plant growth. Nature Chem Biol 5: 301–307.
67. Santino A, Taurino M, De Domenico S, Bonsegna S, Poltronieri P, et al. (2013) Jasmonate signaling in plant development and defense response to multiple (a)biotic stresses. Plant Cell Rep 32: 1085–1098.
68. Bleecker AB (2000) Ethylene: A gaseous signal molecule in plants. Annu Rev Cell Dev Biol 16: 1–18.
69. Licausi F, Ohme-Takagi M, Perata P (2013) APETALA2/Ethylene responsive factor (AP2/ERF) transcription factors: mediators of stress responses and developmental programs. New Phytol 199: 639–649.
70. Teale WD, Paponov IV, Palme K (2006) Auxin in action: signaling, transport and the control of plant growth and development. Nature Rev 7: 847–859.
71. Pérez AC, Goossens A (2013) Jasmonate signaling: a copycat of auxin signaling? Environ 36: 2071–2084.
72. Boatwright J, Pajerowska-Mukhtar K (2013) Salicylic acid: an old hormone up to new tricks. Mol Plant Pathol 14: 623–634.
73. Cutler SR, Rodriguez PL, Finkelstein RR, Abrams SR (2010) Abscisic acid: emergence of a core signaling network. Annu Rev Plant Biol 61: 651–679.
74. Davière JM and Achard P (2013) Gibberellin signaling in plants. Dev 140: 1147–1151.
75. Richards DE, King KE, Ait-ali T, Harberd NP (2001) How gibberellin regulates plant growth and development: a molecular genetic analysis of gibberellin signaling. Annu Rev Plant Physiol 52: 67–88.
76. Harb A, Krishnan A, Ambavaram MMR, Pereira A (2010) Molecular and physiological analysis of drought stress in Arabidopsis reveals early responses leading to acclimation in plant growth. Plant Phys 154: 1254–1271.
77. Damarajua S, Schledeb S, Eckhardta U, Loksteinb H, GrimmaB (2011) Functions of the water soluble chlorophyll-binding protein in plants. J Plant Phys 168: 1444–1451.
78. Gorantla M, Babu PR, Reddy Lachagari VB, Reddy AMM, Wusirika R, et al. (2007) Identification of stress-responsive genes in an indica rice (*Oryza sativa* L.) using ESTs generated from drought-stressed seedlings. J Exp Bot 58: 253–265.
79. Guha A, Sengupta D, Reddy AR (2013) Polyphasic chlorophyll a fluorescence kinetics and leaf protein analyses to track dynamics of photosynthetic performance in mulberry during progressive drought. J Photochem Photobiol B 119: 71–83.
80. Kawasaki S, Miyake C, Kohchi T, Fujii S, Uchida M, et al. (2000) Responses of wild watermelon to drought stress: accumulation of an ArgE homologue and citrulline in leaves during water deficits. Plant Cell Physiol. 41: 864–873.
81. Kusvuran S, Dasgan HY, Abak K (2013) Citrulline is an important biochemical indicator in tolerance to saline and drought stresses in melon. Scientific World J. 2013 ID:253414.
82. Dasgan HY, KusvuranS, Abak K, Leport L, Larher F, et al. (2009) The relationship between citrulline accumulation and salt tolerance during the vegetative growth of melon (*Cucumis melo* L.). Plant Soil Environ 55: 51–57.

4

High-Throughput Phenotyping to Detect Drought Tolerance QTL in Wild Barley Introgression Lines

Nora Honsdorf[1,2], Timothy John March[3], Bettina Berger[4], Mark Tester[5], Klaus Pillen[1]*

1 Chair of Plant Breeding, Institute of Agricultural and Nutritional Sciences, Martin-Luther University Halle-Wittenberg, Halle (Saale), Germany, **2** Interdisciplinary Center for Crop Plant Research (IZN), Halle (Saale), Germany, **3** School of Agriculture, Food and Wine, University of Adelaide, Waite Campus, Adelaide, Australia, **4** The Plant Accelerator, University of Adelaide, Waite Campus, Adelaide, Australia, **5** Center for Desert Agriculture, King Abdullah University of Science and Technology, Thuwal, Saudi Arabia

Abstract

Drought is one of the most severe stresses, endangering crop yields worldwide. In order to select drought tolerant genotypes, access to exotic germplasm and efficient phenotyping protocols are needed. In this study the high-throughput phenotyping platform "The Plant Accelerator", Adelaide, Australia, was used to screen a set of 47 juvenile (six week old) wild barley introgression lines (S42ILs) for drought stress responses. The kinetics of growth development was evaluated under early drought stress and well watered treatments. High correlation (r = 0.98) between image based biomass estimates and actual biomass was demonstrated, and the suitability of the system to accurately and non-destructively estimate biomass was validated. Subsequently, quantitative trait loci (QTL) were located, which contributed to the genetic control of growth under drought stress. In total, 44 QTL for eleven out of 14 investigated traits were mapped, which for example controlled growth rate and water use efficiency. The correspondence of those QTL with QTL previously identified in field trials is shown. For instance, six out of eight QTL controlling plant height were also found in previous field and glasshouse studies with the same introgression lines. This indicates that phenotyping juvenile plants may assist in predicting adult plant performance. In addition, favorable wild barley alleles for growth and biomass parameters were detected, for instance, a QTL that increased biomass by approximately 36%. In particular, introgression line S42IL-121 revealed improved growth under drought stress compared to the control Scarlett. The introgression line showed a similar behavior in previous field experiments, indicating that S42IL-121 may be an attractive donor for breeding of drought tolerant barley cultivars.

Editor: Tianzhen Zhang, Nanjing Agricultural University, China

Funding: This work was supported by the Interdisciplinary Centre for Crop Plant Research (IZN), Halle (Saale) (http://www.uni-halle.de/izn/), the German Plant Genome Research Initiative (GABI) of the Federal Ministry of Education and Research (BMBF, project 0313125B) (www.gabi.de/), and Group of Eight Australia – Germany Joint Research Co-operation Scheme, funded by the German Academic Exchange Service (www.daad.de) and the Group of Eight, Australia (http://www.go8.edu.au/). The funders had no role in study design, data collection and analysis, decision to publish, or preparation of the manuscript.

Competing Interests: The authors have declared that no competing interests exist.

* E-mail: klaus.pillen@landw.uni-halle.de

Introduction

Barley (*Hordeum vulgare* ssp. *vulgare*, hereafter abbreviated with *Hv*) is ranked fourth among the worldwide production of cereals. Due to its multipurpose use as animal feed, human food and substrate for malting it is one of the most important cereals worldwide [1]. Barley is known to be relatively tolerant to abiotic stresses among the major cereal crops and, thus, is often grown in more marginal sites [2]. However, the process of genetic erosion has been under way in barley since its domestication some 10,000 years ago and, in particular, since the advent of intensive modern elite breeding during the last century [3]. As a result of this process, diverse landraces have been replaced by modern elite cultivars with a much narrower gene pool. Therefore there is limited genetic diversity remaining in the elite barley gene pool for abiotic and biotic stress tolerance. The current loss of genetic variation in the elite gene pool tends to limit the breeding success of improved cultivars [4]. To overcome this problem several authors, e.g. Zamir [5], proposed to use wild relatives of crop species as donors of exotic germplasm to enhance elite varieties. Tanksley and Nelson [6] proposed the method of "advanced backcross quantitative trait loci analysis" (AB-QTL) to introduce exotic genes into modern crop varieties. The method combines QTL detection and the introduction of favorable exotic alleles from a wild donor parent. Lines produced by advanced backcrosses ideally contain only one single introgression from the wild parent and are then referred to as introgression lines (ILs). This is achieved by several rounds of backcrossing to the recurrent parent and marker assisted selection (MAS). A set of ILs ideally represents the whole genome of a wild donor plant in the genetic background of a single elite variety [5].

Pillen et al. [7] published the first AB-QTL study in barley. Von Korff et al. [8] developed a BC_2DH population from a cross between the German spring barley cultivar Scarlett (*Hv*) and the Israeli wild barley accession ISR42-8 (*Hordeum vulgare* ssp. *spontaneum*, hereafter abbreviated with *Hsp*). The lines of this S42 population were used in several AB-QTL studies to identify QTL for yield, pathogen resistance and malting quality traits [9–13].

Schmalenbach et al. [14] used the S42 population to develop 59 ILs (S42ILs) by a further round of backcrossing with the recurrent parent Scarlett and subsequent selfing and MAS. Each of the S42ILs contains a single or a small number of *Hsp* introgressions.

Several QTL studies were conducted to verify QTL from AB-QTL studies and to identify new QTL for pathogen resistance, yield and quality parameters [14–16]. Naz et al. [17] studied root architecture of S42ILs and detected QTL for root dry weight and root volume. Later on, the S42IL population was extended to 73 lines and the lines were genotyped with a 1,536-SNP Illumina BOPA1 set [18]. Six hundred thirty-six informative SNPs and their known map order [19] allowed the precise localization of the *Hsp* introgressions. The S42IL set represents 87.3% of the wild barley donor genome. Moreover, Schmalenbach et al. [18] developed segregating high-resolution mapping populations (S42IL-HRs) for 70 S42ILs. Those lines are readily available to facilitate fine mapping and, ultimately, cloning of QTL.

Drought is one of the main factors limiting yield worldwide [20]. Due to climate change extreme weather events are predicted to occur more frequently and an altered pattern of drought occurrence is expected [21]. Therefore maintaining plant growth and yield under drought remains a major objective for plant breeding [22]. Many studies have been conducted on the impacts of terminal drought stress in cereals, while impacts of drought stress at early developmental stages are less well investigated [23]. However, several authors comment that yield may be enhanced by improved early vigor and a rapid development of maximum leaf area [24,25]. López-Castañeda and Richards [26] reported that on average barley has a higher yield in water limited environments compared to wheat, triticale, and oat. As part of a possible explanation, they pinpointed the faster and more vigorous growth of barley during vegetative development. Variation in this trait is, therefore, likely to be in direct relation to drought stress tolerance and yield.

Conventional methods to determine biomass and measure growth are time-consuming and labor intensive. Often they involve destructive harvest of plants and therefore make repeated measurements on the same plant impossible. New developments in plant imaging technologies allow the estimation of biomass and growth parameters as a non-destructive and rapid alternative to more traditional methods [27,28]. New phenotyping facilities enable automated imaging of plants. Several types of plant images can be taken, e.g. with infrared, near infrared, fluorescent and visible light. Scanning with infrared light gives information on plant or leaf temperature, while near infrared imaging sheds light on the plant water content and fluorescent pictures enable conclusions on plant health status. High resolution color pictures (RGB pictures), taken from the top and two side views are used to determine the projected shoot area of the plant. The projected shoot area serves as a measure for biomass. Hence, from RGB images taken at several time points, growth curves as well as growth rates can be calculated.

In fully automated greenhouses plants can be delivered via conveyor belts to watering, weighing and imaging stations. In these high-throughput phenotyping facilities several hundred individual plants can be imaged per day in a fully automated manner. High-throughput phenotyping facilities of this type are currently in use in various research institutes (e.g. The Plant Accelerator, Adelaide, Australia; CropDesign, Gent, Belgium; IPK Gatersleben, Germany, PhenoArch, Montpellier, France).

Such phenotyping facilities are ideal to combine controlled irrigation and phenotyping protocols [29]. A first application was given by Rajendran et al. [30] who used a manual imaging system (LemnaTecScanalyzer3D, Wuerselen, Germany) to screen *Triticum monococcum* accessions for salinity tolerance. They developed high-throughput quantification assays to distinguish sodium exclusion, sodium tissue tolerance and osmotic tolerance as the strategies plants use to establish salinity tolerance.

In this report, "The Plant Accelerator" was used to screen growth of wild barley ILs under well watered and drought treatments during vegetative growth. The aims were (1) to identify wild barley derived QTL within the set of S42ILs that control drought stress responses and (2) to test the use of non-destructive high-throughput imaging to measure vegetative stage drought response in barley.

We could show that high-throughput imaging provides accurate estimates for biomass development over time. Moreover several drought related QTLs were identified and genotypes detected that may be beneficial in future breeding programs.

Materials and Methods

Plant Material

Forty-seven wild barley ILs of the S42IL library and the recipient parent Scarlett were selected for the experiment. The S42ILs are derived from a cross between the German malting barley variety Scarlett and the Israeli wild barley accession ISR42-8. The 47 ILs possess few *Hsp* chromosome segments and were selected based on SSR and SNP genotyping to represent a large portion, 87.3%, of the ISR42–8 genome [18]. Repeated backcrossing and MAS are described in Schmalenbach et al. [14].

Glass House Cultivation

Two drought stress experiments, with duration of six weeks each, were conducted between end of March and mid of July 2011 in The Plant Accelerator greenhouse facilities in Adelaide, Australia (34°58′16.18″S; 138°38′23.88″E). Forty-eight barley genotypes were grown under a well watered and stress treatments with three replicates per genotype and treatment. Each experiment was designed in three randomized blocks. Control and stress treatments of each genotype were placed next to each other (Fig. 1, Table S1).

Single plants were grown in 2.5 L plastic containers with 2.1 kg of soil (50% UC Davis soil mix, 35% Coco-peat, 15% clay-loam). Three seeds per pot were directly sown into the soil and after germination thinned out, leaving one plant per pot. Plants were pre-grown for two weeks in a regular greenhouse and watering was performed manually to allow optimal germination and seedling establishment. Subsequently, the pots were transferred to the "smart house" where each pot was placed onto a cart on a conveyor belt and the two treatments were applied. Every second day, pots were weighed and watered automatically to 22% gravimetric water content for the well watered treatment and 15% for the stress treatment (Fig. 2). Based on the experience of the first

Figure 1. View of experiment 1 with five-week old barley S42IL plants growing in The Plant Accelerator.

Figure 2. Barley plants at the weighing and watering unit after leaving the imaging station.

experiment we adjusted the drought stress in the second experiment to 12% gravimetric water content to slightly increase drought effects. The experiments were carried out under natural lighting with the temperature in the greenhouse kept at a range between 15°C (night) and 22°C (day).

Phenotyping

With the onset of the stress treatment imaging of the plants started. Plant images were captured using a LemnaTec 3D Scanalyzer (LemnaTec, GmbH, Wuerselen, Germany). Every day, three RGB pictures (2056×2454 pixels) were taken of each barley plant, one top view image and two side view images with a 90° horizontal rotation. After background-foreground separation was applied to separate the plant tissue area from the background, pixel numbers per plant were counted and the pixel sum of the three pictures per plant was taken to define the projected shoot area. The shoot area measured over time was used to draw growth curves. For each growth curve, curve fitting with a 6th order polynomial was conducted to adjust for possible missing data points and absolute growth rate [dA/dt] and relative growth rate [(dA/dt)/A] were calculated. For each of the three curves the integral was determined and used as a trait in the statistical analysis. Moreover, six further traits were extracted from the images; caliper length, height, color (as hue angle in the HSI color scheme) and the two parameters shoot area top view and convex hull area to calculate compactness of each plant. At the end of the experiment, barley plants were harvested and above ground biomass, tiller number (TIL), and plant height (HEI) were determined. Fresh biomass was weighed and, subsequently, oven dried to constant weight to determine dry biomass. Water use efficiency (WUE) was calculated by dividing dry biomass at the end of the experiment by the total amount of water added during the four weeks in the "smart house" [mg/g water]. Specific plant weight (SPW) was calculated from the dry weight and the maximum projected shoot area at the end of the experiment. In addition, simple stress indices (SSI) were calculated as follows: SSI = Ts/Tc, where Ts and Tc are the average trait performances of an IL under stress and control conditions, respectively. An overview of trait definitions is given in Table 1.

Genotyping

The S42ILs were genotyped with the 1,536-SNP barley BOPA1 set [19] of the Illumina GoldenGate assay [18]. Six hundred and thirty-six out of the tested 1,536 SNPs were polymorphic and used to characterize the extent of exotic *Hsp* introgressions in each S42IL (see Fig. S1).

Statistical Analysis

Statistical analyses were performed with SAS Enterprise Guide 4.2. [31]. Descriptive statistical parameters (Table S2) were calculated with procedure MEANS. Heritabilities across treatments were calculated as $h^2 = V_G/[V_G + V_{GT}/t + V_{GE}/e + V_{GET}/et + V_R/etr]$, and within treatments as: $h^2 = V_G/[V_G + V_{GE}/e + V_R/er]$. The terms V_G, V_{GT}, V_{GE}, V_{GET} and V_R represent the genotypic, genotype× treatment, genotype×environment, genotype×environment× treatment, and error variance components, respectively, calculated with procedure VARCOMP [31]. The terms t, e, and r indicate the number of treatments, experiments and replicates, respectively. Pearson correlation coefficients between traits were calculated with means across treatments, blocks and experiments and within drought stressed and control treatments, respectively, using the procedure CORR.

Analysis of variance was carried out with the procedure MIXED using model I to test for genotype main effects across treatments and experiments.

Model I:

$$Y_{ijkl} = \mu + L_i + T_j + E_k + L \times T_{ij} + L \times E_{ik} + B\left(E \times T_{kj}\right)_l + \varepsilon_{ijkl}$$

and model II for genotype effects across experiments but within a single treatment.

Model II:

$$Y_{ikl} = \mu + L_i + E_k + L \times E_{ik} + B(E_k)_l + \varepsilon_{ikl}$$

Where μ is the general mean, L_i is the fixed effect of the *i*th line, T_j is the fixed effect of the *j*th treatment, E_k is the fixed effect of the *k*th experiment, $L \times T_{ij}$ is the fixed interaction between the *i*th line and the *j*th treatment, $L \times E_{ik}$ is the fixed interaction between the *i*th line and the *k*th experiment, $B(E \times T_{kj})_l$ is the random effect of the *l*th block nested in the interaction between *k*th experiment and *j*th treatment, $B(E_k)_l$ is the random effect of the *l*th block nested in the *k*th experiment and ε_{ijkl} and ε_{ikl} are the error of Y_{ijkl} and Y_{ikl}, respectively.

Following the mixed model analysis a Dunnett test was conducted where least square means (LSMEANS) of each IL were compared to the control Scarlett. In case an IL revealed a significant (P<0.05) deviation in trait performance from Scarlett, as main effect and/or as line×treatment interaction, a line×trait

Table 1. List of evaluated traits.

Trait	Abbreviation	Unit	Method of measurement
Imaging parameters			
Shoot area integral[a]	SAI	kPix[b]	Calculated from pixel sum of three images per plant per day; A
Absolute growth rate integral	AGRI	kPix/d	Calculated from pixel sum of three images per plant per day; dA/dt
Relative growth rate integral	RGRI	d^{-1}	Calculated from pixel sum of three images per plant per day; (dA/dt)/A
Height integral	HEII	kPix	Max. distance from bottom to top of plant
Caliper length integral	CALI	kPix	Max. distance between two points on the object boundary, top view image
Hull area integral	HULI	kPix	Smallest geometrical object without concave parts that covers whole plant, top view image
Shoot area top view integral	SATVI	kPix	Pixel number
Plant hue integral	HUEI	-	Average hue value calculated from all pixels per plant and day
Harvest parameters			
Tiller number	TIL	-	Number of tillers per pot
Height	HEI	cm	Plant height measured from bottom to leaf tip
Biomass dry	BMD	g	Weight of oven dried biomass per pot
Indices			
Water use efficiency	WUE	mg/g water	Harvested biomass per plant/total amount of irrigation water
Specific plant weight	SPW	mg/kPix	Harvested biomass per plant/pixel number per plant at end of experiment
Compactness integral	COMI	-	SATV/HUL per plant per day
Simple stress index	SSI	-	Trait performance under stress/trait performance under control treatment

[a]Integral: calculated for length of entire experiment, respectively.
[b]kilo Pixel.

association was assumed and the presence of a QTL was accepted. If several ILs with overlapping introgressions showed a similar effect, it was assumed that the ILs contained the same QTL. We consider this QTL as the most likely location of the effect and, thus, define a minimum number of QTL needed to explain all identified trait effects. The relative performance (RP) of an IL was calculated as RP (IL) = [LSMEANS (IL) – LSMEANS (Scarlett)] ×100/LSMEANS (Scarlett), where LSMEANS were calculated with model I across treatments, experiments and blocks or with model II across experiments and blocks, separately for each treatment. The detection of significant line by trait associations was conducted for every trait revealing a heritability with $h^2>0\%$ across treatments or within the two watering treatments, respectively.

Results

Trait Performance of S42ILs

For most traits, means were higher under well watered treatment than under drought stress (Fig. 3 and Fig. 4, Table S2), e.g. 2.2 g of biomass dry (BMD) vs. 0.9 g. There were, however, four exceptions. Compactness integral (COMI) was higher under drought treatment than under well watered treatment. The same was true for plant hue integral (HUEI), SPW, and WUE but differences were marginal.

Coefficients of variation (CV) differed strongly between traits (1.5 to 72.1%). Highest CV was calculated for BMD across treatments (72.1%). The lowest CVs were determined for HUEI, varying from 1.5 to 1.6% for the different treatments and the SSI. CV was generally higher under drought than under well watered treatment. The four exceptions were WUE, SPW and height integral (HEII) where CV under well watered treatment was higher and HUEI where CV was the same under both treatments.

Heritability was generally higher under well watered treatment. HEI, HEII and WUE were exceptions and showed higher heritabilities under drought treatment. Highest heritabilities were found for HEI, HEII and caliper integral (CALI) (between 46.2 and 76.8%). Low heritabilities were determined for WUE and HUEI under drought treatment and across treatments, as well as for BMD under drought treatment (15%). Most of the SSI showed heritabilities equal to 0. SSI (HEII), SSI (HUEI), SSI (SPW), SSI (WUE) revealed heritabilities between 4.7 and 14.5%.

Trait Correlations

Highest correlations were found among measured traits and among stress indices (Table S3). However, correlations between measured traits and stress indices were low. The measured traits showed the highest correlations between shoot area integral (SAI) and absolute growth rate integral (AGRI) ($r = 0.99$), BMD and SAI (Fig. 4), BMD and AGRI, shoot area top view integral (SATVI) and SAI and SATVI and AGRI ($r = 0.98$). Most correlations were positive and statistically significant. HEI showed negative correlation with relative growth rate integral (RGRI) and TIL, but values were not statistically significant. COMI showed negative correlations with all traits but HEII. Among stress indices highest correlations were found for SSI (SAI), SSI (AGRI) and SSI (SAI), as well as between SSI (WUE) and SSI (SPW) (with $r>0.93$). Interestingly, WUE, SPW and BMD showed only low correlations ($r<0.43$). Looking at the simple stress index, however, correlations between SSI (WUE), SSI (SPW) and SSI (BMD) were very high ($r = 0.85$ to 0.96). Autocorrelations between drought and well watered treatments of a single trait were high ($r>0.61$) for most traits. HEI with $r = 0.85$ had the highest correlation between treatments. RGRI showed a low but still significant correlation between the treatments with $r = 0.33$. Autocorrelations for SPW and WUE were not significant.

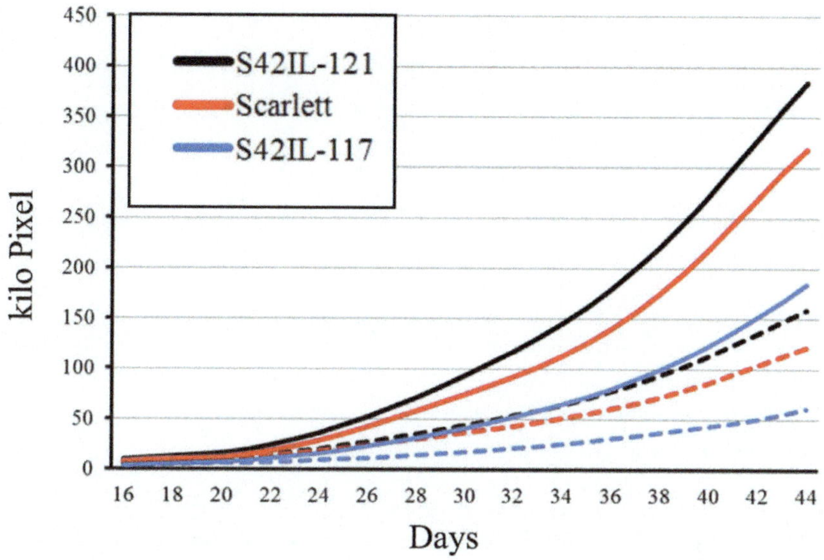

Figure 3. Development of shoot area of S42IL-121, Scarlett and S42IL-117 under well watered (solid line) and drought (dashed line) treatment, respectively.

Mixed Model Analysis of Variance

The mixed model analysis, including fixed line, treatment, and experiment effects (i.e. model I) revealed significant ($P<0.05$) line effects for all investigated traits (Table S4). Treatment had a clear impact on trait performance. For all traits, except WUE and SPW the effect was significant. Line by treatment interaction effects were not significant for any of the measured traits. The experiments had a significant effect on trait performance of all traits except leaf color measured as plant hue integral (HUEI). And line by experiment interaction was significant for all traits but RGRI, COMI, and HEII. In the mixed model analyses for single treatments including fixed line, and experiment effects (i.e. model II), line had a significant effect on trait performance for all traits but HUEI, RGRI, and WUE under well watered and HUEI and SPW under drought conditions (Tables S5 and S6). Also simple stress indices were analyzed with model II (Table S7). The line effect was not significant for any of the simple stress indices.

QTL Detection

QTL were only determined for traits with heritability greater than 0. The Dunnett tests revealed, in total, 63 line effects for eleven out of 14 traits. These effects were detected either across treatments (39), within the drought treatment (15) or within the well watered treatment (9). Several of the measured effects were consistent between the different treatments. Thus, these line effects were summarized to a minimum of 44 QTL (Table 2 and Fig. 5). No QTL were identified for RGRI, SPW, and HUEI. Between two and nine QTL were identified for the traits AGRI, BMD, CALI, COMI, HEI, HEII, hull area integral (HULI), SAI,

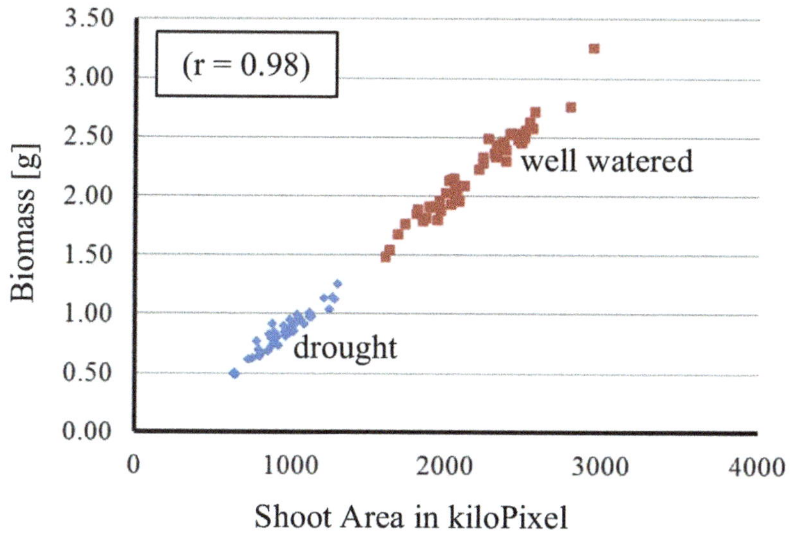

Figure 4. Correlation between biomass and shoot area integral under drought (blue dots) and well watered (red dots) treatment, respectively.

Table 2. List of 44QTL detected for 11 traits in the S42IL-population.

Trait[a]	QTL Name	Position of main introgression[b]	Line	Treatment[c]	LSMEANS Scarlett[d]	LSMEANS IL[e]	Dev. f. Sca	Dev. f. Sca %[g]	Candidate genes[h]	Studies with corresponding QTL[i]
AGRI	QAgri.S42IL-3H	3H; 204.48–255.13	S42IL-115	a	170.84	119.10	−51.74	−30.28		
	QAgri.S42IL-4H	4H; 027.52–064.77	S42IL-117	a, d	170.84	114.69	−56.15	−32.87		
	QAgri.S42IL-6H	6H; 073.90–133.47	S42IL-129	a	170.84	120.89	−49.95	−29.24		
BMD	QBmd.S42IL-3H	3H; 204.48–255.13	S42IL-115	a	1.66	1.10	−0.56	−33.64		
	QBmd.S42IL-4H	4H; 027.52–064.77	S42IL-117	a	1.66	0.99	−0.67	−40.30		
	QBmd.S42IL-4Hb	4H; 074.11–119.06	S42IL-121	a	1.66	2.26	0.60	35.97		
	QBmd.S42IL-6H	6H; 073.90–133.47	S42IL-129	a	1.66	1.10	−0.57	−33.94		IV
CALI	QCali.S42IL-1H	1H; 040.51–089.01	S42IL-103	a, w	19.39	15.83	−3.56	−18.35		
	QCali.S42IL-2H	2H; 102.66–104.81	S42IL-110	a, d	19.39	15.47	−3.92	−20.23		
	QCali.S42IL-3H	3H; 067.01–098.41	S42IL-111	a	19.39	16.52	−2.87	−14.80		
	QCali.S42IL-4H	4H; 027.52–064.77	S42IL-117	a, d	19.39	15.98	−3.42	−17.62		
	QCali.S42IL-4Hb	4H; 074.11–119.06	S42IL-121	a	19.39	22.34	2.94	15.18		
	QCali.S42IL-6H	6H; 073.90–133.47	S42IL-129	a, d, w	19.39	15.02	−4.37	−22.55		
COMI	QComi.S42IL-4H	4H; 027.52–064.77	S42IL-117	a, w	4.10	5.32	1.22	29.77		
	QComi.S42IL-6H	6H; 071.39–132.23	S42IL-128	w	3.82	5.23	1.42	37.13		
HEI	QHei.S42IL-1H	1H; 040.51–089.01	S42IL-103	a, w	49.00	42.75	−6.25	−12.76		
	QHei.S42IL-1Hb	1H; 130.68–173.49	S42IL-143	a	49.00	43.50	−5.50	−11.22	*HvFT-3*[3]	I
	QHei.S42IL-2H	2H; 063.96–110.84	S42IL-109	a	49.00	44.33	−4.67	−9.52	*sdw3*[2], *HvFT4*[3]	I, II, III, IV
	QHei.S42IL-3H	3H; 067.01–098.41	S42IL-111	a	49.00	44.50	−4.50	−9.18		
	QHei.S42IL-3Hb	3H; 154.99–253.73	S42IL-140	a, d, w	49.00	58.17	9.17	18.71	*denso*[1]	I, IV, V
	QHei.S42IL-4H	4H; 074.11–119.06	S42IL-121	a, d, w	49.00	57.75	8.75	17.86		II, IV
	QHei.S42IL-6H	6H; 073.90–133.47	S42IL-129	a	49.00	44.42	−4.58	−9.35		V
	QHei.S42IL-7H	7H; 134.43–193.89	S42IL-137	a, d	49.00	54.67	5.67	11.56		I, II, IV, V
HEII	QHeii.S42IL-1H	1H; 040.51–089.01	S42IL-103	a	11.67	9.85	−1.83	−15.64		
	QHeii.S42IL-1Hb	1H; 130.68–173.49	S42IL-143	a	11.67	10.06	−1.62	−13.85	*HvFT-3*[3]	I
	QHeii.S42IL-4H	4H; 061.15–119.06	S42IL-162	a	11.67	9.99	−1.68	−14.39		
	QHeii.S42IL-4Hb	4H; 171.25–183.54	S42IL-124	a	11.67	10.04	−1.64	−14.02		
	QHeii.S42IL-7H	7H; 176.37–229.66	S42IL-138	a	11.67	10.05	−1.63	−13.93		
HULI	QHuli.S42IL-1H	1H; 040.51–089.01	S42IL-103	a, w	6679	4443	−2236	−33.47		
	QHuli.S42IL-2H	2H; 102.66–104.81	S42IL-110	a, d	6679	4517	−2161	−32.36		
	QHuli.S42IL-3H	3H; 067.01–098.41	S42IL-111	a	6679	4617	−2062	−30.88		
	QHuli.S42IL-3Hb	3H; 204.48–255.13	S42IL-115	a	6679	4616	−2063	−30.89		
	QHuli.S42IL-4H	4H; 027.52–064.77	S42IL-117	a	6679	4327	−2352	−35.21		
	QHuli.S42IL-6H	6H; 073.90–133.47	S42IL-129	a	6679	4179	−2500	−37.42		
SAI	QSai.S42IL-2H	2H; 102.66–104.81	S42IL-110	d	995	625	−369	−37.14		I, (as biomass)

Table 2. Cont.

Trait[a]	QTL Name	Position of main introgression[b]	Line	Treatment[c]	LSMEANS Scarlett[d]	LSMEANS IL[e]	Dev. f. Sca	Dev. f. Sca %[g]	Candidate genes[h]	Studies with corresponding QTL[i]
	QSai.S42IL-4H	4H; 027.52–064.77	S42IL-117	a, d	1654	1116	−538	−32.53		
	QSai.S42IL-6H	6H; 073.90–133.47	S42IL-129	a, d	1654	1151	−504	−30.44		
SATVI	QSatvi.S42IL-2H	2H; 102.66–104.81	S42IL-110	d	398	214	−184	−46.18		
	QSatvi.S42IL-4H	4H; 027.52–064.77	S42IL-117	a	677	437	−240	−35.44		
	QSatvi.S42IL-6H	6H; 073.90–133.47	S42IL-129	a	677	450	−228	−33.64		
TIL	QTil.S42IL-3H	3H; 204.48–255.13	S42IL-115	a	8.17	5.92	−2.25	−27.55		
	QTil.S42IL-4H	4H; 171.25–183.54	S42IL-124	a, d, w	8.17	11.67	3.50	42.86	*VRN-H2*[3]	III, V
WUE	QWue.S42IL-4H	4H; 027.52–064.77	S42IL-117	d	1.97	1.25	−0.72	−36.53		
	QWue.S42IL-6H	6H; 073.90–133.47	S42IL-129	d	1.97	1.17	−0.80	−40.59		

[a]Trait abbreviations are given in Table 1.
[b]By chromosome and cM position.
[c]Treatment under which effect occured, a: main effect, d: under drought treatment, w: under well watered treatment.
[d]LSMEANS Scarlett for the indicated trait and treatment.
[e]LSMEANS for indicated IL for indicated trait and treatment, if under more than one treatment then a.
[f]Deviation from Scarlett = LSMEANS[IL] - LSMEANS [Scarlett].
[g]Deviation from Scarlett in % = (LSMEANS[IL] – LSMEANS [Scarlett])/LSMEANS[Scarlett] *100.
[h]References: [1] Laurie et al. (1995), [2] Gottwald et al. (2004), [3] Wang et al. (2010)
[i]I: Von Korff et al. (2006), II: Schmalenbach et al. (2009), III: Wang et al. (2010), IV: March et al. (in prep.), V: Honsdorf et al. (in prep.).

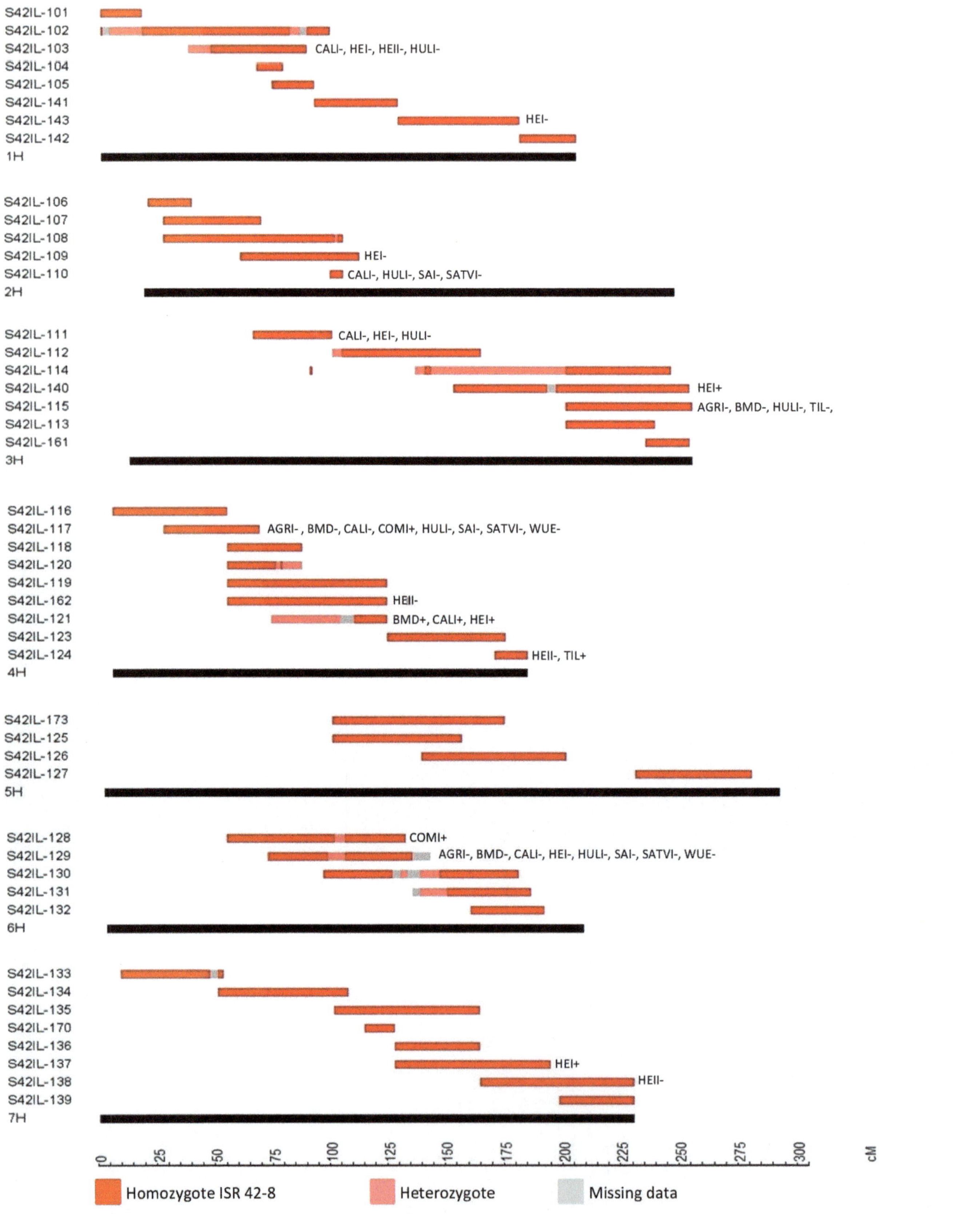

Figure 5. QTL map with indication of S42IL introgressions (Schmalenbach et al. 2011). SNP positions (in cM) are based on Close et al. (2009). QTL are placed right to the S42IL, indicated by trait abbreviations (see Table 1). The sign indicates an increasing (+) or a decreasing (−) *Hsp* effect.

SATVI, TIL, and WUE. In the following, the QTL are presented for each trait separately.

Absolute Growth Rate Integral (AGRI)

Three QTL were identified for AGRI. The QTL are located on chromosomes 3H, 4H and 6H and the *Hsp* allele in all three QTL reduced the trait performance. Across treatments the *Hsp* alleles reduced the integral of the absolute growth rate by approximately 30%. Under drought conditions the *Hsp* allele of QTL QAgri.S42IL-4H reduced the trait performance by almost 40%.

Biomass Dry (BMD)

For biomass four QTL were identified across treatments. The *Hsp* alleles at QTL QBmd.S42IL-3H and QBmd.S42IL-6H on chromosomes 3H and 6H reduced biomass by approximately 33%. The two QTL QBmd.S42IL-4H and QBmd.S42IL-4Hb on chromosome 4H showed contrary effects. While at the first QTL the *Hsp* allele reduced biomass by 40.3%, at the second QTL it increased it by 36.0%.

Caliper Length Integral (CALI)

Six QTL were detected for caliper length on chromosome 1H, 2H, 3H, 4H and 6H. All QTL were detected across treatments, three and two of them also showed effects for drought and well watered treatments, respectively. In five cases the *Hsp* allele had decreasing effects between 14.8 and 22.6% across treatments. In one case the *Hsp* allele at QTL QCali.S42IL-4Hb increased caliper length by 15.2% compared to Scarlett.

Compactness Integral (COMI)

For COMI two QTL were detected on chromosomes 4H and 6H. The effect of the *Hsp* allele at QComi.S42IL-4H was observed across treatments and under well watered treatment. It increased compactness across treatments by 29.8%. In case of QComi.S42IL 6H the effect was solely observed under well watered treatment and lead to an increase of 27.1% by the *Hsp* allele.

Height (HEI)

The highest number of QTL was detected for plant height. Eight QTL were detected across treatments. Of those, four also showed effects under either one or both of well watered and drought treatment. The QTL are located on all chromosomes except 5H. In five cases the *Hsp* alleles reduced plant height (9.2 to 12.8%). In three cases the *Hsp* allele increased plant height by 11.6 to 18.7%.

Height Integral (HEII)

Five QTL across treatments were detected for HEII on 1H, 4H and 7H. Two of those, QHeii.S42IL-1H and QHeii.S42IL-1Hb, coincided with QTL for manual measurement of HEI. At all detected QTL the *Hsp* alleles reduced HEII by between 13.9 to 15.6%.

Hull Area Integral (HULI)

For HULI six QTL were detected on chromosomes 1H, 2H, 3H, 4H, and 6H. In all cases the *Hsp* alleles reduced the hull area by between 30.9 to 37.4% across treatments. In addition, QHuli.S42IL-1H and QHuli.S42IL-2H showed effects under well watered and drought treatments, respectively.

Shoot Area Integral (SAI)

Three QTL were found for the integral of the projected shoot area. The QTL are located on chromosomes 2H, 4H, and 6H. All three QTL were detected under drought treatment, while, in addition, QSai.S42IL-4H and QSai.S42IL-6H showed effects across treatments. The presence of the *Hsp* allele reduced the projected shoot area between 30.4 and 37.1% across treatments.

Shoot Area Top View Integral (SATVI)

Three QTL were detected for shoot area top view that corresponded to the same QTL detected for SAI on chromosomes 2H, 4H, and 6H. However, only QSatvi.S42IL-2H was detected under drought treatment, while the two other QTL were detected across treatments. In all cases the *Hsp* alleles reduced SATVI between 33.6 and 46.2% compared to Scarlett.

Tiller Number (TIL)

For tiller number two QTL were identified on chromosomes 3H and 4H. The *Hsp* allele at QTil.S42IL-3H reduced tiller number by 27.6% across treatments. QTil.S42IL-4H was detected across treatments and separately within the two treatments. Across both treatments the *Hsp* allele increased the tiller number by 42.9%.

Water Use Efficiency (WUE)

Two QTL were detected for WUE. The *Hsp* allele of QTL QWue.S42IL-4H and QWue.S42IL-6H reduced WUE by 36.5 and 40.6% under drought treatment, respectively.

Relative Growth Rate Integral (RGRI) and Plant Hue Integral (HUEI), Specific Plant Weight (SPW), and Simple Stress Index (SSI)

For RGRI, HUEI, and SPW and the SSIs no QTL were detected in this study.

Discussion

The aim of the study was to validate the use of non-destructive high-throughput phenotyping to measure vegetative drought response in barley and to identify QTL derived from wild barley that control physiological traits related to drought stress. To the authors knowledge this is the first QTL report on drought stress that used a high-throughput phenotyping facility.

Plant growth and the biomass parameters tiller number, plant height, and shoot dry weight of 48 barley genotypes were investigated under drought and well watered treatments. Two week old barley plants were transferred into a high-throughput phenotyping greenhouse, where stress and control treatments were applied automatically. During the following four weeks of cultivation, plants were imaged daily in an automated manner. Images were processed and used as a measure for plant height, caliper length, biomass and, consequently, plant growth, and plant color. After a total of six weeks green plants were harvested. Tiller number, plant height, and shoot dry weight were determined for each plant. Moreover, plant compactness, water use efficiency and specific plant weight, as well as stress indices were calculated.

The mixed model ANOVA revealed a clear effect of the treatment on trait expression. Drought stressed plants had a lower growth rate and subsequently produced less biomass (see Fig. 4). However, there was no significant line×treatment interaction, indicating that the S42ILs reacted similar under drought stress and well watered conditions for the investigated traits. This finding is also supported by high autocorrelations for investigation of traits under drought and control treatments, e.g. 0.77 for biomass (see Table S3).

QTL Detection

In this study 44 QTL were identified in 15 ILs for eleven traits. In eight cases the *Hsp* alleles increased the performance of the trait, while in 36 cases there was a decreasing effect. This is to be expected since wild barley is known to carry many unfavorable alleles as well [5]. In six ILs only one QTL was determined, predominantly for HEI and HEII. Multiple QTL effects were found in nine ILs. For S42IL-115,-117, -121, and -129 QTL for BMD were detected in combination with one or more of the traits AGRI, HEI, SAI, and TIL which is in agreement with the high correlations found between those traits. SAI is a measure for biomass, and AGRI is the first derivative of SAI. Therefore, it was expected that ILs show effects for all three traits, simultaneously. However, this was not always the case. Thus, increasing the number of experiments and replications might be useful to increase the power of QTL detection. In the following, the traits or trait complexes are discussed separately.

Absolute Growth Rate Integral, Shoot Area Integral and Biomass Dry (AGRI, SAI, BMD)

The three traits were highly correlated with each other ($r = 0.98$ and $r = 0.99$). Since their relation may be functional, it appears likely that a single pleiotropic QTL may control the three traits AGRI, SAI and BMD, simultaneously.

For line S42IL-129 a biomass reduction of 33.9% was observed. March et al. (in prep.) found a similar decrease in biomass in that line measured under terminal drought stress. This suggests that biomass production may be partly controlled by similar genes during early and late drought stress occurrence.

Three QTL were detected for SAI. Two of those, namely QSai.S42IL-4H and QSai.S42IL-6H, in line S42IL-117 und S42IL-129, respectively, were in accordance with BMD QTL. The *Hsp* allele, in both lines caused a decrease in the projected shoot area. Due to the high correlation of the traits it can be assumed that QTL for biomass correspond to QTL for shoot area. The *Hsp* allele of the QTL QSai.S42IL-2H on chromosome 2H caused a decrease of 37% in projected shoot area, compared to Scarlett. Von Korff et al. [10] described a QTL related to biomass reduction (QMas.S42-2H.a) in the same region in an AB-QTL field trial. Since this AB-QTL population is the parent population of the S42ILs used in this study, it is likely that the same QTL was detected in both the greenhouse and field trials.

AGRI is directly related to SAI. This might be seen as a reason for the detection of similar QTL for AGRI and SAI and, consequently, BMD. Many QTL studies on growth focus on relative growth rate instead of absolute growth rate. In the present study RGRI showed only a weak correlation to AGRI ($r = 0.35$) and other biomass parameters. Poorter et al. [33] pointed out that in their study QTL for RGR rather co-located with QTL for seed mass than with QTL for biomass. This fits well to the weak correlations found between biomass parameters and RGRI. However, two of the QTL detected for AGRI coincided with locations where previous studies mapped QTL for RGR. Yin et al. [34] reported a minor effect for relative growth rate associated with the *denso* locus on chromosome 3H in a spring barley recombinant inbred line (RIL) population of the cross Prisma×A-pex, which may co-localize with QAgri.S42IL-3H. Poorter et al. [33] conducted QTL studies in a F_2 population derived from a cross between two *Hsp* accessions. They mapped QTL for relative growth rate on chromosomes 1H, 2H, 5H and a minor QTL on 6H. The latter one might be in accordance with QAgri.S42IL-6H.

Tiller Number (TIL)

Two QTL were detected for the trait tiller number on chromosomes 3H and 4H. Wang et al. [35] identified the *VRN-H2* gene on chromosome 4H in introgression line S42IL-124. Whereas S42IL-124 carries a dominant winter-type allele, Scarlett carries the recessive and deleted spring type allele at *Vrn-H2*. S42IL-124 showed an increased tiller number compared to Scarlett. Since *Vrn-H2* is known to have a pleiotropic effect on tiller number [36], we assume that this gene explains the underlying effect of the QTL. Studies on other populations revealed QTL for tiller number on chromosomes 4H as well. In a cross between two wild barley accessions Elberse et al. [37] found a QTL for tiller number on that chromosome. Baum et al. [38] identified a QTL on chromosome 4H where the *Hsp* allele increased the number of tillers and a QTL on chromosome 3H where the *Hsp* allele had a decreasing impact in an Arta×*Hsp* 41-1 RIL population. Those effects might correspond to the QTL detected in this study. Both QTL occurred irrespective of the treatment. Especially QTil.S42IL-4H appears to be a very stable QTL. It was detected across and within treatments and was detected in several studies under varying conditions, in field studies as well as under greenhouse conditions. Moreover, von Korff et al. [10] detected QTL for number of ears, which is directly related to tiller number, in the same genomic regions. On 4H the *Hsp* allele increased the number of ears, while on 3H it has a decreasing effect. This supports the observation of a stable QTL.

Height (HEI) and Height Integral (HEII)

Plant height was determined in two ways. First, height (HEII) was modeled from the images taken during four weeks and the integral of the height growth curve was calculated. Second, height (HEI) was measured manually when plants were harvested after six weeks at the end of the experiment. The correlation between HEI and HEII was relatively low with $r = 0.72$, compared to the correlation between SAI and BMD with $r = 0.98$. While SAI shows a constant increase over time, HEII shows an overall increase, but fluctuation between days may be strong. When a new leaf is unfolded the plant grows higher, however, when the leaf becomes too heavy and bends down, the height of the plant appears to be shorter. At the end of the experiment the length of the stretched plant was measured, which is longer than the upright standing plant. Nevertheless two coinciding QTL were found between HEI and HEII on chromosome 1H.

Six out of eight QTL were already identified in previous field studies with the S42 population and/or the S42ILs, exhibiting similar effects of the same direction in all three studies. QTL QHei.S42IL-3Hb was already detected in von Korff et al. [10] and March et al. (in prep.). In this region also the *denso* dwarfing gene was mapped [39], which may be identical with the semi-dwarf gene *sdw1* [40]. The second largest effect, after QHei.S42IL 3Hb, was associated with QHei.S42IL-4H in S42IL-121. This QTL corresponds to QHei.S42IL-4H.a in Schmalenbach et al. [15]. In both studies the *Hsp* allele increased plant height by 18%. March et al. (in prep.) mapped a QTL for height for S42IL-121 as well. A third QTL (QHei.S42IL-7H) with an increasing effect of the *Hsp* allele was detected on chromosome 7H. Here an effect that was already found in the studies of von Korff et al. [10], Schmalenbach et al. [15] and March et al. (in prep.) could be verified. Moreover two QTL where the *Hsp* allele had a decreasing effect on plant height [10] were verified. In S42IL-143 HEI was reduced by 11% (QHei.S42IL-1Hb) and HEII (QHeii.S42IL-1Hb) by 14%. Von Korff et al. [10] detected a QTL in the same region on chromosome 1H. The flowering time gene *HvFT3* is mapped in the same region and known to have a pleiotropic effect on plant height [35]. QHei.S42IL-2H in S42IL-109 had reduced height by 9.5%. March et al. (in prep.) and Schmalenbach et al. [15] found the same effect in their studies. Von Korff et al. [10] found a similar effect in the region where the *Hsp* introgression of S42IL-109 was mapped. Moreover two candidate genes are mapped to the chromosomal region. These are the dwarfing gene *sdw3* [41] and the flowering gene *HvFT4*, which is known to have an effect on plant height [35].

All of the HEI QTL in the present experiments were detected across treatments. Six out of eight QTL were also found in previous field and glasshouse studies. The QTL therefore seem to be very stable across locations as well as across treatments. Moreover they seem to be independent of the developmental stage. The present experiments, thus, allowed the verification of effects after six weeks that were previously screened in field experiments after flowering, indicating that phenotyping juvenile plants may be predictive for adult plant performance, at least in regard to growth parameters. The high heritability of 76.8% supports this finding.

For HEII two QTL coincided with previous studies. Besides QHeii.S42IL-1Hb mentioned above, this was QHeii.S42IL-4Hb where the *Hsp* allele reduced height by 14%. This QTL was also detected by von Korff et al. [10] and Wang et al. [35]. Heritability for digitally determined height was lower (61.4%) than for the manually measured one. Determining height by multiple measurements apparently was not an advantage here. However, this may change at a later stage of plant development. After shooting, the plant height is less subjected to bending of leaves and therefore can be measured more precisely by the imaging technique.

Water Use Efficiency (WUE)

Water use efficiency indicates how much biomass a plant can produce per unit water supplied. Thus, increased WUE has the potential to improve yield under drought stress conditions. Measuring WUE in regular greenhouse experiments is time-consuming. Therefore the high-throughput phenomics facility greatly assisted in scoring of water use efficiency through automated watering of pots to specific weights.

In this study the two S42ILs -117 and -129, with wild barley introgressions on chromosomes 4H and 6H, respectively, showed significant differences in WUE compared to Scarlett. Both ILs showed reduced water use efficiency compared to Scarlett. These ILs also produced less biomass. S42IL-117 and S42IL-129, thus, clearly carry unfavorable alleles for this trait. Chen et al. [42] pointed out that WUE itself is difficult to measure under field conditions and that a suitable tool to measure WUE efficiency is missing. Carbon isotope discrimination is a commonly used technique to measure WUE. Teulat et al. [43] used this method and identified QTL for WUE on chromosome 6H in a set of 167 RILs from a cross between Tadmor and Er/Apm and likewise Diab et al. [44] identified a QTL for the same trait on chromosome 4H. The QTL detected in this study may correspond to the ones found in the studies mentioned before and suggest the results from both techniques are correlated.

Compactness Integral, Shoot Area Top View Integral, Hull Area Integral (COMI, SATVI and HULI)

The compactness of a plant describes how much of the hull area is covered by leaves. It was calculated as the ratio of SATVI to HULI. The more compact a plant is, the more ground cover it has with regard to the hull area. Two QTL were detected for this trait.

SATVI and HULI showed a high correlation of $r = 0.9$, however, correlations between COMI and HULI and between COMI and SATVI were only moderate and negative. This indicates that in general, bigger plants take more space and have a lower compactness compared to smaller plants. In the present experiments this was observed by comparing drought stressed and well watered plants. Drought stressed plants showed on average a higher compactness than well watered plants. Jansen et al. [45] report the same effect on a study in *Arabidopsis thaliana.* Compactness shows negative correlation with all other traits evaluated in this experiment, with the exception of HEII ($r = 0.16$). An example for this is S42IL-117. This introgression line has a higher compactness, but reduced biomass, and other growth parameters compared to Scarlett.

SATVI is one of the three parameters that control SAI and, thus, is highly correlated with this trait as well as with BMD and AGRI. As one may expect, for SATVI the same QTL were detected as for SAI. For HULI a total of six QTL were detected. Three of those may be due to high correlations in accordance with SATVI, AGRI and BMD.

Caliper Length Integral (CALI)

Caliper length describes the maximum diameter of the plant. For this trait six QTL were detected. Those were in accordance with QTL for HULI. This can be explained by the close connection of both traits. Hull area is taken as the basis to calculate caliper length and both traits are highly correlated ($r = 0.94$). CALI also shows positive and high correlations with AGRI, SAI, SATVI, HEI, and BMD. Plants with a large diameter cover a larger area, tend to be bigger, have a higher growth rate and a higher biomass than plants with a smaller diameter. Therefore, a lot of information on plant structure can be deduced from the plant diameter.

Stress Indices

Simple stress indices were calculated for each trait as the ratio of the mean plant performance under drought stress versus well watered treatments. In this study no QTL for a SSI was detected. Additionally the authors used two more complex stress indices (modified after Fischer and Maurer [46]), but were not able to detect QTL with those either. A stress index states how well a genotype performs under stress conditions relative to its performance under control conditions. Therefore, to see differences between genotypes for a stress index, a line by treatment interaction is necessary. If all genotypes show a similar growth reaction under stress and control conditions, the initially existing differences between the genotypes may be drastically reduced. In the present experiments line by treatment interactions were not significant and autocorrelations were high between the treatments. This may be the reason why no significant effect for the stress indices was found. This notion is supported by Wang et al. [47]. In their study on "mathematically-derived traits in QTL mapping" the authors pointed out that an increased complexity of the genetic architecture of derived traits (e.g. stress indices) may reduce the power of QTL detection.

High-throughput Phenotyping using The Plant Accelerator

Determination of biomass by manual harvest is tedious and time-consuming. In addition, destructive harvest makes repeated measurements on the same plant impossible. Visual light imaging technologies applied in this study can solve these problems by utilizing the strong correlation between the projected shoot area and the actual biomass [27]. Imaging technologies have been successfully used in several studies in *Arabidopsis thaliana*, e.g. Granier et al. [48] and Leister et al. [49]. The first study investigated nine accessions under different levels of water deficit in the phenotyping facility "PHENOPSIS". Reaction to water deficits was, amongst other traits, characterized by leaf area growth determined through images. The authors pointed out the importance of the automated watering in their experiment, which enables equal conditions for all plants. A characteristic that was also found very important in the present experiments. Leister et al. [49] described a first approach of using an image based technology for high-throughput growth analysis. They calculated plant area from top view images and found high correlations to plant fresh weight.

In this study the sum of three two-dimensional pictures was used as a measure of plant biomass. In these experiments correlation between SAI and BMD and between AGRI and BMD were very high ($r = 0.98$). The results with six-week old barley plants proved that the sum of three pictures accounts sufficiently for overlapping leaves during early development. Rajendran et al. [30] found the same for *T. monococcum*. However, as Munns et al. [27] pointed out, accuracy may decline when plants become larger and produce multiple shoots. The results of the present study approved that The Plant Accelerator is suited to enable detailed growth analysis of barley plants. The prediction of biomass by the image-based leaf sum gave accurate results when comparing to actual biomass. Growth curves can give detailed information on differences in development of genotypes. For instance, the maximum of the absolute growth rate gives insight into the change from vegetative to generative phase of plant development. The present experiments ran only for six weeks. Therefore not all plants have reached this point. In future experiments this factor should be accounted for by adjusting the duration of the experiment. Automated imaging and the appropriate analysis pipeline make the detection of different developmental stages of plants feasible in high-throughput. With this technique it is possible to detect differences in stress responses between genotypes not only at different time points, but also to account for differences in development at those time points.

Rajendran et al. [30] used non-destructive imaging to screen for different response mechanisms of *T. monococcum* to salt stress. In contrast to conventional salt stress experiments, where tolerance is measured as total biomass production of stressed plants compared to unstressed plants, the growth curves provided through daily imaging gave detailed insight into the tolerance mechanisms of the plants. While osmotically tolerant plants showed a constant growth rate, the growth rate of sodium excluders first dropped than increased after a couple of days. Moreover plant color was analyzed. No stress symptoms occurred on leaves in the present experiments. Due to little variation between the genotypes no QTL was detected for leave color. However, in the experiments by Rajendran et al. [30], color analysis was successfully used to screen leaf damages due to high salt concentrations.

Fluorescence imaging gives information on the health statues of a plant. It allows for detection of leaf senescence and necrosis. However, such symptoms were not observed in the present experiments and, thus, this parameter was not applied. Nonetheless, the technique is readily available. In addition, near infrared (NIR) and infrared (IR) imaging may be useful for future plant growth evaluations and QTL studies. NIR enables the observation of the water status of a plant, while IR is used to determine shoot temperature.

Conclusion

In this study the use of a non-destructive high-throughput phenotyping platform was implemented to map QTL controlling vegetative drought stress responses in barley. Several QTL where the exotic *Hsp* allele had a positive effect on trait performance were detected. In particular, introgression line S42IL-121 showed improved growth under drought stress compared to the recurrent parent Scarlett. The line showed the same behavior in previous field experiments. Thus, this introgression line might be interesting for further breeding.

Moreover, several QTL were detected where the *Hsp* allele had a decreasing effect on trait performance. Especially two QTL for water use efficiency might be interesting for further investigation. In future, interesting effects of S42IL-121 and other S42ILs will be fine mapped with already available high-resolution progeny [18] to further narrow down the QTL region and, ultimately, clone the underlying genes, which caused the observed QTL effects.

Supporting Information

Figure S1 Map of 47 S42ILs, the map contains 636 BOPA1 SNPs.

Table S1 Experimental layout with genotypes referred to positions (1–100) per replication (1–3).

Table S2 Descriptive statistics for S42Ils.

Table S3 Correlations between traits and stress indices.

Table S4 ANOVA (Model I) results of studied traits.

Table S5 ANOVA (Model II drought) results of studied traits.

Table S6 ANOVA (Model II well watered) results of studied traits.

Table S7 ANOVA (Model II SSI) results of studied traits.

Acknowledgments

We are grateful to the team of "The Plant Accelerator, Australian Plant Phenomics Facility" for carrying out the experiments.

Author Contributions

Conceived and designed the experiments: TJM BB MT KP. Performed the experiments: NH TJM BB. Analyzed the data: NH BB. Contributed reagents/materials/analysis tools: BB KP. Wrote the paper: NH KP. Provided comments and corrected the manuscript: TJM BB MT KP.

References

1. Druka A, Sato K, Muehlbauer GJ (2011) Genome Analysis: The State of Knowledge of Barley Genes. In: Ullrich SE, editor. Barley: Production, Improvement, and Uses. 1 ed: Blackwell Publishing Ltd.
2. Jana S, Wilen RW (2005) Breeding for Abiotic Stress Tolerance in Barley. In: Ashraf M, Harris PJC, editors. Abiotic stresses: plant resistance through breeding and molecular approaches: Food Products Press.

3. Tanksley SD, McCouch SR (1997) Seed banks and molecular maps: Unlocking genetic potential from the wild. Science 277: 1063–1066.
4. Zhao J, Sun HY, Dai HX, Zhang GP, Wu FB (2010) Difference in response to drought stress among Tibet wild barley genotypes. Euphytica 172: 395–403.
5. Zamir D (2001) Improving plant breeding with exotic genetic libraries. Nature Reviews Genetics 2: 983–989.
6. Tanksley SD, Nelson JC (1996) Advanced backcross QTL analysis: A method for the simultaneous discovery and transfer of valuable QTLs from unadapted germplasm into elite breeding lines. Theoretical and Applied Genetics 92: 191–203.
7. Pillen K, Zacharias A, Leon J (2003) Advanced backcross QTL analysis in barley (*Hordeum vulgare* L.). Theoretical and Applied Genetics 107: 340–352.
8. von Korff M, Wang H, Leon J, Pillen K (2004) Development of candidate introgression lines using an exotic barley accession (*Hordeum vulgare* ssp. *spontaneum*) as donor. Theoretical and Applied Genetics 109: 1736–1745.
9. von Korff M, Wang H, Leon J, Pillen K (2005) AB-QTL analysis in spring barley. I. Detection of resistance genes against powdery mildew, leaf rust and scald introgressed from wild barley. Theoretical and Applied Genetics 111: 583–590.
10. von Korff M, Wang H, Leon J, Pillen K (2006) AB-QTL analysis in spring barley: II. Detection of favourable exotic alleles for agronomic traits introgressed from wild barley (*H. vulgare* ssp. *spontaneum*). Theoretical and Applied Genetics 112: 1221–1231.
11. von Korff M, Wang H, Leon J, Pillen K (2008) AB-QTL analysis in spring barley: III. Identification of exotic alleles for the improvement of malting quality in spring barley (*H. vulgare* ssp. *spontaneum*). Molecular Breeding 21: 81–93.
12. von Korff M, Leon J, Pillen K (2010) Detection of epistatic interactions between exotic alleles introgressed from wild barley (*H. vulgare* ssp. *spontaneum*). Theoretical and Applied Genetics 121: 1455–1464.
13. Saal B, von Korff M, Leon J, Pillen K (2011) Advanced-backcross QTL analysis in spring barley: IV. Localization of QTL x nitrogen interaction effects for yield-related traits. Euphytica 177: 223–239.
14. Schmalenbach I, Koerber N, Pillen K (2008) Selecting a set of wild barley introgression lines and verification of QTL effects for resistance to powdery mildew and leaf rust. Theoretical and Applied Genetics 117: 1093–1106.
15. Schmalenbach I, Leon J, Pillen K (2009) Identification and verification of QTLs for agronomic traits using wild barley introgression lines. Theoretical and Applied Genetics 118: 483–497.
16. Schmalenbach I, Pillen K (2009) Detection and verification of malting quality QTLs using wild barley introgression lines. Theoretical and Applied Genetics 118: 1411–1427.
17. Naz AA, Ehl A, Pillen K, Leon J (2012) Validation for root-related quantitative trait locus effects of wild origin in the cultivated background of barley (Hordeum vulgare L.). Plant Breeding 131: 392–398.
18. Schmalenbach I, March TJ, Bringezu T, Waugh R, Pillen K (2011) High-Resolution Genotyping of Wild Barley Introgression Lines and Fine-Mapping of the Threshability Locus *thresh-1* Using the Illumina GoldenGate Assay. G3: Genes, Genomes, Genetics 1: 187–196.
19. Close TJ, Bhat PR, Lonardi S, Wu YH, Rostoks N, et al. (2009) Development and implementation of high-throughput SNP genotyping in barley. BMC Genomics 10: 13.
20. Pennisi E (2008) Plant genetics: The blue revolution, drop by drop, gene by gene. Science 320: 171–173.
21. Tester M, Langridge P (2010) Breeding Technologies to Increase Crop Production in a Changing World. Science 327: 818–822.
22. Cattivelli L, Rizza F, Badeck FW, Mazzucotelli E, Mastrangelo AM, et al. (2008) Drought tolerance improvement in crop plants: An integrated view from breeding to genomics. Field Crops Research 105: 1–14.
23. Tyagi K, Park MR, Lee HJ, Lee CA, Rehman S, et al. (2011) Fertile crescent region as source of drought tolerance at early stage of plant growth of wild barley (*Hordeum vulgare* L. ssp. *spontaneum*). Pakistan Journal of Botany 43: 475–486.
24. El Hafid R, Smith DH, Karrou M, Samir K (1998) Root and shoot growth, water use and water use efficiency of spring durum wheat under early-season drought. Agronomie 18: 181–195.
25. Lu ZJ, Neumann PM (1998) Water-stressed maize, barley and rice seedlings show species diversity in mechanisms of leaf growth inhibition. Journal of Experimental Botany 49: 1945–1952.
26. López-Castañeda C, Richards RA (1994) Variation in temperate cereals in rain-fed environments II. Phasic development and growth. Field Crops Research 37: 63–75.
27. Munns R, James RA, Sirault XRR, Furbank RT, Jones HG (2010) New phenotyping methods for screening wheat and barley for beneficial responses to water deficit. Journal of Experimental Botany 61: 3499–3507.
28. Golzarian MR, Frick RA, Rajendran K, Berger B, Roy S, et al. (2011) Accurate inference of shoot biomass from high-throughput images of cereal plants. Plant Methods 7: 11.
29. Berger B, Parent B, Tester M (2010) High-throughput shoot imaging to study drought responses. Journal of Experimental Botany 61: 3519–3528.
30. Rajendran K, Tester M, Roy SJ (2009) Quantifying the three main components of salinity tolerance in cereals. Plant Cell and Environment 32: 237–249.
31. SAS Institute (2008) The SAS Enterprise guide for Windows, release4.2. SAS Institute, Cary, NC, USA.
32. Becker H (2011) Pflanzenzüchtung. Stuttgart: Verlag Eugen Ulmer. 368 p.
33. Poorter H, van Rijn CPE, Vanhala TK, Verhoeven KJF, de Jong YEM, et al. (2005) A genetic analysis of relative growth rate and underlying components in *Hordeum spontaneum*. Oecologia 142: 360–377.
34. Yin X, Stam P, Dourleijn CJ, Kropff MJ (1999) AFLP mapping of quantitative trait loci for yield-determining physiological characters in spring barley. Theoretical and Applied Genetics 99: 244–253.
35. Wang G, Schmalenbach I, von Korff M, Leon J, Kilian B, et al. (2010) Association of barley photoperiod and vernalization genes with QTLs for flowering time and agronomic traits in a BC(2)DH population and a set of wild barley introgression lines. Theoretical and Applied Genetics 120: 1559–1574.
36. Karsai I, Meszaros K, Szucs P, Hayes PM, Lang L, et al. (2006) The influence of photoperiod on the Vrn-H2 locus (4H) which is a major determinant of plant development and reproductive fitness traits in a facultative X winter barley (Hordeum vulgare L.) mapping population. Plant Breeding 125: 468–472.
37. Elberse IAM, Vanhala TK, Turin JHB, Stam P, van Damme JMM, et al. (2004) Quantitative trait loci affecting growth-related traits in wild barley (Hordeum spontaneum) grown under different levels of nutrient supply. Heredity 93: 22–33.
38. Baum M, Grando S, Backes G, Jahoor A, Sabbagh A, et al. (2003) QTLs for agronomic traits in the Mediterranean environment identified in recombinant inbred lines of the cross 'Arta' x *H. spontaneum* 41–1. Theoretical and Applied Genetics 107: 1215–1225.
39. Laurie DA, Pratchett N, Bezant JH, Snape JW (1995) RFLP mapping of 5 major genes and 8 quantitative trait loci controlling flowering time in a winter x spring barley (*Hordeum vulgare* L.) cross. Genome 38: 575–585.
40. Jia Q, Zhang J, Westcott S, Zhang X-Q, Bellgard M, et al. (2009) GA-20 oxidase as a candidate for the semidwarf gene sdw1/denso in barley. Functional & Integrative Genomics 9: 255–262.
41. Gottwald S, Stein N, Borner A, Sasaki T, Graner A (2004) The gibberellic-acid insensitive dwarfing gene *sdw3* of barley is located on chromosome 2HS in a region that shows high colinearity with rice chromosome 7L. Molecular Genetics and Genomics 271: 426–436.
42. Chen J, Chang SX, Anyia AO (2011) Gene discovery in cereals through quantitative trait loci and expression analysis in water-use efficiency measured by carbon isotope discrimination. Plant Cell and Environment 34: 2009–2023.
43. Teulat B, Merah O, Sirault X, Borries C, Waugh R, et al. (2002) QTLs for grain carbon isotope discrimination in field-grown barley. Theoretical and Applied Genetics 106: 118–126.
44. Diab AA, Teulat-Merah B, This D, Ozturk NZ, Benscher D, et al. (2004) Identification of drought-inducible genes and differentially expressed sequence tags in barley. Theoretical and Applied Genetics 109: 1417–1425.
45. Jansen M, Gilmer F, Biskup B, Nagel KA, Rascher U, et al. (2009) Simultaneous phenotyping of leaf growth and chlorophyll fluorescence via GROWSCREEN FLUORO allows detection of stress tolerance in Arabidopsis thaliana and other rosette plants. Functional Plant Biology 36: 902–914.
46. Fischer RA, Maurer R (1978) Drought Resistance in Spring Wheat Cultivars. 1. Grain Yield Responses. Australian Journal of Agricultural Research 29: 897–912.
47. Wang Y, Li HH, Zhang LY, Lu WY, Wang JK (2012) On the use of mathematically-derived traits in QTL mapping. Molecular Breeding 29: 661–673.
48. Granier C, Aguirrezabal L, Chenu K, Cookson SJ, Dauzat M, et al. (2006) PHENOPSIS, an automated platform for reproducible phenotyping of plant responses to soil water deficit in Arabidopsis thaliana permitted the identification of an accession with low sensitivity to soil water deficit. New Phytologist 169: 623–635.
49. Leister D, Varotto C, Pesaresi P, Niwergall A, Salamini F (1999) Large-scale evaluation of plant growth in Arabidopsis thaliana by non-invasive image analysis. Plant Physiology and Biochemistry 37: 671–678.

5

Increased Drought Tolerance through the Suppression of *ESKMO1* Gene and Overexpression of *CBF*-Related Genes in Arabidopsis

Fuhui Xu[1], Zhixue Liu[1], Hongyan Xie[1], Jian Zhu[1], Juren Zhang[3], Josef Kraus[2], Tasja Blaschnig[2], Reinhard Nehls[2], Hong Wang[1,2]*

1 School of Life Sciences and Technology, Tongji University, Shanghai, China, **2** KWS SAAT AG, Einbeck, Germany, **3** School of Life Science, Shandong University, Shandong, China

Abstract

Improved drought tolerance is always a highly desired trait for agricultural plants. Significantly increased drought tolerance in *Arabidopsis thaliana* (Columbia-0) has been achieved in our work through the suppression of *ESKMO1 (ESK1)* gene expression with small-interfering RNA (siRNA) and overexpression of *CBF* genes with constitutive gene expression. *ESK1* has been identified as a gene linked to normal development of the plant vascular system, which is assumed directly related to plant drought response. By using siRNA that specifically targets *ESK1*, the gene expression has been reduced and drought tolerance of the plant has been enhanced dramatically in the work. However, the plant response to external abscisic acid application has not been changed. *ICE1*, *CBF1*, and *CBF3* are genes involved in a well-characterized plant stress response pathway, overexpression of them in the plant has demonstrated capable to increase drought tolerance. By overexpression of these genes combining together with suppression of *ESK1* gene, the significant increase of plant drought tolerance has been achieved in comparison to single gene manipulation, although the effect is not in an additive way. Accompanying the increase of drought tolerance via suppression of *ESK1* gene expression, the negative effect has been observed in seeds yield of transgenic plants in normal watering conditions comparing with wide type plant.

Editor: Haibing Yang, Purdue University, United States of America

Funding: The research fund was provided by KWS SAAT AG (www.kws.com) to FX as her PhD project. KWS participated in the study design, data analysis, decision to publish and preparation of manuscript via HW.

Competing Interests: The work is financially supported by KWS SAAT AG, the employer of JK, TB, RN, and HW. KWS SAAT AG participated in the study design, data analysis, decision to publish and preparation of manuscript via HW. There are no patents, products in development or marketed products to declare.

* Email: hong.wang@kws.com

Introduction

Drought stress is a major limiting factor for crop production worldwide [1]. In 2012, a severe drought in the United States caused heavy losses in crop production, especially in corn, and farmers produced less than three-fourths of the corn that the U.S. Department of Agriculture anticipated [2]. In China, around 20 million hectares of land are at risk of drought each year [3]. Globally, estimated crop losses due to water limitation exceed $10 billion annually [4]. Improving yield under drought, therefore, is a continuous challenge for agriculture, especially for modern breeding.

The development of modern plant biotechnology provides new hope for generating crops with increased drought tolerance. Understanding the response of plants to drought stress is the first step for development of stress tolerance plant through plant biotechnology. Gene expression experiments have identified several hundred genes that are either induced or repressed during drought. Some of those genes encode proteins that play important role in protecting cells from dehydration, such as the enzymes required for biosynthesis of various osmoprotectants, late-embryogenesis–abundant (LEA) proteins, antifreeze proteins, chaperones, and detoxification enzymes [5]. Some others are responsible for gene products including transcription factors, protein kinases, and enzymes involved in phosphoinositide metabolism. C-repeat/dehydration–responsive element binding factors (*CBF*s) are AP2/ERF-type transcription factors, which make up a critical gene cluster of the second group. During in the stress condition, *CBF* genes are rapidly induced in response to abiotic stress, such as dehydration and cold [6,7]. The CBF proteins in turn activate expression of a set of target effector genes by binding to a core sequence in their promoter, C-repeat (CRT) / dehydration response element (DRE) [8,9,10,11]. *CBF* genes appear to be ubiquitous in plant species and almost always present as a gene family [12,13]. In *Arabidopsis*, the three characterized *CBF* genes are *CBF1*, *CBF2*, and *CBF3*, which are organized in tandem on chromosome 4 [14]. *CBF1* and *CBF3* are positive regulators whereas *CBF2* has a negative regulatory role [15]. *CBF* transcription factor genes are induced by the constitutively expressed inducer of *CBF* expression (*ICE1*) by binding to the *CBF* promoter [16,17]. The *ICE1-CBF* cold response pathway is conserved in diverse plant species [14,17,18]. Constitutive overexpression of *CBF* transcription factors in transgenic plants has increased the plant tolerance to freezing, salt, and drought stresses [8,19,20,21,22,23]. This functional conservation has suggested the *ICE1-CBF* genes are important targets for crop improvement for drought tolerance through genetic engineering [24].

ESK1 is a newly discovered member of the second group. It was initially identified as conferring freezing tolerance; a significantly high proline content accumulates in *esk1* mutants [25]. The gene product of *ESK1* belongs to an uncharacterized plant-specific protein family containing 48 members [26]. Bioinformatics analysis of genes whose expression modified by the *eskimo1* mutation showed that a large number of genes were previously reported linking to plant response to salt, osmotic stress, and the stress hormone abscisic acid (ABA) [25,26]. Later work showed that the mutant has a clear advantage in response to drought and salt stress: In standard and drought conditions, transpiration rate of mutant is lower than in wild type (WT) [27]. A biologically relevant parameter is the water required per biomass unit, and with this measure, the *esk1* mutants clearly have shown a higher water use efficiency and photosynthetic rate compared to WT [27]. This higher water use efficiency was independent of stomata closure through ABA biosynthesis. Measurement of root hydraulic conductivity suggests that the *esk1* vegetative apparatus suffers water deficit because of a defect in water transport system [27,28]. *ESK1* promoter-driven reporter gene expression has been observed in xylem and fibers, the vascular tissue which is responsible for the transport of water and mineral nutrients from the soil to the shoots, via the roots. Moreover, in cross sections of hypocotyls, roots, and stems, collapsed xylem vessel has been observed in *esk1* mutant [28]. The *ESK1* gene, therefore, was inferred to play a major role in whole plant water economy. *ESK1* has homologues in numerous species, and it is reasonable to hypothesize that manipulation of *ESK1* in crops could improve water use efficiency.

With an understanding of the molecular mechanism, several gene manipulation approaches have been employed to increase plant drought tolerance. The manipulated genes include those encoding enzymes required for the biosynthesis of various osmoprotectants or enzymes for modifying membrane lipids, LEA protein, and detoxification enzyme [29,30,31]. To date, the *CBF* genes are the most explored genes for improving crop stress tolerance because of the ability of this transcription factor to regulate an entire set of genes in a stress-response pathway [8,32,33]. When rice *OsDREB1A* [34] or corn *ZmDREB1A* is constitutively overexpressed in *Arabidopsis*, the downstream target genes regulated by the *Arabidopsis DREB1* (e.g., *RD29A*) are induced, resulting in desiccation tolerance under 15% humidity [35]. In another study, 35S: *CBF1* transgenic tomato plants are more resistant to water-deficit stress by showing less plant wilting and leaf curling than WT controls after 21-d water withdrawal in the same pot [36]. Constitutive overexpression of two wheat *CBF* factors in barley substantially improves survival under severe drought or cold. In addition, expression of *DREB* factors in wheat and barley under the control of drought-inducible promoters allows for normal development, together with significantly improved survival under severe drought [37].

In spite of the extensive evaluation of *CBF* factors, only a few studies have shown a clear improvement in drought tolerance in crops under field conditions [38]. Considering the complexity of the plant stress response, it has been assumed that better drought tolerance might be obtained if multiple genes involved in different stress response pathways could be manipulated together through molecular stacking. In this work, we selected *ICE1* as well as the *CBF1* and *CBF3* genes involved in one stress-activated pathway and the *ESK1* gene from another stress-regulating pathway as targets of gene manipulation. Here, *ICE1* or *CBF1* and *CBF3* genes were overexpressed under the control of a constitutive promoter whereas the *ESK1* gene was suppressed with siRNA technique. Using gene stacking, we combined the two expression cassettes into one transformation vector for *Agrobacterium*-mediated plant transformation. The results obtained by testing the concept in the model plant *Arabidopsis* showed a significant increase in drought tolerance compared to non-transgenics, indicating that multiple gene manipulation might be a promising strategy for improving stress response and especially drought tolerance in plants.

Results

Manipulation of target gene expression through knock-down and overexpression

To knock down gene expression, a siRNA targeting specifically the *ESK1* gene was designed, and to overexpress the desired *ICE1* or *CBF* gene, the 35S promoter was placed in front of the gene for constitutive expression. The gene suppression and overexpression cassettes were integrated into the *Arabidopsis* genome, either alone or combined, via the flora dipping method. In analysis of gene expression by qRT-PCR, the plant transformed with siRNA cassette that suppresses *ESK1* showed clear knock-down of *ESK1* expression. Although the level of suppression varied from line to line, most plants showed about or more than 50% reduction in gene expression (Fig. 1A). In comparison, the transgenic lines derived via using only the pGPTV-*ESKi* vector showed a better suppression of *ESK1* gene expression than those obtained through use of the vector with the stacked overexpression and suppression cassettes (Figs. 1A, C, D).

Analysis of either *ICE1* or *CBF* overexpression showed the transgenic lines with the relevant transgene cassette had significantly higher *ICE1* or *CBF* expression compared to WT (Figs. 1B–E). For further evaluation of the concept, the best-performing plants based on qRT-PCR results were selected and their T2 seeds produced following the protocol described in the Methods part.

The response of transgenic *Arabidopsis* to osmotic stress *in vitro*

A distinct difference emerged in *in vitro* stress response between transgenic and WT plants. When growing on medium without any applied external stress, the majority of transgenic lines showed an almost identical phenotype to WT plants. However, when growing on medium with 30% polyethylene glycol (PEG), the transgenic plants with suppressed *ESK1* had much better root system growth than WT plants: The main root was longer, and the number of lateral roots was greater (Fig. 2). Leaf growth also differed between the transgenic and WT plants although not as much as the root system: All transgenic lines had eight leaves after 18 days of growth on medium with PEG whereas WT had only six. 15 *ESKi* transgenic lines were tested and the result of response to PEG was similar (data not shown).

Compared to the transgenic lines with the suppressed *ESK1* alone, the plants obtained through using stacking vector that suppress *ESK1* and overexpress either *ICE1* (14 lines) or *CBF* (3 lines) presented a similar performance on medium with PEG. After 14 days of growth with PEG, more and larger leaves were observed in transgenic lines compared to WT (Fig. 2). The growth

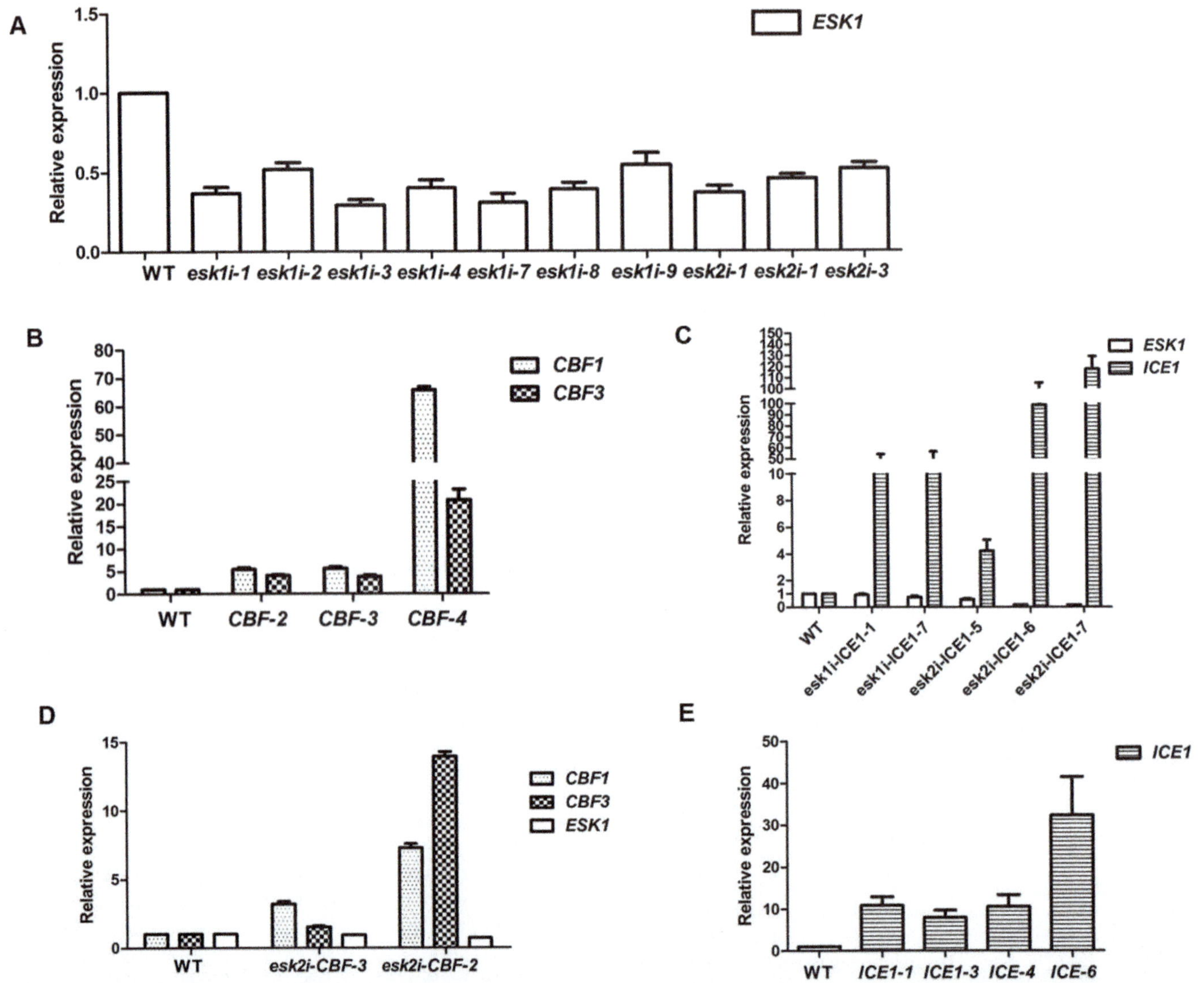

Figure 1. qRT-PCR results show *ESK1*, *ICE1*, *CBF1*, and *CBF3* expression in different transgenic lines. (**A**) *ESK1* was tested in two-week-old *ESKi* transgenic lines. (**B**) *CBF1* and *CBF3* were tested in two-week-old *CBF* transgenic lines. (**C**) *ESK1* and *ICE1* were tested in two-week-old *eski-ICE1* transgenic lines. (**D**) *ESK1*, *CBF1* and *CBF3* were tested in two-week-old *esk2i-CBF* transgenic lines. (**E**) *ICE1* was tested in two-week-old *ICE1* transgenic lines. Values represent means ±SD (error bar) of three replicates.

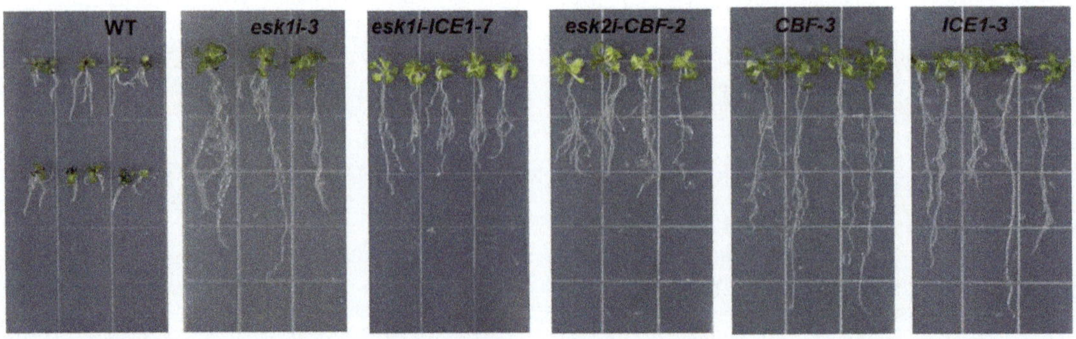

Figure 2. Response of transgenic lines to osmotic stress. Seedlings of different transgenic lines were subjected to osmotic stress: WT, *esk1i*-3, *esk1i-ICE1*-7, *esk2i-CBF*-2, *CBF*-3, *ICE1*-3. Four-day-old WT or transgenic seedlings were transferred to 1/2 Murashige and Skoog (MS) medium previously infused with 30% PEG for 14 days. Experiments were repeated at least three times with similar results. At least 30 seedlings per genotype were measured in each replicate.

of the root system had the same pattern with transgenic lines showing less effect from PEG treatment.

The responses to *in vitro* osmotic stress observed from transgenic plants with only the overexpression for *CBF* (5 lines), or *ICE1* (4 lines) were also similar to those with *ESK1* suppressed (Fig. 2). On normal 1/2 MS medium without PEG, no phenotype difference was observed between transgenics and WT control. On PEG medium, however, most transgenic plants clearly showed better growth than control plants after 14 days: The development of the root system from plants with *CBF* or *ICE1* overexpression was more robust, with a longer main root and more lateral roots. Nevertheless, several lines of *ICE1* transgenic plants showed a stress phenotype like that of WT, although the *ICE1* expression in those plants was much higher than in WT.

The response of transgenic plants to ABA stress *in vitro*

Suppression of *ESK1* gene expression did not significantly influence the sensitivity of plants to external ABA application. On medium without additional ABA, transgenic plants containing *ESK1* suppression cassette appeared almost identical to WT plants in their growth (Fig. 3A). On medium with 20 μM ABA for 14 days, the two still did not differ, and both appeared to suffer the effects of growth on ABA, showing leaves with yellow-brown stress symptoms (Fig. 3B).

The only difference observed between transgenic lines and WT plants was in seed germination time. On medium containing 0.5 μM ABA (Fig. 4B), the transgenic seeds showed delayed germination compared to WT control whereas on medium without ABA (Fig. 4A), the transgenic seeds germinated at the same time as WT.

Similar results were obtained for plant overexpression of *CBF* or *ICE1* in response to ABA stress *in vitro* (Figs. 3, 4). On medium without ABA, both transgenic and control plants showed similar growth with no abnormal phenotype observed in the majority of plants during the first 18 days of growth.

On medium with 20 μM ABA, both transgenic and WT plants were negatively affected. The elongation of the main root was inhibited, the number of lateral roots was reduced, and the leaves showed yellow-brown stress symptoms. No clear differences between transgenic lines and non-transgenic controls were observed (Fig. 3).

The performance of greenhouse transgenic plants in response to drought

To mimic most completely the interaction of plants and their natural environment, the transgenic plants showing the potential to increase drought tolerance via the *in vitro* osmotic test were transferred to greenhouse for evaluation of drought tolerance. The testing protocol was established based on the water-loss rate in greenhouse and the survival rate of plants in the protocol of water withdrawal and re-watering. Plants growing 2 weeks after germination in the described conditions were withdrawn from water for 14 days, so that the soil water content reached around 20%, which is considered a serious drought condition [27]. Following this threshold point, the plants were then fully watered and grown under a normal watering program for two more weeks before survival rates were evaluated.

Non-transgenic plants usually died completely under this protocol; however, transgenic plants obtained by the vector pGPTV-*ESKi* alone showed dramatic improvement in drought tolerance in contrast to the WT control plant (Fig. 5A). After 14 days without water, the transgenic plants still appeared green although with some level of dryness (Figs. 5A, B). Two weeks after re-watering, many of the transgenic plants had re-gained growth, and the survival rate of the best-performing line *esk1i*-3 reached 80% (Fig. 5C). In comparison, almost all WT plants became yellow and dry after 14 days without water, seldom regaining growth after water restoration (survival rate 1.25%, Fig. 5C).

Transgenic plants with *ICE1* or *CBF* gene overexpression vector alone also showed increased drought tolerance under greenhouse conditions (Figs. 5B–E). After 14 days without watering, the transgenic plants still remained green and recovered soon after the re-watering; the non-transgenic controls did not recover. Among the transgenic lines with various vectors, the line transformed with the *ICE1* overexpression vector showed relatively better drought tolerance than those transformed with *CBF* overexpression vector (Figs. 5C, D, E).

The improved drought tolerance test response under greenhouse conditions was also achieved in transgenic plants containing the combined *ESKi* cassette and overexpression cassette, especially those with the combination of the *ESKi* and *ICE1* cassettes; the highest survival rate was obtained from the line transformed with *ESKi-ICE1* stacking vector at 85.25% (Fig. 5C).

To compare the effects of different vectors on improvement of drought tolerance, the transgenic plants were divided into groups according to the vectors used, and the survival rate data was analyzed using *t*-tests. The results showed that the plants performing best were those transformed with *ESKi-ICE1* stacking vector; the lines transformed with the *ESKi* vector showed the second-best improvement while improvement was weakest with overexpression of the *CBF* genes (Figs. 5D, E).

The effect of transgenes on seed biomass

To evaluate potential applications to agriculture, we investigated the effect of transgenes on plant biomass, especially seed mass, under greenhouse conditions, either normal growth conditions or drought test conditions.

Under normal growth conditions in the greenhouse, some *ESK1* siRNA transgenic plants showed a dwarf phenotype with dark leaves (Fig. 6Aa); a delay in flowering was also observed, but most of the plants, otherwise, looked normal. Measuring the mass of seed harvested under normal greenhouse conditions revealed that the *ESK1* siRNA transgenic plants produced less seed than non-transgenic controls. However, upon harvest of the seed from plants that survived the drought test procedure, a clear contrast was evident between the seed yield of transgenic plants and controls, given that most of the non-transgenic plants died of drought. Comparing the seed yield of transgenic plants under drought and normal conditions, a significant reduction was found under drought, with less than 10% of the yield under normal conditions (Fig. 6B).

The impact of transgenes on seed yield showed a similar trend in plants transformed with *ICE1* or *CBF* vector alone and in those transformed with stacking vectors (Fig. 6B). Some dwarf phenotype development was observed during growth under normal conditions whereas in the drought condition, a higher stress tolerance compared to WT was observed, although accompanied by severe seed yield reduction.

Discussion

Plant genetic engineering has presented great potential for improving the drought tolerance of crops, especially since the discovery of *CBF* genes and their functions in stress tolerance [8,19,20,21,22,23]. *ESK1* was recently found to be involved in a stress tolerance pathway that is separate from the *CBF* pathway [25,26]. The *ESK1* pathway originally was identified as being related to plant freezing tolerance, but Bouchabke-Coussa *et al.*

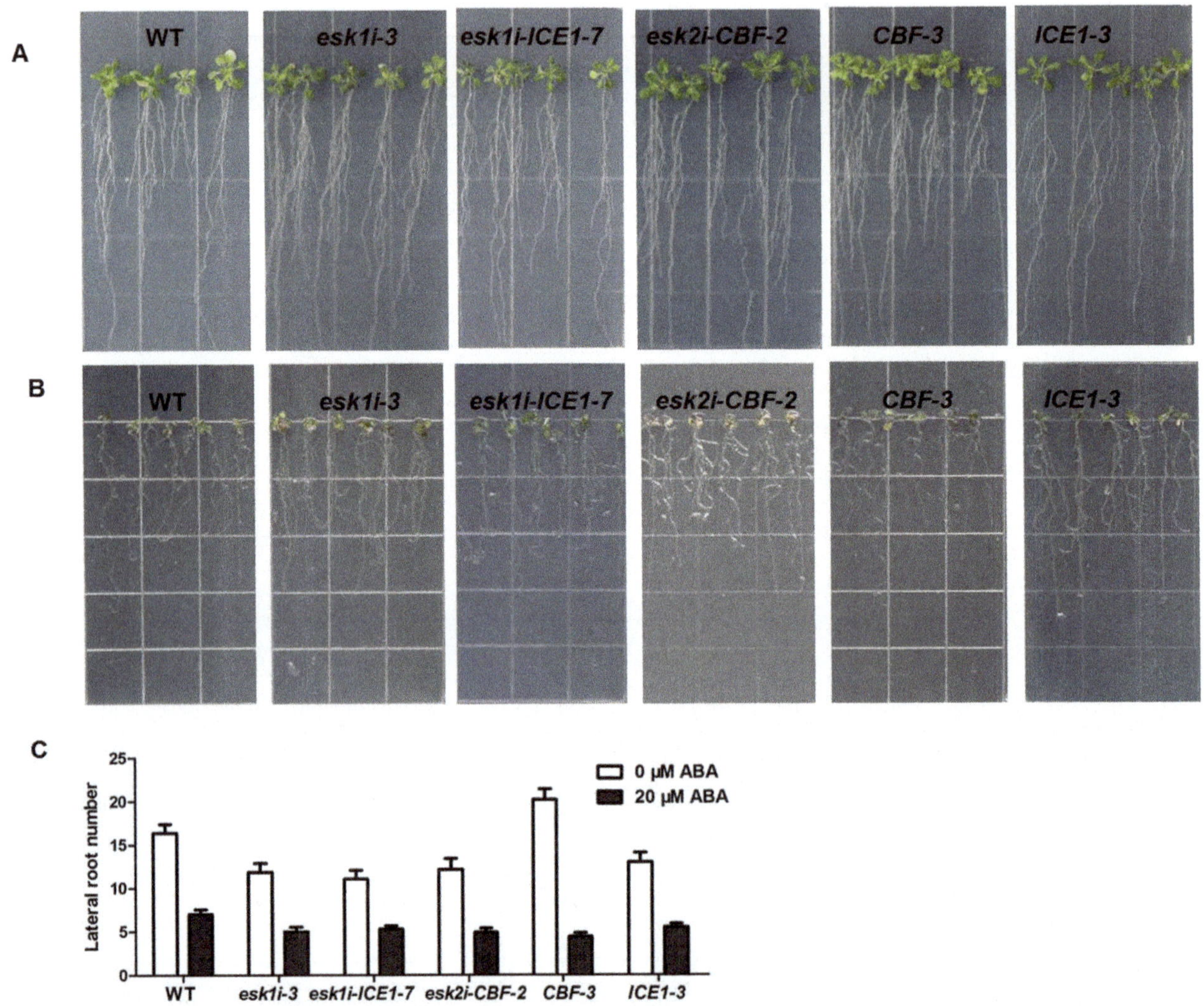

Figure 3. Seedling growth in response to ABA for WT and transgenic plants. (**A**) Four-day-old seedlings grown on 1/2 MS medium were transferred to 1/2 MS medium without ABA. (**B**) Sensitivity of seedlings to ABA. Four-day-old seedlings grown without ABA were transferred to 1/2 MS medium with 20 μM ABA. The photographs were taken 14 d after the transfer. (**C**) Quantification of lateral root number. 30 seedlings were measured in each experiment. Values represent means ±SD (error bar) of three replicates. At least 30 seedlings per genotype were measured in each replicate.

[27] later observed reduced respiration capacity in the *eskl* mutant of *Arabidopsis*, indicating a potential role of *ESK1* in drought tolerance. In our experiment, siRNA specifically targeting *ESK1* was employed to suppress its expression in *Arabidopsis* and the role of *ESK1* was confirmed in drought tolerance: Transgenic plants with efficient *ESK1* expression knock-down showed an obvious increase in osmotic stress tolerance *in vitro* and little effect on root system development by PEG treatment *in vitro* (Fig. 2). Further evaluation confirmed that suppression of the *ESK1* gene significantly enhanced plant drought tolerance (Fig. 5). In ABA *vitro* assay, it has been observed that *ESK1* knock-down transgenic plants has the same sensitivity to ABA change as control (Figs. 3B and 4B), which indicates that acquired stress tolerance of the transgenic plant may not be closely linked to the ABA-regulated network.

The *ESK1* knock-down plants appeared phenotypically normal comparing to non-transgenic plants when grown for 18 days: Plant size, leaf color, and shape were basically same like WT under normal growth conditions (Fig. 3A). Morphological observation also revealed that the number of stomata in the leaves remained unchanged relative to controls (data not shown). The question that arises is how *ESK1* knock-down confers higher tolerance to drought or other stress. One explanation is that the disruption in *ESK1* function may block the normal development of the vascular system. Lefebvre *et al.* [28] observed a kind of collapsed xylem structure in cross sections of hypocotyls, roots, and stems of an *Arabidopsis esk1* mutant. In the current study, we also observed that the vascular system appeared abnormal in both leaf and stem (data not shown). These observations may explain why the respiration rate of a transgenic plant is lower compared to WT and why drought tolerance is higher than in controls.

Measuring the weight of total seeds showed that knocking down the *ESK1* gene can negatively affect seed production under normal growth conditions but that under the drought condition, the yield is much less affected, in contrast to complete loss in the non-transgenic control (Fig. 6B). Linking the yield penalty and the vascular distortion of the *ESK1* knock-down plant, we can infer that manipulation of *ESK1* gene expression can indeed increase the drought tolerance of plants, but expecting no reduction in biomass is not realistic considering the physical distortion of the vascular structure. Adoption of an inducible promoter or tissue-specific promoter may alleviate some negative consequences, but

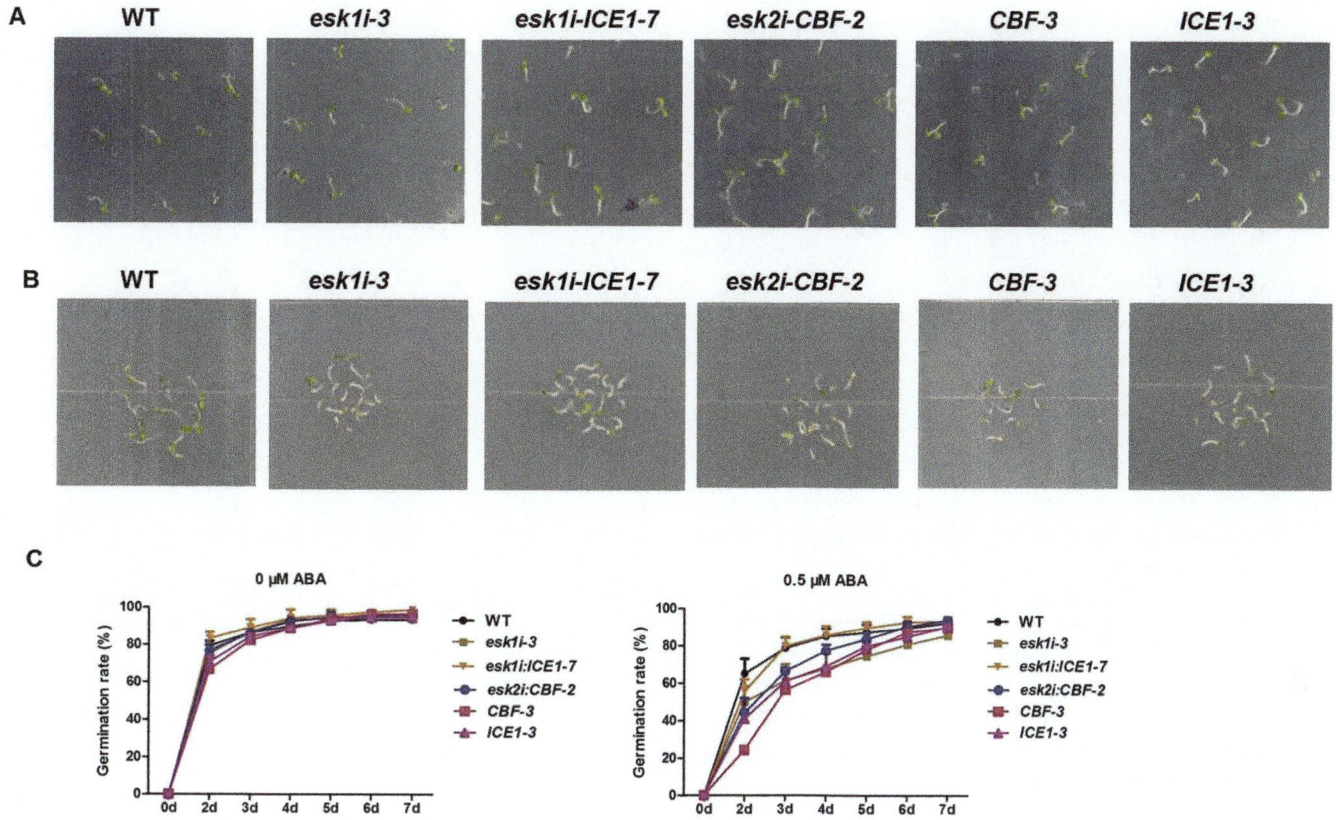

Figure 4. Seed germination in response to ABA for WT and transgenic plants. (**A**) and (**B**) Photographs of young seedlings at 5 d after the end of stratification. Seeds were germinated and allowed to grow on horizontal agar medium containing 0 or 0.5 μM ABA. (**C**) Seed germination time course of the six genotypes grown on medium without ABA or with 0.5 μM ABA. Values represent means ±SD (error bar) of three replicates. At least 100 seeds per genotype were measured in each replicate.

the enhancement of drought tolerance may be compromised at the same time. Therefore, use of *ESK1* for stress tolerance should be carefully balanced. One potential application may be in improvement of ornamental plants, for which biomass is not important.

Many studies have illustrated the potential of manipulating *CBF/DREB* genes to confer improved drought tolerance [20,21,37,39]. For example, overexpression of *CBF1/DREB1B* from *Arabidopsis* improves tolerance to water-deficit stress in tomato, but few plants clearly show enhanced drought tolerance under natural conditions [36]. We speculate that the complexity of the stress response pathway could be the reason and that modification of a single gene in a complicated pathway might not be sufficient to alter plant drought tolerance dramatically. Gene discovery and functional genomics projects have revealed many mechanisms and gene families that confer improved adaptation to abiotic stresses. These gene families can be manipulated into novel combinations, expressed ectopically, or transferred to species in which they do not naturally occur or vary [40,41,42]. Therefore, we have designed a gene stacking strategy by combining manipulation of the *CBF* and *ESK1* genes, hoping to obtain at least an additive effect on drought tolerance.

The result showed that gene stacking can indeed further improve drought tolerance in *Arabidopsis*, but an additive effect was not observed. One possibility for the lack of additive effect is that genes stacked in the same transformation vector may not work as efficiently as those in completely independent transformation vectors. qRT-PCR analysis in our work showed that the suppression effect of siRNA targeting *ESK1* was less significant when the same vector was used for the stacked RNAi cassette and the overexpression cassette. In the future, the effect of combined gene manipulation will be evaluated through stacking the gene cassette by crossing the transgenic lines with individual vectors.

Materials and Methods

Plant materials and growth conditions

Arabidopsis thaliana ecotype Columbia (Col-0) and transgenic plants in Col-0 background were grown under long-day conditions (16 h light/8 h dark) with light intensity at 100 μE m^{-2} s^{-1}.

For plate-grown plants, *Arabidopsis thaliana* seeds were surface sterilized with 20% (v/v) bleach and sown on medium containing half-strength Murashige and Skoog (MS) salts [43], 1% sucrose, and 0.8% agar. After stratification for 3 d at 4°C, the plates were kept in a growth incubator under a long-day photoperiod (16 h light, 8 h darkness) at 24°C for 10 d. For transgenic seeds, medium was supplemented with 50 μg/mL kanamycin sulfate.

Gene isolation and binary vector construct

The leaf of 2-week-old *Arabidopsis* Col-0 was used for DNA and RNA extraction. The Col-0 cDNA sequences were obtained from GenBank (http://www.ncbi.nlm.nih.gov/). Total DNA was extracted from the samples using the DNeasy Plant Mini Kit (Qiagen, Germany), and total RNA was extracted using the RNeasy Plant Kit (Qiagen, Germany). cDNA was synthesized as described in the Thermo Scientific protocol (#K1631).

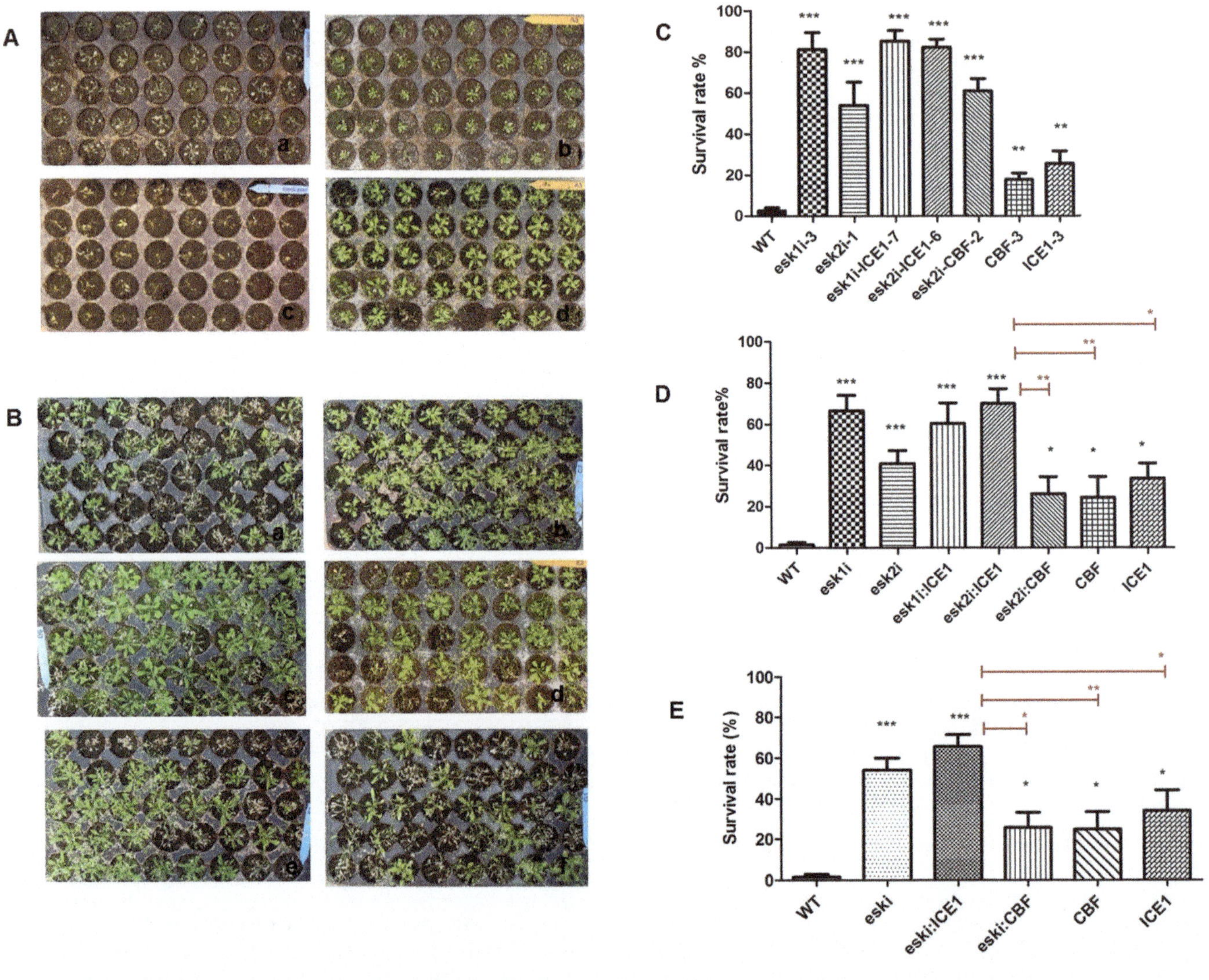

Figure 5. Drought responses of WT and transgenic plants. (**A**) and (**B**) Drought tolerance assay. 7-d seedlings were transferred to soil for another 1 week, subjected to drought by water withholding for 14 d, and then re-watered for 7 d. WT (Aa and Ac), *esk1i*-3 (Ab and Ad), *esk2i*-1 (Ba), *esk1i-ICE1*-7 (Bb), *esk2i-ICE1*-6 (Bc), *esk2i-CBF-2* (Bd), *CBF*-3 (Be), and *ICE1*-3 (Bf) plants. (**C**) Survival rate of plants from (A) and (B) after re-watering. (**D**) Survival rate of plants from three individual transgenic lines of each vector. (**E**) Survival rate of plants from all transgenic lines as description in methods and materials. SD (error bars) was calculated from results of three independent experiments (n>30 for each experiment). Asterisks indicate significant differences from the corresponding WT values as determined by Student's *t*-tests (*0.01≤P≤0.05, **P≤0.01, ***P≤0.001). Experiments were repeated at least three times with similar results.

The *AtCBF1* (AT4G25490) and *AtCBF3* (AT4G25480) were amplified from *Arabidopsis* Col-0 DNA by PCR. *AtCBF1* was amplified with a forward primer containing a BamHI restriction site at the 5′ end (5′-CGGGATCCCTCTGATCAATGAACTCATT-3′) and a reverse primer containing a SacI site at the 3′ end (5′-GCGAGCTCTTAGTAACTCCAAAGCGACA-3′). *AtCBF3* was amplified with a forward primer containing an ApaI site at the 5′ end (5′-GGGCCCGATCAATGAACTCATTTTCTGC-3′) and a reverse primer containing a XbaI site at the 3′ end (5′-GCTCTAGATTAATAACTCCATAACGATACGTCG-3′).

The *AtICE1* (AT3G26744) and *AtESK1* (AT3G55990) cDNA fragments were amplified from *Arabidopsis* Col-0 RNA by RT-PCR. *AtICE1* was amplified with a forward primer containing a XhoI site at the 5′ end (5′-GCCTCGAGGCGATGGGTCTTGACGGAAACAATGGTG-3′) and a reverse primer containing a XbaI site at the 3′ end (5′-GCTCTAGATCAGATCATACCAGCATACCCTGCTGTATCG-3′).

For the RNA interference (RNAi) construct, for a 341-bp specific fragment of *ESK1*, *esk1i*, the sense fragment was amplified by PCR using a forward primer (5′-GCCTCGAGTTGCTAGCATGTCTCCTCTT-3′) and a reverse primer (5′-GCGAGCTCATTCCACGTGTCAGGTAAAC-3′), as was the antisense fragment (forward primer: 5′-CGCCCGGGATTCCACGTGTCAGGTAAAC-3′; reverse: 5′-CGGTCGACTTGCTAGCATGTCTCCTCTT-3′). For a 284-bp specific fragment of *ESK1*, *esk2i*, the sense fragment also was amplified using forward and reverse primers (5′-GCCTCGAGTCAAGTGTGCATTAGAGACG-3′ and 5′-GCGAGCTCATTCCACGTGTCAGGTAAC-3′, respectively), as was the antisense fragment (forward: 5′-CGCCCGGGATTCCACGTGTCAGGTAAAC-3′; reverse: 5′-CGGTCGACTCAAGTGTGCATTAGAGACG-3′). The

Figure 6. Seed biomass for WT and transgenic plants. (**A**) Phenotype of WT, *esk1i*-3, *esk1i-ICE1*-7, *esk2i-ICE1*-6, *CBF*-3, and *ICE1*-3 in normal watering environment. a, eight-week-old seedlings, b, six-week-old seedlings. (**B**) Average seed biomass. In the normal watering environment, SD (error bars) calculated from 30 plants of each phenotype and the results of three independent experiments, transgenic lines were as description in methods and materials. After drought treatment, SD (error bars) calculated from 10 plants of each phenotype and the results of three independent experiments, transgenic lines were as description in methods and materials. Asterisks indicate significant differences from the corresponding WT values determined by Student's *t*-tests (*0.01≤P≤0.05, **P≤0.01, ***P≤0.001). Experiments were repeated at least three times with similar results.

PCR products were sequenced to ensure that they encoded the expected gene products.

The pGPTV-SdaI vector was used to construct a binary expression vector. The pRNAi-vector was used to construct an intron-spliced hairpin RNA (RNAi construct), and the isolated gene or gene fragment was constructed into the relevant vector to yield pGPTV-*CBF1-CBF3* (*CBF*), pGPTV-*ICE1* (*ICE1*), pGPTV-*esk1i* (*esk1i*), pGPTV-*esk2i* (*esk2i*), pGPTV-*esk2i-CBF1-CBF3* (*esk2i-CBF*), pGPTV-*esk1i-ICE1* (*esk1i-ICE1*), and pGPTV-*esk2i-ICE1* (*esk2i-ICE1*). The transcription of each gene or gene fragment was under the control of the 35S promoter and 35S terminator.

Agrobacterium-mediated gene transformation

The floral dip method [44] was applied to stably transform the *Arabidopsis* plant by using the *Agrobacterium* strain ATHV containing the designed binary vector. Seeds from treated plants were germinated on media containing 1/2 MS salts, 1% sucrose, 0.8% agar and 50 μg/mL kanamycin. Resistant seedlings were transferred to soil and grown under 16 h light /8 h dark at 24°C in a growth chamber. After PCR confirmation, positive seedlings were used to produce T2 and T3 seeds, which were always subjected to kanamycin selection and PCR confirmation; only the positive plants were used for seed production of the next generation. The kanamycin-tolerant and PCR-positive T3 or T4 plants were used in all experiments. Ecotype Col-0 served as control.

Quantitative real-time PCR (qRT-PCR)

Total RNA from different *Arabidopsis* plants was isolated using the RNeasy plant mini kit (Qiagen, Germany) according to the manufacturer's instructions. To eliminate any residual genomic DNA, total RNA was treated with ribonuclease-free DNase I (Thermo Scientific). Two micrograms of the total RNA was used as template to synthesize cDNA employing the RevertAid H Minus First Strand cDNA Synthesis Kit (Thermo Scientific). Real-time PCR was performed according to Kant *et al.* [45]. The PCR reaction was performed with three replicates and repeated with three biological samples. Relative quantification values for each target gene were calculated by the 2 (-Delta Delta C(T)) method [46]. For normalizing the amount of total RNA in all *Arabidopsis* samples, *Actin2* (AT3G18780) and *beta tubulin* (AT5G23860) were used as internal reference genes to compare data from different PCR runs or cDNA samples. The GenScript online tool was used to design the primers (https://www.genscript.com/), and primers used in q-PCR are listed in Supplemental Table S1.

In vitro assay for plant response to ABA and PEG

For germination, seeds were planted on the plate containing 1/2 MS salts, 1% sucrose, and 0.8% agar with 0 or 0.5 μM ABA as indicated. Plates were chilled at 4°C in the dark for 3 d and moved to 24°C with a 16 h light/8 h dark cycle. The percentage of seed germination was scored at the indicated times. Germination was defined as an obvious emergence of the radicle through the seed coat [47].

To study the inhibition effect of ABA in post-germinative growth, seeds were sown on 1/2 MS medium with 50 μg/mL kanamycin for 4 d after 3 d stratification, then transferred to 1/2 MS medium containing 0 or 20 μM ABA.

For response to *in vitro* dehydration, 3-day-old seedlings were transferred to 1/2 MS medium previously infused with 30% PEG8000 for 14 d [48]. After growing for 14 d on the treatment medium vertically, seedlings were photographed with a digital camera.

15 *ESKi* transgenic lines, 14 *eski-ICE1* transgenic lines, 3 *eski-CBF* transgenic lines, 5 *CBF* transgenic lines and 5 *ICE1* transgenic lines were used in both PEG and ABA *in vitro* assay. For each treatment, 30 plants of each line have been subjected to the treatment. Each test has been repeated three times. The influence of the tested element was concluded based the result pooled together from three repeats.

Greenhouse test for drought tolerance

15 *ESKi* transgenic lines, 14 *eski-ICE1* transgenic lines, 3 *eski-CBF* transgenic lines, 5 *CBF* transgenic lines and 5 *ICE1* transgenic lines, which showed desired gene manipulation result in qRT-PCR, were used for drought tolerance test. All seeds from transgenic lines and Col-0 together were sown in 40 5 cm×5 cm pots. Plant growth conditions were as described [49]. All seeds were cold-treated (4°C) for 3 d immediately after planting to ensure uniform germination. Plants were grown in controlled environment chambers at 24°C under sunshine illumination or supplementary illumination from cool-white fluorescent lights (100–150 μmol m^{-2} s^{-1}) and irrigated with water three times a week, ensuring soil saturation; the last watering was on day 14 after germination. Two-week-old seedlings were used for drought treatment, in which watering was withdrawn for two weeks. Afterwards, all plants were re-watered according to the pre-drought treatment. Tolerance to drought was determined by the capacity of plants to resume growth after 7 d of re-watering. The test was repeated at least three times. The data obtained was subjected to statistical analysis described as below.

Measurement of drought effect on seed biomass

14 *ESKi* transgenic lines, 12 *eski-ICE1* transgenic lines, 3 *eski-CBF* transgenic lines, 4 *CBF* transgenic lines and 3 *ICE1* transgenic lines were used for the test. At least 30 plants from each transgenic line and wild type were exposed to both control and drought treatment described as above. After the treatment, the survived plants were grown in normal water condition for seed setting. The seeds were harvested by bag that had been covered to plant in reproductive period. The harvested seeds were cleaned and then dried at room temperature. The weight of seed was measured after 2 weeks drying. The average weight (total harvested seed divided by the number of tested plant) was used as result to compare the drought effect on each line of plant. The data was analyzed based the statistical methods described below.

Statistics

The results shown are representative of three independent experiments, and within each experiment, treatments were replicated three times. Data were statistically analyzed using SigmaPlot 10.0 and GraphPad Prism 5.

Acknowledgments

We thank Professor Juren Zhang (School of Life Science, Shandong University) for thoughtful suggestions. This research was supported by KWS SAAT AG.

Author Contributions

Conceived and designed the experiments: HW FX ZL RN. Performed the experiments: FX TB HX. Analyzed the data: FX HW ZL JZ JZ JK. Contributed reagents/materials/analysis tools: FX TB. Contributed to the writing of the manuscript: HW FX.

References

1. Bohnert HJ, Nelson DE, Jensen RG (1995) Adaptations to Environmental Stresses. Plant Cell 7: 1099–1111.
2. U.S. Drought 2012: Farm and Food Impacts. Available: http://www.ers.usda.gov/topics/in-the-news/us-drought-2012-farm-and-food-impacts.aspx. Accessed 2013 July 26.
3. Jenifer Huang McBeath, Jerry McBeath (2010) Environmental Change and Food Security in China. New York: Springer.pp. 112.
4. Sergey M, Govorushko (2012) Natrual Processes and Human Impacts: Interactions between Humanity and the Environment. New York: Springer.pp. 113.
5. Umezawa T, Fujita M, Fujita Y, Yamaguchi-Shinozaki K, Shinozaki K (2006) Engineering drought tolerance in plants: discovering and tailoring genes to unlock the future. Curr Opin biotechnol 17: 113–122.
6. Shinozaki K, Yamaguchi-Shinozaki K (2007) Gene networks involved in drought stress response and tolerance. J Exp Bot 58: 221–227.
7. Zhang X, Fowler SG, Cheng H, Lou Y, Rhee SY, et al. (2004) Freezing-sensitive tomato has a functional CBF cold response pathway, but a CBF regulon that differs from that of freezing-tolerant Arabidopsis. Plant J 39: 905–919.
8. Liu Q, Kasuga M, Sakuma Y, Abe H, Miura S, et al. (1998) Two transcription factors, DREB1 and DREB2, with an EREBP/AP2 DNA binding domain separate two cellular signal transduction pathways in drought- and low-temperature-responsive gene expression, respectively, in Arabidopsis. Plant Cell 10: 1391–1406.
9. Miller AK, Galiba G, Dubcovsky J (2006) A cluster of 11 CBF transcription factors is located at the frost tolerance locus Fr-A^m2 in Triticum monococcum. Mol Genet Genomics 275: 193–203.
10. Lata C, Bhutty S, Bahadur RP, Majee M, Prasad M (2011) Association of an SNP in a novel DREB2-like gene SiDREB2 with stress tolerance in foxtail millet [Setaria italica (L.)]. J Exp Bot 62: 3387–3401.
11. Roychoudhury A, Paul S, Basu S (2013) Cross-talk between abscisic acid-dependent and abscisic acid-independent pathways during abiotic stress. Plant Cell Rep 32: 985–1006.
12. Skinner JS, von Zitzewitz J, Szucs P, Marquez-Cedillo L, Filichkin T, et al. (2005) Structural, functional, and phylogenetic characterization of a large CBF gene family in barley. Plant Mol Biol 59: 533–551.
13. El Kayal W, Navarro M, Marque G, Keller G, Marque C, et al. (2006) Expression profile of CBF-like transcriptional factor genes from Eucalyptus in response to cold. J Exp Bot 57: 2455–2469.
14. Medina J, Bargues M, Terol J, Pérez-Alonso M, Salinas J (1999) The Arabidopsis CBF gene family is composed of three genes encoding AP2 domain-containing proteins whose expression is regulated by low temperature but not by abscisic acid or dehydration. Plant Physiol 119: 463–470.
15. Novillo F, Medina J, Salinas J (2007) Arabidopsis CBF1 and CBF3 have a different function than CBF2 in cold acclimation and define different gene classes in the CBF regulon. Proc Natl Acad Sci USA 104: 21002–21007.
16. Zarka DG, Vogel JT, Cook D, Thomashow MF (2003) Cold induction of Arabidopsis CBF genes involves multiple ICE (inducer of CBF expression)

promoter elements and a cold-regulatory circuit that is desensitized by low temperature. Plant Physiol 133: 910–918.
17. Chinnusamy V, Ohta M, Kanrar S, Lee BH, Hong X, et al. (2003) ICE1: a regulator of cold-induced transcriptome and freezing tolerance in Arabidopsis. Genes Dev 17: 1043–1054.
18. Chinnusamy V, Zhu J, Zhu JK (2007) Cold stress regulation of gene expression in plants. Trends Plant Sci 12: 444–451.
19. Jaglo-Ottosen KR, Gilmour SJ, Zarka DG, Schabenberger O, Thomashow MF (1998) Arabidopsis CBF1 overexpression induces COR genes and enhances freezing tolerance. Science 280: 104–106.
20. Haake V, Cook D, Riechmann JL, Pineda O, Thomashow MF, et al. (2002) Transcription factor CBF4 is a regulator of drought adaptation in Arabidopsis. Plant Physiol 130: 639–648.
21. Siddiqua M, Nassuth A (2011) Vitis CBF1 and Vitis CBF4 differ in their effect on Arabidopsis abiotic stress tolerance, development and gene expression. Plant Cell Environ 34: 1345–1359.
22. Orabya H, Ahmad R (2012) Physiological and biochemical changes of CBF3 transgenic oat in response to salinity stress. Plant Sci 185–186: 331–339.
23. Zhang L, Li Z, Li J, Wang A (2013) Ectopic overexpression of SsCBF1, a CRT/DRE-binding factor from the nightshade plant Solanum lycopersicoides, confers freezing and salt tolerance in transgenic Arabidopsis. PLoS One 8: e61810.
24. Boyer JS (1982) Plant productivity and environment. Science 218: 443–448.
25. Xin Z, Browse J (1998) Eskimo1 mutants of Arabidopsis are constitutively freezing tolerant. Proc Natl Acad Sci U S A 95: 7799–7804.
26. Xin Z, Mandaokar A, Chen J, Last RL, Browse J (2007) Arabidopsis ESK1 encodes a novel regulator of freezing tolerance. Plant J 49: 786–799.
27. Bouchabke-Coussa O, Quashie ML, Seoane-Redondo J, Fortabat MN, Gery C, et al. (2008) ESKIMO1 is a key gene involved in water economy as well as cold acclimation and salt tolerance. BMC Plant Biol 8: 125.
28. Lefebvre V, Fortabat MN, Ducamp A, North HM, Maia-Grondard A, et al. (2011) ESKIMO1 disruption in Arabidopsis alters vascular tissue and impairs water transport. PLoS One 6: e16645.
29. Zhang M, Barg R, Yin M, Gueta-Dahan Y, Leikin-Frenkel A, et al. (2005) Modulated fatty acid desaturation via overexpression of two distinct omega-3 desaturases differentially alters tolerance to various abiotic stresses in transgenic tobacco cells and plants. Plant J 44: 361–371.
30. Duan J, Cai W (2012) OsLEA3-2, an abiotic stress induced gene of rice plays a key role in salt and drought tolerance. PLoS One 7: e45117.
31. Badawi GH, Kawano N, Yamauchi Y, Shimada E, Sasaki R, et al. (2004) Overexpression of ascorbate peroxidase in tobacco chloroplasts enhances the tolerance to salt stress and water deficit. Physiol Plant 121: 231–238.
32. Stockinger EJ, Gilmour SJ, Thomashow MF (1997) Arabidopsis thaliana CBF1 encodes an AP2 domain-containing transcriptional activator that binds to the C-repeat/DRE, a cis-acting DNA regulatory element that stimulates transcription in response to low temperature and water deficit. Proc Natl Acad Sci U S A 94: 1035–1040.
33. Nakashima K, Ito Y, Yamaguchi-Shinozaki K (2009) Transcriptional regulatory networks in response to abiotic stresses in Arabidopsis and grasses. Plant Physiol 149: 88–95.
34. Dubouzet JG, Sakuma Y, Ito Y, Kasuga M, Dubouzet EG, et al. (2003) OsDREB genes in rice, Oryza sativa L., encode transcription activators that function in drought-, high-salt- and cold-responsive gene expression. Plant J 33: 751–763.
35. Qin F, Sakuma Y, Li J, Liu Q, Li YQ, et al. (2004) Cloning and functional analysis of a novel DREB1/CBF transcription factor involved in cold-responsive gene expression in Zea mays L. Plant Cell Physiol 45: 1042–1052.
36. Hsieh TH, Lee JT, Charng YY, Chan MT (2002) Tomato plants ectopically expressing Arabidopsis CBF1 show enhanced resistance to water deficit stress. Plant Physiol 130: 618–626.
37. Morran S, Eini O, Pyvovarenko T, Parent B, Singh R, et al. (2011) Improvement of stress tolerance of wheat and barley by modulation of expression of DREB/CBF factors. Plant Biotechnol J 9: 230–249.
38. Xiao BZ, Chen X, Xiang CB, Tang N, Zhang QF, et al. (2009) Evaluation of seven function-known candidate genes for their effects on improving drought resistance of transgenic rice under field conditions. Mol Plant 2: 73–83.
39. Gutha LR, Reddy AR (2008) Rice DREB1B promoter shows distinct stress-specific responses, and the overexpression of cDNA in tobacco confers improved abiotic and biotic stress tolerance. Plant Mol Bio 68: 533–555.
40. Halpin C (2005) Gene stacking in transgenic plants-the challenge for 21st century plant biotechnology. Plant Biotechnol J 3: 141–155.
41. Ramana Rao MV, Parameswari C, Sripriya R, Veluthambi K (2011) Transgene stacking and marker elimination in transgenic rice by sequential Agrobacterium-mediated co-transformation with the same selectable marker gene. Plant Cell Rep 30: 1241–1252.
42. Ye X, Al-Babili S, Klöti A, Zhang J, Lucca P, et al. (2000) Engineering the provitamin A (beta-carotene) biosynthetic pathway into (carotenoid-free) rice endosperm. Science 287: 303–305.
43. Murashige T, Skoog F (1962) A revised medium for rapid growth and bio-assays with tobacco tissue cultures. Physiol Plant 15: 473–497.
44. Clough SJ, Bent AF (1998) Floral dip: a simplified method for Agrobacterium-mediated transformation of Arabidopsis thaliana. Plant J 16: 735–743.
45. Kant S, Kant P, Raveh E, Barak S (2006) Evidence that differential gene expression between the halophyte, Thellungiella halophila, and Arabidopsis thaliana is responsible for higher levels of the compatible osmolyte proline and tight control of Na^+ uptake in T. Halophila. Plant Cell Environ 29: 1220–1234.
46. Livak KJ, Schmittgen TD (2001) Analysis of relative gene expression data using real-time quantitative PCR and the 2 (-Delta Delta C(T)) method. Methods 25: 402–408.
47. Li H, Jiang H, Bu Q, Zhao Q, Sun J, et al. (2011) The Arabidopsis RING finger E3 ligase RHA2b acts additively with RHA2a in regulating abscisic acid signaling and drought response. Plant Physiol 156: 550–563.
48. van der Weele CM, Spollen WG, Sharp RE, Baskin TI (2000) Growth of Arabidopsis thaliana seedlings under water deficit studied by control of water potential in nutrient-agar media. J Exp Bot 51: 1555–1562.
49. Novillo F, Alonso JM, Ecker JR, Salinas J (2004) CBF2/DREB1C is a negative regulator of CBF1/DREB1B and CBF3/DREB1A expression and plays a central role in stress tolerance in Arabidopsis. Proc Natl Acad Sci U S A 101: 3985–3990.

6

On the Use of Leaf Spectral Indices to Assess Water Status and Photosynthetic Limitations in *Olea europaea* L. during Water-Stress and Recovery

Pengsen Sun[1], Said Wahbi[2], Tsonko Tsonev[3], Matthew Haworth[4], Shirong Liu[1], Mauro Centritto[4]*

1 Institute of Forest Ecology, Environment and Protection, Chinese Academy of Forestry, Beijing, P. R. China, **2** Laboratoire de Biotechnologie et Physiologie Végétale, Faculté des Sciences Semlalia, Université Cadi Ayyad, Marrakech, Morocco, **3** Institute of Plant Physiology and Genetics, Bulgarian Academy of Sciences, Sofia, Bulgaria, **4** Trees and Timber Institute, National Research Council, Sesto Fiorentino, Florence, Italy

Abstract

Diffusional limitations to photosynthesis, relative water content (RWC), pigment concentrations and their association with reflectance indices were studied in olive (*Olea europaea*) saplings subjected to water-stress and re-watering. RWC decreased sharply as drought progressed. Following rewatering, RWC gradually increased to pre-stress values. Photosynthesis (*A*), stomatal conductance (g_s), mesophyll conductance (g_m), total conductance (g_t), photochemical reflectance index (PRI), water index (WI) and relative depth index (RDI) closely followed RWC. In contrast, carotenoid concentration, the carotenoid to chlorophyll ratio, water content reflectance index (WCRI) and structural independent pigment index (SIPI) showed an opposite trend to that of RWC. Photosynthesis scaled linearly with leaf conductance to CO_2; however, *A* measured under non-photorespiratory conditions ($A_{1\%O2}$) was approximately two times greater than *A* measured at 21% [O_2], indicating that photorespiration likely increased in response to drought. $A_{1\%O2}$ also significantly correlated with leaf conductance parameters. These relationships were apparent in saturation type curves, indicating that under non-photorespiratory conditions, CO_2 conductance was not the major limitations to *A*. PRI was significant correlated with RWC. PRI was also very sensitive to pigment concentrations and photosynthesis, and significantly tracked all CO_2 conductance parameters. WI, RDI and WCRI were all significantly correlated with RWC, and most notably to leaf transpiration. Overall, PRI correlated more closely with carotenoid concentration than SIPI; whereas WI tracked leaf transpiration more effectively than RDI and WCRI. This study clearly demonstrates that PRI and WI can be used for the fast detection of physiological traits of olive trees subjected to water-stress.

Editor: Pilar Hernandez, Institute for Sustainable Agriculture (IAS-CSIC), Spain

Funding: Financial support was provided by the National Natural Science Foundation of China (Grant No. 31290223), the Key Laboratory of Forest Ecology and Environment of State Forestry Administration of China, and the Ministero dell□Istruzione dell□Universit□ □ della Ricerca of Italy: PRIN 2010-2011 PRO-ROOT and Progetto Premiale 2012 Aqua. The funders had no role in study design, data collection and analysis, decision to publish, or preparation of the manuscript.

Competing Interests: The authors have declared that no competing interests exist.

* Email: mauro.centritto@cnr.it

Introduction

The effective utilisation of renewable water resources in agriculture is a significant challenge in many areas of the globe. To mitigate the effects of increasing chronic water shortages, there is a need to develop and expand irrigation management practices. However, considering the unprecedented pressure on water resources for agriculture caused by rapidly growing water demand for urban and industrial uses, the expansion of irrigation may only become possible through the development of precision water saving irrigation techniques [1]. These are based on the real time detection of crop physiological status, and require the development of advanced, non-invasive phenotyping methods to monitor water relations and photosynthetic status in plants experiencing water-stress [2]. To this end, remotely sensed vegetation indices are increasingly being used as reliable cost-effective plant-based indicators to assess physiological traits associated with plant water status [3]. Furthermore, the real-time detection of plant physiological changes in regions subjected to drought is important for precision crop management and the estimation of terrestrial productivity. Therefore, improved knowledge of the relationship between leaf spectral and physiological responses under variable water conditions is of crucial importance.

Water deficit constrains all physiological processes involved in plant growth and development. These changes are part of a cascade of responses to drought that affect primary processes including tissue water relations and gas exchange mechanisms. It is well known that one of the earliest responses to water deficit is diminished stomatal conductance (g_s), because stomata act as control valves in the exchange of water vapor between leaf and the atmosphere to match soil water uptake rate with transpiration rate (E) to maintain plant water balance [4]. However, as g_s declines, CO_2 diffusion into the leaves also decreases, leading to reduced substomatal CO_2 concentration (C_i) [5]. This diffusional limitation to photosynthesis (A) in the gas phase is frequently associated with a coordinated change limiting CO_2 transport through the mesophyll (i.e., mesophyll conductance, g_m) [6,7]. As a consequence, during the early stages of drought stress, photosynthetic limitations have been shown to be predominantly caused by decreased total diffusive conductance (g_t, which is related to g_s and

g_m) leading to low chloroplastic CO_2 concentration (C_c) [2,7]. Alteration of photosynthetic metabolism generally becomes more prominent as water-stress progresses [5]. Nevertheless, there is still a degree of controversy on the relative importance of g_s, g_m or metabolic impairments in the limitation of A under drought (Lawlor and Cornic 2002) and during recovery upon re-watering [5].

Alterations in leaf water status, photosynthetic pigment concentrations and photosynthetic activity in turn lead to changes in spectral reflectance properties [3]. Reflectance in specific wavelength bands in the visible and near-infrared (NIR) region have potential applications in the estimation of plant water status [3]. Many spectral reflectance indices have been proposed in the direct monitoring of plant water-status; expressed as relative water content (RWC), leaf water potential and photosynthetic status. Among the water spectral indices, the NIR-based water index (WI) [3] is increasingly employed in monitoring water status; as recently demonstrated by Gutierrez et al., [8] to assess water relations in contrasting wheat genotypes, and by both Serrano et al., [9] and Marino et al., [10] to estimate vine water status at leaf and canopy levels. Furthermore, Sun et al., [11] developed a new index, the water content reflectance index (WCRI), which was inversely related to RWC. However, this relationship was only clear at low RWC values, suggesting that this water content-based spectral index may not be effective in detecting moderate stress.

The photochemical reflectance index (PRI) was originally developed to estimate rapid changes in the epoxidation state of xanthophyll pigments [12], a major component of non-photochemical quenching. However, PRI has been increasingly used to detect photosynthetic activity at different scales from plant biochemistry to the canopy level (see the review by Garbulsky et al., [13]. Dobrowsky et al., [14] found that heat and water-stress induced changes in steady-state chlorophyll fluorescence that were tracked by PRI; suggesting that PRI was a more effective real time indicator of photosynthetic function than indices based upon leaf water content and pigment concentration. Consistent with these findings, Ripullone et al., [15] observed strong relationships between photosynthetic activity and PRI in individual forest tree species subjected to water-stress. Similarly, significant linear relationships between PRI and A, and between PRI and g_s, were also demonstrated in *Solanum lycopersicum* [16] and *Quercus ilex* [17] subjected to water-stress, and at the tree canopy-level in irrigated and rainfed adult olive trees grown under field condition [10]. The PRI is not sensitive to sudden drought, but did track A, g_s and leaf water content during slow developing drought stress in *Olea europaea* [11,18]. However, to the best of our knowledge there is only one study that investigates the possibility of using PRI to track photosynthesis diffusional limitations (g_s and g_m) during the early stages of drought stress and during recovery upon re-watering [17].

In this study, diffusional limitations to photosynthesis, leaf water status, pigment concentrations and spectral reflectance indices were evaluated in olive (*Olea europaea* L.) plants during water-stress and recovery. The objective was to understand the links between simultaneous variations in photosynthetic diffusional limitations and PRI alongside leaf water relation parameters and WI by examining time course changes in response to water availability. Additionally, the constancy of other water content and pigment related spectral indices were also evaluated in response to water-stress and recovery.

Materials and Methods

Growing conditions and drought treatment

Three-year-old plants of olive (*Olea europaea* L., cv Leccino) were grown in 15 dm^3 pots filled with commercial soil in the National Research Council greenhouse in Montelibretti, Rome, Italy. Temperature within the greenhouse was maintained at 25–27°C, photosynthetically active radiation followed the natural light regime and relative humidity ranged between 60–70%. Large volume 15 dm^3 pots relative to the *Olea europaea* plants were chosen to avoid any potential effects of root restriction generating root to shoot signals that may interfere with the response of the plants to drought [19,20]. Furthermore, the g_s and PRI values of well-watered and drought treated plants were indistinguishable in the early stages of drought treatment and the values recorded in the drought treatment plants subsequently returned to pre-stress levels following re-watering in the recovery period, indicating that potential root restriction did not influence the effect of drought on simultaneous measurement of reflectance and gas exchange under conditions of water-deficit. The plants were regularly watered to pot water capacity and fertilized with Hoagland solution once a week to supply mineral nutrients at free access rates [21,22] for the first two months after the onset of the growing season. On the afternoon preceding the initiation of the experiment (Day 0), plants were fully irrigated and allowed to drain the excess water over-night. Then half of the plants were water-stressed by withholding water until the seedlings showed symptoms of severe water-stress, while the remaining half of the seedlings continued to be well-watered to pot capacity. On the afternoon of day 18 water-stressed plants were irrigated with 500 cm^3 of water as supplementary irrigation. On the evening of the Day 23, the water-stressed seedlings were then re-watered daily to pot capacity over a 7-day recovery period. Drought stress treatment was performed during May - June. The plants were fully randomized and five replicate plants per treatment were analyzed for all measurements.

Gas exchange and fluorescence measurements

Simultaneous measurement of gas exchange and fluorescence were conducted on five leaves per plant (these values were then averaged to produce a mean value for the individual plant, and then the mean of five replicate plants taken for a given treatment value) using a LI-6400-40 portable infrared gas-analyser (Li-Cor, Lincoln, NE, USA). All gas exchange measurements were made between 11.00 and 15.00 h, with the leaf chamber set to a saturating photosynthetic photon flux density (PPFD) (1300 μmol m^{-2} s^{-1}), relative humidity of the air ranging between 45–55% and at a leaf temperature of 25°C. To reduce diffusion leaks through the chamber gasket [23], a supplementary external chamber gasket composed of the same polymer foam was added to create an interspace between the two gaskets (i.e. a double-gasket design with a 5 mm space separating the internal and external gaskets). Then the CO_2 and H_2O gradients between the in-chamber air and pre-chamber air were minimized by feeding the IRGA exhaust air into the interspace between the chamber and the pre-chamber gaskets [24]. Instantaneous measurement of steady-state photosynthesis (A), transpiration (E), stomatal conductance (g_s), and $\Delta F/F_m$' (i.e. the quantum yield of PSII in the light) were made on five plants per treatment after removal of stomatal limitation of A by lowering the external atmospheric [CO_2] (C_a) to 50 ppm as described by Centritto et al., [6]. Measurements of dark respiration (R_d) were performed at ambient [CO_2] concentration in the dark on the same leaves by switching off the light in the leaf cuvette; when CO_2 release from the leaf had become stable for approximately five to ten minutes this was recorded and considered to represent R_d. Mesophyll conductance to CO_2 diffusion (g_m), the inverse of the total resistance encountered by CO_2 across the leaf mesophyll, was calculated using the variable J method. One percent oxygen for the measurement of A under non-photorespiratory conditions was produced using a Brooks Instruments mass flow controllers (Brooks Instruments, Hatfield, PA, USA) and a gas mixing system connected to cylinders of compressed pure nitrogen and oxygen gas, this was then connected to the gas input line of the LiCor Li6400. Then g_m in olive leaves was calculated following Aganchich et al., [30], whereas total diffusion conductance (g_t) was calculated as: $g_t = g_s\, g_m/(g_s+g_m)$.

Determination of relative water content and pigment concentrations

Leaf samples for relative water content (RWC) and pigment concentrations were taken immediately after the gas exchange and spectral measurements. Fully-developed leaves were detached and weighed to determine leaf fresh mass (F_M). After measuring leaf area, using a leaf area meter (LI3100, LI-COR Inc., Lincoln, NE, USA), leaves were then covered with a plastic bag and allowed to rehydrate with the cut-end under water in a dark cold room at 5°C for18 h. Following rehydration, each leaf was weighed to determine saturated mass (S_M), and then oven-dried at 80°C for 48 hours to determine dry mass (D_M). RWC was finally calculated as follows: $RWC = (F_M - D_M)/(S_M - D_M)$.

The carotenoid and total chlorophyll (*Chl*, a and b) concentration was measured in intact leaf tissues by immersion in N,N-dimethylformamide (DMF). Leaf discs with a total area of 1.5 cm^2 were placed into glass vials containing 2 cm^3 DMF and immediately placed in darkness at 4°C in an orbital shaker set to 100 rpm for 4 h. The absorbance (A_B) of the solution was then read on a spectrophotometer (Perkin Elmer, Norwalk, CT) at 663.8, 646.8, and 480 nm, using DMF as a blank. The pigment concentrations were calculated according to the following equations [25]:

$$\text{Total chlorophyll } (Chl_a + Chl_b) = (12A_{B663.8} - 3.11A_{B646.8}) + (20.78A_{B646.8} - 4.88A_{B663.8})$$

$$\text{Carotenoid} = (1000A_{B480} - 1.12Chl_a - 34.7Chl_b)/245.$$

Spectral measurements and indices

Spectral measurements were carried out in the laboratory, by a portable spectroradiometer (Field spec FR 350–2500 nm, ASD Inc., Boulder, CO, USA). As the reflectance properties of leaves alter with age [26], all of the leaves analysed were of approximately the same age, less than one year-old from the uppermost part of the canopy. The instrument measures spectral reflectance between 350 and 2500 nm to a resolution of 3 nm. The instrument automatically calculates the reflectance value as the ratio between the incident radiation reflected from the surface target and the incident radiation reflected by a reference spectral panel. The reference spectral panel material can be regarded as a Lambertian reflector. The Field spec FR may be operated with different lenses that control the field of view, in this study an 8° lens was used to restrict the field of view of the instrument. Reflectance spectra (R) were collected from a distance of 5.0 cm from the sample, which was fixed on a dark platform. To reduce instrument noise each spectral signature was calculated from the average of 100 scans. A further check of the stability of the reflected signal was performed by taking the white reflectance at the beginning and end of each sample. Reflectance spectra were preprocessed using the View Spec Pro (version 5.6, ASD, Inc.) software. A summary of the spectral measurements used within the study are presented in Appendix S1.

Statistics

Data were tested using a simple factorial ANOVA (two-way maximum interactions) and Tukey *post-hoc* test. The statistical analysis was performed by Sigma Plot 11.0 (Systat software, Inc., San Jose, CA, USA). Linear and non-linear curve-fitting (Sigma Plot 11.0), which minimizes the difference between observed and predicted values, were used for regression analyses.

Ethics Statement

No permits were required for the experimental analysis of *Olea europaea* plants that were bought commercially and complied with all relevant regulations and did not involve an endangered or protected species.

Results

Figure 1 shows the time-course of RWC, PRI, A, g_s, g_m and g_t during the study period of soil drying and subsequent recovery. As expected, relative water content (Fig. 1a) was strongly affected by the water-regime ($P<0.001$), decreasing first slowly (Days 1–8) and then sharply (Days 10–16) as drought progressed. RWC was significantly increased ($P<0.05$) by supplementary irrigation (Day 20) and then following re-watering to pot water capacity (Days 23–30). The whole experimental process resulted in a "W-shaped" time course in RWC. In fact, upon relief of the water-stress, RWC increased very rapidly to pre-stress levels. The response of PRI to the drought and recovery cycle (Fig. 1b) closely followed the pattern of RWC.

Photosynthesis (Fig. 1c), g_s, g_m, and g_t (Fig. 1d) were also strongly affected ($P<0.001$) by water availability. Water-stressed saplings exhibited a significant decrease in A even during the early phase of the drying cycle (Fig. 1c). When water-stress became very severe (Day 16), A measured in ambient air was almost completely inhibited and only slightly, but not significantly, increased following the supplementary irrigation (Day 20). At low [O_2], A measured under non-photorespiratory conditions ($A_{1\%O2}$), was increased by ~75% in well-watered plants at the beginning and at the end of the experiment. In comparison to control values of $A_{1\%O2}$, significant decreases in $A_{1\%O2}$ were only observed in severely stressed plants. It is interesting to note that the ratio between $A_{1\%O2}$ and A increased during water-stress, reaching a value of about two on Day 20 following the supplementary irrigation, and a similar value one day after re-watering on Day 24. Mild water-stress also induced significant decreases in g_s (Fig. 1d); in a similar pattern to A, g_s approached zero when soil water decreased to a level where there was no longer water available within the pot to sustain transpiration, and then did not significantly respond to supplementary irrigation. Upon relief of water-stress, g_s and A increased rapidly to pre-stress levels; but their recovery was less rapid than that of RWC. Interestingly, seven days after re-watering, A and g_s significantly exceeded pre-stress values by about 25% and 31%, respectively. Mesophyll conductance to CO_2 diffusion and, in turn, g_t mirrored the trend in g_s during the drought and recovery cycle (Fig 1d).

Stomatal and mesophyll conductance values were not only similar in magnitude but also linearly correlated ($P<0.001$)

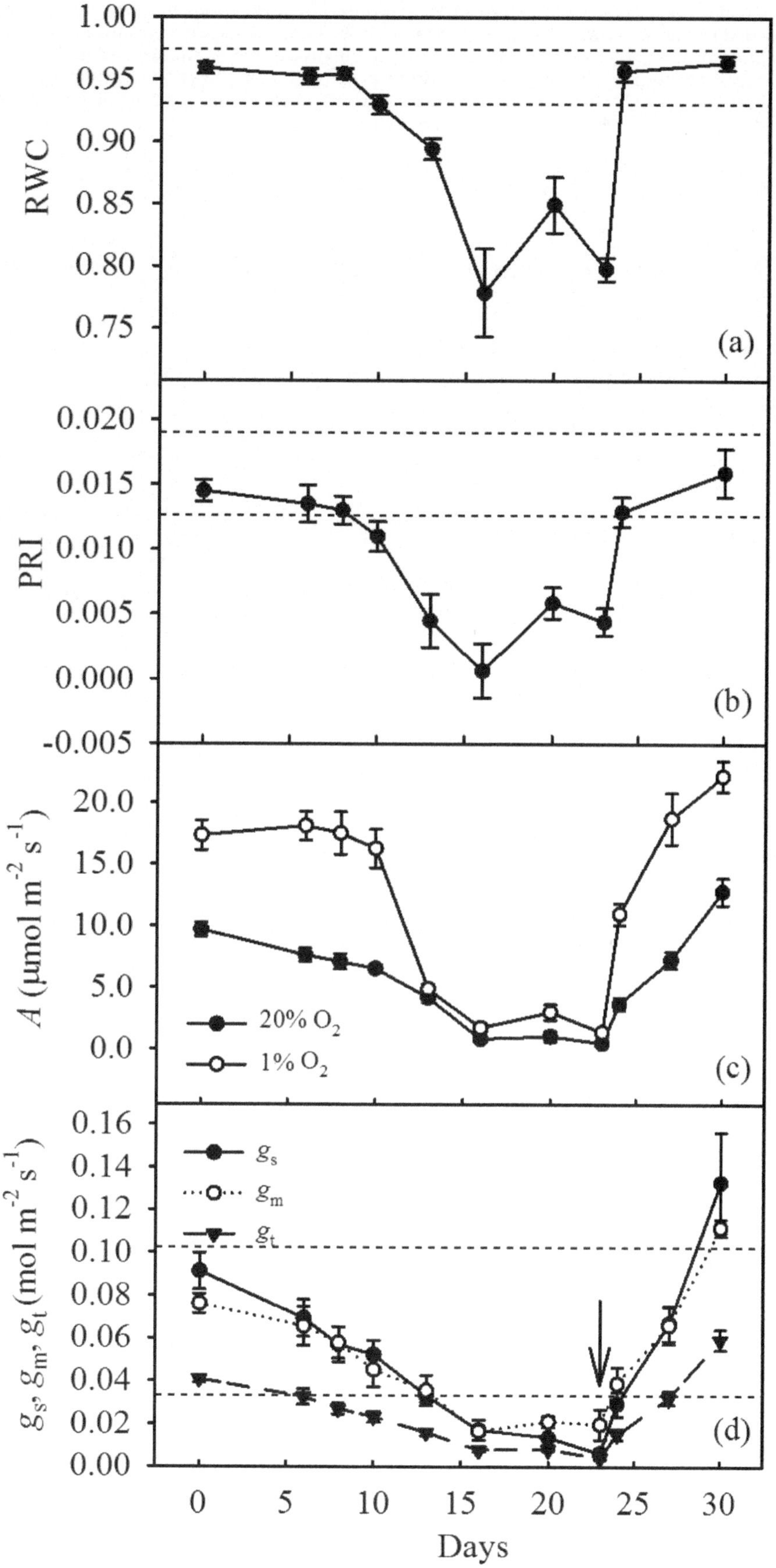
RWC
1.00
0.95
0.90
0.85
0.80
0.75
(a)
PRI
0.020
0.015
0.010
0.005
0.000
-0.005
(b)
A (µmol m-2 s-1)
20.0
15.0
10.0
5.0
0.0
20% O2
1% O2
(c)
gs, gm, gt (mol m-2 s-1)
0.16
0.14
0.12
0.10
0.08
0.06
0.04
0.02
0.00
gs
gm
gt
(d)
0
5
10
15
20
25
30
Days

Figure 1. Time courses of leaf (a) relative water content (RWC), dashed horizontal lines indicate the range of values (mean ± one standard deviation) recorded in the well-watered plants over the duration of the experiment; (b) photochemical reflectance index (PRI) (dashed horizontal lines as in 1a); (c) photosynthesis (*A*) (measured in both ambient air and in air with 1% [O_2] (*A*1%O2), and; (d) stomatal conductance (g_s), mesophyll conductance (g_m) and total conductance (g_t) of olive saplings grown during and after the drought cycle (days 1–23), dashed horizontal lines indicate the range of g_s values recorded in the well-watered plants. ↓ = end of the drying cycle. Data points are means of five plants (five leaves per plant) ±1 SEM.

(Fig. 2). There was a significant linear correlation between A and g_s ($r^2 = 0.966$, $P<0.001$), when measurements conducted on control and water-stressed leaves were pooled (Fig. 3a). Similar linear relationships were also found between A and g_m ($r^2 = 0.959$, $P<0.001$) (Fig. 3b) and A and g_t ($r^2 = 0.969$, $P<0.001$) (Fig. 3c). Furthermore, $A_{1\%O2}$ scaled significantly with leaf conductance parameters (Fig. 3d,e,f). In contrast to A, the responses of $A_{1\%O2}$ to g_s ($r^2 = 0.945$, $P<0.001$), g_m ($r^2 = 0.965$, $P<0.001$), and g_t ($r^2 = 0.955$, $P<0.001$) took the form of a saturating curve.

After pooling together of all measurements during the drought and recovery cycle, there were significant exponential relationships between the gas-exchange parameters A (Fig. 4a), g_s, g_m and g_t (Fig. 4b), when plotted against RWC. The pre-stress and recovery (i.e., high RWC) gas-exchange parameters showed higher variation than those recorded during water-stress (Fig.4a,b). Stomatal conductance increased more rapidly than g_m and g_t during the late water-stress relief period (i.e., at leaf RWC ranging from 0.9 to 0.96) (Fig. 1d, Fig. 4b). The relationship between PRI and RWC was also non-linear ($r^2 = 0.935$, $P<0.001$), with RWC varying from 0.77 to 0.96 during the study period (Fig. 4c). The PRI showed less variability than the gas exchange parameters as RWC recovered to pre-stress levels. The PRI was also significantly correlated with to A (Fig. 4d), g_m (Fig. 4e) and g_t (Fig. 4e).

The water content-related reflectance indices, RDI (Fig. 5a) and WI (Fig 5c), showed a similar time-course response to that of RWC (Fig. 1a); they decreased significantly as water-stress progressed and significantly increased following the relief of water-stress. In contrast, WCRI (Fig. 5b) increased significantly as water-stress became severe, and recovered quickly and completely upon re-watering to pot water capacity. The supplementary irrigation (Day 20) altered RDI significantly ($P<0.05$), but had a less apparent effect on both WI and WCRI. Overall, RDI (Fig. 5d), WCRI (Fig. 5e) and WI (Fig. 5f) showed significant linear relationships with RWC. However, WCRI differed from the other two indices by being inversely related to RWC. The RDI and WCRI were found to exhibit comparatively stronger correlations with RWC than WI.

The RDI showed significant non-linear relationships with g_s (Fig. 6a) and E (Fig. 6b); whereas, WI scaled linearly with g_s (Fig. 6c) and E (Fig. 6d). In contrast, there was an inverse curvilinear relationship between WCRI and both g_s (Fig. 6e) and E (Fig. 6f). It is noteworthy that RDI and WCRI were highly sensitive to both g_s and E under severe stress conditions ($g_s <$ 0.06 mol m^{-2} s^{-1}), whereas this sensitivity decreased sharply in mild-stress or optimal conditions ($g_s >$0.06 mol m^{-2} s^{-1}).

Carotenoid concentration (Fig. 7a) was significantly affected ($P<0.05$) by water-stress. Carotenoid concentration increased as water-stress progressed; resulting in a 34% increase when water-stress was most severe (Days 16 and 23), and then decreased significantly following both the supplementary irrigation (Day 20) and re-watering to pot water capacity (Day 24). Chlorophyll concentration was inversely associated to that of carotenoids (data not shown), and this led to a progressive decrease in the carotenoid to chlorophyll ratio as water-stress progressed, and to a correspondingly rapid increase in response to the supplementary irrigation and relief of water-stress (Fig. 7b). The time-course of structural independent pigment index (SIPI, i.e. a measure of carotenoid to chlorophyll *a* ratio) (Fig. 7c) mirrored that of the overall carotenoid to chlorophyll ratio. Finally, significant linear relationships to carotenoid concentration were found between both PRI (Fig. 8a) and SIPI (Fig. 8b).

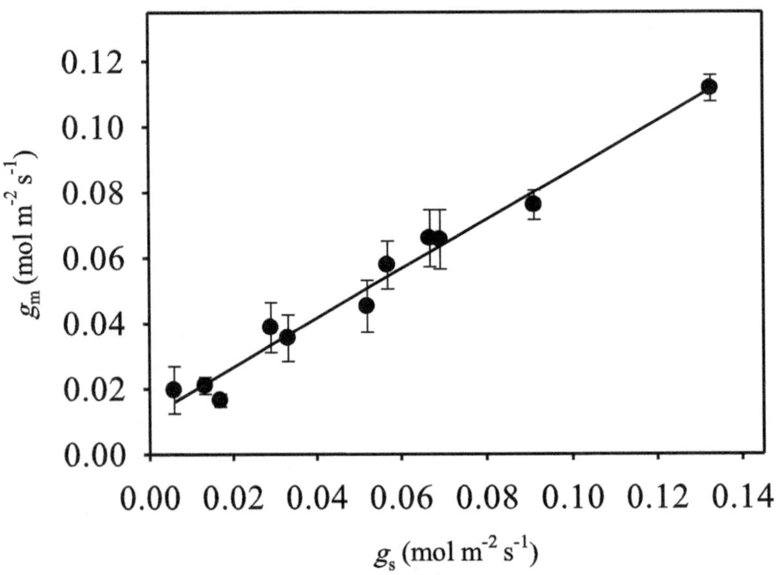

Figure 2. Linear relationships between mesophyll conductance (g_m) and stomatal conductance (g_s) ($r^2 = 0.978$, $P<0.001$) in olive saplings grown during and after the drought cycle.

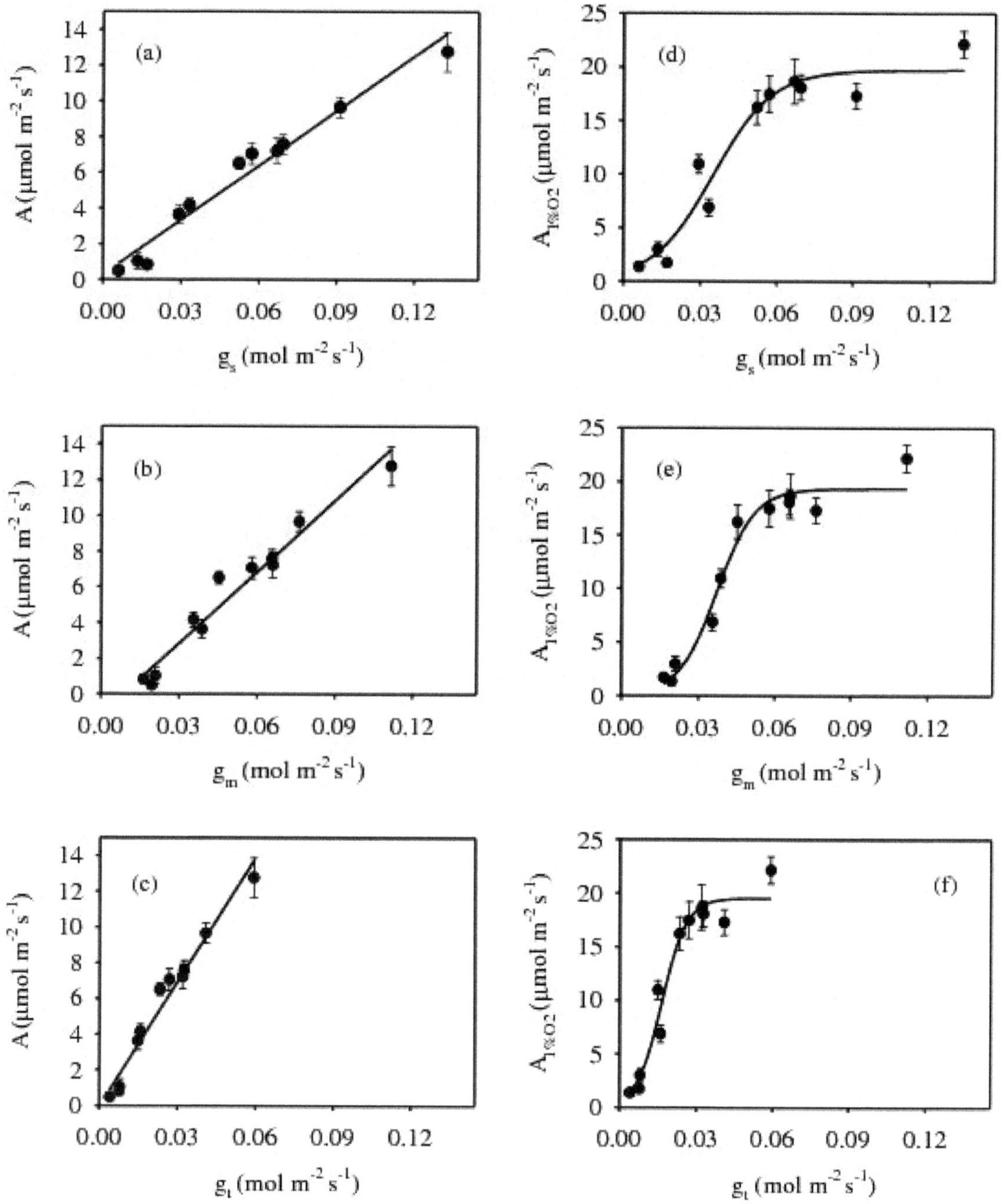

Figure 3. Relationships of photosynthesis (*A*), measured in ambient air (a, b, c) and in air with 1% [O_2] (*A*1%O2) (d, e, f), and (a, d) stomatal conductance (g_s), (b, e) mesophyll conductance (g_m), and (c, f) total conductance (g_t) in olive saplings grown during and after the drought cycle.

Discussion

Water shortage is a crucial problem limiting plant productivity across large areas of the globe. Changes in biogeophysical cycles (that is increased evaporative demand and water holding capacity of the air) associated with global warming are expected to further worsen the water crisis in arid and semi-arid zones, and to also increasingly affect temperate regions [27]. To mitigate the effects of drought on plant growth it essential to improve our knowledge of plant functional responses to soil drying. In this study, we assessed a cascade of changes in leaf water relations, pigment concentrations, photosynthetic CO_2 diffusional limitations in olive saplings subjected to water-stress and recovery by using spectral reflectance indices to monitor the water-status and photosynthetic function of olive trees.

The time-course in leaf water status, as expressed by RWC (Fig. 1a) and PRI (Fig. 1b), in response to water availability was mirrored by that of the photosynthetic parameters. Furthermore, as observed in other studies [2], the decline of both *A* (Fig. 4a) and leaf conductance parameters (Fig. 4b) was significantly correlated to RWC. We observed a gradual decrease of *A* (Fig. 1c) in water-stressed saplings during the early period of the drought treatment

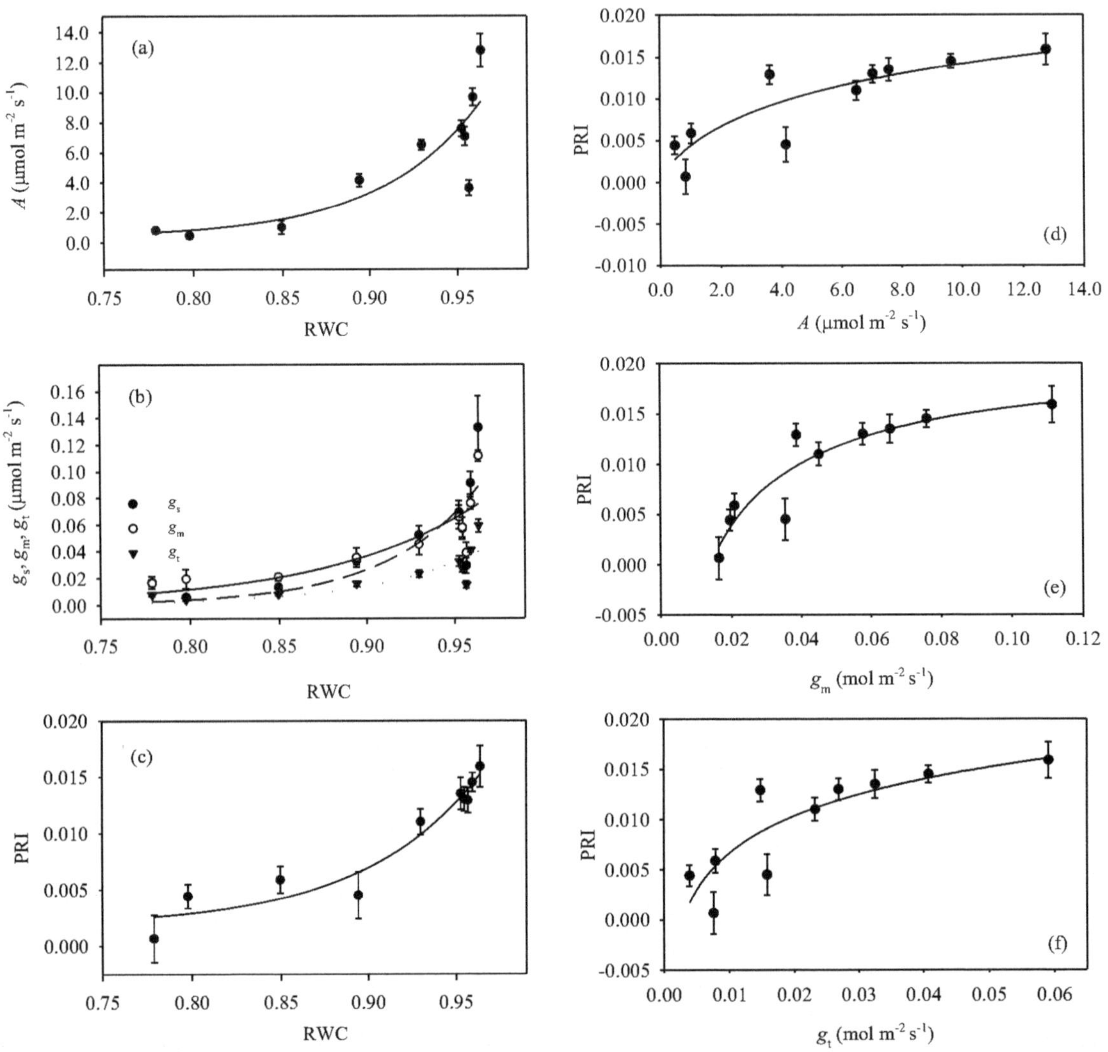

Figure 4. Relationships between relative water content (RWC) and (a) photosynthesis (*A*) ($r^2=0.742$, $P=0.008$), (b) stomatal conductance (g_s, —) ($r^2=0.667$, $P=0.005$), mesophyll conductance (g_m, ----) ($r^2=0.748$, $P=0.003$) and total conductance (g_t, ·····) ($r^2=0.653$, $P<0.02$), and (c) photochemical reflectance index (PRI) ($r^2=0.935$, $P<0.001$), and between PRI and (d) *A* ($r^2=0.762$, $P=0.001$, (e) g_m ($r^2=0.844$, $P<0.001$), and (f) g_t ($r^2=0.725$, $P=0.002$) in olive saplings grown during and after the drought cycle.

(Day 10). The reduction in conductance values of CO_2 diffusion (g_s, g_m and g_t) (Fig. 1d) followed similar patterns, suggesting that during the early phase of the drought stress *A* was limited by reduced C_c [5,28,29]. There are numerous studies suggesting that under water-stress, *A* is mostly limited by reduced C_i due to stomatal closure [5]. However, information about the contribution of g_m during water-stress is still lacking [28,29]; although new evidence indicates that g_m is linearly correlated to g_s (Fig. 2) [2,6], and an important component in the determination of water use efficiency [7]. When water-stress reached severe levels (Day 16), *A* was almost fully inhibited, and upon relief of the water-stress *A* gradually recovered to the pre-stress level in parallel with g_s, g_m and g_t. The *A* and g_s values observed in this study, in both well-watered and stressed conditions, are consistent with the findings of previous work performed on young pot-grown olive plants [11,30] and mature olive trees grown under field conditions [10,31,32,33]. Moriana et al., [31] and Marino et al., [10] observed that some degree of metabolic limitation to *A* occurred in olive trees exposed to severe summer drought. However, the rapid increase in *A* and its complete restoration to pre-stress levels after one week of stress-relief, in addition to the linear relationships between *A* and leaf conductance parameters (Fig. 3a,b,c), may indicate that *A* was likely mainly inhibited by impaired CO_2 uptake [34].

Photosynthesis measured under non-photorespiratory conditions ($A_{1\%O2}$) changed in parallel with *A* (Fig. 1c). Assuming that the difference between $A_{1\%O2}$ and *A* is an approximate measure for photorespiration, it is evident that this parameter was stimulated dramatically at the onset of water stress. This is may have been caused by reductions in C_c as a result of decreases in g_t (Fig. 1d). Reduced C_c may then lead to increased oxygenation of ribulose-1,5-bisphosphate (RuBP) by Rubisco, thus increasing photorespiration, not only relative to photosynthesis, but also in absolute terms [35,36]. For example, photorespiration in four grape varieties increased under moderate water-stress and helped maintain a relatively high photochemical efficiency of PSII and rapid recovery of *A* after re-watering [37]. However, as photorespiration depends directly on RuBP recycling in the Benson-Calvin cycle, which in turns depends on C_c, severe drought may also result in reduced rates of photorespiration [35]. This may explain the progressive reduction of $A_{1\%O2}$ under

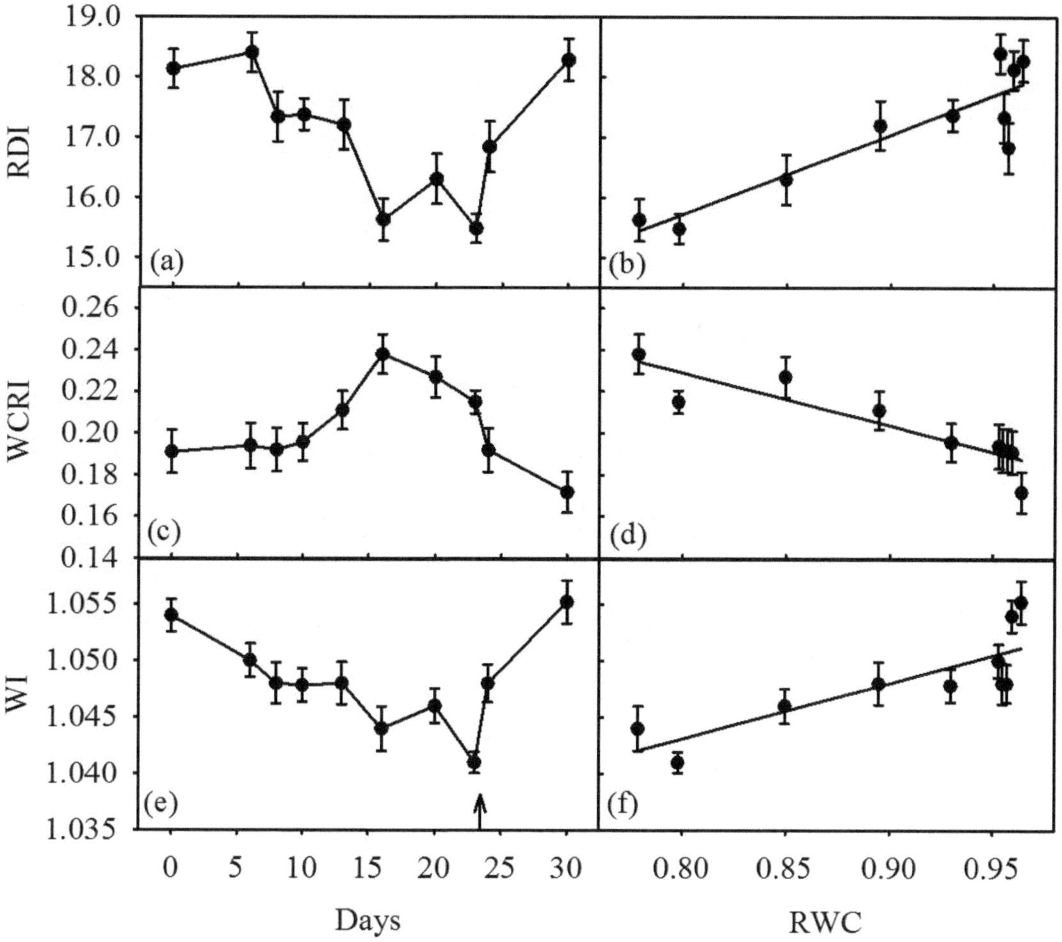

Figure 5. Time courses of (a) relative depth index (RDI), (b) water correlated reflectance index (WCRI) and (c) water index (WI), and relationships between relative water content (RWC) and (d) RDI (r^2=0.803, P<0.001), (e) WCRI (r^2=0.814, P<0.001) and (f) WI (r^2=0.675, P=0.004) in olive saplings grown during and after the drought cycle (days 1–23). ↑ = end of the drying cycle. Data points are means of five plants (five leaves per plant) per measurement ±1 SEM.

conditions of severe water-stress. Conversely, the increase in $A_{1\%O2}$ observed after the supplementary irrigation and relief of water-stress (Days 20 and 24, respectively), when $A_{1\%O2}$ was approximately two-fold higher than A, may likely be attributed to the corresponding increase in g_m and C_c. Furthermore, the saturating responses of $A_{1\%O2}$ to g_m (Fig. 3d), g_s (Fig. 3e), and g_t (Fig. 3f) imply that under non-photorespiratory conditions, conductance of CO_2 is not the major limitation to A. Thus, because broad-leaved sclerophyll plants, such as olive, possess inherently low leaf conductance, they may have a competitive advantage over mesophyllous vegetation in the future as atmospheric [CO_2] rises [38].

As CO_2 transport from the ambient air to the active site of Rubisco decreases in response to progressive water-stress, A and the quantum yield of photosystem II (PSII) become progressively inhibited. In turn, the proportion of the PPFD energy absorbed by the photosynthetic antenna and dissipated as heat increases, while that used for photochemistry declines. This photo-protective reaction is associated with the de-epoxidation state of xanthophylls [39]. PRI is based on a normalized difference of the 531 and 570 nm bands where xanthophyll pigments absorption occurs and, thus can be used as a robust proxy of radiation use efficiency [12] and photosynthetic activity [13]. The PRI, was dramatically affected by water availability (Fig. 1b), and exhibited close correlation to RWC (Fig. 4c) and A (Fig. 4d). The strong relationship between PRI and RWC (P<0.001) observed in this study is in keeping with previous studies showing that PRI is a good indicator of leaf water status [15]. Furthermore, PRI detected at tree canopy level, has been recently found to be correlated to A in field-grown mature olive plants under rainfed and well-watered conditions [10]. Significant correlations between PRI and A have also been found in *Arbutus unedo* [15], *Ceratonia siliqua* [40], *Solanum lycopersicum* [16] and *Quercus ilex* [15,17] subjected to water-stress in different environmental conditions. Photosynthesis is mainly limited by diffusive limitations to CO_2 uptake [7,28,29], unless water-stress becomes particularly severe to the point where there is virtually no water available to support transpiration [4,34]; consequently, the direct detection of diffusional limitations by spectral indices may be of great importance in the detection of plant function under stress conditions. In this study, significant relationships between g_t (Fig. 4f) and g_m (Fig. 4e) were established. Similar results were also observed in a study of *Q. ilex* [17] response to water-stress and recovery. These relationships, which were similar to that observed between A and PRI (Fig. 4d), can be viewed as supportive indications of the suitability of PRI in tracking photosynthetic activity.

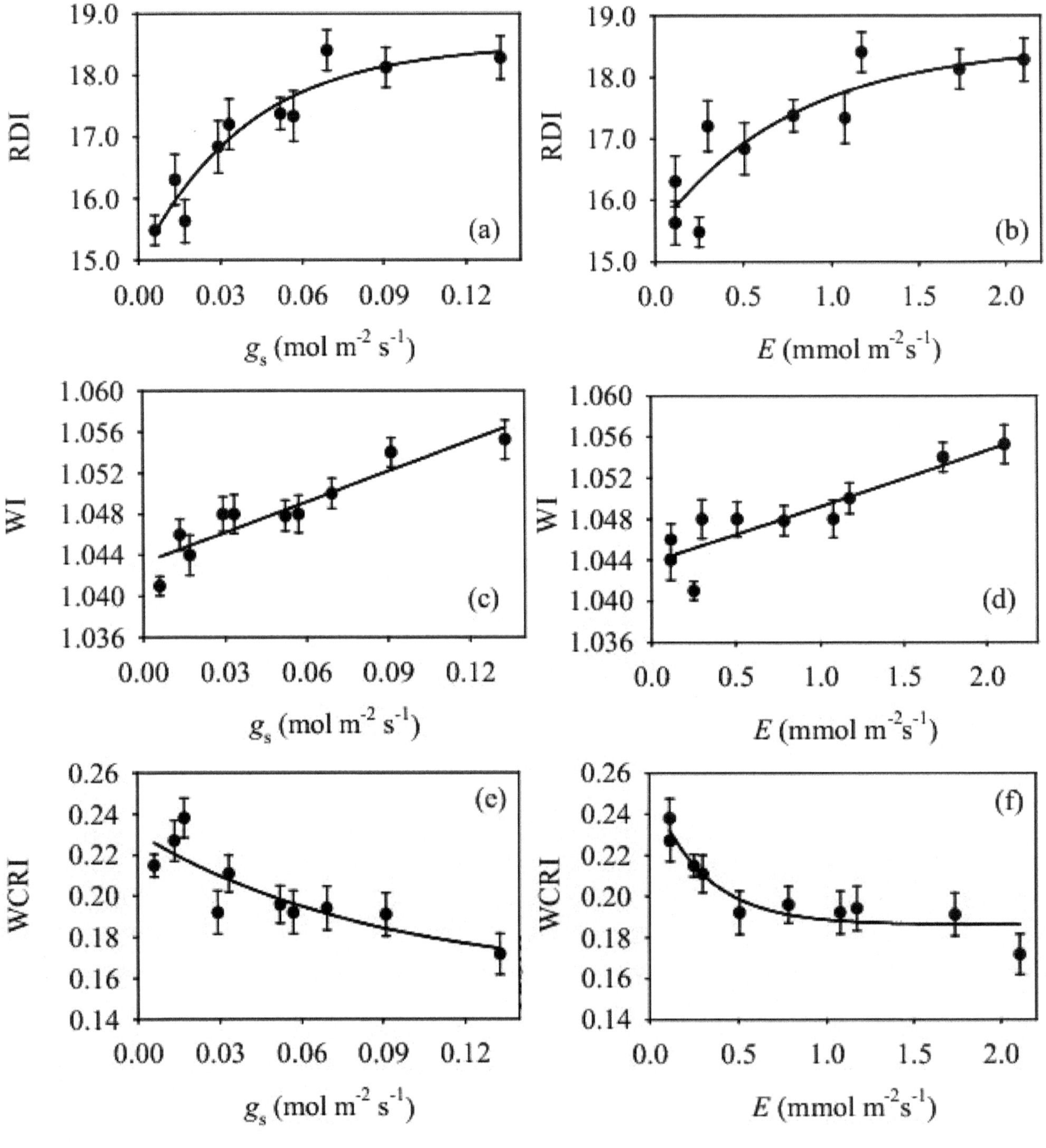

Figure 6. Relationships between relative depth index (RDI) and (a) leaf stomatal conductance (g_s) ($r^2 = 0.908$, $P<0.001$) and (b) leaf transpiration (E) ($r^2 = 0.796$, $P<0.01$), between WI and (c) g_s ($r^2 = 0.863$, $P<0.001$) and (d) E ($r^2 = 0.801$, $P<0.001$), and between WCRI and (e) g_s ($r^2 = 0.717$, $P<0.012$) and (f) E ($r^2 = 0.902$, $P<0.002$) in olive saplings grown during and after the drought cycle (days 1–23).

Variation in PRI can be affected by short-term changes in xanthophyll epoxidation state and over longer time-scales by shifts in the size of carotenoid and chlorophyll pools [41,42,43]. Similarly, time-course variations in SIPI, a pigment-based reflectance index [3], in response to water status (Fig. 7c) may reflect corresponding changes in the ratio of carotenoids to chlorophyll. In this study, carotenoid concentration (Fig. 7a) and the carotenoid to chlorophyll ratio (Fig. 7b) were responsive to water availability, as they significantly increased as the water-stress progressed and then decreased following supplementary irrigation and re-watering to pot water capacity. This may indicate the ability of olive trees to up- and down-regulate photo-protective thermal dissipation mechanisms by xanthophyll cycle carotenoids in response to water status [39]. The PRI (Fig. 8a) and SIPI (Fig. 8b) both inversely reflected changes in carotenoid concentrations. Interestingly, SIPI was less effective than PRI in gauging carotenoid concentration. Unlike PRI, which is estimated in the green-yellow region of the radiation spectrum (i.e., within the

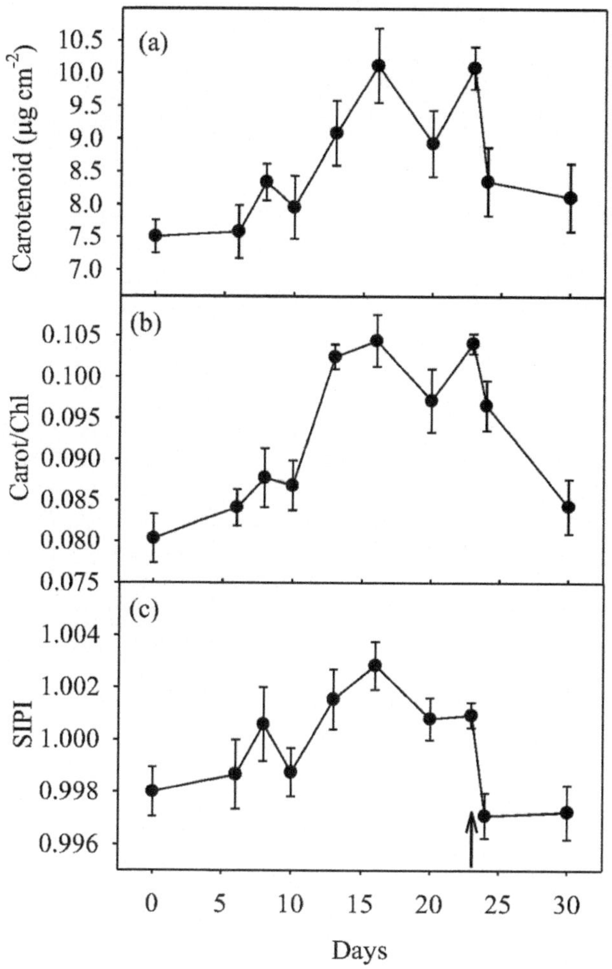

Figure 7. Time courses of (a) leaf carotenoid concentration (μg cm^{-2}), (b) carotenoid to chlorophyll ratio (Carot/Chl), and (c) structural independent pigment index (SIPI) in olive saplings grown during and after the drought cycle (days 1–23). ↑ = end of the drying cycle. Data points are means of five plants (five leaves per plant) per measurement ±1 SEM.

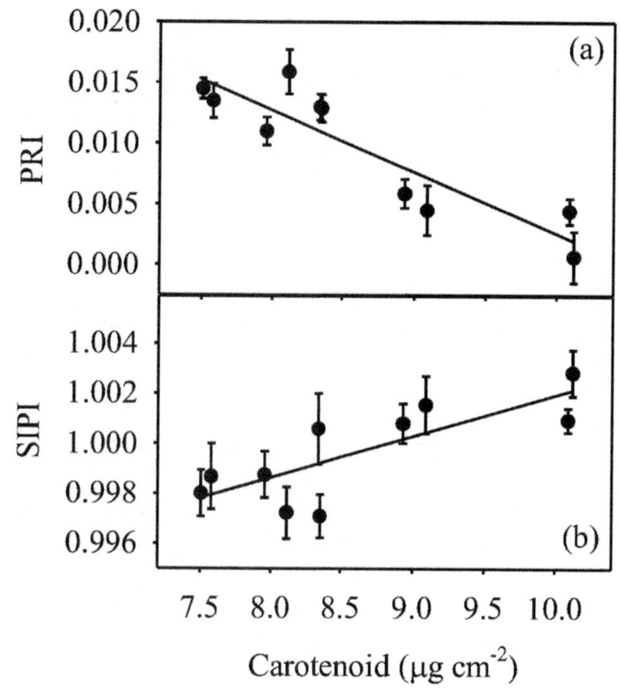

Figure 8. Linear relationships between carotenoid concentration (μg cm^{-2}) and (a) photochemical reflectance index (PRI) (r^2 = 0.813, P<0.001) and (b) structural independent pigment index (SIPI) (r^2 = 0.622, P<0.007) in olive saplings grown during and after the drought cycle.

carotenoid absorption spectra), SIPI is estimated using blue and red bands that are sensitive to both carotenoid and chlorophyll pigment changes. Therefore, the combined variations in both carotenoid and chlorophyll pools may have contributed to the higher variability of SIPI in comparison to that of PRI. However, taken together, these results reinforce the growing body of literature showing that PRI [10,13,41], and to a lesser extent SIPI, [10,44] are correlated to photosynthetic parameters.

The three water-related reflectance indices, RDI, WI and WCRI, were significantly affected during the water-stress cycle and re-watering. The RDI (Fig. 5a) and WI (Fig. 5c) mirrored the time course of RWC (Fig.1a); whereas WCRI (Fig. 5b) increased as water-stress progressed and decreased upon rewatering. These differential responses resulted in significant direct positive relationships between RWC and both RDI (Fig. 5d) and WI (Fig. 5f), and a significant negative relationship between RWC and WCRI (Fig. 5e). As all these indices are formulated by several water absorption wavelengths, these linear relationships indicate the consistent sensitivity of RDI, WI and WCRI to mild or severe drought conditions. Furthermore, these three spectral indices were also significantly correlated to g_s and E (Fig. 6), due to the close dependence of stomata behavior and transpiration upon leaf water status. The RDI (Fig. 6a,b) and WCRI (Fig. 6e,f) had curvilinear relationships with g_s and E, and were consequently more sensitive as E and g_s decreased, and less sensitive under mild-stress and optimal conditions. Conversely, WI was linearly related to both g_s (Fig. 6c) and E (Fig. 6d), and as a consequence was more sensitive to these two physiological parameters than RDI and WCRI. This differential sensitivity may result from the different wavelengths used in calculating RDI, WI and WCRI, as middle infrared wavelengths are used to estimate RDI and WCRI while near-infrared wavelengths are used to calculated WI. It is noteworthy that Marino et al., [10] demonstrated in a recent study of mature olive trees that WI, estimated at tree canopy level, was the most accurate predictive index of plant water status and whole-plant transpiration. Similarly, Serrano et al., [9] found that WI was a good indicator of vineyard water status. We therefore suggest that WI is the most reliable tool in predicting parameters of plant water-status.

Conclusion

Drought is a complex syndrome affecting several leaf biophysical and biochemical properties that subsequently influence leaf reflectance spectra. In the present study, we examined the reliability of water absorption and pigment based spectral indices to assess changes in leaf water status, pigment concentration and photosynthetic traits during water-stress and recovery. In general, PRI was very sensitive to leaf water status, pigments concentrations and A. It is noteworthy that PRI also significantly tracked leaf CO_2 conductance. These relationships may reflect the tight control that diffusional limitations exert on A under water-stress and then following relief from water-stress. The WI was the most reliable index in the prediction of plant water-status. Finally, A was likely mainly inhibited by leaf CO_2 conductance parameters during water-stress; whereas under non-photorespiratory conditions, CO_2 conductance was not the major limitation to photosynthesis. In consideration of a predicted future elevated atmospheric [CO_2], it is possible that sclerophyll plants will likely out-perform mesophyllous vegetation.

Acknowledgments

The comments of two anonymous reviewers significantly improved this manuscript. The authors are grateful to the Scientific Cooperation Agreements between CNR and the Chinese Academy of Forestry; CNR and the Centre National pour la Recherche Scientifique et Technique of Morocco.

Author Contributions

Conceived and designed the experiments: PS SW TT SL MC. Performed the experiments: PS SW TT. Analyzed the data: MC MH. Contributed reagents/materials/analysis tools: MC TT. Wrote the paper: PS MC MH.

References

1. Fereres E, Orgaz F, Gonzalez-Dugo V (2011) Reflections on food security under water scarcity. Journal of Experimental Botany 62: 4079–4086
2. Centritto M, Lauteri M, Monteverdi MC, Serraj R (2009) Leaf gas exchange, carbon isotope discrimination, and grain yield in contrasting rice genotypes subjected to water deficits during the reproductive stage. Journal of Experimental Botany 60: 2325–2339
3. Peñuelas J, Filella L (1998) Visible and near-infrared reflectance techniques for diagnosing plant physiological status. Trends in Plant Science 3, 151–156
4. Centritto M, Brilli F, Fodale R, Loreto F (2011a) Different sensitivity of isoprene emission, respiration, and photosynthesis to high growth temperature coupled with drought stress in black poplar (*Populus nigra*). Tree Physiology 31: 275–286
5. Lawlor DW, Cornic G (2002) Photosynthetic carbon assimilation and associated metabolism in relation to water deficits in higher plants. Plant, Cell and Environment 25: 275–294
6. Centritto M, Loreto F, Chartzoulakis K (2003) The use of low [CO_2] to estimate diffusional and non-diffusional limitations of photosynthetic capacity of salt-stressed olive saplings. Plant Cell and Environment 26: 585–594
7. Flexas J, Niinemets U, Gallé A, Barbour MM, Centritto M, et al. (2013) Diffusional conductances to CO_2 as a target for increasing photosynthesis and photosynthetic water-use efficiency. Photosynthesis Research, 117: 45–59
8. Gutierrez M, Reynolds MP, Klatt AR (2010) Association of water spectral indices with plant and soil water relations in contrasting wheat genotypes. Journal of Experimental Botany 61: 3291–3303
9. Serrano L, González-Flor C, Gorchs G (2010) Assessing vineyard water status using the reflectance based water index. Agriculture, Ecosystems and Environment 139: 490–499
10. Marino G, Pallozzi E, Cocozza C, Tognetti R, Giovannelli A, et al. (2014) Assessing gas exchange, sap flow and water relations using tree canopy spectral reflectance indices in irrigated and rainfed *Olea europaea* L. Environmental and Experimental Botany 99: 43–52
11. Sun P, Grignetti A, Liu S, Casacchia R, Salvatori R, et al. (2008) Associated changes in physiological parameters and spectral reflectance indices in olive (*Olea europaea* L.) leaves in response to different levels of water stress. International Journal of Remote Sensing 29: 1725–1743
12. Gamon JA, Peñuelas J, Field CB (1992) A narrow-waveband spectral index that tracks diurnal changes in photosynthetic efficiency. Remote Sensing of Environment 41: 35–44
13. Garbulsky MF, Peñuelas J, Gamon JA, Inoue Y, Filella I (2011) The photochemical reflectance index (PRI) and the remote sensing of leaf, canopy and ecosystem radiation use efficiencies: a review and meta-analysis. Remote Sensing of Environment 115: 281–297
14. Dobrowski SZ, Pushnik JC, Zarco-Tejada PJ, Ustin SL (2005) Simple reflectance indices track heat and water stress-induced changes in steady-state chlorophyll fluorescence at the canopy scale. Remote Sensing of Environment 97: 403–414
15. Ripullone F, Rivelli AR, Baraldi R, Guarini R, Guerrieri R, et al. (2011) Effectiveness of the photochemical reflectance index to track photosynthetic activity over a range of forest tree species and plant water statuses. Functional Plant Biology 38: 177–186
16. Sarlikioti V, Driever SM, Marcelis LFM (2010) Photochemical reflectance index as a mean of monitoring early water stress. Annals of Applied Biology 157: 81–89
17. Tsonev T, Wahbi S, Sun P, Sorrentino G, Centritto M (2014) Gas exchange, water relations and their relationships with photochemical reflectance index in *Quercus ilex* plants during water stress and recovery. International Journal of Agriculture and Biology 16: 335–341
18. Suárez L, Zarco-Tejada PJ, Sepulcre-Cantó G, Pérez-Priego O, Miller JR, et al. (2008) Assessing canopy PRI for water stress detection with diurnal airborne imagery. Remote Sensing of Environment 112: 560–575
19. Arp WJ (1991) Effects of source-sink relations on photosynthetic acclimation to elevated CO_2. Plant, Cell & Environment 14: 869–875
20. Ismail A, Hall A, Bray E (1994) Drought and pot size effects on transpiration efficiency and carbon isotope discrimination of cowpea accessions and hybrids. Functional Plant Biology 21: 23–35.
21. Magnani F, Centritto M, Grace J (1996) Measurement of apoplasmic and cell-to-cell components of root hydraulic conductance by a pressure-clamp technique. Planta 199: 296–306
22. Centritto M, Lee HSJ, Jarvis PG (1999) Long-term effects of elevated carbon dioxide concentration and provenance on four clones of Sitka spruce (*Picea sitchensis*) I. Plant growth, allocation and ontogeny. Tree Physiology 19: 799–806
23. Flexas J, Díaz-Espejo A, Berry J, Cifre J, Galmés J, et al. (2007) Analysis of leakage in IRGA's leaf chambers of open gas exchange systems: quantification and its effects in photosynthesis parameterization. Journal of Experimental Botany 58: 1533–1543
24. Rodeghiero M, Niinemets U, Cescatti A (2007) Major diffusion leaks of clamp-on leaf cuvettes still unaccounted: how erroneous are the estimates of Farquhar, et al. model parameters? Plant, Cell and Environment 30: 1006–1022
25. Wellburn AR (1994) The spectral determination of Chlorophylls *a* and *b*, as well as total carotenoids, using various solvents with spectrophotometers of different resolution. Journal of Plant Physiology 144: 307–313
26. Peñuelas J, Filella I, Gamon JA (1995) Assessment of photosynthetic radiation-use efficiency with spectral reflectance. New Phytologist 131: 291–296
27. Dai A (2010) Drought under global warming: a review. WIREs Climate Change 2: 45–65
28. Flexas J, Ribas-Carbó M, Diaz-Espejo A, Galmés J, Medrano H (2008) Mesophyll conductance to CO_2: current knowledge and future prospects. Plant, Cell and Environment 31: 602–621
29. Loreto F, Centritto M (2008) Leaf carbon assimilation in a water-limited world. Plant Biosystems 142: 154–161
30. Aganchich B, Wahbi S, Loreto F, Centritto M (2009) Partial root zone drying: regulation of photosynthetic limitations and antioxidant enzymatic activities in young olive (*Olea europaea*) saplings. Tree Physiology 29: 685–696
31. Moriana A, Villalobos FJ, Fereres E (2002) Stomatal and photosynthetic responses of olive (*Olea europaea* L.) leaves to water deficits. Plant, Cell and Environment 25: 395–405
32. Centritto M, Wahbi S, Serraj R, Chaves MM (2005) Effects of partial rootzone drying (PRD) on adult olive tree (*Olea europaea*) in field conditions under arid climate. II. Photosynthetic responses. Agriculture, Ecosystems and Environment 106: 303–311
33. Diaz-Espejo A, Nicolas E, Fernández JE (2007) Seasonal evolution of diffusional limitations and photosynthetic capacity in olive under drought. Plant, Cell and Environment 30: 922–933
34. Brilli F, Tsonev T, Mahmood T, Velikova V, Loreto F, et al. (2013) Ultradian variation of isoprene emission, photosynthesis, mesophyll conductance and optimum temperature sensitivity for isoprene emission in water-stressed *Eucalyptus citriodora* saplings. Journal of Experimental Botany 64: 519–528
35. Wingler A, Quick WP, Bungard RA, Bailey KJ, Lea PJ, et al. (1999) The role of photorespiration during drought stress: an analysis utilizing barley mutants with reduced activities of photorespiratory enzymes. Plant, Cell and Environment 22: 361–373
36. Sage RF (2013) Photorespiratory compensation: a driver for biological diversity. Plant Biology 15: 624–638
37. Guan XQ, Zhao SJ, Li DQ, Shu HR (2004) Photoprotective function of photorespiration in several grapevine cultivars under drought stress. Photosynthetica 42: 31–36
38. Centritto M, Tognetti R, Leitgeb E, Střelcová K, Cohen S (2011). Above ground processes - Anticipating climate change influences. In: Bredemeier M, Cohen S, Godbold DL, Lode E, Pichler V, Schleppi P, eds. Forest Management and the Water Cycle: An Ecosystem-Based Approach. Ecological Studies 212, Springer, pp 31–64
39. Demmig-Adams B, Adams WWIII (2000) Harvesting sunlight safely. Nature 403: 371–374

40. Osório J, Osório ML, Romano A (2012) Reflectance indices as nondestructive indicators of the physiological status of *Ceratonia siliqua* seedlings under varying moisture and temperature regimes. Functional Plant Biology, 39: 588–597
41. Sims DA, Gamon JA (2002) Relationships between leaf pigment content and spectral reflectance across a wide range of species, leaf structures and developmental stages. Remote Sensing of Environment 81: 337–354
42. Guo J, Trotter CM (2004) Estimating photosynthetic light-use efficiency using the photochemical reflectance index: variations among species. Functional Plant Biology 31: 255–265
43. Filella I, Porcar-Castell A, Munne-Bosch S, Back J, Garbulsky MF, et al. (2009) PRI assessment of long-term changes in carotenoids/chlorophyllratio and short-term changes in de-epoxidation state of the xanthophyllcycle. International Journal of Remote Sensing 30: 4443–4455
44. Ollinger SV (2011) Sources of variability in canopy reflectance and the convergent properties of plants. New Phytologist 189: 375–394.

7

Modeling the Impact of Soil and Water Conservation on Surface and Ground Water Based on the SCS and Visual Modflow

Hong Wang[1,3], Jian-en Gao[1,2,4,5*], Shao-long Zhang[4], Meng-jie Zhang[5], Xing-hua Li[4]

1 Institute of Soil and Water Conservation, Chinese Academy of Sciences and Ministry of Water Resources, Yangling, Shaanxi Province, China, **2** Institute of Soil and Water Conservation, Northwest A&F University, Yangling, Shaanxi Province, China, **3** University of Chinese Academy of Sciences, Beijing, China, **4** College of Water Resources and Architectural Engineering, Northwest A&F University, Yangling, Shaanxi Province, China, **5** College of Natural Resources and Environment, Northwest A&F University, Yangling, Shaanxi Province, China

Abstract

Soil and water conservation measures can impact hydrological cycle, but quantitative analysis of this impact is still difficult in a watershed scale. To assess the effect quantitatively, a three-dimensional finite-difference groundwater flow model (MODFLOW) with a surface runoff model–the Soil Conservation Service (SCS) were calibrated and applied based on the artificial rainfall experiments. Then, three soil and water conservation scenarios were simulated on the sand-box model to assess the effect of bare slope changing to grass land and straw mulching on water volume, hydraulic head, runoff process of groundwater and surface water. Under the 120 mm rainfall, 60 mm/h rainfall intensity, 5 m^2 area, 3° slope conditions, the comparative results indicated that the trend was decrease in surface runoff and increase in subsurface runoff coincided with the land-use converted from bare slope to grass land and straw mulching. The simulated mean surface runoff modulus was 3.64×10^{-2} $m^3/m^2/h$ in the bare slope scenario, while the observed values were 1.54×10^{-2} $m^3/m^2/h$ and 0.12×10^{-2} $m^3/m^2/h$ in the lawn and straw mulching scenarios respectively. Compared to the bare slope, the benefits of surface water reduction were 57.8% and 92.4% correspondingly. At the end of simulation period (T = 396 min), the simulated mean groundwater runoff modulus was 2.82×10^{-2} $m^3/m^2/h$ in the bare slope scenario, while the observed volumes were 3.46×10^{-2} $m^3/m^2/h$ and 4.91×10^{-2} $m^3/m^2/h$ in the lawn and straw mulching scenarios respectively. So the benefits of groundwater increase were 22.7% and 60.4% correspondingly. It was concluded that the soil and water conservation played an important role in weakening the surface runoff and strengthening the underground runoff. Meanwhile the quantitative analysis using a modeling approach could provide a thought for the study in a watershed scale to help decision-makers manage water resources.

Editor: Jose Luis Balcazar, Catalan Institute for Water Research (ICRA), Spain

Funding: This paper was supported by National Technology Support Project (2011BAD31B05), by Natural Science Foundation of China (41371276), by the Subject of National Science and Technology Major Project (2009ZX07212-002-003-02) and by Knowledge Innovation Project of Institute of Soil and Water Conservation, CAS & MWR (Soil and Water Conservation Project) (A315021304). The funders had no role in study design, data collection and analysis, decision to publish, or preparation of the manuscript.

Competing Interests: The authors have declared that no competing interests exist.

* E-mail: gaojianen@126.com

Introduction

In recent years, with more and more officials and professionals paying attention to the problems such as soil erosion and water loss, the contradiction between water supply and demand, whether soil and water conservation measures on upper regions would have effect on the amount of water resources of the lower reaches is a scientific issue to be discussed and probed into urgently [1]. For many years, the work of soil and water conservation as same as the Yellow River management, not only obtains great achievement, but also brings some problems [2]. The biggest achievement was no more than soil erosion control in the watershed, and thereby reducing the amount of sediment flowed into the river. Took the Yellow River basin for example, changes in land use and vegetation in the Loess Plateau had a decisive impact on sediment transport of Yellow River [3], even about 60% sediment yield since 10 ka BP occurred during the last 1040 years of the period of estrepement in Loess Plateau. The sediment monitoring at Sanmenxia hydrological station of Yellow River showed the annual average sediment discharge was 1040 million tons during 1950–2010, 589 million tons during 1987–2010, while it was only 198 and 351 million tons in 2009 and 2010 separately [4].

The reduction in the quantity of the sediment discharge by soil and water conservation is no doubt pleasing, but it also brings a new eco-environmental problem for its surface water reduction function. Some studies found that land use/land cover (LULC) change could alter hydrological cycles by affecting ecosystem evaportranspiration, soil infiltration capacity, surface and subsurface flow regimes [5–8]. Just as the study of the relation between control in Loess Plateau and no-flow in the Yellow River showed that the influence of water reduction on no-flow in the lower Yellow River for the comprehensive control changed the condition underlying surface and influenced the water cycle was nonnegligible [9]. Results of soil and water conservation method showed

Figure 1. The experimental flume.

that the mean flood reduced 0.5456 billion tons in He-Long section of middle Yellow River, Jinghe, Beiluohe and Weihe watershed from 1970 to 1996 [10]. The average runoff reduction benefits of soil and water conservation measures for Chabagou, Dalihe and Wudinghe basins in the 1970s were 14.47%, 20.22% and 20.78% respectively [11]. Jain and Mishra et al [12] evaluated the suitability to particular land use, soil type and combination a quantitative evaluation of the existing Soil Conservation Service Curve Number (SCS-CN) model using a large set of rainfall-runoff data from small to large watersheds of the U.S.A. Shi and Yi et al [13] evaluated the effect of land use/cover change on surface runoff by SCS model in Shenzhen region, China. Guo and Wang et al [14] calculated the volume of surface runoff during 5 rainfalls on 5 different kinds of land use types in sloping runoff plots by SCS model.

Soil and water conservation could reduce surface runoff and it is bound to affect groundwater recharge, which is the entry into the saturated zone of water made available at the table surface, together with the associated flow away from the water table within the saturated zone [15]. But less direct study has previously been undertaken to assess this effect quantitatively. It had been shown that groundwater recharge was closely related to land-use types very much. The impact of land-use on distributed groundwater recharge and discharge in the western Jilin, China, using MODFLOW, WetSpass, the Seepage packages, and ArcGIS showed that forest vegetation had the highest recharge, followed by agricultural farmlands and the recharge generally decreased when vegetated forests deteriorate to be other landforms (bush, grassland or bare-land) [16]. A study by Cho and Barone using MODFLOW indicated that subsurface flow regimes could be negatively affected by urbanization due to increased withdrawal and reduced recharge [17]. A number of studies showed the effects of changes in forest cover on groundwater recharge [18–22]. These studies showed a range in the reduction of groundwater recharge beneath trees, from 15% to 90%, compared to that under grass [18]. Thus the impacts of land-use on the atmospheric components of the hydrologic cycle (regional and global) are increasingly being recognized, though those on the subsurface components of the hydrologic cycle, particularly groundwater recharge are not equally known [23].

Therefore special attention should be given to the effect of soil and water conservation on the hydrologic cycle, especially recharge and discharge. Both monitoring and modeling approaches are used for conjunctive investigation of surface water and groundwater. The monitoring approach is expensive and time demanding yet measuring actual changes in stream and groundwater levels over time may lead to more direct estimates of the impact of land development on both surface and subsurface flows [17]. Through the model coupling approach, interaction between surface water and groundwater has become a trend [24,25]. Linkage between MODFLOW and existing surface models such as the SCS method to consider surface water and groundwater interactions could validate the outputs of the model.

The overall goal of this study is to assess the impact of soil and water conservation on the surface and subsurface flow by mathematical models combined with artificial rainfall simulation experiments and gives a quantitative analysis thought simultaneously.

Materials and Methods

SCS Curve Number Method

The runoff equation of the U.S. Soil Conservation Service, commonly called the curve number method, for estimating runoff from rainfall, came into common use in the mid - 1950s [26]. Although the SCS method for runoff estimation has changed little since the 1960s, its popularity has been maintained over the years [27]. The SCS-CN equation, in the typical form [28], is given as:

Table 1. Mechanical composition of experimental materials (%).

Types	1~0.5	0.5~0.25	0.25~0.05	0.05~0.01	0.01~0.005	0.005~0.001	<0.001
Lou soil	0.07	0.65	5.86	49.04	12.18	13.72	18.48
Mixed soil	1.83	24.34	30.91	25.29	4.82	5.53	7.27

$$\bar{Q}_{surf}=\frac{(P-I_a)^2}{(P-I_a+S)} \text{ for } P>I_a;\ \bar{Q}_{surf}=0 \text{ for } P\leq I_a \quad (1)$$

Where $\bar{Q}_{surf}$ is the surface runoff (mm), in this study it is the average runoff depth per hour; P is the precipitation (mm), in this study it is the precipitation per hour; S is the amount of water storage available in the soil profile or the maximum storage (mm), and I_a is the initial abstraction (mm). To reduce the numbers of variables, the empirical relationship $I_a = 0.2S$ was adopted, which then gives the most familiar form of the runoff equation [29]:

$$\bar{Q}_{surf}=\frac{(P-0.2S)^2}{(P+0.8S)} \text{ for } P>0.2S;\ \bar{Q}_{surf}=0 \text{ for } P\leq 0.2S \quad (2)$$

Table 2. Calibrated parameter value and the relative error (RE) (%).

Period	Calibration period	Verification period			
Rainfall Intensity (mm/h)	75	45	90	105	120
RE (%)	0.015	3.41	0.40	2.20	1.77

Because the range of the S value is too large to obtain a suitable value, the dimensionless parameter, curve number (CN), was introduced into this formula. The parameter is related to CN by the relationship:

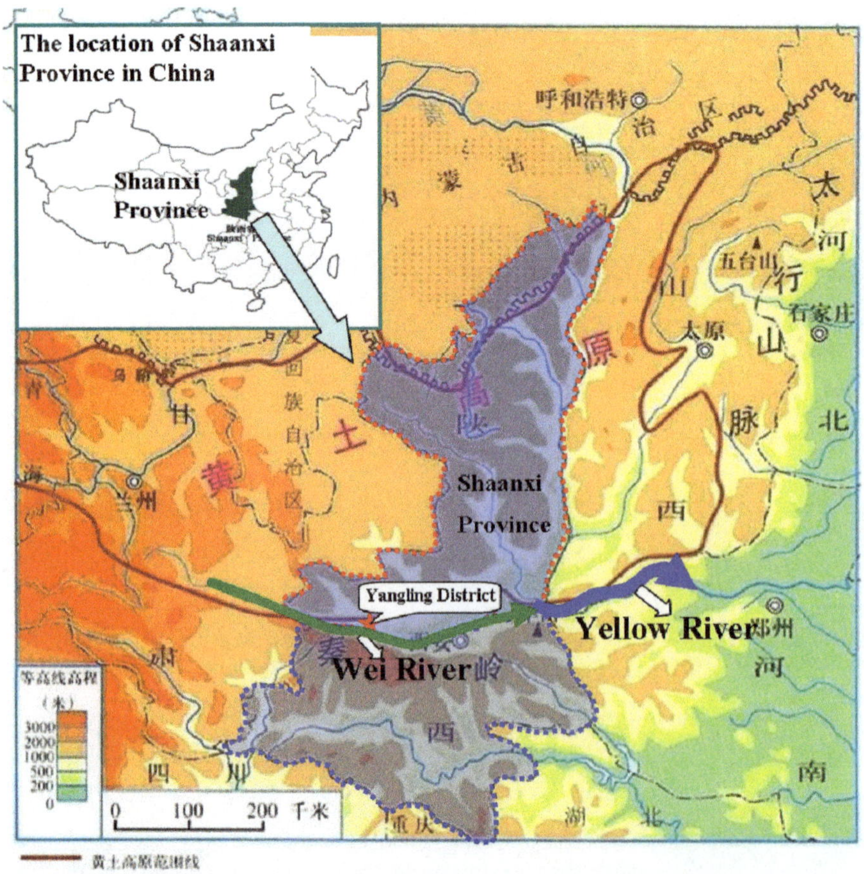

Figure 2. The schematic map of Wei River and Yangling District location.

Table 3. The main calibrated parameters of the MODFLOW model.

Parameters	Kx1(m/s)	Kz1(m/s)	Kx2(m/s)	Kz2(m/s)	Kx3(m/s)	Kz3(m/s)	Sy	Ss
Minimum	1.16×10^{-8}	1.16×10^{-9}	5.79×10^{-7}	5.79×10^{-8}	5.79×10^{-5}	5.79×10^{-6}	0.05	1×10^{-5}
Maximum	5.79×10^{-6}	5.79×10^{-6}	5.79×10^{-5}	5.79×10^{-5}	5.21×10^{-4}	5.21×10^{-4}	0.5	1×10^{-3}
Initial Value	2.90×10^{-6}	2.90×10^{-7}	3.18×10^{-5}	3.18×10^{-6}	2.2×10^{-4}	2.2×10^{-5}	0.2	1×10^{-4}
Calibrated Value	1.12×10^{-7}	2.52×10^{-9}	1.50×10^{-6}	7.47×10^{-7}	3.1×10^{-4}	7.60×10^{-5}	0.36	1.385×10^{-4}

$$CN=\frac{25400}{(S+254)};\ S=25.4\left(\frac{1000}{CN}-10\right) \tag{3}$$

The CN is a comprehensive parameter which related to initial soil moisture, slope, vegetation, soil type and land use status etc [30]. It appeared that the curve numbers were used as a proxy for the retention parameter S in order to scale the curves to a convenient range between zero and 100 [29].

Meanwhile, the runoff amount could be calculated combined the rainfall duration by the relationship:

$$Q_{surf}=\bar{Q}_{surf}\times h \tag{4}$$

Where Q_{surf} is the total runoff of rainfall (120 mm) (mm); h is the rainfall duration (h).

Relative error (RE) was selected as the statistical evaluation index of SCS and it was defined by the equation below:

$$RE=\frac{Q_{cal}-Q_{obs}}{Q_{cal}} \tag{5}$$

Where Q_{cal} was the calculated runoff by SCS model (mm); Q_{obs} was the observed runoff obtained from the rainfall experiment (mm).

Visual Modflow Model

MODFLOW (Modular Three-dimensional Finite-difference Ground-water Flow Model)) is a modular finite-difference ground-water flow model published by the U.S. Geological Survey, which can simulate ground-water flow in a three-dimensional heterogeneous and anisotropic medium. Using the finite-difference method, the domain in which flow is to be simulated is divided into a rectilinear mesh of rows, columns, and layers [31–33]. Visual MODFLOW is also a three-dimensional groundwater flow and contaminant transport modeling application that integrates MODFLOW, MODPATH, MT3DMS, WinPEST, Zone Budget, and so on. Applications include well head capture zone delineation, pumping well optimization, aquifer storage and recovery, groundwater remediation design, simulating natural attenuation, and saltwater intrusion [32].

The three-dimensional movement of ground water of constant density through porous earth material may be described by the partial-differential equation [31]:

$$\frac{\partial}{\partial x}\left(Kxx-\frac{h}{x}\right)+\frac{\partial}{\partial y}\left(Kyy-\frac{h}{y}\right)+\frac{\partial}{\partial z}\left(Kzz-\frac{h}{z}\right)+W = Ss\frac{\partial h}{\partial t} \tag{6}$$

where Kxx, Kyy, and Kzz are values of hydraulic conductivity along the x, y, and z coordinate axes, which are assumed to be parallel to the major axes of hydraulic conductivity (m/s); h is the potentiometric head (m); W is a volumetric flux per unit volume representing sources and/or sinks of water, with $W<0$ for flow out of the ground water system, and $W>0$ for flow into the system; Ss is the specific storage of the porous material (1/m); and t is time (min).

The calibration statistics can be reported in the result plots and the statistical evaluation indexes, including Calibration Residual (CM), Residual Mean (RM), Absolute Residual Mean (ARM), Standard Error of the Estimate (SEE), Root Mean Squared (RMS), Normalized Root Mean Squared (NRMS) and Correlation Coefficient (Cor) [32]. Based on the principle of the modeling and simulated rainfall experiments, a transient three-dimensional groundwater flow model was built using Visual MODFLOW 4.1. The model domain was divided into a 50×265 array of 4×10^{-4} square meter cells uniformly. The model consisted of three individual layers with gradient of 3° and the thicknesses of layers were respectively 0.5, 1 and 98.5 cm from top to bottom. The transient flow simulation was selected as the flow type and the time unit for all parameters was minute. The edge of the model domain was modeled as a no-flow boundary. Actual quantities of groundwater abstraction from drain pipe were treated as flux boundary condition in form of pumping well. Groundwater recharge from rainfall was modeled as recharge package in MODFLOW. The initial ground water level was 0.39 m high and paralleled to the bottom of the model. The initial value and the range of hydraulic conductivity and storage coefficient values at different layers were assigned to each active grid cell by interpolation of discrete property data derived from water releasing test analysis and regional geology data.

Simulated Rainfall Experiments

Experimental conditions and equipment. The simulated rainfall experiments were carried out in the Rainfall Simulation Hall of the State Key Laboratory of Soil Erosion and Dryland Farming on the Loess Plateau during the period of June to October of the year 2012. The simulated rainfall system, with automatic simulation device of under sprinkler, could ensure the kinetic energy of simulated precipitation close to the natural rainfall for the mean fall-height of 18 meters. And the calibration tests showed that rainfall uniformity was greater than 85%. The experiments were conducted in the sand-box model [34,35] (figure 1), 5.3 m×1 m×1 m at the Rainfall Simulation Hall. The

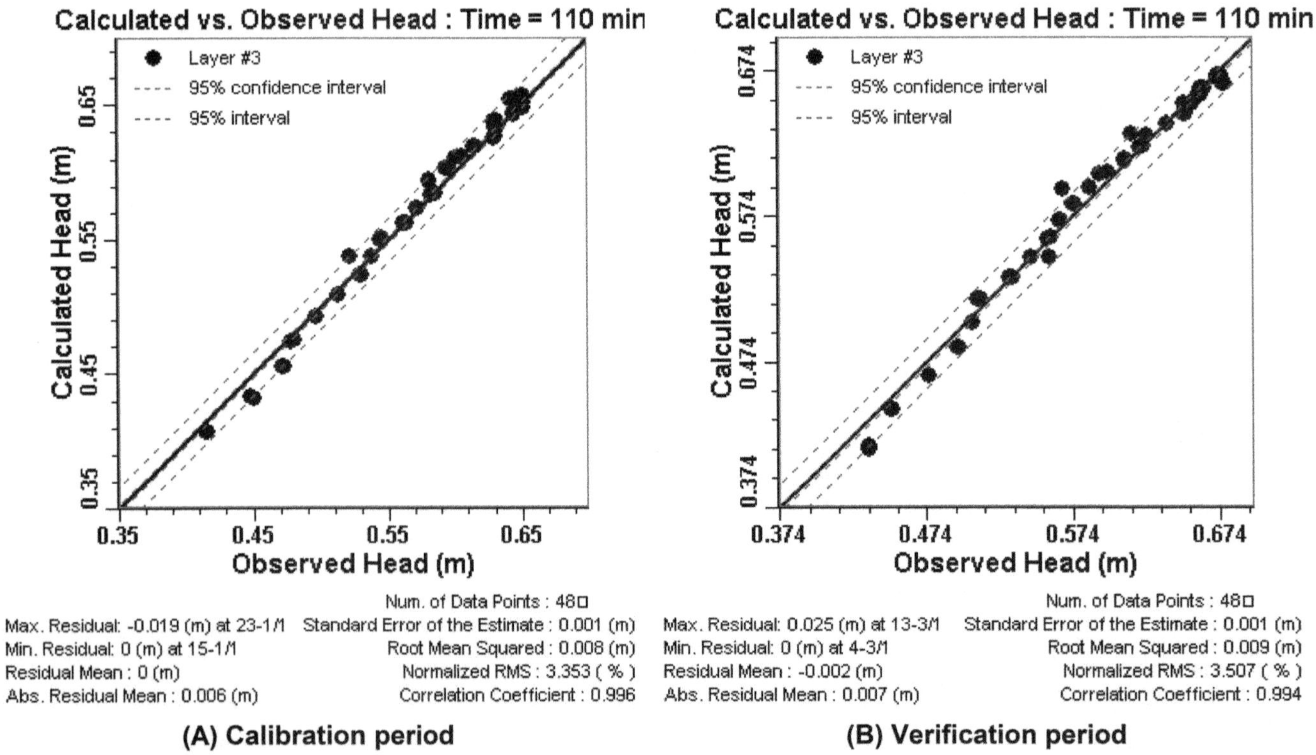

Figure 3. The scatter graph of calculated vs. observed values.

slope of the model could be adjusted manually from 0° to 35°. There were two water tanks, 0.15 m×1 m×1 m in front and back of the model, for the regulation of the groundwater level. Above the front side of the water tank there was one surface water groove and there was one drainage pipe of groundwater in the 0.39 m high at the front side of the water tank.And one hundred and twenty 0.2 m×0.2 m sets of piezometric tubes of level observation were installed in left side of the sand-box model. And two neutron probe access tubes down to a depth of 0.9 m in experimental flume for soil moisture control.

The experimental flume was fixed at an angle of 3° in this study. Three water and soil conservation measures (bare slope, straw mulching and grass land) were considered and the soil surface condition should be roughen to keep the beginning condition consistent every time before experiment. The precipitation for a control to be equal was 120 mm. Six gradient rainfall intensities (45~120 mm/h) with uniform rainfall conditions were simulated for 160 to 60 min correspondingly. The initial ground water level was parallel to the bed of the flume and the distance between two lines was 0.39 m.

Experimental materials and monitoring methods. The test materials included riversand and Lou soil. The riversand samples were dug from the middle and lower reaches of the Wei River bank in Yangling District and the Lou soil was also collected from Yangling District, Shaanxi Province, China. Figure 2 is the schematic map of Wei River and Yangling District location. The mechanical composition of soil particles was shown in table 1. The samples were air-dried for about ten days and sieved through a series of corresponding magnitude sieves. Then the test materials were packed into the flume layer by layer. The experimental flume consists of three individual layers and the thickness was respectively 0.5, 1 and 98.5 cm from top to bottom. The medium sand (0.25 mm ~ 0.5 mm) was paved in the third 98.5 cm deep layer and the soil bulk density was 1.4~1.5 g/cm^3. In the second layer the riversand particle size was less than 0.25 mm and the average soil bulk density was about 1.6 g/cm^3. On the top layer, the soil used was the composite sandy loam, composed of riversand (<0.25 mm) and Lou soil, the weight ratio of this composite soil was about 2:5 and the average soil bulk density was also about 1.6 g/cm^3. The setting of water and soil conservation measures as follows: for the straw mulching, the quantity of arid straw mulching was designed as 0.4 kg/m^2 uniformly each experiment with coverage of 85~90%; for the grassland, the grass was grew on the soil in the flume and its species was ophiopogon japonicus with coverage of 65%~70% and the planting structure was, 10 cm row spacing×5 cm plant spacing×8.2 cm average plant height.

The main monitoring items measured during the experiments were: the surface runoff amount, the surface runoff in the process, underground runoff and groundwater level etc. Surface runoff was collected by collecting buckets at the beginning of runoff yield. The surface runoff in the process was collected for 30 seconds at 5-min intervals during experiments and the runoff amount was collected during the whole process. After groundwater level adjustment, recorded the initial level first and the dynamic changes every 10 minutes with the changes begin for about 4

Table 4. Error indexes of calculated vs. observed values.

Evaluation Indexes	RM/m	ARM/m	SEE/m	RMS/m	NRMS/%	Cor
Calibration	0.0002	0.00727	0.0012	0.0097	4.146	0.996
Verification	0.000375	0.00843	0.00119	0.00981	4.259	0.994

Table 5. The benefits of water reduction in field controlled and uncontrolled small watersheds %.

Water System	Watershed Name	Area (km^2)	Condition	Representative Type Area	Year	Benefits δ(%)
Kuye River	Mengjiagou	2.03	Controlled	Earth-rocky Mountainous Area	1959~1961	42.64
	Yangyagou	1.88	Uncontrolled			
Yuxi River	Qingcaogou	0.373	Controlled	Half sandstorm area	1959~1960	88.19
	Wangjiagou	0.434	Uncontrolled			
	Tiaogou	0.7677	Controlled		1959~1961	16.51
	Lijiagou	0.693	Uncontrolled			
Wuding River	Jiuyuangou	70.1	Controlled	Loess Hilly and Gully Region	1959~1969	16.88
	Peijiamaogou	41.2	Uncontrolled			
	Xiangtagou	0.454	Controlled		1958~1961	23.70
	Tuanyuangou	0.491	Uncontrolled			
	Wangmaozhuanggou	5.967	Controlled		1962~1963	43.70
	Lijiazhaigou	5.45	Uncontrolled			
Yanhe River	Dabiangou	3.7	Controlled		1963~1967	39.63
	Xiaobiangou	3.925	Uncontrolled			
Luohe River	Sigou	4.37	Controlled	gully region of loess plateau	1959~1961	57.98
	Nangou	5.11	Uncontrolled			
Juhe River	Guanzhuanggou	3.39	Controlled		1959~1961	57.98
	Yuanguzhuanggou	2.82	Uncontrolled			
Luohe River	Xingshugou	0.522	Controlled	Loess Hilly and Gully Region	1958~1961	83.62
	Beilougou	0.334	Uncontrolled			
Jinghe River	Fengyugou	1.176	Controlled		1958~1960	14.71
	Wangjiagou	0.874	Uncontrolled			
Chanhe River	Yaojiagou	7.815	Controlled	Terrace region of loess plateau	1960~1961	64.99
	Dicungou	4.4	Uncontrolled			
Average						44.09

hours. Meanwhile, underground runoff was collected by measuring cylinder per 2000 ml uninterruptedly for about 8 hours. Then according to the condition of the underground runoff, the monitoring interval was lengthened gradually until the water level dropped to the initial control water level. Soil moisture (neutron probe method) measurements were taken immediately before and after precipitation to ensure replicability of the initial soil surface conditions. And water temperature was also recorded during each simulation.

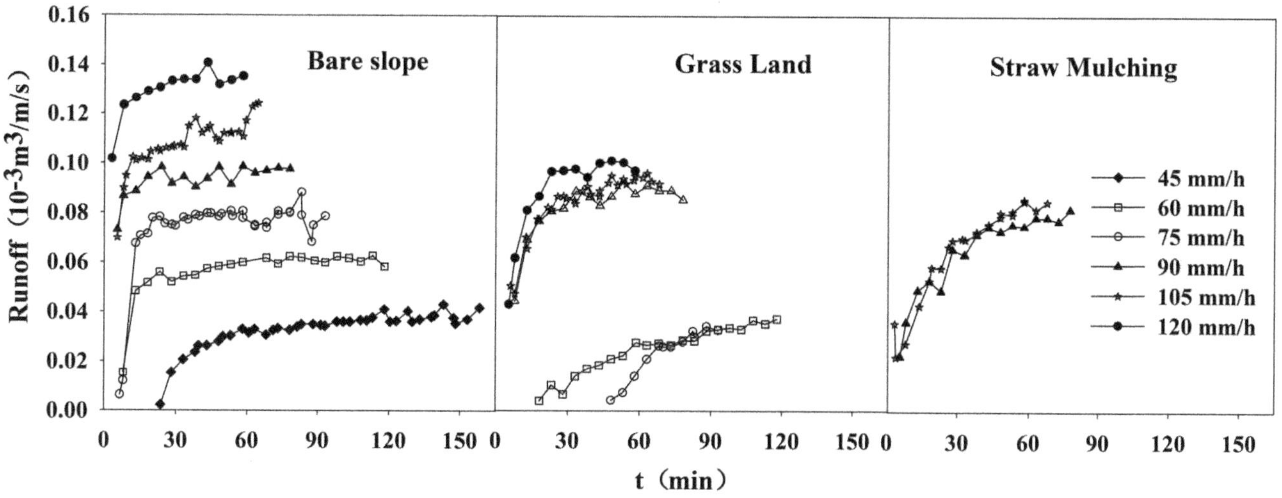

Figure 4. The time-series graphs of the surface runoff under three simulation scenarios.

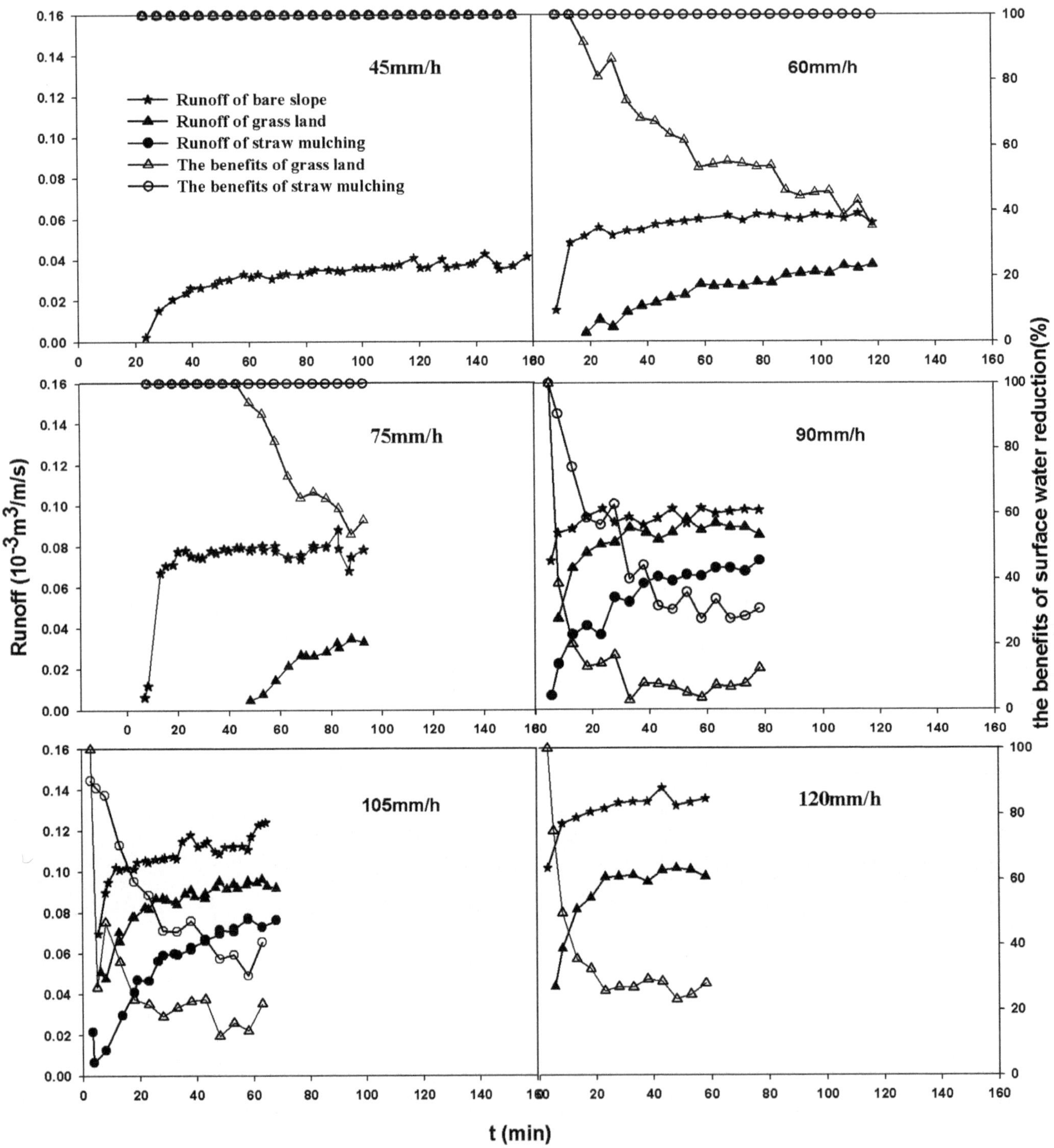

Figure 5. The time-series graphs of the surface runoff and the benefits of surface water reduction.

Ethics Statement

No specific permissions were required for these sampling activities because the location is not privately-owned or protected in any way and the field activities did not involve endangered or protected species.

The Benefits of Water Quantity Change for Soil and Water Conservation Measures

The quantity of surface water and groundwater could be changed by adoption of conservation management practices. Quantitative analysis of this change could be computed by the relationship:

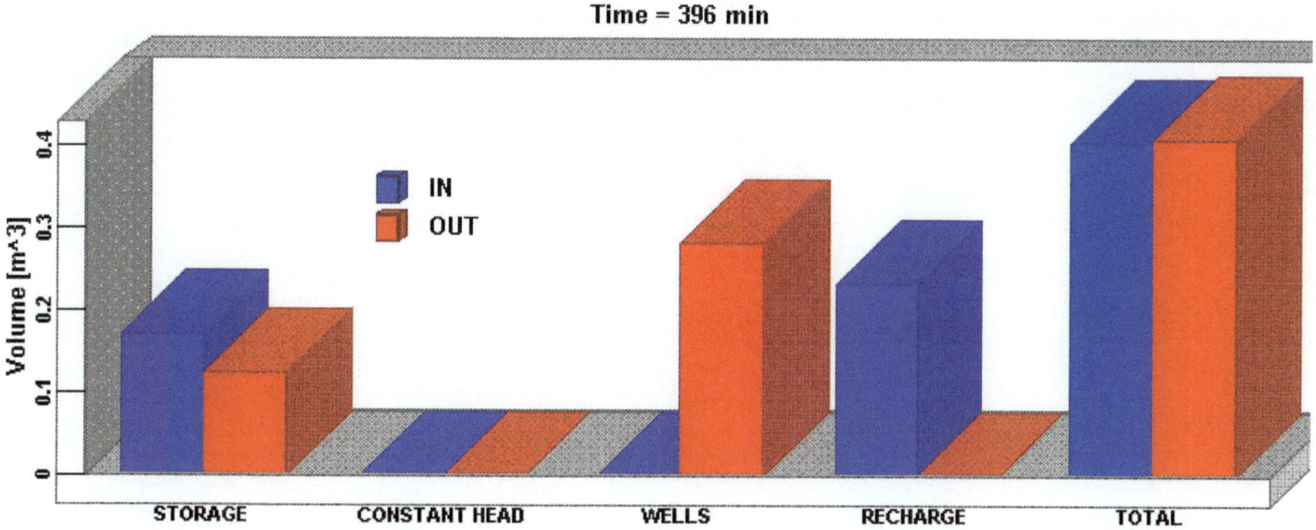

Figure 6. The flow mass balance graph.

$$\delta = \frac{Q_c - Q_{uc}}{Q_{uc}} \times 100\% \tag{7}$$

Where δ is the benefit of water change contributed by soil and water conservation measures, %; Q_c is the runoff modulus in the scenario controlled by soil and water conservation; Q_{uc} is the runoff modulus in the bare slope scenario.

Results and Discussions

Model Calibration and Validation

Model calibration is the process whereby selected model input parameters are adjusted within reasonable limits to produce simulation results that best match the known or measured values [32]. It is the most critical process in building a model, because the quality of the calibration and validation inevitably determines the reliability of any conclusions and recommendations made using the simulation results.

Calibration and Validation of SCS Model

Where concurrent rainfall and runoff data are available, an 'optimal' curve number can be found by calibrating the curve number to the data [29]. Based on the rainfall runoff data obtained from simulated rainfall experiment for the rainfall intensity 75 mm/h of bare slope scenario, the SCS model was calibrated. Observed $\bar{Q}_{surf}$ and P were 50.01 mm and 75 mm respectively. According to eq. (2) and (3), S and CN values were calculated successively. The calibrated CN value was 90.33. Then based on precipitation data from rainfall intensity 45, 90, 105, 120 mm/h of bare slope scenarios respectively, the SCS model were verified by comparing the estimated runoff with in situ measured data.

Based on the statistical analysis of model results, the relative error (RE, eq. 5) in the estimated runoff ranged from 0.40 to 3.41% (table 2) during the verification period respectively. The calibration and verification statistics showed the model did a relatively good job for surface water predicting. Hence, the SCS model was available for the predicting in this study. So surface runoff of bare slope scenario could be simulated by the verified

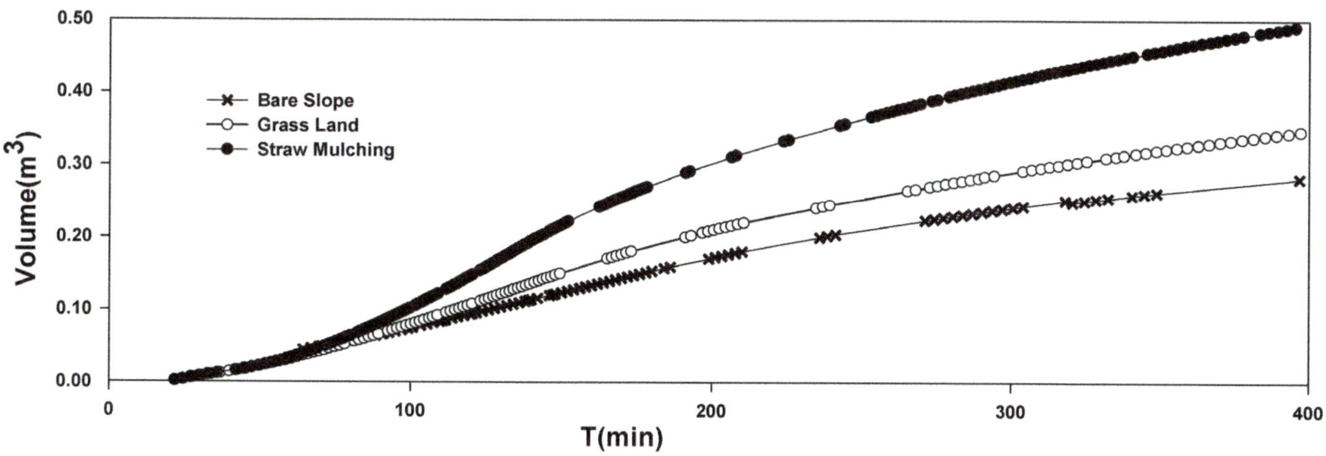

Figure 7. The time-series graphs of accumulated underground water volume under three simulation scenarios.

Table 6. Differences of calculated heads of bare slope vs. observed values of grass land/straw mulching.

Scenario	Average change(cm)	CM_{max}/cm	RM/cm	ARM/cm	RMS/cm	NRMS/%
Grass Land	3.91	−11.8	−3.6	4.1	4.5	14.1
Straw Mulching	7.63	−20.5	−7.3	7.4	7.8	23.4

SCS model and the calculated result was compared with the corresponding observed runoff of lawn and straw mulching experiments to study the impact of soil and water conservation on surface water.

Calibration and Validation of Visual MODFLOW Model

The main parameters of the model were: hydraulic conductivity (Kxx, Kyy, and Kzz), storage (specific storage (Ss), specific yield (Sy)). The initial values and the range of these parameters at different layers were provided by the water releasing test analysis combined with the regional geology data. These data were imported into the constructed MODFLOW model as the initial parameters and adjusted during the process of model calibration. Then based on the simulated rainfall experiment of the bare slope, which intensity was also 75 mm/h and rainfall was 120 mm, a uncertainty analysis was carried out after calibration to quantify the uncertainty in the calibrated model caused by uncertainty in the estimates of aquifer parameters, recharge boundary conditions, initial head conditions and river-aquifer interactions [18]. The uncertainty analysis showed that horizontal conductivity in the third layer (Kx3), specific yield (Sy) were the most sensitive parameters in this model. The main parameters calibrated were shown in table 3. Then to verify the model used another simulated rainfall experiment data of the bare slope, which intensity was 45 mm/h and rainfall was also 120 mm.

The scatter graph of calculated vs. observed values was the default calibration graph. This graph represents a snap-shot in time of the comparison between the values calculated by the model (Y-axis), and the values observed or measured in the field (X-axis) [32]. For example, the scatter graphs at t = 110 min of the calibration period and the verification period (figure 3) showed that most of the data points intersect the 45 degree line on the graph where X = Y. These represented an ideal calibration that simulated hydraulic heads were consistent with the observed heads. The calibration statistics were reported in the footer of the calibration plots window when the calculated vs. observed scatter graph were displayed [32]. Based on these statistical analysis of model results, error indexes including the residual mean (RM), the absolute residual mean (ARM), the standard error of the estimate (SEE), the root mean squared error (RMS), the normalized root mean squared (NRMS) and the correlation coefficient (Cor) for groundwater levels at the location of 48 observation boreholes during the calibration period and the verification period respectively were shown in table 4.

Based upon comparisons the scatter graphs and the calibration statistics between the observed and simulated hydraulic head, we

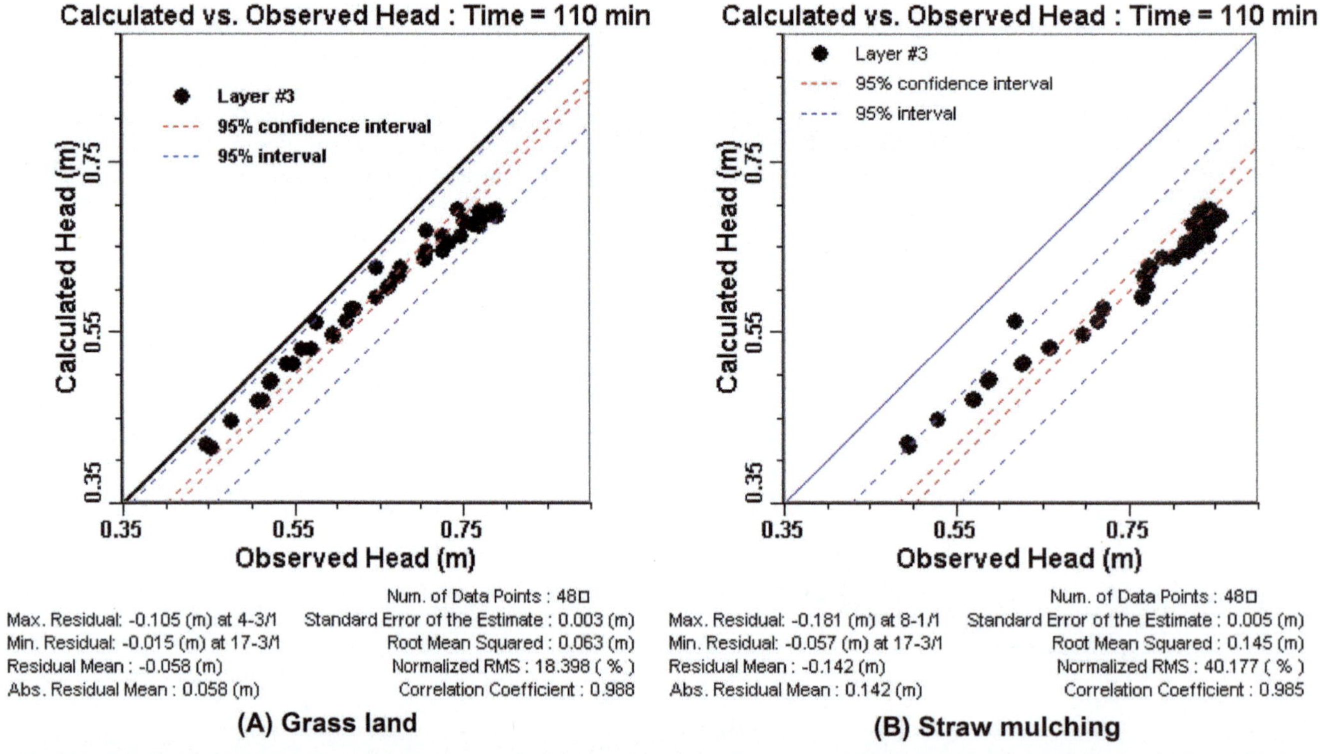

Figure 8. The scatter graph of calculated vs. observed values.

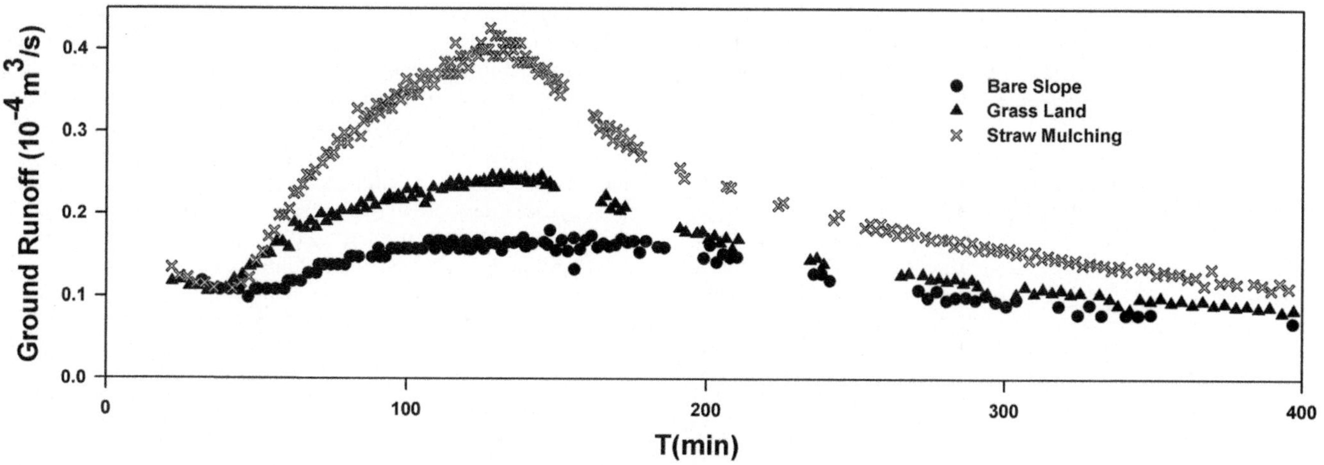

Figure 9. The groundwater runoff of bare slope vs. the discharge of lawn and straw mulching.

could see the groundwater levels simulated by the calibrated groundwater flow model were generally consistent with the physical system represented.

The Impact of Soil and Water Conservation on Surface and Ground Water

Then the calibrated and verified SCS and MODFLOW models were used to simulate rainfall intensity 60 mm/h of bare slope scenario to research the hydrologic impact of soil and water conservation. First, surface runoff amount was simulated by the SCS model and was compared with the corresponding results of lawn and straw mulching experiments to study the impact for the surface water. Second, the calibrated groundwater transient flow model with the input levels obtained from corresponding lawn and straw mulching experiments was used to assess the impact of land change on groundwater.

The Impact of Soil and Water Conservation on Surface Water

The surface runoff modulus amount calculated by SCS model was 0.364 m^3 in bare slope scenario, while the surface runoff observed from the rainfall experiments were 0.154 m^3 and 0.012 m^3 in the lawn and straw mulching scenarios respectively. Compared to the bare slope scenario, the benefits of surface water reduction by these two measures were 57.8% and 92.4% correspondingly. These results indicated that the soil and water conservation played an important role in the benefits of water reduction. These were consistent with the field observation results of the small watersheds. For example, by comparing the 12 groups (24) of controlled and uncontrolled small watersheds (table 5) in Shaanxi Province, China [36], we could see the benefits of surface water reduction by the comprehensive control of soil and water conservation varied from 14.71% to 88.19%, while the corresponding average value was 44.09%.

Not only the amount of surface runoff, but also the runoff process was important for studying the hydrologic impact of soil and water conservation. In this research, the SCS model could reliably estimate runoff amount, but it could not predict runoff process. So the influence for the process of surface runoff was analyzed using simulated rainfall experiments. The time-series graphs (figure 4) were used to evaluate and compare temporal trends in the runoff under three simulation scenarios. The process of runoff would appear to increase first and then stabilize gradually with the passage of rainfall time under three simulation scenarios. This result was consistent with the research that rainfall infiltration rate was decreased first and then stabilize gradually with the passage of rainfall time [37].

Compared with the bare slope scenario, the reduced range of runoff was different in the lawn scenario and in the straw mulching scenario. When rainfall intensity increased from 75 mm/h to 90 mm/h, the latter average runoff was 7.3 times larger than that of the former in the lawn scenario and it started runoff in the straw mulch scenario. During the whole runoff process of the experiments, the benefits of surface water reduction by the grassland varied from 24.3% to 100% and its average value was 58.9%, while the range was 55.3% to 100% and the average value was 84.2% in the straw mulching scenario (figure 5).

The Impact of Soil and Water Conservation on Ground Water

As mentioned above, the calibrated and verified MODFLOW model was then applied to simulating the impact of soil and water conservation scenarios on groundwater recharge and level. The mass balance which is one of the key indicators of a successful simulation [32] was analyzed first. The flow mass balance graph (figure 6) showed the volume of water entering and leaving the system through the flow boundary conditions, and from aquifer storage at the end of simulation period. The total volume flow into the entire system was 0.4059 m^3 which included 0.1719 m^3 storage and 0.2345 m^3 recharge. The total volume flow out-of the model was 0.4070 m^3 which consisted of 0.1252 m^3 storage and 0.2818 m^3 well. The mass balance error for the simulation inflow and outflow was 1.47% $<2\%$. The results of the simulation may generally be considered acceptable, provided the model was also calibrated [32].

Figure 7 compared the changes in accumulated underground runoff with scenarios converted from bare slope to grass land and straw mulching. As a result, the cumulative curve of straw mulching was the highest, the grassland took second place, and the bare slope was the lowest. Hence, there was a significant increase for straw mulching, contrasting to the changes for grassland. Then the impact of groundwater cumulant was analyzed quantitatively. At the end of simulation period, the accumulated underground runoff was 0.282 m^3 in the bare slope scenario, 0.346 m^3 in the lawn scenario, and 0.491 m^3 in the straw mulching scenario. Compared to the bare slope scenario, the amount and the benefit of groundwater recharge increased by 0.064 m^3 and 22.7% respectively in the lawn scenario. Accordingly, they were 0.209 m^3 and 60.4% in the straw mulching scenario. It was explicit that groundwater recharge very strongly depended upon land-use type [16].

The groundwater levels responded to the changes in groundwater recharge described as follows. Figure 8 compared the influence of soil and water conservation scenarios on groundwater levels at t = 110 min. Figure 8 (A) was the scatter plots of calculated levels of bare slope scenario vs. observed heads of grass land scenario. And figure 8 (B) was the graph of calculated levels of bare slope scenario vs. observed values of straw mulching scenario. As shown in the figures, the data points were under the X = Y line, so the calculated values were less than the observed values, especially in figure 8 (B). These indicated that groundwater levels increased with land-use converted from bare slope to grassland and even more with land-use converted to straw mulching. To quantify the impact on groundwater heads, the changes between the calculated heads of bare slope scenario and observed levels of grass land and straw mulching scenarios were showed in table 6.

It was explicit from the curve of figure 9 that the soil and water conservations had great influence upon the groundwater runoff. Straw mulching showed the higher runoff, followed by grassland. It was illustrated by further analysis that average flow was 0.14×10^{-4} m^3/s in the bare slope scenario, 0.17×10^{-4} m^3/s in the lawn scenario, and 0.26×10^{-4} m^3/s in the straw mulching scenario, during the simulation period. So the average runoff increased by 1.23 and 1.87 times through measures of lawn and straw mulching.

In summary, both grass and straw mulching scenarios played an important role in reducing the surface runoff and increasing the underground runoff. The effects of measures on redistribution of rainfall runoff was mainly due to the rainfall infiltration rata was changed. The grass and straw mulching dispersed the large raindrops into small raindrops, which reduced the actual rainfall intensity on the ground [38–42] and made more precipitation meet the condition of infiltration. But different soil and water conservation measures had different effects. In this study, straw mulching has more significant hydrological effects than lawn. The main difference between the two scenarios was that the straw mulching scenario had less bare land area compared to lawn scenario.

Conclusions

Based on the principle of water balance, a linked approach for SCS model and Visual MODFLOW was conducted to assess the impact of soil and water conservation on the surface water and groundwater runoff. The calibration results showed that the predicted results matched well with the observed data. Therefore, three land use management scenarios were simulated on the sandbox model to assess the effect of bare slope converted to grass land and straw mulching on water volume, hydraulic head, runoff process of groundwater and surface water.

Soil and water conservation measures could reduce surface runoff effectively. Under the 120 mm rainfall, 60 mm/h rainfall intensity, 5 m^2 area, 3° slope conditions, the comparative results indicated that decrease in surface runoff and increase in subsurface runoff coincided with the land-use converted from bare slope to grass land and straw mulching. Compared to the bare slope, the benefits of surface water reduction by these two measures were 57.8% and 92.4% correspondingly. Not only the individual soil and water conservation measure but also the comprehensive management of the small watershed had a significant benefit of surface water reduction. The comparative results of 12 groups (24) of controlled and uncontrolled small watersheds showed that the benefits of surface water reduction varied from 14.71% to 88.19% while the corresponding average value was 44.09%.

Soil and water conservation measures could promote rainfall recharge groundwater. Under the same condition, compared to the bare slope, the amount and the benefit of groundwater recharge increased by 0.064 m^3 and 22.7% respectively in the lawn scenario. Accordingly, they were 0.209 m^3 and 60.4% in the straw mulching scenario. For the runoff, the average flow of straw mulching was highest, the grassland took second place, the bare slope was the lowest, and the average runoff increased by 1.23 and 1.87 times through measures of lawn and straw mulching.

It was concluded that the soil and water conservation played an important role in weakening the surface runoff and strengthening the underground runoff. Meanwhile the groundwater flow model coupled with the surface water model used in this study, when properly validated, could be used as a tool in the evaluation of soil and water conservation measures on surface water and groundwater resources, and this approach could provide a thought for the study in a watershed scale to help decision-makers manage groundwater resources.

Acknowledgments

We would like to thank Mr. Xiu-quan XU, Mr. Li-zhi JIA, and Mr. Tong ZHANG for their assistant in experiment processing and Miss. Chun-hong ZHAO, Mr. Hong-jie WANG and Mr. Yuan-xing ZHANG for their advices in writing the paper. At the same time, the authors are very grateful for the editors' and reviewer's hard work in reviewing the paper.

Author Contributions

Conceived and designed the experiments: JEG HW. Performed the experiments: HW SLZ MJZ XHL. Analyzed the data: HW JEG. Contributed reagents/materials/analysis tools: JEG HW. Wrote the paper: HW JEG.

References

1. Li ZJ, Zhou PX, Mao LH (2006) General Review on Effects of Soil and Water Conservation Measures on Water Resources in China. Progress in Geography 25: 49–57.
2. Zhu XM, Jiang DS, Zhou PH, Jin ZS (1984) Discussion on the strategic issues of soil and water conservation in the loess region. Bulletin of soil and water conservation 1: 15–18.
3. Ren ME (2006) Sediment Discharge of the Yellow River, China: Past,Present and Future–A Synthesis. Advances in earth science 6: 551–563.
4. Yellow River Conservancy Commission of MWR (2010) Yellow river sediment bulletin, http://www.yellowriver.gov.cn/nishagonggao/2010/index.html.Accessed 2013 March 6.
5. Kim J, Choi J, Choi C, Park S (2013) Impacts of changes in climate and land use/land cover under IPCC RCP scenarios on streamflow in the Hoeya River Basin, Korea. Science of The Total Environment 452: 181–195.
6. Skaggs RW, Amatya DM, Chescheir GM, Blanton CD, Gilliam JW (2006) Effect of drainage and management practices on hydrology of pine plantation. Notes.
7. Sun G, Riedel M, Jackson R, Kolka R, Amatya D, et al. (2004) Influences of management of Southern forests on water quantity and quality. Southern Forest Science, General Technical Report SRS-75 198.
8. Qi S, Sun G, Wang Y, McNulty SG, Moore Myers JA (2009) Streamflow response to climate and landuse changes in a coastal watershed in North Carolina. Trans ASABE 52: 739–749.
9. Li YS (1997) Relation Between Control in Loess Plateau and No-flow in the Yellow River. Bulletin of Soil and Water Conservation 17: 41–45.
10. Ran DC (2006) Water and Sediment Variation and Ecological Protection Measures in the Middle Reach of the Yellow River. Resources Science 28: 93–100.

11. Qi JY Cai QG, Cai L, Sun LY (2011) Scale Effect of Runoff and Sediment Reduction Effects of Soil and Water Conservation Measures in Chabagou, Dalihe and Wudinghe Basins. Progress in Geography 30: 95–102.
12. Jain MK, Mishra SK, Singh VP (2006) Evaluation of AMC-dependent SCS-CN-based models using watershed characteristics. Water Resources Management 20: 531–552.
13. Shi PJ, Yuan Y, Zheng J, Wang JA, GeY, etal. (2007) The effect of land use/ cover change on surface runoff in Shenzhen region, China. Catena 69: 31–35.
14. Guo XJ, Wang DJ, Zhuang JQ (2010) Application of SCS model on simulation of the slope -runoff process in dry - hot valley. Science of Soil and Water Conservation 8: 14–18.
15. Freeze RA (1969) The Mechanism of Natural Ground - Water Recharge and Discharge: 1. One - dimensional, Vertical, Unsteady, Unsaturated Flow above a Recharging or Discharging Ground - Water Flow System. Water Resources Research 5: 153–171.
16. Paul MJ (2006) Impact of land-use patterns on distributed groundwater recharge and discharge. Chinese Geographical Science 16: 229–235.
17. Cho J, Barone VA, Mostaghimi S (2009) Simulation of land use impacts on groundwater levels and streamflow in a Virginia watershed. Agricultural water management 96: 1–11.
18. Zhang H, Hiscock KM (2010) Modelling the impact of forest cover on groundwater resources: A case study of the Sherwood Sandstone aquifer in the East Midlands, UK. Journal of Hydrology 392: 136–149.
19. Cooper JD (1980) Measurement of moisture fluxes in unsaturated soil in Thetford Forest. Research Report 66. Institute of Hydrology, Wallingford, UK.
20. Bosch JM, Hewlett JD (1982) A review of catchment experiments to determine the effect of vegetation changes on water yield and evapotranspiration. Journal of Hydrology 55: 3–23.
21. Dams J, Woldeamlak ST, Batelaan O (2008) Predicting land-use change and its impact on the groundwater system of the Kleine Nete catchment, Belgium. Hydrology and Earth System Sciences 12: 1369–1385.
22. Allen A, Chapman D (2001) Impacts of afforestation on groundwater resources and quality. Hydrogeology Journal 9: 390–400.
23. Pouget L, Escaler I, Guiu R, Mc Ennis S, Versini PA (2012) Global Change adaptation in water resources management: The Water Change project. Science of The Total Environment 440: 186–193.
24. Ferrer J, Pérez-Martín MA, Jiménez S, Estrela T, Andreu J (2012) GIS-based models for water quantity and quality assessment in the Júcar River Basin, Spain, including climate change effects. Science of The Total Environment 440: 42–59.
25. Candela L, Tamoh K, Olivares G, Gomez M (2012) Modelling impacts of climate change on water resources in ungauged and data-scarce watersheds. Application to the Siurana catchment (NE Spain). Science of The Total Environment 440: 253–260.
26. Boughton WC, Stone JJ (1985) Variation of runoff with watershed area in a semi-arid location. Journal of Arid Environments 9: 13–25.
27. Yu B (1998) Theoretical justification of SCS method for runoff estimation. Journal of irrigation and drainage engineering 124: 306–310.
28. Rallison RE (1980) Origin and evolution of the SCS runoff equation. Symposium on Watershed Management 1980. ASCE, New York. N, Y. II: 912–924.
29. Boughton WC (1989) A review of the USDA SCS curve number method. Soil Research 27: 511–523.
30. Wang ML, Guo SL, Yi Y (2004) SCS Monthly hydrological model building and its application. Journal of Water Resources Research 2: 002.
31. Harbaugh AW, Banta ER, Hill MC, McDonald MG (2000) MODFLOW-2000, the US Geological Survey modular ground-water model: User guide to modularization concepts and the ground-water flow process. Reston: US Geological Survey.
32. Hydrogeologic Water (2005) Visual MODFLOW v. 4.1 User's Manual. Waterloo, Ontario.
33. Harbaugh AW, McDonald MG (1996) User's documentation for MODFLOW-96, an update to the US Geological Survey modular finite-difference ground-water flow model. United States, Geological Survey.
34. Zhang WZ (1983) The Calculation of the Groundwater Unsteady Flow and Appraisal of Groundwater Resources. Beijing: Science Press.
35. Liang W, Liu L, PAN P (2007) Simulation of nitate-nitrogen transfer in trough scale. Lake Science 19: 710–717.
36. Shaanxi Provincial Bureau of Soil and Water Conservation (1976) Shaanxi Provincial runoff and sediment Data of soil and Water conservation. Xi'an: Shaanxi Provincial Bureau of Soil and Water Conservation.
37. Li YY, Shao MA (2004) Experimental study on characteristics of water transformation on slope land. Journal of hydraulic engineering 4: 48–53.
38. Laflen JM, Colvin TS (1981) Effect of crop residue on soil loss from continuous row cropping. Transactions of the ASAE 24: 1472–1475.
39. West LT, Miller WP, Langdale GW, Bruce RR, Laflen JM, et al. (1991) Cropping system effects on interrill soil loss in the Georgia Piedmont. Soil Science Society of America Journal 55: 460–466.
40. Zhu XM (1960) The impact of vegetation factors on soil and water erosion in Loess area. Acta pedologica sinica 8: 110–121.
41. Tang T, Hao MD, Shan FX (2008) Effects of Straw Mulch Application on Water Loss and Soil Erosion Under Simulated Rainfall. Research of Soil and Water Conservation 15: 9–11.
42. Li M, Yao WY, Li ZB (2005) Progress of the Effect of Grassland and Vegetation for Conserving soil and water on Loess Plateau. Advances in earth science 20: 74–80.

8

Novel Microbiological and Spatial Statistical Methods to Improve Strength of Epidemiological Evidence in a Community-Wide Waterborne Outbreak

Katri Jalava[1]*¤, Hanna Rintala[2], Jukka Ollgren[1], Leena Maunula[3], Vicente Gomez-Alvarez[4], Joana Revez[3], Marja Palander[1], Jenni Antikainen[5], Ari Kauppinen[6], Pia Räsänen[1], Sallamaari Siponen[1], Outi Nyholm[1], Aino Kyyhkynen[1], Sirpa Hakkarainen[2], Juhani Merentie[2], Martti Pärnänen[2], Raisa Loginov[5], Hodon Ryu[4], Markku Kuusi[1], Anja Siitonen[1], Ilkka Miettinen[6], Jorge W. Santo Domingo[4], Marja-Liisa Hänninen[3], Tarja Pitkänen[6]

1 Department of Infectious Disease Surveillance and Control, National Institute for Health and Welfare, Helsinki, Finland, **2** Siilinjärvi municipality, Finland, **3** Department of Food Hygiene and Environmental Health, University of Helsinki, Helskinki, Finland, **4** Office of Research and Development, United States Environmental Protection Agency, Cincinnati, Ohio, United States of America, **5** Laboratory HUSLAB, Helsinki University Hospital, Helsinki, Finland, **6** Department of Environmental Health, National Institute for Health and Welfare, Kuopio, Finland

Abstract

Failures in the drinking water distribution system cause gastrointestinal outbreaks with multiple pathogens. A water distribution pipe breakage caused a community-wide waterborne outbreak in Vuorela, Finland, July 2012. We investigated this outbreak with advanced epidemiological and microbiological methods. A total of 473/2931 inhabitants (16%) responded to a web-based questionnaire. Water and patient samples were subjected to analysis of multiple microbial targets, molecular typing and microbial community analysis. Spatial analysis on the water distribution network was done and we applied a spatial logistic regression model. The course of the illness was mild. Drinking untreated tap water from the defined outbreak area was significantly associated with illness (RR 5.6, 95% CI 1.9–16.4) increasing in a dose response manner. The closer a person lived to the water distribution breakage point, the higher the risk of becoming ill. Sapovirus, enterovirus, single *Campylobacter jejuni* and EHEC O157:H7 findings as well as virulence genes for EPEC, EAEC and EHEC pathogroups were detected by molecular or culture methods from the faecal samples of the patients. EPEC, EAEC and EHEC virulence genes and faecal indicator bacteria were also detected in water samples. Microbial community sequencing of contaminated tap water revealed abundance of *Arcobacter* species. The polyphasic approach improved the understanding of the source of the infections, and aided to define the extent and magnitude of this outbreak.

Editor: Martyn Kirk, The Australian National University, Australia

Funding: No external funding execpt routine governmental work funding was obtained for this work. The funders had no role in study design, data collection and analysis, decision to publish, or preparation of the manuscript.

Competing Interests: The authors have declared that no competing interests exist.

* Email: katri.jalava@thl.fi

¤ Current address: Environmental Health Department, City of Tampere, Finland

Introduction

Community-wide waterborne outbreaks are characterized by a large number of exposed people with high attack rates [1–6]. Waterborne outbreaks are frequently associated with large number of symptomatic cases in a point source manner. Such outbreaks may be caused by a failure in the drinking water distribution system [1,3,4] or water treatment breakthrough of contaminating agents due to heavy rainfall or other excess weather conditions [2,5]. The water distribution system can be potentially contaminated with multiple pathogens during a relatively short period of time as the result of intrusion from surface or waste water [7]. Indeed, waterborne outbreaks with multiple causative organisms, e.g. *Campylobacter* spp., norovirus-like organisms, *Shigella* and enterohaemorrhagic *Escherichia coli* (EHEC) have been described [1,3,4,6,8,9]. In particular, when norovirus and sapovirus types are implicated in large scale waterborne outbreaks, this strongly indicates drinking water distribution system contamination by a human faecal sources [10,11]. Sapovirus usually causes sporadic infections [12] but has been isolated from cases of waterborne outbreaks [12,13].

Waterborne outbreaks may be classified according to the level of evidence indicating that the drinking water was the cause of the outbreak. Evidence may be found by microbiological and/or epidemiological studies and the level of evidence can be assessed according to standardized criteria [14]. During the years 1998–2009 there have been 3–10 waterborne outbreaks in Finland annually and the outbreaks have typically been detected in small community groundwater plants with fewer than 500 consumers [7]. The implicated technical failures for the groundwater contamination in Finland have been flooding and surface run-off caused by heavy rains or rapid melting of snow. Also intrusion of contaminated water and cross-connections in the water distribution system play important role as a cause of Finnish waterborne outbreaks [15,16]. Most common causative microbes have been norovirus and *Campylobacter* [7].

In waterborne outbreak investigations, the delay between environmental investigation and the original presence of the pathogen within a water body has often hampered the detection of causative microbiological agents [15]. Special methods suitable for concentration of large water volumes, such as the use of large diameter membrane filters or ultrafiltration apparatus have been developed to increase the sensitivity of microbial detection in dilute water samples [17–19]. For the identification of a faecal contamination source, various detection methods targeting specific pathogens and microbial water quality indicators [15] and faecal source host specific molecular assays are available [20]. The use of RNA-based methods may not only help detecting but also identify active and thus potentially infective pathogens, providing a better estimate of public health risks than current DNA-based methods [21]. Additionally, next generation sequencing (NGS) can provide a path towards detecting multiple microbial taxa within complex microbial communities including rare members [17]. The high number of reads per sample in NGS applications also enables the use of these techniques for source tracking and identification of the transmission route of an outbreak [22].

Epidemiological cohort or case control studies may be applied in drinking water outbreaks due to the point source nature of the contamination and a well-defined population [3,11,23]. Web-based questionnaires are now increasingly used in outbreak investigations [24], especially in large and/or widely spread outbreaks [25]. Additionally, novel spatial methods have shown potential to define the source and location of the outbreak [26–28]. Modelling may be used to describe the person-to-person transmission or infection spreading from the environment [28]. Spatial variation in disease incidence has been studied in detail [27]. Also distance from water tanks has been evaluated by statistical methods with respect to diarrhoeal incidence [26].

The outbreak

The main water pipe was accidentally broken on 4th July, 2012 during road construction work in Vuorela, a community of 3000 inhabitants within the municipality of Siilinjärvi in Eastern Finland. The pipe breakage caused the contents of the upper drinking water storage reservoir to leak into the road construction pit. The pipe breakage was fixed within 14 hours, flushed and quality of the water was shown to fulfill the hygienic quality criteria. On the 16th July, the local environmental health authorities were informed by the health care centre of an excess number of patients with gastrointestinal symptoms. Given the recent pipe break in the area, a waterborne outbreak was suspected. The local outbreak control team of health and environmental authorities and waterworks personnel was activated, national outbreak awareness team was informed and consulted. On the following day, the results from tap water samples revealed faecal contamination of the water further confirming the waterborne nature of the outbreak. An immediate boil water notice was issued on 17th July and collection of patient and drinking water samples was initiated. The water distribution pipeline was subsequently flushed, the water storage reservoir was cleaned and the whole distribution system was disinfected with chlorine. The boil water notice was cancelled when the outbreak was declared over on 3rd August 2012.

The aim of our study was to reveal the role of contaminated water as a cause of a community wide outbreak detected in a small Finnish municipality during July 2012. We used a polyphasic approach carrying out advanced epidemiological, microbiological and environmental investigation to verify the source and scale of the outbreak. We included advanced and novel statistical, epidemiological and molecular microbiology methods, not previously applied to outbreak investigations according to our knowledge.

Materials and Methods

This study was part of public health response. According to Finnish legislation, no ethical approval is needed in this type of response. This study and related sampling and modelling were part of an official waterborne outbreak investigation, which is on the responsibility of municipal health and public health authorities (many of the coauthors are representatives of respective authorities).

Epidemiological investigation

Data collection. The contaminated part of the water distribution system provides drinking water for the area of Vuorela and Toivala in the municipality of Siilinjärvi. The total population of this area was 5934 and the exposed population who were served by the contaminated water distribution system in the defined outbreak area was 2931 persons (source population). The age, sex and living coordinates for the population were obtained from the National Population Register. A case was defined as a person staying or living in the Vuorela area during July 2012 with diarrhoea or two of the following symptoms: nausea, vomiting, stomach ache or fever. We excluded persons who were absent from the outbreak area during the whole study period and those who travelled abroad during July 2012. Based on the geographical coordinates we also determined the number of households or blocks of houses with unique water delivery points in the affected area.

Study design. A retrospective cohort study was conducted. A web based questionnaire was designed to define the extent and cause of the outbreak. The exposed population was informed by the local newspaper and press releases on the municipality website to participate in the study. All inhabitants of the Vuorela and Toivala area were invited as well as those visiting or working in this area. Data was collected between 19th July and 1st August, 2012, and any person living or spending time during the study period in the defined outbreak area was eligible to participate in the study. Study participants were asked about their basic demographic characteristics, clinical symptoms (from 4th to 20th July, 2012) and habits of consumption of tap water in the defined outbreak area (from 1st to 30th July, 2012).

Data analysis. We used a commercial web based questionnaire from Webropol (www.webropol.fi). The data was analysed as univariate factors calculating the risk ratios for all risk factors asked and frequencies of the illness as a retrospective cohort sample using R [29]. Subsequently, a binary log and logistic regression models with case status as the outcome variable and those explanatory variables that were significant in the univariate analysis were included, the analysis were performed in R. Furthermore, as we had the information of all the inhabitants, we compared the age and sex between the non-cases and the source population using a standard Wald's statistic for calculating the confidence intervals for the observed difference in percentages between the groups.

Spatial analysis and regression. For the spatial analysis, only those persons with address information available from the National Population Register and who replied to the cohort study were included. We obtained information on name, address, date of birth and living location co-ordinates. Sampling locations in the water distribution pipeline were plotted on a schematic map obtained from the local construction office (Figure 1). The shortest distance via the water pipe to the pipe breakage was determined from the obtained digital map using an R package gdistance. The distance data was subsequently allocated to each person living in the area. We also included a spatial correlation variable for the model to explain the possible transmission of the infection within households and closely living contacts and due to the other possibly unmeasured spatially correlated factors. We evaluated the model by residual diagnostics, residuals were plotted against the predicted values and Cook's diagnostic values were calculated [30]. We subsequently categorised the distance to four equal groups and compared the proportion of ill persons in each.The statistical calculations were done in R using both routine and special spatial packages [29]. The code for calculating the spatial distance via the water pipe in R which is presented in (File S1).

Clinical microbiology investigation

Microbiological analysis of patient samples. A total of 25 patients were sampled between 17th July and 2nd August, 2012. The age of the sampled patients had a median of 43 years (range 5–85 years). The potential causative agents were tested broadly, each from 5–21 samples aiming of qualitative detection of possible pathogens (Table 1). The initial samples were analysed in the local clinical laboratory with routine tests for common pathogens. As no major pathogens were detected, more samples were collected and analysed in national reference or specialized laboratories. A part of the patient samples were tested for enteric bacterial pathogens using standard clinical microbiology culture methods and PCR for the target genes of *Campylobacter jejuni/coli* (*rimR, gyrB*), *Salmonella* spp. (*invA*), *Shigella* spp./enteroinvasive *E. coli* (*ipaH, invE*) and *Yersinia* spp. (*virF, rumB*) [31]. The presence of diarrheagenic *E. coli* (i.e. EHEC, EPEC, ETEC, EIEC/*Shigella* spp. and EAEC) virulence genes (*pic, bfpB, invE, hlyA, elt, ent, escV, eae, ipaH, aggR,* stx_1, stx_2, *estlb, estla* and *astA*) were determined using multiplex PCR and qPCR techniques from mixed preliminary cultures grown on CLED (cystine lactose electrolyte deficient) agar medium plates [31,32]. The *ipaH* and *invG* genes are specific for both *Shigella* spp. and EIEC and the tests do not distinguish between these two organisms. If PCR for EHEC genes was positive, the specific colony was picked out when possible from the mixed culture plate for a single PCR of the *stx* genes. Electron microscopy was performed to detect enteric viruses, such as noro-, rota-, adeno-, entero-, sapo- and astroviruses. In addition, for norovirus reverse transcription (RT)-qPCR [33,34] was performed. For sapovirus analysis, two different PCR protocols were used [35,36] and nucleic acid sequences were determined from the amplicons of the polymerase region [37]. Samples were also tested for enteric parasites *Giardia* and *Cryptosporidium* using enzyme immunoassay method for antigen detection [38]. Seven frozen faecal samples were tested in a retrospective testing scenario for the presence of *Arcobacter* spp. by culture method as described previously [39] and the plates were inspected regularly for up to 3 weeks [40]. A species-specific multiplex-PCR was performed for detection of *A. buzleri, A. cryaerophilus* and *A. skirrowii* as described previously [41]. The samples were also tested using an additional *Arcobacter* genus specific PCR method [42].

Environmental investigation

Environmental sampling and analysis. The drinking water in the distribution network of Vuorela and Toivala area is UV-disinfected groundwater produced in the nearby Jälaniemi waterworks. The Siilinjärvi municipality owns and operates this public drinking waterworks and its distribution network. The employees of the Siilinjärvi municipality, including health protection authorities and personnel of the waterworks were responsible of environmental sampling and delivery of samples to expert laboratories. The water quality management routines include testing for coliform bacteria, heterotrophic plate counts and *Clostridium perfringens* at the waterworks and around the

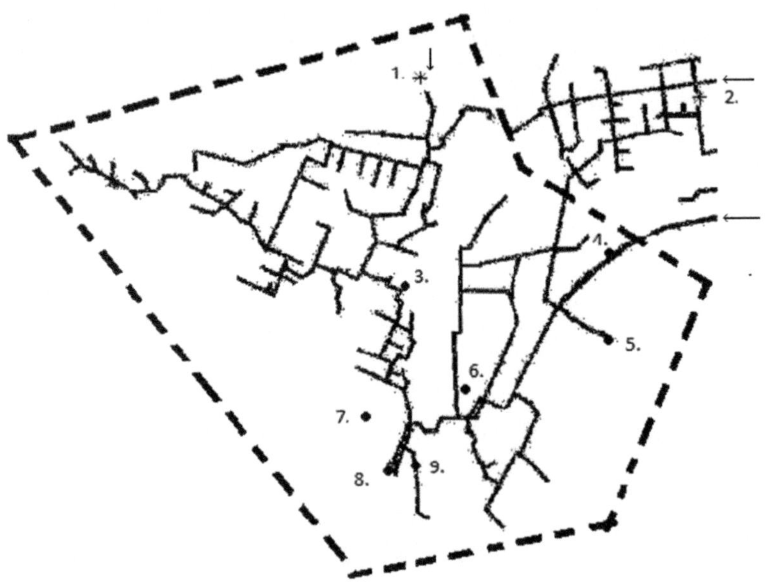

Figure 1. Schematic map of the water pipe of a defined outbreak area in Vuorela, July 2012. The outbreak (boil water notice) area is indicated by dashed line (- - -), the water sampling points (1–9) are coded as (•) with a positive culture finding and (*) with a negative finding. Arrows indicate the inflow points of the water from the water plant (outside the figure). Points 1,2,3,6,7 and 9 are tap water sampling locations, point 4 is the water pipe line breakage point (surface water), point 5 represent drinking water from the upper water storage reservoir and point 8 is the municipal effluent sampling location.

Table 1. Microbiological results of faecal samples from symptomatic patients of a waterborne outbreak in Vuorela, July 2012.

Microbial pathogen and methods used	Number of patients tested positive/number of tested
Campylobacter jejuni/coli[1,2]	1[5]/21
Salmonella spp.[1,2]	0/21
Shigella spp.[1,2]	0/21
Yersinia spp.[1,2]	0/21
E. coli	
- EHEC[1,2]	2[6]/12
- EPEC[2]	6/12
- ETEC[2]	0/12
- EIEC/*Shigella*[1,2]	0/12
- EAEC[2]	2/12
Arcobacter spp.[1,2]	0/7
Norovirus[2,3]	0[2]/12, 0[3]/17
Adenovirus[3]	0/17
Enterovirus[3]	1/17
Sapovirus[2,3]	5[2,7]/12, 3[3]/17
Astrovirus[2,3]	0[2]/7, 0[3]/17
Giardia[4]	0/19
Cryptosporidium[4]	0/19

Methods used were cultivation[1], PCR/RT-PCR[2] (polymerase chain reaction/reverse transcripition-PCR), EM[3] (electron microscopy) and/or EIA[4] (Enzyme Immunoassay). [5]*Campylobacter jejuni* and [6]EHEC O157:H7 (from one sample) and [7]sapovirus GII.P3 were isolated from the samples.

water distribution network according to European regulation [43]. The water quality results were acceptable prior to the outbreak with the last results dating back to 7th June and 2nd July, 2012. After accidental pipe breakage at 4th July, the pipeline was fixed within 14 hours, then flushed and sampled from two points (locations before the breakage point outside the schematic map and after the breakage point, point 6 in Figure 1) in 5th July, 2012 for indicator bacteria testing. The water quality was shown to comply the legal microbiological quality criteria (i.e. absence of coliform bacteria, *E. coli* and intestinal enterococci with no abnormal change in heterotrophic plate counts, Table 2).

A total of 65 water samples were collected between 5th July and 29th August (Figure 1). The majority of samples (n = 54) were tap water samples collected from various parts of the distribution system (1–10 sampling events/location) (Table 2). Also a raw water sample from a water production well, two biofilm samples from water meters and a sample from the community wastewater influent were collected. Surface water from a rainwater collection well and ditches at the pipeline breakage location (n = 5) were sampled as potential contamination sources. Both small (500–2000 mL) and large (10 L) scale water samples were analysed for pathogens. On-site large volume sampling (more than 100 L) using tangential flow ultrafiltration was conducted 17th July at one of the distribution system sites where high counts of faecal bacteria were detected at the previous day small scale sampling (point 7 represented in Figure 1). The ultrafiltration was conducted without sodium polyphosphate using a semi-automated system [17,18] and the bovine-serum pre-treated ultrafilter was eluted after sampling with 9:1 retentate/elution solution and 500 ml elution solution [17,44].

Culture methods were used to analyse *E. coli* and coliform bacteria [45,46], intestinal enterococci and heterotrophic plate counts [46], and vegetative cells and spores of *Clostridium perfringens* [47] from water samples. The presence/absence of faecal bacterial pathogens, thermotolerant *Campylobacter* spp. and *Salmonella* spp., was determined using culture-based selective enrichment methods [46,48], and EHEC was analysed using serotype O157 specific immunomagnetic separation coupled with the selective culture enrichment [49]. Enteric viruses were concentrated as previously described [50], except using Sartolon polyamide membranes (diameter 47 mm, pore size 45 μm; Sartorius, Goettingen, Germany). Norovirus, rotavirus and adenovirus were analysed using previously described RT-qPCR and qPCR methods [48,50,51]. For sapovirus analysis, the faecal protocol was applied to water as described earlier. Protozoan parasites *Giardia* and *Cryptosporidium* were determined from a tap water concentrate using immunomagnetic separation and epifluorescence microscopy according to the international standard method [52] and from water meter biofilm samples using molecular techniques [53]. The concentration of free and total chlorine in the drinking water distribution was measured on-site using the Palintest Micro 1000 photometer and in the laboratory using ISO 7393 standard method [54].

Further identification of pathogens. *E. coli* isolates from seven tap water samples (n = 19) and DNA extracts (n = 4) were screened for presence of virulence genes of EHEC, EPEC, ETEC, EIEC and EAEC [31,32]. In addition, the same virulence genes were tested from mixed cultures (tap water and rainwater sample) collected from membranes during the *E. coli*/coliform analysis. PFGE-profiles of *C. jejuni* isolates (n = 6) from patient and environmental samples were produced using *Kpn*I and *Sma*I restriction enzymes [55].

Extracted nucleic acids from total (DNA) and active bacterial fraction (RNA) in water samples were used as template to amplify faecal source identifiers and the bacterial 16S rRNA gene for microbial source tracking (MST) and NGS applications, respectively. MST assays were performed as previously described [21]. The samples were obtained from the upper storage reservoir water

Table 2. The counts of water quality indicator bacteria (range of MPN or CFU/100 mL), occurrence of faecal pathogens and chlorine concentrations in the water samples taken in July and August, 2012 in Vuorela, Finland.

Date (number of samples)	*E. coli*	Coliform bacteria	Entero-coccus	Faecal pathogens	Chlorine (mg/l)
Drinking water distribution system samples from the clean area					
5–19 Jul (12)	0	0–1	0	Not detected/3–6 samples[1]	ND, 0.4–1.8[2]
Drinking water distribution system samples from the boil water notice area					
5 Jul (1)	0	0	ND	ND	ND
16 Jul (4)	0–150	0–150	0–17	*Arcobacter* spp./1–4 samples[3]	ND
17 Jul (4)	0–21	0–34	0–2	EHEC, EPEC and EAEC virulence genes[4]	ND
18–20 Jul (10)	0[5]	0	0	Not detected/2–4 samples[1]	<0.1–1.6
23–30 Jul (13)	ND[6]	ND	ND	Not detected/3 samples[7]	<0.1–2.0
1–29 Aug (9)	ND[6]	ND	ND	ND	0.4–1.9
Water storage reservoir					
17 July (1)	110	190	15	Norovirus and adenovirus[8]	ND
21 July (1)	0	0	ND	Not detected[9]	ND
Biofilms from the water meters removed from the boil water notice area					
1 Aug (2)	ND	ND	ND	*Arcobacter* spp.[10]	ND
Raw water at the groundwater abstraction plant					
9 August (1)	0	0	ND	Not detected[11]	ND
Surface water samples (contaminant source)					
23 Jul (1)	86	450	44	*Campylobacter jejuni* 10–100 cfu/l, EPEC virulence genes[12]	ND
29 Aug (5)	0–94	ND	5–80	ND	ND
Community wastewater					
25 Jul (1)	ND	ND	ND	Sapovirus	ND

ND; not determined.
[1]A portion of the samples were selected for *Salmonella*, *Campylobacter*, enterohaemorragic *E. coli* (EHEC) culture analyses and for norovirus analysis.
[2]Measured from three locations at 19 July.
[3]Samples were tested for noro-, adeno-, rota- and sapoviruses, *Campylobacter* and *E. coli* virulence genes. *Arcobacter* was tested from DNA extracts and genus specific PCR was positive in one sample (point 9 in Figure 1).
[4]*E. coli* virulence genes were detected after ultrafiltration from one sampling location (point 5 in Figure 1). *Salmonella*, *Campylobacter*, EHEC (culture method), noro-, rota- and sapovirus, *Giardia* and *Cryptosporidium* were not detected (1–4 samples tested/method).
[5]One colony of *Clostridium perfringens* was found from 1 000 mL of tap water sample taken from the most contaminated area.
[6]*Clostridium perfringens* was analyzed and not detected from 5 000 mL samples.
[7]Samples were tested for sapovirus.
[8]Sample was tested for noro-, adeno-, rota- and sapoviruses, *Salmonella*, *Campylobacter* and *E. coli* virulence genes. *Clostridium perfringens* was detected (10 CFU/L).
[9]Sample was tested for *Campylobacter* and noroviruses. *Clostridium perfringens* was detected (2 CFU/L).
[10]Sample tested for *Campylobacter*, *Arcobacter*, *Giardia* and *Cryptosporidium*.
[11]Sample tested for *Campylobacter*.
[12]Sample tested for *Campylobacter*, *E. coli* virulence genes, noro- and adenoviruses, *Giardia* and *Cryptosporidium*. *Clostridium perfringens* was detected (40 CFU/L).

before and after cleaning (point 5 in Figure 1) and from a tap water sample collected during the contamination episode (point 7 in Figure 1). The MST assays included the analysis of faecal bacterial groups *Bacteroidales* spp. (GenBac3 assay [56]) and human-specific *Bacteroidales* (HF183 assay [57]).

For Illumina MiSeq NGS, we utilized barcoded primers 515F and 806R to produce 250 bp pair-ended sequences [58]. Reads were processed and analyzed using the software MOTHUR v1.30.2 ([59]; http://www.mothur.org) as described previously [60]. Briefly, fastq files with forward and reverse reads were used to form contigs. The reads that met the following criteria were excluded from further analysis: the length was no greater than 255 bp; contained ambiguous bases (N) or homopolymers greater than 7 bases; identified as chimera; or classified as Chloroplasts or Mitochondria. Reads were aligned and sorted with >97% similarity into operational taxonomic units (OTUs). Taxonomic classification was obtained using the tool Classifier in the Ribosomal Database Project II release 10.28 [61]. The raw reads were deposited in the NCBI Sequence Read Archive (SRA) under accession number SRP041117. Using MEGA v5.2 [62], a phylogenetic tree based on the aligned 16S rRNA gene sequences (~255 bp) was constructed with the Maximum Likelihood (ML) method using Tamura-Nei model [63] with 1,000 bootstrap replicates. The tree was used to infer the phylogenetic relationship among sequences classified as *Arcobacter* obtained in this study. *Sulfurospirillum deleyianum* (NR_074378) and *Campylobacter fetus* (L04314) were used as outgroup. Three water samples and two water meter samples were subsequently analysed for *Arcobacter* spp. using genus specific PCR [42] and species specific multiplex-PCR [41] from the DNA extracted as previously described [50].

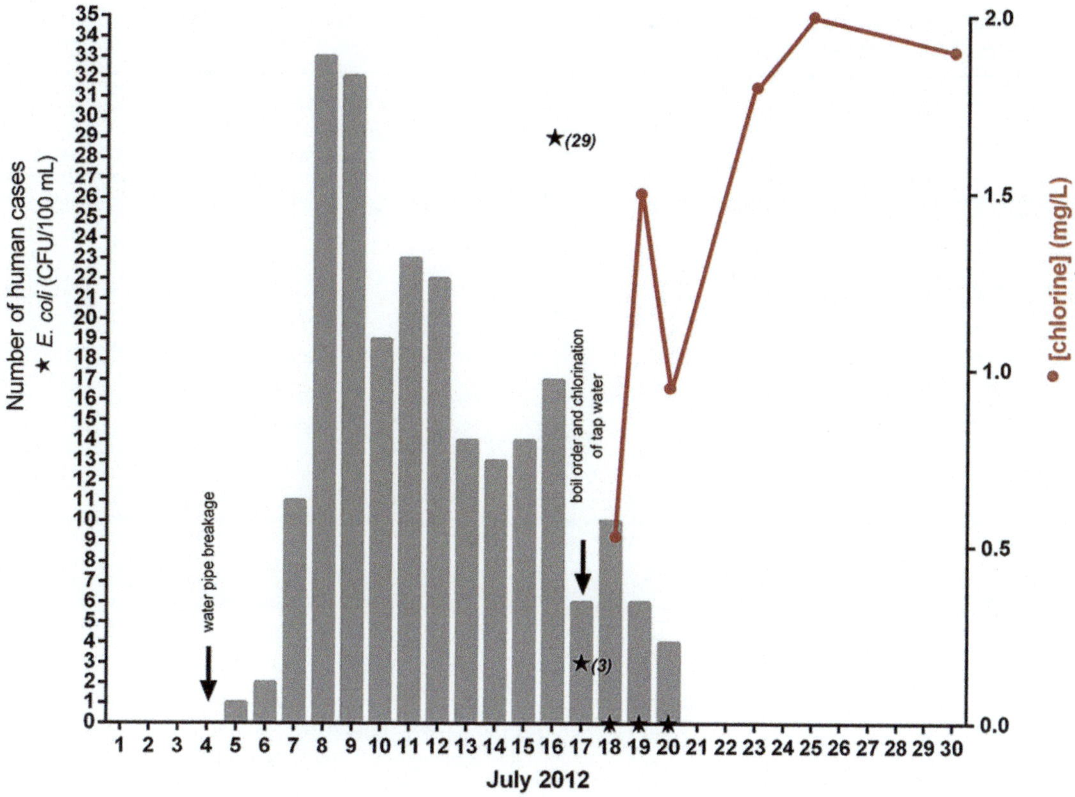

Figure 2. Epidemic curve of a waterborne outbreak in Vuorela, July 2012 based on the reported onset date of illness of the cases, and *E. coli* bacteria counts and chlorine levels in the point 7 (See Fig. 1) of the water distribution network.

Results

Epidemiological investigation

Descriptive analysis. Of all 2931 inhabitants (source population) of the defined outbreak area, 473 (16%) persons participated in the study (study population). We excluded 19 persons absent from the outbreak area during the whole study period and 23 due to travelling abroad. In total, we identified 225 cases and 206 healthy persons from the cohort study according to the case definition. However, for the individual risk factors data contained a few missing values. During the four week period after the water distribution breakage, an estimated 800 excess persons visited either a nurse or GP in the local health care centre. The outbreak curve presented in Figure 2 implicates a point source outbreak.

Demographic characteristics. Of the 431 respondents, 33% (135/408) were male. The mean age was 41.2 years (range 1–80 years). As there were few children among the respondent (10 in total, 3 among non-cases), we excluded the children from comparison analysis. The proportion of male (33%, 60/182) among the non-cases in the study population was lower to that of the source population (48%, 1074/2252). This difference for the male proportions of 14.5% (7.4%–21.6%) was statistically significant. Persons were slightly older among the source population compared to the study population, Table 3. Also, there were 50.2% (232/462) households (unique water delivery points) with at least one person under 18 years of age in the source population compared to 68.5% (126/184) among the study population. This difference was significant, CI for the difference was 18.3% (10.2%–26.4%). Therefore, households with children were more likely to respond to the study.

The course of the illness was mild, only one person was admitted to hospital. The main symptoms were stomach ache 88% (199/225), nausea 85% (191/225) and diarrhoea 82% (185/225). The length of the illness had a median of 3 days (range 1–30), 16% (35/225) of the cases sought for medical assistance and one patient was hospitalized (37-year old female with no underlying medical condition). Absence from work due to outbreak illness was

Table 3. Comparison of the study population to source population with respect to age groups in a waterborne outbreak in Vuorela, July 2012.

Age group	Study population (non-cases)% of population (number of persons)	Source population % of population (number of persons)	% difference (95% confidence intervals)
20–39 years	40% (78/195)	29% (655/2252)	11% (3.9%–18.1%)
≥40 years	60% (117/195)	71% (1597/2252)	−11% (−18.1%–−3.9%)

Table 4. The Univariate and multivariate results for individual risk factors and the generalized additive model risk ratios with the spatial term of a waterborne outbreak in Vuorela, July 2012.

	Explanatory variable	Risk ratio or univariate log regression exp (β-values), (95% confidence intervals) for individual risk factors	Multivariable generalized additive logit model, exp (β-values), (p-value)
Personal characteristic	Age (continuous in years)	0.99 (0.98–0.99)	0.975 (0.0061)
Drinking at home	Tap water	2.2 (1.2–4.1)	5.90 (0.0037)
	Water from own well	1.0 (0.57–1.87)	n/a
	Bottled water	0.86 (0.69–1.07)	n/a
	Boiled water	0.69 (0.53–0.90)	n/a
Drinking water in Vuorela (outside home)	Tap water	1.6 (1.2–2.0)	n/a
	Water from own well	0.83 (0.56–1.22)	n/a
	Bottled water	0.90 (0.72–1.12)	n/a
	Boiled water	0.80 (0.56–1.15)	n/a
Spatial variables	Distance from the breakage by waterpipe (metres)	0.99950 (0.99930–0.99969)	0.998 (0.060)
	Spatial variable (coordinates)	n/a	n/a (0.002)

reported by 31% (133/431) of the study participants, the total number of working days lost because of illness was 398. Taking care of a sick child was not included.

Univariate analysis of the cohort. Of all possible cases occurring in the area, 225 cases responded to the study in an area with a population of 2931 persons. Of the water related risk factors, drinking untreated tap water in the outbreak area had a risk ratio of 5.6 (95% CI 1.9–16.4), also drinking untreated tap water at home RR 2.2 (95% CI 1.2–4.1) and outside home RR 1.6 (95% CI 1.2–2.0) in the outbreak area were associated with illness, Table 4. Drinking boiled water at home was a protective factor, RR 0.69 (95% CI 0.53–0.90), drinking well water or bottled water were not significant either at home or outside home. It was observed that the risk ratio increased as the number of glasses of water consumed increased. This further strengthened the role of the contaminated tap water as a cause of the outbreak in a dose response manner, table 5.

Multivariate and spatial analysis of the cohort. Of those fulfilling eligibility criteria, 20 participants had no address information, 81 lived outside the defined boil water notice area and 17 could not be found in the national population register, leaving 313 persons for the spatial analysis. After these exclusions, 154 cases and 159 non-cases were used for the spatial analysis. Age, drinking tap water at home and distance from the breakage point were significant in the multivariable model, Table 4. Younger persons were more likely to become ill ($p = 0.0061$), risk also increased if tap water was consumed at home ($p = 0.0037$). Distance from the leakage point was inversely associated with becoming a case ($p = 0.060$). Also spatial variable was significant in the generalized additive model ($p = 0.002$), this is likely to reflect the person to person transmission within households or neighbourhoods (Table 4). The distance via the water pipe was shorter for the cases compared with the non-cases (Table 4). This indicates that the closer one lived to the water breakage point, the more contaminated the drinking water was and therefore the likelihood of the illness was higher. The diagnostics of the model for randomized quantile residuals were normally distributed to suggest correct specification of the model. Between the categorised groups, the proportion of ill persons was higher among those living closer to water breakage point, Table 6.

Clinical microbiology investigation

Patient faecal samples. The microbiological analysis of faecal samples from the patients identified several pathogens (Table 1). The pathogens identified included sapovirus (detected from 5 patients of 12 tested, 2 sapoviruses were genotyped as

Table 5. The dose-response between the illness and the amount of water consumed at home in a waterborne outbreak in Vuorela, July 2012.

Number of glasses of water consumed at home per day	Cases/total number of persons in the implicated group (%)	Risk ratio (95% confidence intervals)	p-value
0	6/28	reference	
1–3	49/114 (43%)	2.01 (1.06–4.81)	0.065
4–6	112/193 (58%)	2.71 (1.48–6.40)	0.0066
7–9	37/48 (77%)	3.60 (1.94–8.54)	0.00054
10 or more	9/19 (47%)	2.21 (0.96–5.68)	0.068

Those not drinking water at home served as a control group.

Table 6. Categorized distance and proportion of cases within those groups in a waterborne outbreak in Vuorela, July 2012.

Distance categories	% (Cases/total)
Distance 1 (<2332)	72.4% (63/87)
Distance 2 (2332–2713)	45.5% (35/77)
Distance 3 (2713–3202)	43.5% (37/85)
Distance 4 (>3202)	29.7% (19/64)

GII.P3), enterovirus (detected from one patient out of 17 tested) and *C. jejuni* (isolated from one patient out of 21 tested). Two sapovirus nucleic acid sequences were submitted to Genbank with accession numbers KJ200380 and KJ200381. In addition, specific virulence genes of pathogenic *E. coli* were detected, including genes from EHEC (2/12), EPEC (6/12) and EAEC (2/12). In subsequent culturing of the virulence gene positive samples, one *E. coli* O157:H7 positive culture was found. No suspect *Arcobacter* spp. were detected by culture methods or by PCR.

Environmental investigation

Faecal microbes in the drinking water distribution. Faecal indicator bacteria were detected in four out of 11 tap water samples taken from different sampling points of the distribution on 16th and 17th July, 2012 (Table 2). The bacterial counts from the drinking water reservoir sample (point 5 in Figure 1) taken at the same time were high for total coliform bacteria, *E. coli* and enterococci. In addition, low numbers of adenoviruses and noroviruses (genogroup II) were detected in the storage reservoir sample. Water samples tested negative for sapovirus, although the community waste water influent was shown to contain sapovirus. Specific virulence genes of EHEC, EPEC and EAEC were detected from the large volume tap water sample (point 7 in Figure 1).

Cause of the water contamination. Technical investigations revealed that the water storage reservoir had been rapidly filled with contaminated water after the pipe breakage repair. The microbial contaminants unintentionally funnelled to the storage reservoir could remain viable as no disinfection was used in the distribution system (Table 2). Subsequently the contaminated water was introduced to the Vuorela and Toivala distribution system due to the water usage. The water level in the storage reservoir varies depending on water usage and pumping of the fresh water from the groundwater abstraction plant into the reservoir. The water quality at the groundwater abstraction plant was tested acceptable and free of microbial contaminants shortly after the breakage (Table 2). De-contamination of the reservoir was ensured by emptying and then mechanical cleaning, washing and chlorination of the inner surfaces of the storage reservoir. Indicator bacteria were no longer detected from the water distribution system after flushing and successful chlorination (Table 2). However, it took approximately a week before *C. perfringens* spores were absent in the water samples from the contaminated distribution system and chlorine levels reached the minimum target concentration (0.5 mg/l) at the most distant parts of the distribution.

Microbial source tracking. GenBac3 and HF183 assays showed the presence of *Bacteroides* spp. and human-specific *Bacteroidales*, respectively, in the contaminated tap water and in the storage reservoir (Table S1). There were more rRNA markers present compared to the rRNA gene (DNA) markers (rRNA:rDNA ratios were higher) in the storage reservoir sample than in the contaminated tap water sample. After cleaning of the storage reservoir, the GenBac3 and HF183 marker concentrations declined to below the quantification and detection limits, respectively (Table S1). A reactive surface water sample from the rainwater well was collected one week after the outbreak notification to evaluate the potential environmental sources near the pipe breakage location (Table 2). *C. jejuni* and EPEC virulence genes were identified from the surface water sample. Also indicator bacteria; coliforms, *E. coli* and intestinal enterococci were abundant. Five *C. jejuni* isolates from this suspected contamination source had identical *Kpn*I and *Sma*I profiles in PFGE, but different from those of the *C. jejuni* patient isolate (data not shown).

Bacterial communities in water. Community analysis based on the bacterial 16S rRNA region showed a higher diversity for contaminated sampling sites, while upper storage reservoir after the cleaning showed a lower diversity (Table S2). It is noteworthy that the community diversity measured as a total number of OTUs in the community was higher within the DNA reads than within the RNA reads (Table S2). Taxonomic analysis of contaminated water samples indicated a high presence of the class Beta-proteobacteria including the family *Comamonadaceae* and genus *Zoogloea*, and Gamma-proteobacteria including affiliated family *Methylococcaceae* and genera *Methylobacter* and *Pseudomonas* (Figure 3, A and B; Table S3).

Interestingly, also a significant share of reads generated from the contaminated water samples affiliated with Epsilon-proteobacteria genus *Arcobacter* (Table S3 and Table S4). Further sequence analysis revealed close relationship to *A. cryaerophilus* group 1A LMG9865 (OTU00012 in Figure 4). However, the corresponding tap water sample was negative for the species-specific *Arcobacter* multiplex-PCR (for *A. butzleri*, *A. cryaerophilus*, *A. skirrowii*) and the storage reservoir samples were no longer available for species-specific testing. *Arcobacter* genus specific PCR was positive for a tap water sample and a water meter sample collected from another location (point 9 in Figure 1) but these samples remained negative by culture. Although the sequences from both samples (Genbank with accession number KJ196910) showed high sequence similarity (>98%) with *A. cryaerophilus* (uncultured), *A. skirrowii* and *A. cibarius*, no species attribution can be made.

After the cleaning and chlorination of the upper storage reservoir, the relative abundance of Beta- and Gamma-proteobacteria decreased and the abundance of *Sphingobacteria* including genus *Pedobacter* increased (Figure 3, C; Table S3). The reads associated *Bacteroides*, *Escherichia* and *Clostridia* in the contaminated water samples were absent from the storage reservoir sample after cleaning and chlorination (Table S4).

Discussion

This study describes a large-scale municipality area wide outbreak due to breakage of drinking water distribution pipeline during a road construction work. The illness in the community became apparent only two weeks after the incident but the cause of the illness was obvious. This was confirmed using a polyphasic approach applying microbiological, epidemiological and statistical methods. Multiple causative microorganisms were isolated from patient samples. No major definite causative pathogen was identified but sapovirus was most frequently detected in patient samples. Also various types of virulent *E. coli*, *C. jejuni* and *Arcobacter* spp. might have played a role in the onset of gastrointestinal symptoms. All these organisms have been previously associated with waterborne transmission [5,9,12,13,23]. By

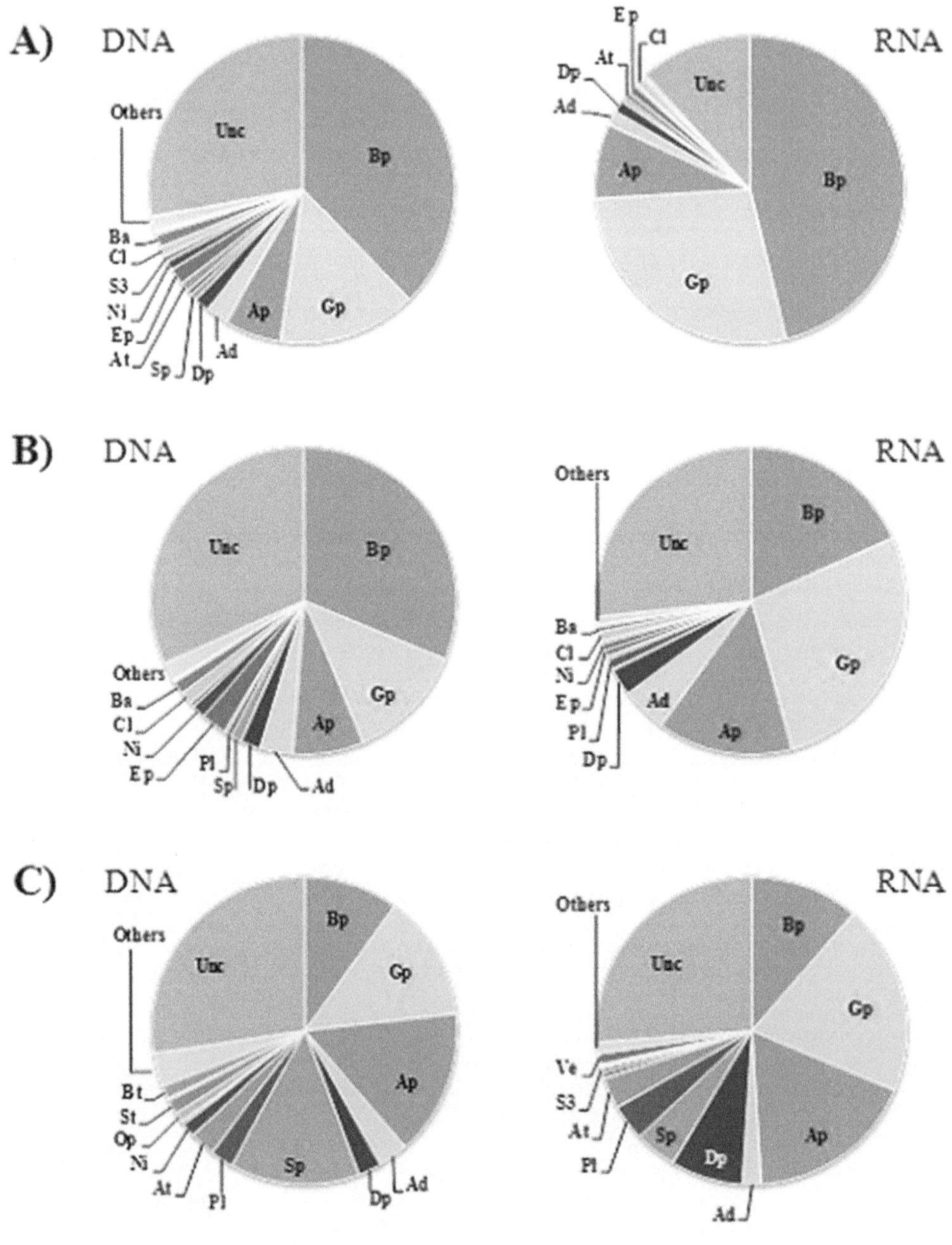

Figure 3. Distribution of the Bacteria domain as determined by taxonomic identification of partial 16S rRNA gene sequencing (at class level) in a waterborne outbreak in Vuorela, July 2012. Charts and tables represent the cumulative distribution of total DNA and RNA extracted from samples of A) the upper storage reservoir before cleaning (point 5 in Figure 1), B) tap water during contamination (point 7 in Figure 1) and C) the upper storage reservoir after cleaning. Legend: Beta-proteobacteria (Bp); Gamma-proteobacteria (Gp); Alpha-proteobacteria (Ap); Acidobacteria (Ad); Delta-proteobacteria (Dp); Sphingobacteria (Sp); Planctomycetacia (Pl); Actinobacteria (At); Epsilon-proteobacteria (Ep); Nitrospira (Ni); Verrucomicrobia Subdivision3 (S3); Clostridia (Cl); Bacteroidia (Ba); Opitutae (Op); Spartobacteria (St); Verrucomicrobiae (Ve); Bacteroidetes (Bt); Others (classes each representing $<1\%$); unclassified (Unc).

applying the spatial and microbiological analysis, we could identify the contamination route for this outbreak. The distance to the water breakage point was inversely associated with illness both by epidemiological and spatial methods. We further confirmed the role of the contaminated water as a vehicle by novel spatial analysis. The spatial method provided here could be used to compare the likelihoods between possible candidate point source locations. Effective control measures and rapid and continuous communication with the media were essential to ensure appropriate public health measures.

The overall clinical illness in most patients was mild with only 16% accessing medical assistance and only one patient being hospitalized. Majority of the cases had gastrointestinal symptoms with mild course and relatively short length. Overall, the number of cases was likely to be an underestimate for the whole population. The case definition used in the present study was relatively sensitive due to the mild overall nature of the illness. The

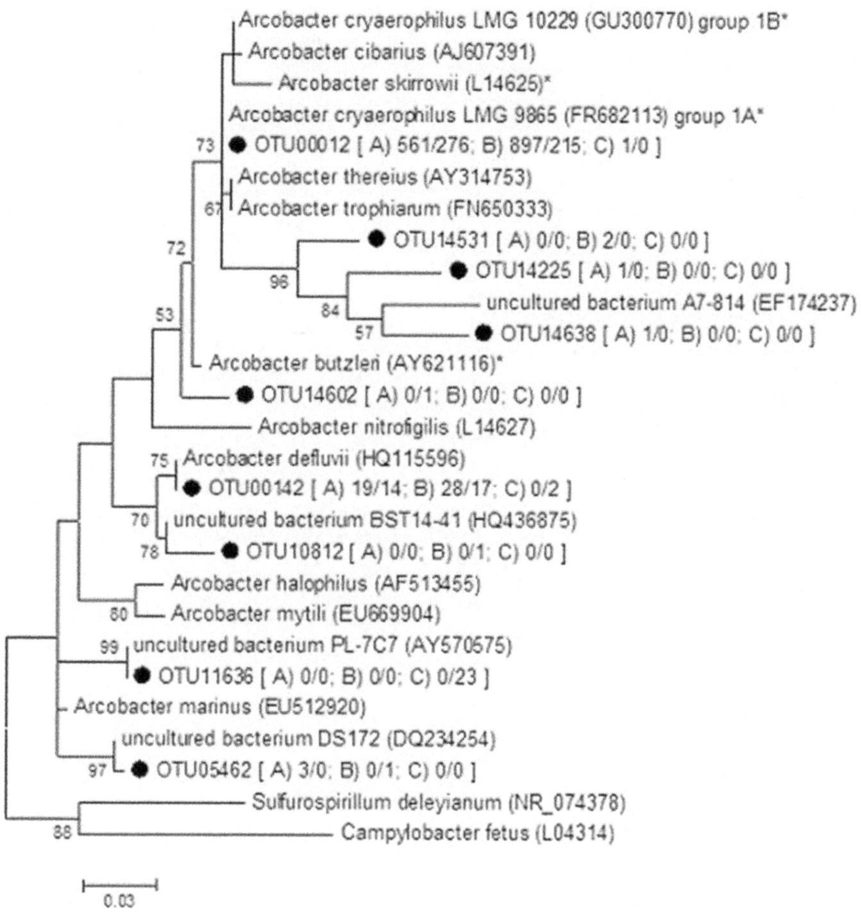

Figure 4. Phylogenetic relationships among OTUs (•) of the genus *Arcobacter* in a waterborne outbreak in Vuorela, July 2012. The tree was inferred from a maximum likelihood analysis of aligned 16S rRNA gene sequence (≈255 bp) and nodes with a bootstrap value ≥50% of 1 000 replicates are identified. *Sulfurospirillum deleyianum* (NR_074378) and *Campylobacter fetus* (L04314) were used as outgroup. Number in bracket represents the total amount of DNA/RNA reads identified in samples from A) the upper storage reservoir before cleaning, B) tap water during contamination and C) the upper storage reservoir after cleaning. **A. butzleri*, *A. cryaerophilus* and *A. skirrowii* have been associated with gastrointestinal diseases [76].

contaminated water consumed whether at or outside home in the defined outbreak area was associated with illness. We further showed by epidemiological methods that drinking contaminated tap water was associated with illness in a dose-response manner. The effect levelled off at the highest dose, often observed in food or waterborne outbreaks [64,65], this has been observed also in previous studies [1,3,9]. Furthermore, drinking boiled water was a protective factor indicating the effectiveness of the implemented control measures.

It is challenging to estimate the true attack rate in the present outbreak. It is quite likely that people showing symptoms responded more actively in the study compared to non-cases. We had a unique opportunity to compare the study population to the source population and found that women responded more actively to the study. This has been found also previously [66–68]. As we also found slightly more younger people among the non-cases compared to the source population, therefore, the observed effect of age should be interpreted with caution. The crude attack rate obtained from the questionnaire study (225/431, 52%) is within the same range to what has been observed in similar outbreaks, namely between 31–88% [2–4]. The true number of ill persons is difficult to estimate, but the educated evaluation of the local general practitioner (800 cases) is a fair estimate.

By applying the spatial and microbiological analysis, we could identify the contamination route for this outbreak. The contaminated water from the breakage initially filled the water storage reservoir and subsequently this water was distributed to the community over a number of days. The drinking water storage reservoirs have contributed to the transmission of waterborne infections also elsewhere, usually due to the improper maintenance and structure of the reservoirs [46,69,70]. In our case, the exact cause of the contamination was pipeline breakage at the road construction site. The storage reservoir prolonged the contaminant transport time to the water consumers. The likelihood of illness was higher closer to the water breakage point as measured by water pipeline length. Also the higher amount of water consumed in the boil water notice area increased the likelihood of illness. Additionally, younger persons (i.e. parents of small children) might have been more likely to become infected presumably due to the fact that they had more contact to young infants.

The source of the illness was obvious from the beginning of the outbreak, but we initially searched for one major pathogen as a causative agent yet aware that this type of outbreak is often caused by multiple pathogens. The sapovirus finding was novel and has rarely been detected in previous Finnish waterborne outbreaks [13]. According to recent reports, sapoviruses can be commonly found in environmental waters [10,71,72] highlighting the importance of including sapovirus detection into the outbreak investigations. Sapoviruses cause acute gastroenteritis primarily among young children [36], and indeed, young age was suspected to be one risk factor in this study. This result, together with the presence of sapoviruses in communal wastewater influent, support the assumption that sapovirus was commonly circulating in the community. We did not detect sapoviruses in the drinking water samples, presumably due to delayed sampling and low viral concentration in the drinking water. We also found that adults had symptomatic sapovirus infections. Adults have been found to have sapovirus infections in other studies [12,73,74]. The occurrence of human infecting sapoviruses among the patient cases together with the detection of human-specific genetic marker HF183, adenoviruses and noroviruses from the contaminated drinking water suggests that human faecal material was present at the pipe breakage site. No leakage was identified in a waste-water pipe line located in the same construction pit as the drinking water pipeline. Instead, it was concluded that surface water runoff from the nearby recreational area was the potential source of the faecal microbes. The abundant pathogenic *E. coli* findings from patient and water samples were not surprising bearing in mind the nature of the contamination. One positive *C. jejuni* finding in a patient may have had sporadic origin as the environmental isolate was of different type.

We used a novel NGS approach to study the microbial communities in the tap water distributed in the contaminated water pipe network. To our knowledge, this was the first time that high-throughput sequencing was employed for detection of potential causative bacterial agents in a waterborne outbreak case. By using the novel microbial community analysis for the water samples, we could show an abundance of *Arcobacter* spp. in the drinking water distribution during the contamination. Further analysis found high sequence similarity with potential pathogenic *Arcobacter* spp. (*A. cryaerophilus*, *A. skirrowii* and *A. cibarius*). Although the abundance of reads of *Arcobacter* correlated with the quality of water (increase in contaminated waters), the resolution of *Arcobacter* species lineage was not possible due to ambigious nature of the 16S rRNA sequences [75]. Since *Arcobacter* spp. have been suspected as a cause of gastroenteritis in humans [5,76], we aimed to isolate these from the remaining patient samples. No *Arcobacter* spp. were isolated perhaps due to the small quantity, long storage time (more than one year) or storage conditions of the samples. Nevertheless, considering that there is no gold standard for the isolation of these putative pathogens, their presence cannot be ruled out.

The NGS approach also provided information on the bacterial communities in the non-chlorinated and faecally contaminated drinking water. Until now, the analysis of indigeneous microbial community composition in non-chlorinated drinking water distribution systems has been scarce [77]. The detected change in the storage reservoir bacterial community structure after the cleaning and chlorination proves the effectiveness of these measures. Moreover, our result showing higher community diversity within the DNA reads than within the RNA reads suggest that DNA-based methods may not effectively discriminate between active and dormant populations. However, more work is needed to better characterize the bacterial community changes in the different drinking water distribution systems. For example, understanding the role of drinking water retention time in the network [77] and the effect of drinking water treatment processes such as filtration on the bacterial communities in the distribution [78] would facilitate the sustainable management of the microbial water quality in the distribution networks.

We further confirmed the role of the contaminated water as a vehicle by novel spatial analysis. By calculating the distance for each household to the water breakage point via the water pipe, we could show that the probability of illness decreased by the increasing length of the water pipe. However, it should be noted that non-symptomatic persons in the same households and neighbourhoods participated more actively in the cohort study. We used all cases in the model, including potential secondary cases. As the spatial term describes just secondary spread and was significant in the spatial logistic regression model, it suggests that there were indeed a number of secondary cases among the case patients. In addition to distance from the breakage point, age was inversely associated with illness and consumption of water in the Vuorela area was positively associated in a spatial logistic regression.

The study was limited by a slow response due to notification delay as it took a relatively long time before the outbreak was detected and confirmed. Therefore, the causative agents could not be isolated from the water samples. The microbial community analysis also warrants further studies. We used 16S rRNA as a target of the NGS studies. In the future, with careful design of pathogen specific primers for sequencing purposes, it might be possible to abandon the requirement of culture isolation for genotyping purposes. In successful NGS applications, the huge number of reads per sample potentially enables the identification phylogenetic relatedness between the causative pathogenic strains in different samples [79]. Subsequently, this might allow the use of NGS approach for source tracking and understanding of the transmission route of an outbreak.

Conclusions

We used novel and existing statistical, microbiological and spatial methods to characterize a community wide waterborne outbreak. These methods may be applied to wide range of food and waterborne outbreak investigations and beyond. In particular, microbial community analysis in combination with traditional culture and PCR based methods aided to clarify the potential causative agents. In addition, the cohort study was carried out as a rapid web based application added with a novel spatial method to show the statistical association to the suspected point source leakage of the water pipe network. We also confirmed previous observations that women and younger people were more likely to respond to this type of study. In the present outbreak the event associated with the outbreak was fairly obvious, but in many outbreak situations the source cannot be easily identified [3]. Failures in the water distribution networks are common causes of waterborne outbreaks [2]. The method provided here could be used to compare the likelihoods between possible candidate point source locations.

Supporting Information

Table S1 Numbers of the GenBac3 and HF183 markers ($\log_{10}$ copies 100 mL^{-1}) in TaqMan rRNA-targeted RT-qPCR (rRNA) and rRNA gene-targeted qPCR (rDNA) assays.

Table S2 Community diversity estimates (±CI) of the domain *Bacteria*.

Table S3 Taxonomic affiliation of the most abundant *Bacteria* domain representatives.

Table S4 Abundance of reads associated with faecal bacteria and/or disease agents detected in the Vuorela and Toivala drinking water distribution system.

File S1 R code for calculating the shortest direct distance and the distance via the water pipe between each inhabitant location and water pipe breakage point.

Acknowledgments

We thank Anne-Mari Rissanen, Paula Muona and Mirja Rissanen and all the other personnel at the water works, in the local analytical laboratories and municipality of Siilinjärvi for their help in the outbreak investigation. We thank Pekka Pere from the University of Helsinki in helping with the statistical analysis. The opinions expressed in the paper are those of the authors and do not necessarily reflect the official positions and policies of the US Environmental Protection Agency or any other public authorities mentioned in this study. Any mention of product or trade names does not constitute recommendation for use by the US Environmental Protection Agency or any other public authorities mentioned in this study.

Author Contributions

Conceived and designed the experiments: KJ JO LM VGA JR H. Rintala AS JWSD MLH TP. Performed the experiments: KJ H. Rintala LM VGA JR M. Palander JA A. Kauppinen PR SS ON A. Kyyhkynen SH JM M. Pärnänen RL H. Ryu TP. Analyzed the data: KJ JO LM VGA JR A. Kauppinen A. Kyyhkynen SH M. Pärnänen MK AS JWSD MLH TP. Contributed reagents/materials/analysis tools: KJ JO LM VGA JR M. Palander JA A. Kyyhkynen SH JM M. Pärnänen RL H. Ryu MK AS IM JWSD MLH TP. Wrote the paper: KJ JO LM VGA JR A. Kauppinen MK AS JWSD MLH TP.

References

1. Maurer AM, Sturchler D (2000) A waterborne outbreak of small round structured virus, *Campylobacter* and *Shigella* co-infections in la Neuveville, Switzerland, 1998. Epidemiol Infect 125: 325–332.
2. Laursen E, Mygind O, Rasmussen B, Ronne T (1994) Gastroenteritis: A waterborne outbreak affecting 1600 people in a small Danish town. J Epidemiol Community Health 48: 453–458.
3. Jakopanec I, Borgen K, Vold L, Lund H, Forseth T, et al. (2008) A large waterborne outbreak of campylobacteriosis in Norway: The need to focus on distribution system safety. BMC Infect Dis 8: 128-2334-8-128.
4. Laine J, Huovinen E, Virtanen MJ, Snellman M, Lumio J, et al. (2011) An extensive gastroenteritis outbreak after drinking-water contamination by sewage effluent, Finland. Epidemiol Infect: 1–9.
5. Fong TT, Mansfield LS, Wilson DL, Schwab DJ, Molloy SL, et al. (2007) Massive microbiological groundwater contamination associated with a waterborne outbreak in Lake Erie, South Bass Island, Ohio. Environ Health Perspect 115: 856–864. 10.1289/ehp.9430.
6. Bopp DJ, Sauders BD, Waring AL, Ackelsberg J, Dumas N, et al. (2003) Detection, isolation, and molecular subtyping of *Escherichia coli* O157:H7 and *Campylobacter jejuni* associated with a large waterborne outbreak. J Clin Microbiol 41: 174–180.
7. Zacheus O, Miettinen IT (2011) Increased information on waterborne outbreaks through efficient notification system enforces actions towards safe drinking water. J Water Health 9: 763–772. 10.2166/wh.2011.021.
8. Dev VJ, Main M, Gould I (1991) Waterborne outbreak of *Escherichia coli* O157. Lancet 337: 1412.
9. Gubbels SM, Kuhn KG, Larsson JT, Adelhardt M, Engberg J, et al. (2012) A waterborne outbreak with a single clone of *Campylobacter jejuni* in the Danish town of Koge in May 2010. Scand J Infect Dis 44: 586–594. 10.3109/00365548.2012.655773.
10. Nenonen NP, Hannoun C, Larsson CU, Bergstrom T (2012) Marked genomic diversity of norovirus genogroup I strains in a waterborne outbreak. Appl Environ Microbiol 78: 1846–1852. 10.1128/AEM.07350-11.
11. Parshionikar SU, Willian-True S, Fout GS, Robbins DE, Seys SA, et al. (2003) Waterborne outbreak of gastroenteritis associated with a norovirus. Appl Environ Microbiol 69: 5263–5268.
12. Svraka S, Vennema H, van der Veer B, Hedlund KO, Thorhagen M, et al. (2010) Epidemiology and genotype analysis of emerging sapovirus-associated infections across Europe. J Clin Microbiol 48: 2191–2198. 10.1128/JCM.02427-09.
13. Rasanen S, Lappalainen S, Kaikkonen S, Hamalainen M, Salminen M, et al. (2010) Mixed viral infections causing acute gastroenteritis in children in a waterborne outbreak. Epidemiol Infect 138: 1227–1234. 10.1017/S0950268809991671.
14. Tillett HE, de Louvois J, Wall PG (1998) Surveillance of outbreaks of waterborne infectious disease: Categorizing levels of evidence. Epidemiol Infect 120: 37–42.
15. Pitkänen T (2013) Review of *Campylobacter* spp. in drinking and environmental waters. J Microbiol Methods 95: 39–47. 10.1016/j.mimet.2013.06.008.
16. Miettinen IT, Zacheus O, von Bonsdorff CH, Vartiainen T (2001) Waterborne epidemics in Finland in 1998–1999. Water Sci Technol 43: 67–71.
17. Hill VR, Kahler AM, Jothikumar N, Johnson TB, Hahn D, et al. (2007) Multistate evaluation of an ultrafiltration-based procedure for simultaneous recovery of enteric microbes in 100-liter tap water samples. Appl Environ Microbiol 73: 4218–4225. 10.1128/AEM.02713-06.
18. Rhodes ER, Hamilton DW, See MJ, Wymer L (2011) Evaluation of hollow-fiber ultrafiltration primary concentration of pathogens and secondary concentration of viruses from water. J Virol Methods 176: 38–45. 10.1016/j.jviromet.2011.05.031.
19. Hijnen WA, Biraud D, Cornelissen ER, van der Kooij D (2009) Threshold concentration of easily assimilable organic carton in feedwater for biofouling of spiral-wound membranes. Environ Sci Technol 43: 4890–4895.
20. Toledo-Hernandez C, Ryu H, Gonzalez-Nieves J, Huertas E, Toranzos GA, et al. (2013) Tracking the primary sources of fecal pollution in a tropical watershed in a one-year study. Appl Environ Microbiol 79: 1689–1696. 10.1128/AEM.03070-12.
21. Pitkänen T, Ryu H, Elk M, Hokajarvi AM, Siponen S, et al. (2013) Detection of fecal bacteria and source tracking identifiers in environmental waters using rRNA-based RT-qPCR and rDNA-based qPCR assays. Environ Sci Technol 47: 13611–13620. 10.1021/es403489b.
22. Di Bella JM, Bao Y, Gloor GB, Burton JP, Reid G (2013) High throughput sequencing methods and analysis for microbiome research. J Microbiol Methods. 10.1016/j.mimet.2013.08.011.
23. Riera-Montes M, Brus Sjolander K, Allestam G, Hallin E, Hedlund KO, et al. (2011) Waterborne norovirus outbreak in a municipal drinking-water supply in Sweden. Epidemiol Infect 139: 1928–1935. 10.1017/S0950268810003146.
24. de Jong B, Ancker C (2008) Web-based questionnaires - a tool used in a *Campylobacter* outbreak investigation in Stockholm, Sweden, October 2007. Euro Surveill 13: 18847.
25. Oxenford CJ, Black AP, Bell RJ, Munnoch SA, Irwin MJ, et al. (2005) Investigation of a multi-state outbreak of *Salmonella hvittingfoss* using a web-based case reporting form. Commun Dis Intell Q Rep 29: 379–381.
26. Bessong PO, Odiyo JO, Musekene JN, Tessema A (2009) Spatial distribution of diarrhoea and microbial quality of domestic water during an outbreak of diarrhoea in the Tshikuwi community in Venda, South Africa. J Health Popul Nutr 27: 652–659.
27. Dangendorf F, Herbst S, Reintjes R, Kistemann T (2002) Spatial patterns of diarrhoeal illnesses with regard to water supply structures–a GIS analysis. Int J Hyg Environ Health 205: 183–191. 10.1078/1438-4639-00151.
28. Tuite AR, Tien J, Eisenberg M, Earn DJ, Ma J, et al. (2011) Cholera epidemic in haiti, 2010: Using a transmission model to explain spatial spread of disease and identify optimal control interventions. Ann Intern Med 154: 593–601. 10.1059/0003-4819-154-9-201105030-00334.
29. R_Core_Team R (2013) R: A language and environment for statistical computing. Vienna, Austria.
30. Hosmer DWJ, Lemeshow S, Strudivant RX (2013) Applied logistic regression. 3rd edition, Wiley.
31. Antikainen J, Kantele A, Pakkanen SH, Laaveri T, Riutta J, et al. (2013) A quantitative polymerase chain reaction assay for rapid detection of 9 pathogens directly from stools of travelers with diarrhea. Clin Gastroenterol Hepatol 11: 1300–1307.e3. 10.1016/j.cgh.2013.03.037.
32. Antikainen J, Tarkka E, Haukka K, Siitonen A, Vaara M, et al. (2009) New 16-plex PCR method for rapid detection of diarrheagenic *Escherichia coli* directly from stool samples. Eur J Clin Microbiol Infect Dis 28: 899–908. 10.1007/s10096-009-0720-x.
33. Summa M, von Bonsdorff CH, Maunula L (2012) Pet dogs–a transmission route for human noroviruses? J Clin Virol 53: 244–247. 10.1016/j.jcv.2011.12.014.
34. Ronnqvist M, Ratto M, Tuominen P, Salo S, Maunula L (2013) Swabs as a tool for monitoring the presence of norovirus on environmental surfaces in the food industry. J Food Prot 76: 1421–1428. 10.4315/0362-028X.JFP-12-371.
35. Oka T, Katayama K, Hansman GS, Kageyama T, Ogawa S, et al. (2006) Detection of human sapovirus by real-time reverse transcription-polymerase chain reaction. J Med Virol 78: 1347–1353. 10.1002/jmv.20699.
36. van Maarseveen NM, Wessels E, de Brouwer CS, Vossen AC, Claas EC (2010) Diagnosis of viral gastroenteritis by simultaneous detection of adenovirus group F, astrovirus, rotavirus group A, norovirus genogroups I and II, and sapovirus in two internally controlled multiplex real-time PCR assays. J Clin Virol 49: 205–210. 10.1016/j.jcv.2010.07.019.

37. Reuter G, Zimsek-Mijovski J, Poljsak-Prijatelj M, Di Bartolo I, Ruggeri FM, et al. (2010) Incidence, diversity, and molecular epidemiology of sapoviruses in swine across Europe. J Clin Microbiol 48: 363–368. 10.1128/JCM.01279-09.
38. Rimhanen-Finne R, Hanninen ML, Vuento R, Laine J, Jokiranta TS, et al. (2010) Contaminated water caused the first outbreak of giardiasis in Finland, 2007: A descriptive study. Scand J Infect Dis 42: 613–619. 10.3109/00365541003774608.
39. Revez J, Huuskonen M, Ruusunen M, Lindstrom M, Hanninen ML (2013) Arcobacter species and their pulsed-field gel electrophoresis genotypes in Finnish raw milk during summer 2011. J Food Prot 76: 1630–1632. 10.4315/0362-028X.JFP-13-083.
40. Merga JY, Leatherbarrow AJ, Winstanley C, Bennett M, Hart CA, et al. (2011) Comparison of *Arcobacter* isolation methods, and diversity of *Arcobacter* spp. in Cheshire, United Kingdom. Appl Environ Microbiol 77: 1646–1650. 10.1128/AEM.01964-10.
41. Houf K, Tutenel A, De Zutter L, Van Hoof J, Vandamme P (2000) Development of a multiplex PCR assay for the simultaneous detection and identification of *Arcobacter butzleri*, *Arcobacter cryaerophilus* and *Arcobacter skirrowii*. FEMS Microbiol Lett 193: 89–94.
42. Harmon KM, Wesley IV (1996) Identification of *Arcobacter* isolates by PCR. Lett Appl Microbiol 23: 241–244.
43. European U (1998) Council directive 98/83/EC of 3rd November on the quality of water intended for human consumption. Official Journal of European Communities L330: 32–54.
44. Polaczyk AL, Narayanan J, Cromeans TL, Hahn D, Roberts JM, et al. (2008) Ultrafiltration-based techniques for rapid and simultaneous concentration of multiple microbe classes from 100-L tap water samples. J Microbiol Methods 73: 92–99. 10.1016/j.mimet.2008.02.014.
45. ISO_9308-2 A (2012) Water quality - enumeration of Escherichia coli and coliform bacteria – part 2: Most probable number method. International Organization for Standardization, Geneva, Switzerland.
46. Pitkänen T, Miettinen IT, Nakari UM, Takkinen J, Nieminen K, et al. (2008) Faecal contamination of a municipal drinking water distribution system in association with *Campylobacter jejuni* infections. J Water Health 6: 365–376.
47. ISO/CD 6461-2 (2002) Water quality – Detection and enumeration of the spores of sulfite-reducing anaerobes (clostridia) – part 2: Method by membrane filtration. Revised committee draft. International Organization for Standardization, Geneva, Switzerland.
48. Pradhan SK, Kauppinen A, Martikainen K, Pitkänen T, Kusnetsov J, et al. (2013) Microbial reduction in waste water treatment using Fe^{3+} and Al^{3+} coagulants and PAA disinfectant. J Water Health 11: 581–589.
49. NMKL 164 A (2005) *Escherichia coli* O157. Detection in food and feeding stuffs, Nordic committee on food analysis.
50. Kauppinen A, Ikonen J, Pursiainen A, Pitkänen T, Miettinen IT (2012) Decontamination of a drinking water pipeline system contaminated with adenovirus and *Escherichia coli* utilizing peracetic acid and chlorine. J Water Health 10: 406–418. 10.2166/wh.2012.003.
51. Kauppinen A, Martikainen K, Matikka V, Veijalainen AM, Pitkänen T, et al. (2014) Sand filters for removal of microbes and nutrients from wastewater during a one-year pilot study in a cold temperate climate. J Environ Manage 133: 206–213. 10.1016/j.jenvman.2013.12.008.
52. ISO_15553 (2006) Water quality - isolation and identification of *Cryptosporidium* oocysts and *Giardia* cysts from water. International Organization for Standardization, Geneva, Switzerland.
53. Guy RA, Payment P, Krull UJ, Horgen PA (2003) Real-time PCR for quantification of *Giardia* and *Cryptosporidium* in environmental water samples and sewage. Appl Environ Microbiol 69: 5178–5185.
54. ISO 7393 (1985) Water quality – Determination of free chlorine and total chlorine. International Organization for Standardization, Geneva, Switzerland.
55. Nakari UM, Hakkinen M, Siitonen A (2011) Identification of persistent subtypes of *Campylobacter jejuni* by pulsed-field gel electrophoresis in Finland. Foodborne Pathog Dis 8: 1143–1145. 10.1089/fpd.2011.0882.
56. Siefring S, Varma M, Atikovic E, Wymer L, Haugland RA (2008) Improved real-time PCR assays for the detection of fecal indicator bacteria in surface waters with different instrument and reagent systems. J Water Health 6: 225–237. 10.2166/wh.2008.022.
57. Kildare BJ, Leutenegger CM, McSwain BS, Bambic DG, Rajal VB, et al. (2007) 16S rRNA-based assays for quantitative detection of universal, human-, cow-, and dog-specific fecal *Bacteroidales*: A bayesian approach. Water Res 41: 3701–3715. 10.1016/j.watres.2007.06.037.
58. Caporaso JG, Lauber CL, Walters WA, Berg-Lyons D, Lozupone CA, et al. (2011) Global patterns of 16S rRNA diversity at a depth of millions of sequences per sample. Proc Natl Acad Sci U S A 108 Suppl 1: 4516–4522. 10.1073/pnas.1000080107.
59. Schloss PD, Westcott SL, Ryabin T, Hall JR, Hartmann M, et al. (2009) Introducing mothur: Open-source, platform-independent, community-supported software for describing and comparing microbial communities. Appl Environ Microbiol 75: 7537–7541. 10.1128/AEM.01541-09.
60. Kozich JJ, Westcott SL, Baxter NT, Highlander SK, Schloss PD (2013) Development of a dual-index sequencing strategy and curation pipeline for analyzing amplicon sequence data on the MiSeq illumina sequencing platform. Appl Environ Microbiol 79: 5112–5120. 10.1128/AEM.01043-13.
61. Cole JR, Wang Q, Cardenas E, Fish J, Chai B, et al. (2009) The ribosomal database project: Improved alignments and new tools for rRNA analysis. Nucleic Acids Res 37: D141–5. 10.1093/nar/gkn879.
62. Tamura K, Peterson D, Peterson N, Stecher G, Nei M, et al. (2011) MEGA5: Molecular evolutionary genetics analysis using maximum likelihood, evolutionary distance, and maximum parsimony methods. Mol Biol Evol 28: 2731–2739. 10.1093/molbev/msr121.
63. Tamura K, Nei M, Kumar S (2004) Prospects for inferring very large phylogenies by using the neighbor-joining method. Proc Natl Acad Sci U S A 101: 11030–11035. 10.1073/pnas.0404206101.
64. ter Waarbeek HL, Dukers-Muijrers NH, Vennema H, Hoebe CJ (2010) Waterborne gastroenteritis outbreak at a scouting camp caused by two norovirus genogroups: GI and GII. J Clin Virol 47: 268–272. 10.1016/j.jcv.2009.12.002 [doi].
65. Teunis P, Van den Brandhof W, Nauta M, Wagenaar J, Van den Kerkhof H, et al. (2005) A reconsideration of the *Campylobacter* dose-response relation. Epidemiol Infect 133: 583–592.
66. Volken T (2013) Second-stage non-response in the swiss health survey: Determinants and bias in outcomes. BMC Public Health 13: 167-2458-13-167. 10.1186/1471-2458-13-167 [doi].
67. Johnson TP, Wislar JS (2012) Response rates and nonresponse errors in surveys. JAMA 307: 1805–1806. 10.1001/jama.2012.3532 [doi].
68. Martikainen P, Laaksonen M, Piha K, Lallukka T (2007) Does survey non-response bias the association between occupational social class and health? Scand J Public Health 35: 212–215. 10.1080/14034940600996563.
69. Miettinen IT, Pitkänen T, Nakari UM, Hakkinen M, Wermudsen K, et al. (2006) 5th nordic drinking water conference, 8.-10. June, 2006.
70. Richardson G, Thomas DR, Smith RM, Nehaul L, Ribeiro CD, et al. (2007) A community outbreak of *Campylobacter jejuni* infection from a chlorinated public water supply. Epidemiol Infect 135: 1151–1158. 10.1017/S0950268807007960.
71. Kitajima M, Oka T, Haramoto E, Katayama H, Takeda N, et al. (2010) Detection and genetic analysis of human sapoviruses in river water in Japan. Appl Environ Microbiol 76: 2461–2467. 10.1128/AEM.02739-09.
72. Murray TY, Mans J, Taylor MB (2013) First detection of human sapoviruses in river water in South Africa. Water Sci Technol 67: 2776–2783. 10.2166/wst.2013.203.
73. Mikula C, Springer B, Reichart S, Bierbacher K, Lichtenschopf A, et al. (2010) Sapovirus in adults in rehabilitation center, upper austria. Emerg Infect Dis 16: 1186–1187. 10.3201/eid1607.091789.
74. Lee LE, Cebelinski EA, Fuller C, Keene WE, Smith K, et al. (2012) Sapovirus outbreaks in long-term care facilities, Oregon and Minnesota, USA, 2002–2009. Emerg Infect Dis 18: 873–876. 10.3201/eid1805.111843.
75. Vandamme P, Pot B, Gillis M, de Vos P, Kersters K, et al. (1996) Polyphasic taxonomy, a consensus approach to bacterial systematics. Microbiol Rev 60: 407–438.
76. Collado L, Figueras MJ (2011) Taxonomy, epidemiology, and clinical relevance of the genus *Arcobacter*. Clin Microbiol Rev 24: 174–192. 10.1128/CMR.00034-10.
77. Lautenschlager K, Hwang C, Liu WT, Boon N, Koster O, et al. (2013) A microbiology-based multi-parametric approach towards assessing biological stability in drinking water distribution networks. Water Res 47: 3015–3025. 10.1016/j.watres.2013.03.002.
78. Pinto AJ, Xi C, Raskin L (2012) Bacterial community structure in the drinking water microbiome is governed by filtration processes. Environ Sci Technol 46: 8851–8859. 10.1021/es302042t.
79. Cox MJ, Cookson WO, Moffatt MF (2013) Sequencing the human microbiome in health and disease. Hum Mol Genet 22: R88–94. 10.1093/hmg/ddt398.

9

Detecting Leaf Pulvinar Movements on NDVI Time Series of Desert Trees: A New Approach for Water Stress Detection

Roberto O. Chávez[1]*, Jan G. P. W. Clevers[1], Jan Verbesselt[1], Paulette I. Naulin[2], Martin Herold[1]

1 Laboratory of Geo-Information Science and Remote Sensing, Wageningen University, Wageningen, The Netherlands, 2 Laboratorio de Biología de Plantas, Departamento Silvicultura y Conservación de la Naturaleza, Universidad de Chile, Santiago, Chile

Abstract

Heliotropic leaf movement or leaf 'solar tracking' occurs for a wide variety of plants, including many desert species and some crops. This has an important effect on the canopy spectral reflectance as measured from satellites. For this reason, monitoring systems based on spectral vegetation indices, such as the normalized difference vegetation index (NDVI), should account for heliotropic movements when evaluating the health condition of such species. In the hyper-arid Atacama Desert, Northern Chile, we studied seasonal and diurnal variations of MODIS and Landsat NDVI time series of plantation stands of the endemic species *Prosopis tamarugo* Phil., subject to different levels of groundwater depletion. As solar irradiation increased during the day and also during the summer, the paraheliotropic leaves of Tamarugo moved to an erectophile position (parallel to the sun rays) making the NDVI signal to drop. This way, Tamarugo stands with no water stress showed a positive NDVI difference between morning and midday ($\Delta NDVI_{mo\text{-}mi}$) and between winter and summer ($\Delta NDVI_{W\text{-}S}$). In this paper, we showed that the $\Delta NDVI_{mo\text{-}mi}$ of Tamarugo stands can be detected using MODIS Terra and Aqua images, and the $\Delta NDVI_{W\text{-}S}$ using Landsat or MODIS Terra images. Because pulvinar movement is triggered by changes in cell turgor, the effects of water stress caused by groundwater depletion can be assessed and monitored using $\Delta NDVI_{mo\text{-}mi}$ and $\Delta NDVI_{W\text{-}S}$. For an 11-year time series without rainfall events, Landsat $\Delta NDVI_{W\text{-}S}$ of Tamarugo stands showed a positive linear relationship with cumulative groundwater depletion. We conclude that both $\Delta NDVI_{mo\text{-}mi}$ and $\Delta NDVI_{W\text{-}S}$ have potential to detect early water stress of paraheliotropic vegetation.

Editor: Guy J-P. Schumann, NASA Jet Propulsion Laboratory, United States of America

Funding: This work has been funded by the CONICYT (Chile)-Wageningen University scholarship (Res. Ex. N. 281/2009). The funders had no role in study design, data collection and analysis, decision to publish, or preparation of the manuscript.

Competing Interests: The authors have declared that no competing interests exist.

* Email: roberto.chavez@wur.nl

Introduction

Heliotropism or 'solar tracking' is the ability of many desert plant species and crops to move leaves and flowers as a response to changes in the position of the sun throughout the day [1]. There are two types of heliotropic movements: diaheliotropic movements in which leaves adjust the leaf lamina to face direct solar irradiation, and paraheliotropic movements, in which leaves adjust to avoid facing incoming radiation by contraction of pulvinar structures located at the base of leaves [1–3]. Paraheliotropic movements are triggered by directional solar irradiation and allow partial regulation of the incident irradiance on the leaves. For desert species, the regulation of the solar irradiation intensity on the leaves is an important adaptation to avoid photosynthesis saturation (photoinhibition) and to enhance the water use efficiency [4–6].

Diurnal paraheliotropic movements have a direct impact on the canopy reflectance properties of vegetation [7–9]. Therefore, spectral vegetation indices derived from remote sensing data, like the normalized difference vegetation index (NDVI), can significantly vary during the day and during the year due to leaf movement as solar irradiation changes. Nevertheless, no studies have quantified the effects of solar tracking by plants on the NDVI signal recorded from satellites. Diurnal leaf movements of Tamarugo plants (*Prosopis tamarugo* Phil.) were first described by Chávez et al. [9] under laboratory conditions, and later by Chávez et al. [8] for adult trees in the field. These diurnal leaf movements corresponded to paraheliotropic movements since the leaves moved to an erectophyle leaf distribution (facing away from the sun) around midday when solar irradiation was maximum. The paper of Chávez et al. [8] showed that leaf pulvinar movements caused diurnal changes of Tamarugo's canopy spectral reflectance and NDVI signal, which was negatively correlated to diurnal solar irradiation values.

In the present study, we hypothesize that the effects of Tamarugo's diurnal leaf pulvinar movements on the NDVI can also be recorded by remote sensors from space, since the acquisition time of the different sensors differ. A high solar irradiation at midday is assumed to cause a lower NDVI than in the morning as indicated in Figure 1a. In this context the MODIS (Moderate Resolution Imaging Spectroradiometer) sensor seems to be especially suitable to capture this difference in NDVI between morning (low solar irradiation, high NDVI) and midday (high solar irradiation, low NDVI), since the MODIS sensor on board of the Terra satellite acquires data for the study site at 10 a.m. (local

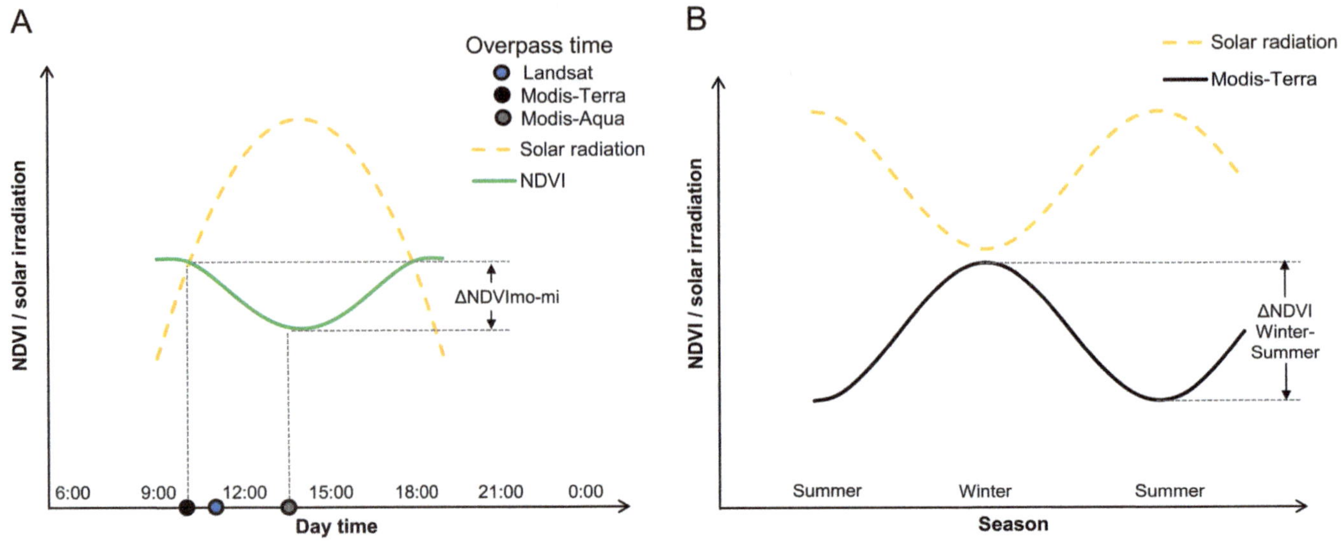

Figure 1. Conceptual diagram of the effect of leaf pulvinar movement on the NDVI signal. (A) NDVI difference between morning and midday ($\Delta NDVI_{mo\text{-}mi}$) occuring as solar irradiation changes during the day, and (B) NDVI difference between winter and summer ($\Delta NDVI_{W\text{-}S}$) occuring as solar irradiation varies between seasons. The time at which the Landsat (5–7), MODIS-Terra, and MODIS-Aqua satellites acquire data is displayed to illustrate the impact of pulvinar movements on the NDVI retrieved from these platforms.

time) and the MODIS sensor on board of the Aqua satellite acquires data at 1.30 p.m. (local time). Thus, the NDVI difference between morning and midday ($\Delta NDVI_{mo\text{-}mi}$) can be calculated as the difference between the NDVI MODIS Terra and the NDVI MODIS Aqua.

Considering the negative correlation between diurnal NDVI measurements and solar irradiation reported by Chávez et al. [8], we expect also seasonal NDVI variations associated with seasonal changes in solar irradiation with peaks in winter when the solar irradiation is the lowest. We hypothesize that this effect can also be recorded by sensors from space as indicated in Figure 1b. In this case, the Landsat TM (Thematic Mapper) and ETM (Enhanced Thematic Mapper) catalogue seems to be very suitable for studying NDVI seasonal variations since it offers one of the longest existing time series of systematically recorded satellite data worldwide [10]. Besides, the Landsat catalogue is considered the most relevant satellite dataset for ecological applications and environmental monitoring [11,12]. MODIS data might also be used to study seasonal effects of the pulvinar movements on the NDVI signal, providing images with a coarser spatial resolution (250 meters vs the 30 meters of Landsat), but with a higher temporal resolution (daily) enabling near real time vegetation monitoring [13]. However, the MODIS time series is considerably shorter than the Landsat time series (only since 2000).

Tamarugo is an endemic tree of the hyper-arid Atacama Desert, Northern Chile, a location considered among the most extreme environments for life [14,15]. The Tamarugo forest, locally known as Pampa del Tamarugal, sustains a biodiversity of about 40 species of plants and animals, some of them endemic for this particular ecosystem [16–19]. Precipitation events are very rare and the only source of water supply for vegetation is the groundwater (GW), from which Tamarugo is completely dependent. However, not only Tamarugo trees are demanding water: the main economic activity in Atacama is mining, which is also demanding water for human consumption and for many industrial processes. This has led to an overexploitation of the GW sources and a progressive depletion of the GW over the whole Pampa del Tamarugal [20].

The natural Tamarugo forest was almost extinct in the 19th century and during the 1970's an enormous reforestation effort was carried out by the Chilean government and 13,000 hectares of Tamarugo were planted in the Pampa del Tamarugal basin [21]. Currently, the Pampa del Tamarugal is under threat due to GW overexploitation. Chilean policy makers, scientists and private companies have debated intensively about defining environmentally safe GW extractions. To achieve this, good indicators of the Tamarugo water condition are needed and remote sensing, and specifically the NDVI, has proved to be useful for assessing Tamarugo's water condition [8,9]. Nevertheless, time series of NDVI have not been directly related to GW depletion yet and to do so, the effect of the leaf pulvinar movements must be considered to understand a) the natural NDVI dynamic in the absence of water stress, and b) how this dynamic may be altered by GW depletion. In this paper, we use MODIS and Landsat NDVI time series to study both the natural and the altered NDVI dynamics of Tamarugo stands located in the Pampa del Tamarugal basin. Furthermore, we explore other biological (phenology) and environmental factors (precipitation) with potential effects on the NDVI signal.

Material and methods

2.1 Species description

Tamarugo is a phreatophytic desert tree that is highly specialized to survive the hyper-arid conditions of the Atacama Desert. This species belongs to the Leguminoseae family, Mimosaceae subfamily and it can reach up to 25 meters height, 20–30 meters crown size and 2 meters stem diameter [22,23]. The branches are arched and twigs flexuous with composite leaves, often bipinnate with 6–15 pairs of folioles (Figure 2b, c, f) [24]. The Tamarugo petioles have a distinctive structure of motor cells in the pulvinus, responsible for the leaf paraheliotropic movements (Figure 2d, e, f). Differential turgor changes of the pulvinus cells make the leaves to stand up and orientate the leaf lamina parallel to the incoming sun rays. The composite leaves of Tamarugo have three levels of pulvinar structures: the first at the base of the

Figure 2. Pulvinar structures of Prosopis tamarugo leaves. (A) Tamarugo trees, (B) leaf angle randomly distributed during the morning when the solar radiation is low, (C) leaf angle in erectophyle position to avoid facing high solar irradiation at midday, (D) transversal section of a closed pulvinus (empty of water) during the morning, (E) transversal section of an open pulvinus (filled with water), which allows leaves to stand up and reach the erectophyle position, and (F) detail of the base of a Tamrugo pinna showing the three levels of pulvinar structures (at the base of the bipinna, of each pinna and each foliole).

bipinna, the second at the base of each pinna, and the third at the base of each of the folioles. This pulvinar mechanism at the three levels allows the Tamarugo canopy to adjust its internal structure to avoid facing excessive solar irradiation. Tamarugos are phreatophytic species [25,26] presenting a dual root system consisting of a deep taping root and a dense superficial root mat [27]. This dual system would allow Tamarugos to move water from the deep groundwater table to the superficial root mat layer during the night to ensure water supply during the growing season when the water demand at the capillary fringe increases [26].

2.2 Study area

The study area is located in the Atacama Desert (Northern Chile), specifically in the southern part of the Pampa del Tamarugal basin, where most of the remaining Tamarugo population is concentrated (Figure 3). The Tamarugo forest is practically the only ecosystem of the Absolute Desert eco-region [16], and it is characterized by almost null precipitation, high day-night temperature oscillation, and high potential evapotranspiration [28,29]. Most of the plantation stands (Pintados and Bellavista) are in the southern part of the basin and within the study area. Just little natural patches of Tamarugo remain in the

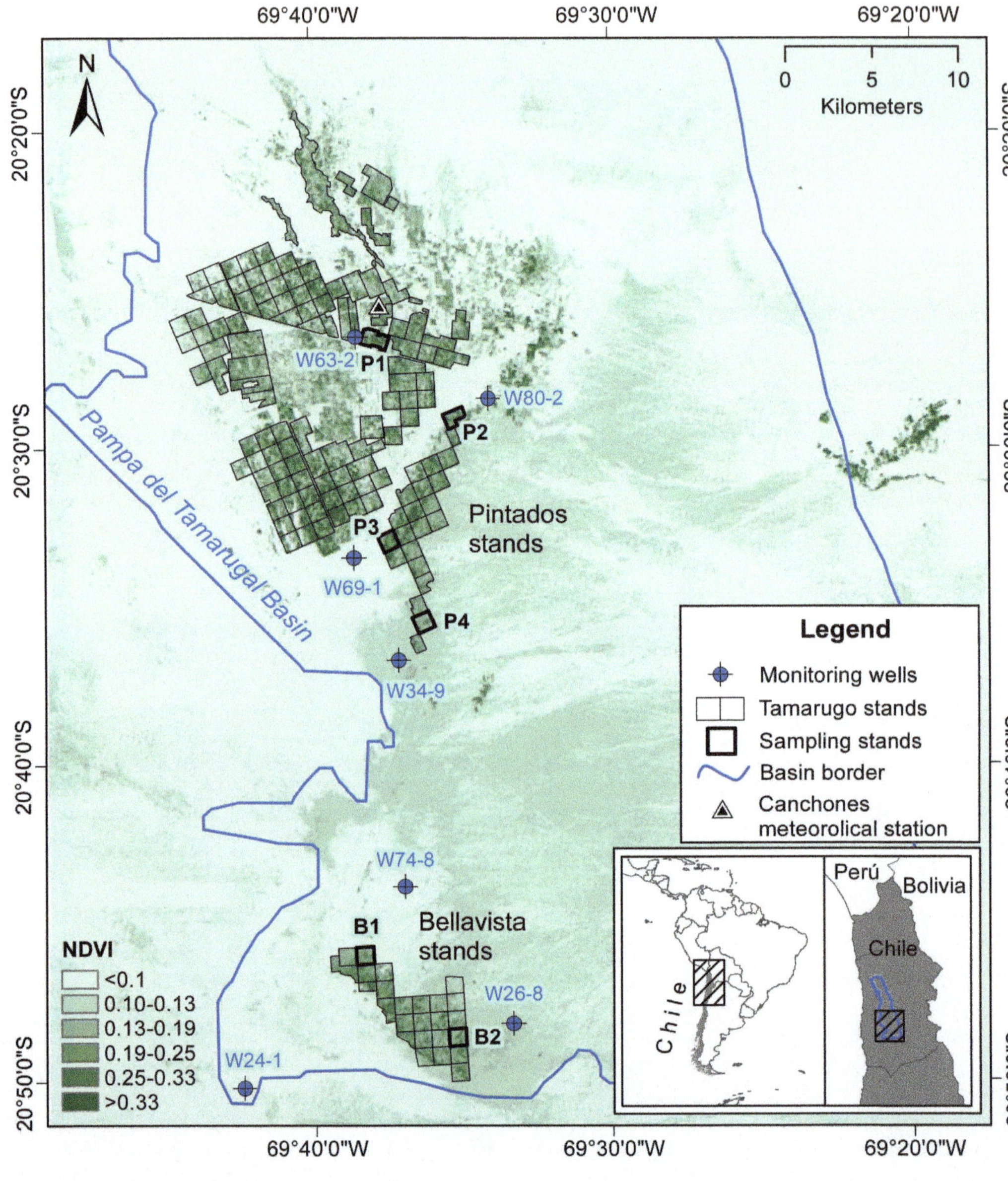

Figure 3. Landsat NDVI image showing the location of the Tamarugo stands (Winter 2007).

northern portion of the Pintados plantation (Figure 3). Although the oldest plantation stands were established as early as 1936, most of the existing plantation stands were planted between 1968 and 1972 [30]. The plantation scheme consisted of squared 1×1 kilometres stands and trees separated 10×10 meters. Besides Tamarugo plantations, there are plantations of other *Prosopis* species, sometimes mixed with Tamarugo. Only pure Tamarugo plantation stands and some natural forest patches were considered in this study and they can be identified in Figure 3 as the green areas highlighted in black.

2.3 Landsat and MODIS NDVI time series

We used all available Landsat 5 TM and Landsat 7 ETM data (referred from here onwards in the text as 'Landsat' data) as well as MODIS-Terra and MODIS-Aqua data of the study area covering the period 1989–2012. We selected this time frame since this is the period of time with available GW depth records for most of the monitoring wells located in the study area (Figure 3). For the Landsat NDVI time series we used cloud free L1T images of 30 meters pixel resolution (471 scenes) corresponding to path 1 and row 34 and pre-processed using the Landsat Ecosystem Disturbance Adaptive Processing System (LEDAPS) to obtain surface reflectance values for all spectral bands [31]. Finally, we used the surface reflectance values of red and NIR to compute the NDVI for each date as follows: NDVI = (NIR-Red)/(NIR+Red). For the MODIS-Terra and MODIS-Aqua NDVI time series we used the MODIS 16-day composites at 250 meters pixel resolution (MOD13Q1 and MYD13Q1 data products). MODIS pixel reliability showed that 85% of the observations can be used with confidence (reliability =0) and 15% were considered useful (reliability=1) of which MODIS vegetation index quality indicated average aerosol quantity. MODIS pixels with reliability 0 and 1 were considered in this study and showed consistent values for the NDVI time series of all forest stands. Both MODIS and Landsat data were downloaded from the USGS Earth Explorer website. Complementary, we used a panchromatic WorldView2 image of

Table 1. Plantation stands close to monitoring wells in the Pampa del Tamarugal basin.

					Groundwater depth (m)			
Stand	Plantation year	Canopy coverage (%)	Closest monitoring well (DGA code)	Distance to well (km)	1989	1997	2007	2012
B1	1968–1969	17	017000-74-8	4.7	10.99*	11.19*	11.64*	11.90*
			017000-26-8	9.5				
			017000-24-1	10.1				
B2	1968–1969	21	017000-74-8	9.3	14.75*	14.87*	15.45*	15.82*
			017000-26-8	3.4				
			017000-24-1	12.8				
P1	unknown	22	017000-63-2	0.4	7.97	9.03	9.93	10.26
P2	1972	25	017000-80-2	1.6	9.90	11.91	12.97	13.11
P3	1972	27	017000-69-1	1.7	5.70	6.26	6.87	6.89
P4	1972	11	017000-34-9	1.9	7.93	8.65	9.54	9.55

(*)Groundwater depth and depletion estimated using inverse distance weighted interpolation of 3 neighbouring wells.

0.6 meters pixel resolution to quantify the tree coverage of each plantation stand. This was carried out by using object-based image classification and the eCognition software following the procedure used by Chávez et al. [8].

2.4 Groundwater and climatic data

GW records were obtained from the monitoring network of the Dirección General de Aguas (DGA), the Chilean Water Service. From this network, seven wells were close to the Tamarugo stands and had enough records to establish a direct relationship between the groundwater table and the forest status (Figure 3). We averaged the (three to twelve) records of each year to obtain annual values of groundwater depth for the seven monitoring wells. Figure 3 shows the location of the monitoring wells used in this study as well as the forest stands located close to each well. Basic data of each monitoring well are provided in Table 1. This way we obtained representative groundwater data for six Tamarugo stands for the period 1989–2012. In the case of stands B1 and B2, the groundwater depth was estimated using an inverse distance weighted interpolation of records from three wells (see Table 1).

Although groundwater is the main water source of the Tamarugo forest, sporadic precipitation may occur in the Atacama Desert, having a positive impact on the water status of the trees and the NDVI signal. For this reason, we included in our analysis precipitation records from the DGA meteorological station Huara en Fuerte Baquedano (20°07'51"S, 69°44'59"W) located about 30 km north from the study area and at a similar altitude (1,100 m). Solar irradiation records were obtained from the Canchones Experimental Station of the Universidad Arturo Prat (Chile), located next to the Tamarugo stand P1 in the northern part of the study area (Figure 3).

2.5 Data analysis

2.5.1 NDVI signal in the absence of water stress (natural dynamic). Finding Tamarugo vegetation without nearby GW depletion in the Atacama Desert was a difficult task. We identified a Tamarugo forest stand (B1) and a time frame (2005-2008) with almost null GW depletion and no precipitation events in the southern part of the study area (see Figure 3, Bellavista stand). We assumed the NDVI time series of this three year period was not strongly influenced by the growth of trees. The Landsat, MODIS-Terra and MODIS-Aqua NDVI time series for the stand B1 were calculated using the median value of the pixels inside the 1 × 1 km stand. This aggregation enabled direct comparison of Landsat and MODIS NDVI time series. In the case of the Landsat time series we excluded pixels with NDVI values lower than 0.13, which were considered as no forest pixels. This threshold was set by considering the NDVI values observed outside the plantation stands, which correspond to completely bare areas (Figure 3). We first analysed the time series without any level of temporal aggregation, and then we aggregated the values to monthly averages in order to study the relationship between the NDVI and the monthly mean solar irradiation. For the latter purpose we used simple linear regression.

2.5.2 NDVI signal under water stress. After studying the natural dynamic of the NDVI time series for the three satellite sensors, we analysed the relationship between the average annual records of GW depletion and different metrics derived from the NDVI signal. To achieve this we used simple linear regression between the cumulative GW depletion and the NDVI derived metrics of the period between 1997 and 2007 with no precipitation. For the Landsat NDVI time series, these metrics were: annual NDVI average ($NDVI_{av}$), NDVI in winter ($NDVI_{W}$), and the NDVI difference between winter and summer ($\Delta NDVI_{W\text{-}S}$). For MODIS NDVI time series, these metrics were the $\Delta NDVI_{W\text{-}S}$ and the NDVI difference between morning and midday ($\Delta NDVI_{mo\text{-}mi}$). We calculated $\Delta NDVI_{W\text{-}S}$ and $\Delta NDVI_{mo\text{-}mi}$ for each year as follows:

1) $MODIS\ \Delta\ NDVI_{W-S} = MODIS\ Terra\ NDVI_{W} - MODIS\ Terra\ NDVI_{S}$,

2) $Landsat\ \Delta\ NDVI_{W-S} = Landsat\ NDVI_{W} - Landsat\ NDVI_{S}$, and

3) $\Delta\ NDVI_{mo-mi} = MODIS\ Terra\ NDVI_{W} - MODIS\ Aqua\ NDVI_{W}$.

Where:

$NDVI_{W}$
$= average\ of\ all\ NDVI\ scenes\ of\ May,\ June,\ July (winter)$

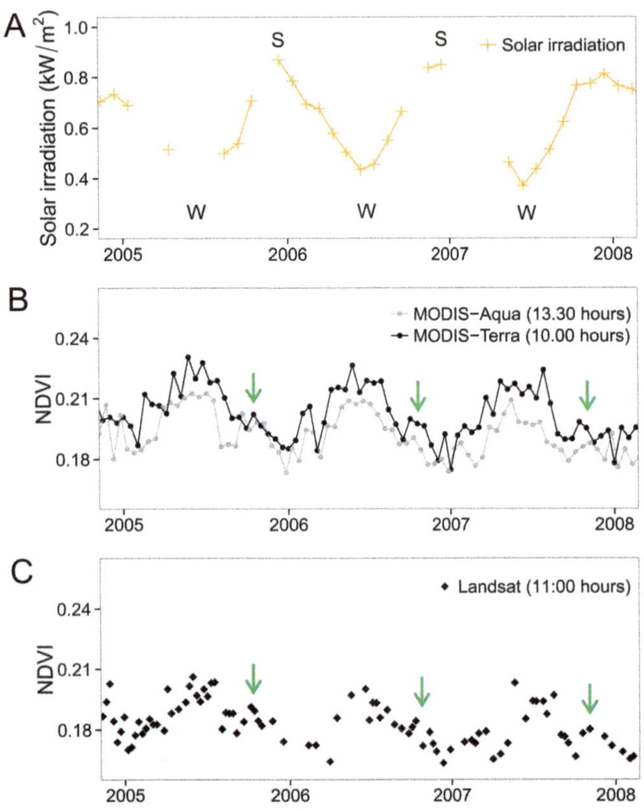

Figure 4. Time series of solar irradiation and NDVI for the B1 site (low groundwater depletion). (A) Solar irradiation, (B) MODIS 16 days composite NDVI, and (C) Landsat NDVI of the B1 site. Arrows indicate the peak of Tamarugo's vegetative period. S = summer, W = winter.

$$NDVI_S = \textit{average of all NDVI scenes of November, December, January(summer)}$$

In the case of the Landsat NDVI time series, we considered a minimum of three scenes for the summer and winter period to obtain a representative value of the respective season. For all Landsat and MODIS NDVI scenes we used the median value of the pixels inside the 1 × 1 km stands.

Results

3.1 Leaf pulvinar movement and the NDVI natural dynamic

In the absence of GW extraction or precipitation events, the NDVI signal of the Tamarugo stand B1 presented a strong seasonal variation for the period 2005–2008, mainly explained by the seasonal variation of the monthly average solar irradiation (Figure 4) influencing the pulvinar movement of paraheliotropic plants (Figure 1). The R^2 for the linear relationship between NDVI and solar irradiation was 0.66 for the Landsat NDVI time series, 0.65 for the MODIS-Terra NDVI, and 0.41 for the MODIS-Aqua NDVI. Partial foliage loss during the period May-September and the peak of the vegetative period occurring around October seemed to have only a marginal effect on the NDVI time series, noticeable as a small drop followed by a peak around October (green arrows in Figure 4). Overall, the seasonal variation is the main feature of the annual NDVI signal, and therefore the $\Delta NDVI_{W\text{-}S}$ may be used to detect the leaf pulvinar movement occurring in the Tamarugo canopy under natural conditions.

Besides the seasonal variation, the MODIS-Terra and MODIS-Aqua NDVI time series allowed to identify the $\Delta NDVI_{mo\text{-}mi}$ reported by Chávez et al. [8] on single Tamarugo trees, for instance for the Tamarugo stand B1 (Figure 4b). Although the $\Delta NDVI_{mo\text{-}mi}$ was clearly noticeable during winter, it was close to zero in summer. This was expected since both the morning and midday solar irradiation in the Atacama Desert are much higher in summer than in winter. For example, the average solar irradiation of June 2007 (winter) was 0.21 kW/m^2 at 10.00 hours and 0.62 kW/m^2 at 13.30 hours while the average of December 2006 (summer) was 0.87 kW/m^2 at 10.00 hours and 0.94 kW/m^2 at 13.30 hours. As a result, in winter the leaves will only have an erectophile position at midday, but in summer this occurs already half way the morning (yielding a small $\Delta NDVI_{mo\text{-}mi}$). Based on what we observed in Figure 4 for a Tamarugo stand without water stress we can expect that it has a positive $\Delta NDVI_{mo\text{-}mi}$ in winter as well as a positive $\Delta NDVI_{W\text{-}S}$. Both NDVI derived metrics can be quantified and mapped using Landsat and MODIS images as shown in Figure 5.

Figure 5 displays the NDVI values at pixel level of all Bellavista plantation stands (including the stand B1) in the winter of 2007 (first row), the summer of 2006–2007 (second row), and the $\Delta NDVI_{W\text{-}S}$ of 2007 (third row) obtained from Landsat images (first column), MODIS-Terra images (second column), and MODIS-Aqua images (third column). The fourth column corresponds to the $\Delta NDVI_{mo\text{-}mi}$ in winter (Figure 5d) and summer (Figure 5h) based on Terra (morning) and Aqua (midday). This figure confirms that the $\Delta NDVI_{mo\text{-}mi}$ in winter and the $\Delta NDVI_{W\text{-}S}$ of 2007 was positive for the forested area. On the other hand, the $\Delta NDVI_{mo\text{-}mi}$ in summer was zero or close to zero. In a similar way, and as a consequence of the diurnal pulvinar movements, the $\Delta NDVI_{W\text{-}S}$ was higher when using MODIS-Terra images than when using MODIS-Aqua images. Thus, the most promising indicators of pulvinar movement seemed to be the $\Delta NDVI_{mo\text{-}mi}$ in winter and the $\Delta NDVI_{W\text{-}S}$ in the morning (MODIS-Terra). When using the NDVI as a potential indicator of Tamarugo's water status, the signal in winter was stronger. The canopy coverage can also play an important role in the strength of the NDVI signal and its effect has to be considered when using these NDVI derived metrics for monitoring purposes. We will discuss this issue further in the next section where more Tamarugo stands, with different canopy coverage, were analysed.

3.2 Groundwater depletion: the NDVI signal under water stress

Figure 6 displays the annual time series of GW depth and the Landsat $NDVI_W$ and $\Delta NDVI_{W\text{-}S}$ for the six Tamarugo stands analysed in this study. The precipitation events are indicated with arrows. Only four precipitation events were recorded in the 24 years period analysed: 3.0 mm in 1996, 1.8 mm in 2008, 7.9 mm in 2011, and 2.2 mm in 2012, three of them during the last five years. The Landsat $NDVI_W$ signal reacted to the precipitation event of 1996 by showing a short recovering phase (about one year) and quickly returned to the general decreasing trend. For the precipitation events in the last years this effect was difficult to observe since they occurred close to each other in time. These precipitation events did not have any impact on the groundwater table, so we assumed this water was only available for the trees in the superficial soil layers. Although these precipitation events contributed little water to the basin, we assume the moisture added

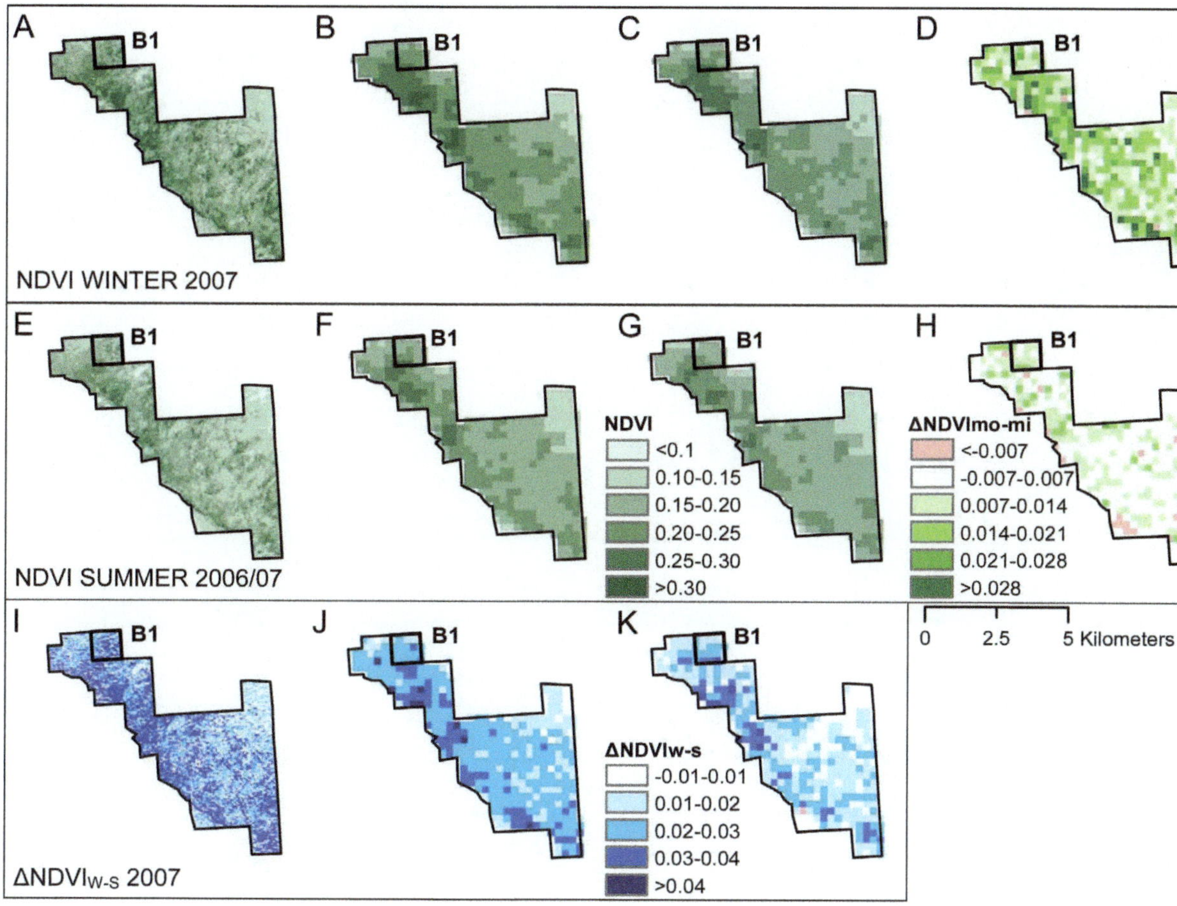

Figure 5. ΔNDVI morning-midday and ΔNDVI winter-summer of the Bellavista plantation in 2007. Winter 2007: (A) Landsat NDVI, (B) MODIS-Terra NDVI (morning), (C) MODIS-Aqua NDVI (midday), (D) $\Delta NDVI_{mo-mi}$ = B–C; **Summer 2006–07**: (E) Landsat NDVI, (F) MODIS-Terra NDVI (morning), (G) MODIS-Aqua NDVI (midday), and (H) $\Delta NDVI_{mo-mi}$ = F–G. Graphs I, J and K display the **$\Delta NDVI_{W-S}$ 2007**, where (I) Landsat $\Delta NDVI_{W-S}$ = A–E, (J) MODIS-Terra $\Delta NDVI_{W-S}$ = B–F, and (K) the MODIS-Terra $\Delta NDVI_{W-S}$ = C–G.

to the superficial root mat of Tamarugo trees temporally had a positive impact on the growth of the trees. Apart from the precipitation events, the analysed NDVI metrics seemed to follow the GW depth trend for all stands. To quantify this relationship, we calculated the R^2 for the linear regression between each of the NDVI metrics and the cumulative GW depletion for the period without precipitation (1997–2007). We also included in this analysis the Landsat annual $NDVI_{av}$ values to check whether the $NDVI_W$ was a better indicator than the simple annual NDVI average. The results are given in Table 2.

The Bellavista Tamarugo stands (B1 and B2) are located in the southern part of the basin and far from the area where the pumping wells are concentrated, which is towards the north and east of the Pintados stands (Figure 3). For this reason, the GW depletion in the stands B1 and B2 was less in comparison to the stands of the Pintados sector (P stands), especially in the case of P2. The stand B1 showed the lowest cumulative depletion (0.45 m) for the period 1997–2007 as well as the lowest R^2 (<0.1) for the relationship between Landsat $NDVI_{av}$ and GW depletion. Furthermore, the R^2 of the GW depletion - Landsat $NDVI_W$ relationship was also the lowest, but higher than the GW depletion - Landsat $NDVI_{av}$ relationship. In fact, this was the case for almost all stands. Thus, the Landsat $NDVI_W$ was more sensitive to changes in GW depth than the Landsat $NDVI_{av}$. This was also the case when comparing Landsat $NDVI_W$ with Landsat $\Delta NDVI_{W-S}$. Only for the stand B1, the R^2 of the Landsat $\Delta NDVI_{W-S}$ - GW depletion relationship was higher than for the $NDVI_W$ - GW depletion relationship.

The rest of the stands showed GW depletions between 0.58 and 1.06 meters between 1997 and 2007 and R^2 values for the Landsat $NDVI_W$ - GW depletion relationship higher than 0.75 except for the stand P2 with an R^2 of 0.29. The stand P2 is located close to the pumping area, and therefore the GW depletion could have been influenced by short-term changes of the pumping rate. If the intra-annual GW values fluctuated too rapidly, the depletion may not have had an effect on the NDVI signal. However, this is difficult to detect in annually averaged records. Overall the Landsat $NDVI_W$ was the most sensitive NDVI derived metric to the 11-year changes in GW depletion.

In the case of the MODIS NDVI derived metrics, the R^2 values presented in Table 2 were difficult to interpret since the time series without precipitation events was very short (2003–2007). We found R^2 values as high as 0.70 when using the MODIS $\Delta NDVI_{W-S}$ (stand B2) and the MODIS $\Delta NDVI_{mo-mi}$ (stand P2), but also <0.1 (stands P1 and P4) for $\Delta NDVI_{W-S}$ or $\Delta NDVI_{mo-mi}$ (stand P3).

3.3 Mapping water stress using Landsat $\Delta NDVI_{W-S}$

The $NDVI_W$ and $\Delta NDVI_{W-S}$ showed good potential to assess the effect of GW depletion on the water status of Tamarugo trees.

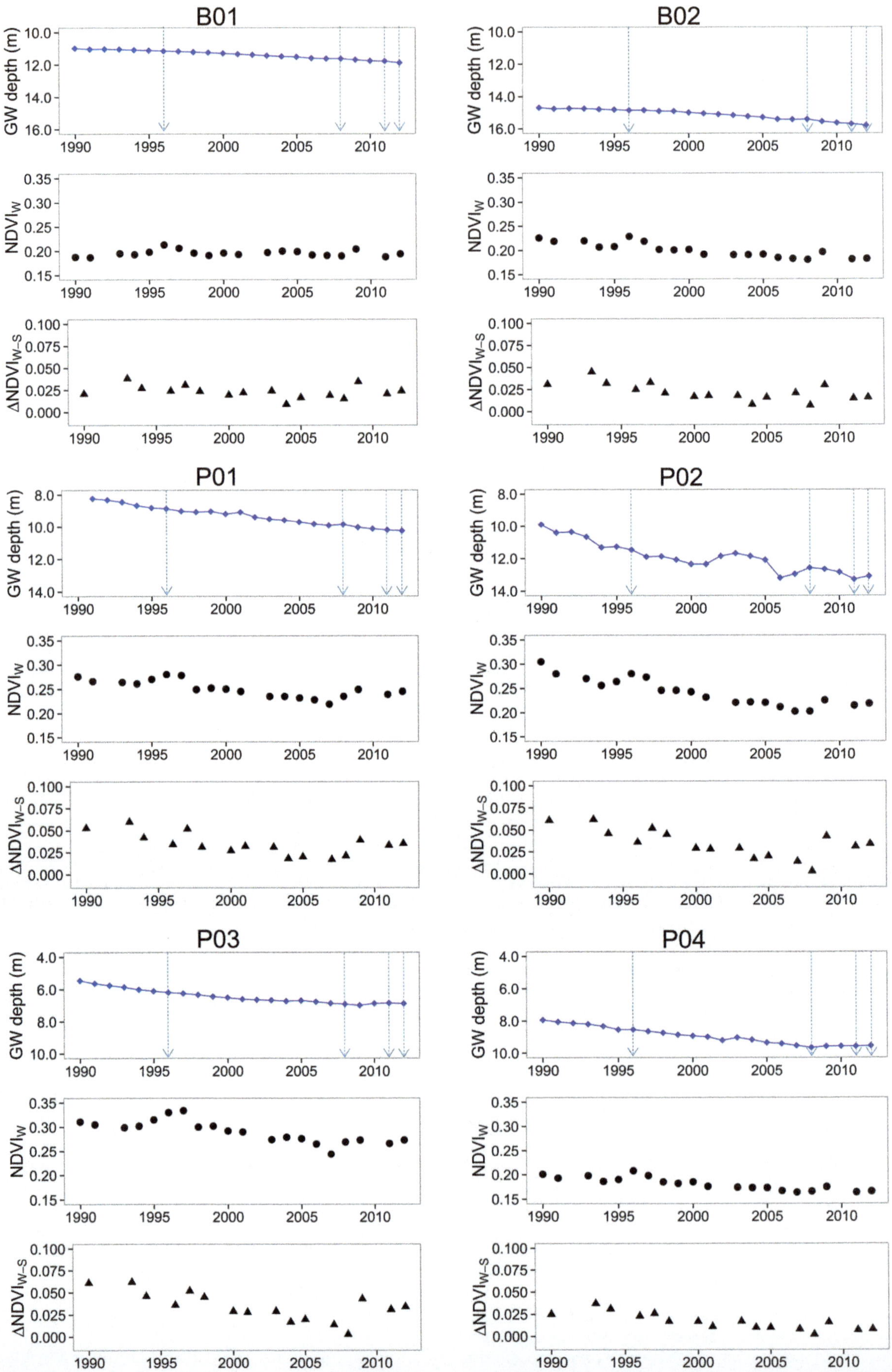

Figure 6. Time series of groundwater depth, Landsat $NDVI_W$, and Landsat $\Delta NDVI_{W-S}$ for six Tamarugo plantation stands. Blue arrows indicate precipitation events.

Table 2. R^2 of the linear model of cumulative groundwater depletion v/s $NDVI_{av}$, $NDVI_W$, $\Delta NDVI_{W-S}$ and $\Delta NDVI_{mo-mi}$ for the period 1997–2007 (no precipitation events).

Stand	Cumulative GW depletion (1997–2007)	Landsat $NDVI_{av}$	Landsat $NDVI_W$	Landsat $\Delta NDVI_{W-S}$	MODIS $\Delta NDVI_{W-S}$	MODIS $\Delta NDVI_{mo-mi}$
		n = 11	n = 10	n = 8	n = 4	n = 5
B1	0.45	<0.1	0.13	0.44*	0.55	0.29
B2	0.58	0.74***	0.76***	0.26	0.76	0.35
P1	0.90	0.58***	0.75***	0.60**	<0.1	<0.1
P2	1.06	0.27*	0.29	0.24	0.70	0.70
P3	0.61	0.82***	0.90***	0.70***	0.66	0.14
P4	0.89	0.77***	0.85***	0.74***	<0.1	0.52

Significative linear relationship with ***P<0,01; **P<0.05, *P<0.1.
av = average; W = winter; W-S = winter-summer; mo-mi = morning-midday.

We selected the Landsat $\Delta NDVI_{W-S}$ to map this effect in the study area because we believe it senses water stress earlier than NDVIw (see Discussion section for more details). We mapped the $\Delta NDVI_{W-S}$ for three different years: 1997, 2007, and 2011 as shown in Figure 7 averaged to a 1×1 km grid. The $\Delta NDVI_{W-S}$ difference between 1997 and 2007 can be explained by groundwater depletion since no precipitation events occurred in this period. For most of the stands, the $\Delta NDVI_{W-S}$ values in 2011 showed a recovery of the forest after the precipitation event of 2011 (7.9 mm), the most intense rain recorded in the last 25 years in Pampa del Tamarugal. The stands with more stable $\Delta NDVI_{W-S}$ through time were those located at the west border of the Bellavista plantation, close to the well W24-1, which reported very shallow GW depths in 1997 (2.6 m), 2007 (3.0 m), and 2011 (3.1 m). Furthermore, a $\Delta NDVI_{W-S}$ gradient can be observed in the Bellavista sector from east to west, showing a good spatial agreement with the increasing GW depletion towards the east (GW depth in well W26-8 was about 19.6 m in 1997, 20.3 m in 2007, and 21 m in 2011).

Discussion

Early stages of water stress in plants are associated with a lower leaf water potential, reduction in transpiration rate, and foliage water loss while late stages are associated with pigment degradation, biomass loss, and finally dying plants [32–34]. Although Tamarugo trees are naturally adapted to the predominant water scarcity of the Atacama desert, they can be affected by water stress due to GW depletion as shown in this paper. From previous papers [8,9], we know that Tamarugos show the typical water stress symptoms of most plants, but additionally water stress limits the normal functioning of the leaf pulvinar mechanism. These pulvinar movements are typical for heliotropic species. Furthermore, we have shown that leaf pulvinar movement can be remotely sensed by different metrics derived from the NDVI signal, allowing to understand both the temporal natural dynamic of the Tamarugo forest and the effects of GW depletion. In Table 3 we give an overview of the water stress symptoms of Tamarugo trees, the temporal scale at which they occur, and the NDVI derived metrics we can use to study these symptoms.

Diurnal leaf movements can be studied using $\Delta NDVI_{mo-mi}$ from MODIS Terra and Aqua satellites as shown in this paper. No significant differences have been found for the MODIS NDVI Terra and Aqua for other non-solar tracker vegetation [35,36]. Since these two satellites acquire data on a daily basis, it would be possible to map the $\Delta NDVI_{mo-mi}$ of the Pampa del Tamarugal basin every day at a spatial resolution of 250×250 m. This way, the effects of an abrupt GW depletion could be identified using MODIS data if the forest is dense enough to provide a sufficiently strong signal as well as large enough to cover one or more MODIS pixels [37,38]. In this paper we analysed averaged $\Delta NDVI_{mo-mi}$ values for the winter seasons and its relationship with annual records of GW depth. This time series was rather short, sometimes resulting in low R^2 values. Perhaps better results can be achieved when using the full temporal resolution (daily or 16 days) of the MODIS NDVI products and more detailed records of the water availability. This is an interesting topic for further research and not only for Tamarugo plants, but also for detecting short-term water stress in, e.g., bean crops, which also have documented paraheliotropic behaviour [5,6].

Seasonal differences of leaf pulvinar adjustments of Tamarugo vegetation can be studied at a large scale using the $\Delta NDVI_{W-S}$ as measured from Landsat (Figure 7) and MODIS Terra satellites. The advantage of using Landsat images is the possibility to map this variable at 30 meters pixel resolution and the disadvantage is that these satellites (Landsat 5, 7 and 8) have a revisit time of 16 days, increasing the chance of missing dates due to cloud cover. Although cloud cover is not such as problem in deserts, missing data can have an important impact on the calculation of the $\Delta NDVI_{W-S}$ if the NDVI values of winter or summer are not well represented by sufficient images. In this paper, we considered a minimum of three Landsat scenes for calculating a representative value of the summer or winter period. The NDVI signal of Tamarugo showed a strong seasonality (Figure 4c) and, for example, a calculation of the $NDVI_W$ using one or two images in May and a calculation of the $NDVI_S$ using one or two images in December may lead to a serious underestimation of the $\Delta NDVI_{W-S}$. This is not a problem for MODIS 16-day composites, which provide five or six images for the winter and summer period systematically distributed within the three months' timeframe. Therefore, there is a trade-off between temporal and spatial resolution when choosing Landsat or MODIS to detect the $\Delta NDVI_{W-S}$.

If the water stress persists, Tamarugo trees will react by selectively shutting down leaves, twigs and entire branches to reduce the transpiration surface while keeping the remaining foliage green with hydric parameters within normal ranges [8]. Foliage loss has been successfully assessed using NDVI for a wide range of vegetation types and it is especially accurate for LAI values <2 [39]. Such assessments are usually carried out at the

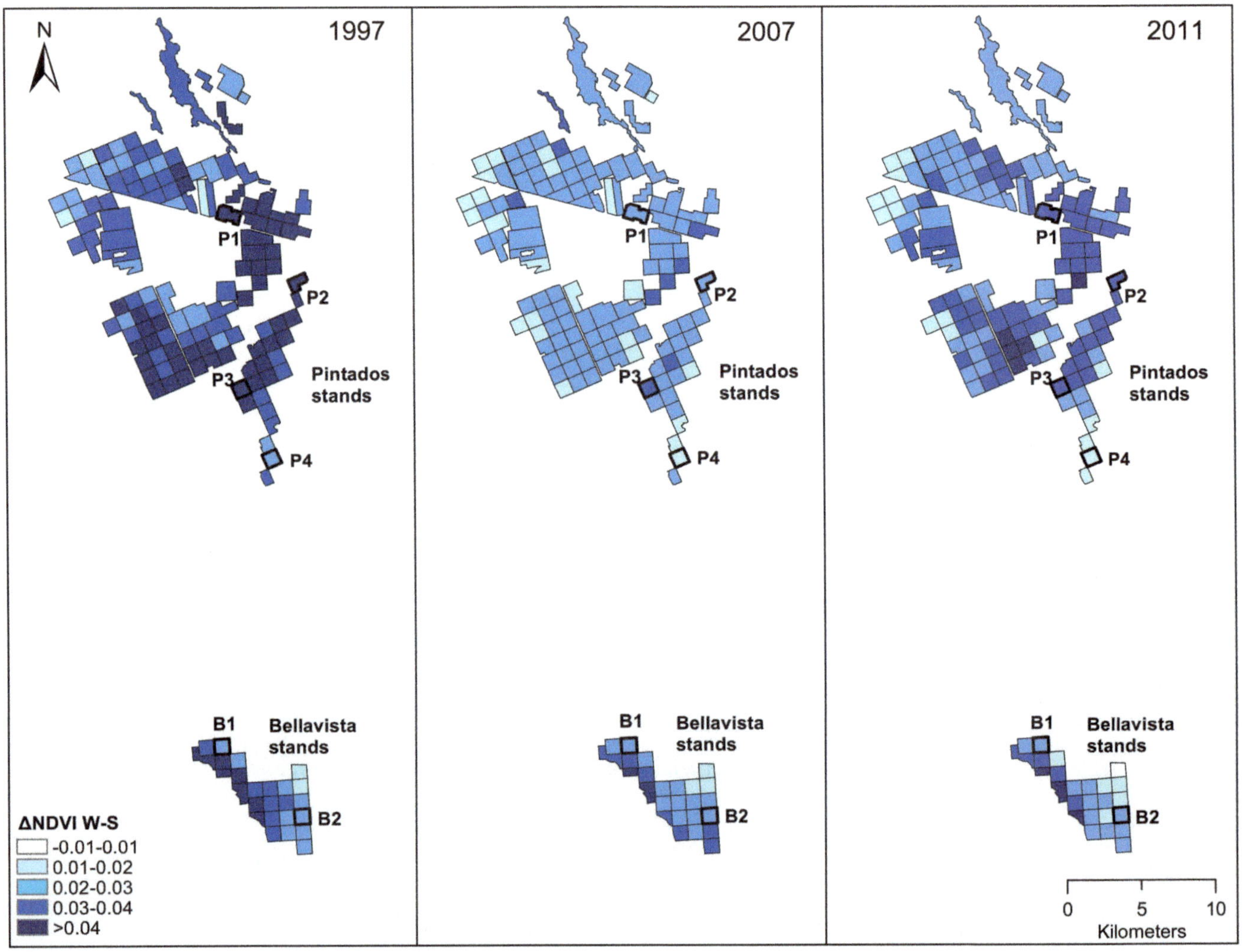

Figure 7. Landsat $\Delta NDVI_{W-S}$ of all plantation stands in 1997, 2007, and 2011 (after a precipitation event).

peak of the vegetative period, usually in spring. In the case of Tamarugo the seasonal variation of the NDVI signal is mainly driven by the pulvinar movements, which are primarily driven by seasonal changes in solar irradiation. Thus, the 'pulvinar effect' on the NDVI signal is minimum in winter and therefore this is the best time to retrieve the NDVI for inter annual foliage loss estimations (Table 3).

The strong relationship between $NDVI_W$ and cumulative GW depletion observed for most of the Tamarugo stands is an indication that foliage is decreasing in the study area as a consequence of water extraction, in other words, the forest is reaching an advanced stage of water stress (Table 3). However, it was not possible to discriminate whether the decreasing $NDVI_W$ signal was because some trees were dying while others remained alive (intra species competition) or all trees were losing foliage gradually. The tree coverage played also an important role in the absolute value of the $NDVI_W$ signal and, therefore, it was not possible to directly compare different stands at a single point in time. In order to better interpret the Landat and MODIS $NDVI_W$ signal, we believe that high spatial resolution remote sensing data can provide complementary information about the actual tree coverage of the forest as well as the water status of single trees. This will be the topic for further research.

A recent publication entitled 'Remote sensing: A green illusion' [40] has drawn the attention of the scientific community and policy makers on the issue of the correct interpretation of remote sensing derived products for environmental applications. The authors reflected on this issue based on the results of Morton et al. [41] showing how the apparent canopy greenness of the Amazon forest, interpreted as a positive response to more sunlight in the dry season, was caused by a bidirectional reflectance effect. In other words, it was caused by an optical artefact due to seasonal changes of the sun-sensor geometry. In this paper, we also discussed the correct interpretation of remote sensing derived products, but this time for paraheliotropic vegetation. As shown for the case of Tamarugo in this study, the seasonal changes in NDVI were related to leaf pulvinar movements causing a change in the canopy structure. This change in canopy structure explained the observed seasonal changes (Figure 4). Three pieces of evidence support the hypothesis that pulvinar movements are responsible for NDVI diurnal and seasonal changes of Tamarugo vegetation and that this is not an optical artefact due to bidirectional reflectance effects:

i. As shown in a previous paper [9], canopy spectral reflectance of Tamarugo plants simulated with the Soil-Leaf-Canopy

Table 3. Water stress stages of Tamarugo desert trees and NDVI based variables to assess their effects using satellite remote sensing.

Water stress			Remote sensing	
Stage	**Effects on Tamarugo vegetation**	**Temporal scale to perceive the effects**	**Monitoring variable**	**Sensor useful to retrieve this variable**
Instantaneous	Limitation of the diurnal photoinhibition control via pulvinar movements	Diurnal	$\Delta NDVI_{mo\text{-}mi}$	MODIS Terra and Aqua
Early	Limitation of the seasonal photoinhibition control via pulvinar movements	Seasonal	$\Delta NDVI_{W\text{-}S}$	MODIS Terra; Landsat
Advanced	Foliage loss	More than 1 year	$NDVI_W$	MODIS Terra; Landsat
Irreversible	Partial crown death and tree death	Several years	$NDVI_W$	MODIS Terra; Landsat, combined with very high spatial resolution sensors: Quickbird2, WorldView2 or GeoEye

(SLC) radiative transfer model showed that the SLC parameter LIDF (leaf inclination distribution function) could explain diurnal changes in canopy reflectance measured empirically with a spectroradiometer under laboratory conditions (lamp-sensor geometry was fixed). Thus, leaf movements, set in the SLC simulations as a 'random' LIDF in the morning and as an 'erectophile' LIDF after midday, explained diurnal changes in canopy reflectance in the absence of water stress.

ii. Another previous paper [8] showed a negative empirical relationship between diurnal values of NDVI, measured for single Tamarugo trees with a spectroradiometer, and solar irradiation under field conditions. In that study, the authors observed a predominantly erectophyle position of Tamarugo leaves around midday, corresponding to the diurnal peak of solar irradiation and the lowest values of NDVI. In the current paper, we showed a negative empirical relationship between seasonal NDVI values, measured by Landsat and MODIS satellites for Tamarugo stands, and solar irradiation (Figure 4). Furthermore, it is a known botanical fact that paraheliotropic movements are a response to increasing solar irradiation on the leaves [1]. Thus, pulvinar movements activated by changes in solar irradiation govern diurnal and seasonal changes in the NDVI signal of Tamarugo vegetation.

iii. This study provided evidence that the amplitude of the seasonal NDVI trend ($\Delta NDVI_{W\text{-}S}$) of Tamarugo stands declined with water stress (Figure 6). If the NDVI seasonal trend measured by satellite remote sensing was governed by a sun-sensor artefact, there is no reason why water stress would cause the amplitude of the NDVI signal to decline significantly.

In the southern hemisphere, more internal shadowing in satellite images (captured at nadir) is expected to occur in winter at lower solar elevation, and therefore, the bidirectional reflectance effect should cause an 'apparent greening' towards spring/summer [41]. However, as shown in Figure 4, the peak of the NDVI signal of Tamarugo stands does not occur in summer, but in winter. Although bidirectional reflectance effects may also occur in the case of Tamarugo vegetation, we believe that such effects are obscured by the stronger effect of seasonal pulvinar movement.

Conclusions

1. Monthly values of solar irradiation were negatively correlated to NDVI measured by the MODIS-Terra and Landsat satellites. Previous studies have shown that pulvinar movement causes the NDVI signal to drop from morning to midday as solar irradiation increases, and therefore, in the absence of water stress the seasonal variation of NDVI is also expected to be controlled by pulvinar movement.
2. The NDVI difference between midday and morning ($\Delta NDVI_{mo\text{-}mi}$), as measured by the difference of the NDVI signal from the MODIS Terra and Aqua satellites, can be used to detect the diurnal leaf pulvinar movement of Tamarugo plantation stands. This has not been reported in literature before, and therefore, this paper constitutes a proof of concept that MODIS images can be used to detect diurnal movements of paraheliotropic vegetation.
3. Similarly, the NDVI difference between winter and summer ($\Delta NDVI_{W\text{-}S}$), as measured by the Landsat or the MODIS Terra satellites, can be used to detect differences in seasonal pulvinar movements, associated to photoinhibition regulation.
4. Leaf pulvinar movements are triggered by changes in cell turgor and they can be limited by water stress. Thus, water stress in Tamarugo vegetation caused by groundwater overexploitation can be assessed and monitored using $\Delta NDVI_{mo\text{-}mi}$ and $\Delta NDVI_{W\text{-}S}$. For long time series (more than 10 years), Landsat $\Delta NDVI_{W\text{-}S}$ of Tamarugo stands showed a positive linear relationship with cumulative groundwater depletion.
5. Under water stress, a limitation of the pulvinar movement occurs in Tamarugo trees before they start losing foliage. For this reason, changes in $\Delta NDVI_{mo\text{-}mi}$ and $\Delta NDVI_{W\text{-}S}$ are expected to occur before NDVI decreases due to foliage loss, and therefore, $\Delta NDVI_{mo\text{-}mi}$ and $\Delta NDVI_{W\text{-}S}$ have potential for early water stress detection.

Acknowledgments

The authors would like to thank Digital Globe for providing the WorldView2 imagery and UNAP (Chile) for providing the solar irradiation records. Especial thanks to L. Dutrieux, B. de Vries, M. Schultz, and J. González de Tanago from Wageningen University for the valuable contribution of R code for the time series analysis. Finally, we thank V.

Urra and G. Valenzuela from Universidad de Chile for their collaboration on the anatomical description of Tamarugo leaves.

Author Contributions

Conceived and designed the experiments: ROC JGPWC MH. Performed the experiments: ROC JV PIN. Analyzed the data: ROC JGPWC JV PIN MH. Contributed reagents/materials/analysis tools: ROC JV PIN. Wrote the paper: ROC JGPWC.

References

1. Ehleringer J, Forseth I (1980) Solar tracking by plants. Science 210: 1094–1098.
2. Koller D (1990) Light-driven leaf movements*. Plant Cell Environ 13: 615–632.
3. Moran MS, Pinter Jr PJ, Clothier BE, Allen SG (1989) Effect of water stress on the canopy architecture and spectral indices of irrigated alfalfa. Remote Sens Environ 29: 251–261.
4. Koller D (2001) Solar navigation by plants. In: Comprehensive Series in Photosciences. pp.833–895.
5. Pastenes C, Porter V, Baginsky C, Norton P, González J (2004) Paraheliotropism can protect water-stressed bean (*Phaseolus vulgaris* L.) plants against photoinhibition. J Plant Physiol 161: 1315–1323.
6. Pastenes C, Pimentel P, Lillo J (2005) Leaf movements and photoinhibition in relation to water stress in field-grown beans. J Exp Bot 56: 425–433.
7. Kimes DS, Kirchner JA (1983) Diurnal variations of vegetation canopy structure. Int J Remote Sens 4: 257–271.
8. Chávez RO, Clevers JGPW, Herold M, Acevedo E, Ortiz M (2013) Assessing water stress of desert tamarugo trees using in situ data and very high spatial resolution remote sensing. Remote Sens 5: 5064–5088.
9. Chávez RO, Clevers JGPW, Herold M, Ortiz M, Acevedo E (2013) Modelling the spectral response of the desert tree *Prosopis tamarugo* to water stress. Int J Appl Earth Obs Geoinf 21: 53–65.
10. Williams DL, Goward S, Arvidson T (2006) Landsat: Yesterday, today, and tomorrow. Photogramm Eng Remote Sensing 72: 1171–1178.
11. Birdsey R, Angeles-Perez G, Kurz WA, Lister A, Olguin M, et al. (2013) Approaches to monitoring changes in carbon stocks for REDD+. Carbon Manage 4: 519–537.
12. Cohen WB, Goward SN (2004) Landsat's role in ecological applications of remote sensing. Bioscience 54: 535–545.
13. Verbesselt J, Zeileis A, Herold M (2012) Near real-time disturbance detection using satellite image time series. Remote Sens Environ 123: 98–108.
14. McKay CP, Friedmann EI, Gómez-Silva B, Cáceres-Villanueva L, Andersen DT, et al. (2003) Temperature and moisture conditions for life in the extreme arid region of the atacama desert: Four years of observations including the El Niño of 1997–1998. Astrobiology 3: 393–406.
15. Navarro-González R, Rainey FA, Molina P, Bagaley DR, Hollen BJ, et al. (2003) Mars-Like Soils in the Atacama Desert, Chile, and the Dry Limit of Microbial Life. Science 302: 1018–1021.
16. Gajardo R (1994) La vegetación natural de Chile. Clasificación y distribución geográfica. Santiago (Chile): Editorial Universitaria. 165 p.
17. CONAMA (2008) Biodiversidad de Chile, patrimonio y desafíos. Santiago de Chile: Ocho Libros Editores. 640 p.
18. Estades CF (1996) Natural history and conservation status of the Tamarugo Conebill in northern Chile. Wilson Bull 108: 268–279.
19. Ramírez-Leyton G, Pincheira-Donoso D (2005) Fauna del Altiplano y Desierto de Atacama. Vertebrados de la Provncia de El Loa. Calama, Chile: Phrynosaura Ediciones. 396 p.
20. Rojas R, Dassargues A (2007) Groundwater flow modelling of the regional aquifer of the Pampa del Tamarugal, Northern Chile. Hydrogeol J 15: 537–551.
21. Zelada L (1986) The influence of the productivity of *Prosopis tamarugo* on livestock production in the Pampa del Tamarugal - a review. For Ecol Manage 16: 15–31.
22. Altamirano H (2006) *Prosopis tamarugo* Phil. Tamarugo. In: C. Donoso, editor. Las especies arbóreas de los bosques templados de Chile y Argentina Autoecología. Valdivia (Chile): Marisa Cuneo Ediciones. pp.534–540.
23. Riedemann P, Aldunate G, Teillier S (2006) Flora nativa de valor ornamental. Chile, Zona Norte. Identificación y propagación. Santiago, Chile: Productora Gráfica Andros Ltda. 404 p.
24. Trobok S (1985) Fruit and seed morphology of Chilean *Prosopis* (Fabaceae-Mimosoidae). In: M. Habit, editor. The current state of knowledge on *Prosopis tamarugo*. Rome: F.A.O. Available: http://www.fao.org/docrep/006/ad316e/AD316E13.htm#ch3.2. Accessed 2014 February 28.
25. Aravena R, Acevedo E (1985) The use of environmental isotopes Oxigen-18 and deuterium in the study of water relations of *Prosopis tamarugo* Phil. In: M. Habit, editor. The current state of knowledge on *Prosopis tamarugo*. Rome: F.A.O. Available: http://www.fao.org/docrep/006/ad316e/AD316E14.htm#ch3.4. Accessed 2014 February 28.
26. Mooney HA, Gulmon SL, Rundel PW, Ehleringer J (1980) Further observations on the water relations of *Prosopis tamarugo* of the northern Atacama Desert. Oecologia 44: 177–180.
27. Sudzuki F (1985) Environmental moisture utilization by *Prosopis tamarugo* Phil. In: M. Habit, editor. The current state of knowledge on *Prosopis tamarugo*. Rome: F.A.O. Available: http://www.fao.org/docrep/006/ad316e/AD316E04.htm#ch1.2. Accessed 2014 February 28.
28. Houston J (2006) Evaporation in the Atacama Desert: An empirical study of spatio-temporal variations and their causes. J Hydrol 3: 402–412.
29. Houston J, Hartley AJ (2003) The central andean west-slope rainshadow and its potential contribution to the origin of hyper-aridity in the Atacama desert. Int J Climatol 23: 1453–1464.
30. CONAF (1997) Plan de manejo reserva nacional Pampa del Tamarugal. Corporación Nacional Forestal (CONAF). Ministerio de Agricultura. Gobierno de Chile. 110 p.
31. Masek JG, Vermote EF, Saleous NE, Wolfe R, Hall FG, et al. (2006) A landsat surface reflectance dataset for North America, 1990–2000. IIEE Geosci Remote S 3: 68–72.
32. Baret F, Houlès V, Guérif M (2007) Quantification of plant stress using remote sensing observations and crop models: The case of nitrogen management. J Exp Bot 58: 869–880.
33. Alpert P, Oliver MJ (2002) Drying without dying. In: M. . Black and H. W. Pritchard, editors. Desiccation and survival in plants: drying without dying. Wallingford (UK): CABI. pp.4–31.
34. Taiz L, Zeiger E (2010) Plant physiology. Sunderland, MA: Sinauer Associates. 782 p.
35. Wang J, Guo N, Wang X, Yang J (2007) Comparisons of normalized difference vegetation index from MODIS Terra and Aqua data in northwestern China. Int Geosci Remote Se pp.3390–3393.
36. Wu A, Xiong X, Cao C (2008) Terra and Aqua MODIS inter-comparison of three reflective solar bands using AVHRR onboard the NOAA-KLM satellites. Int J Remote Sens 29: 1997–2010.
37. Verbesselt J, Robinson A, Stone C, Culvenor D (2009) Forecasting tree mortality using change metrics derived from MODIS satellite data. For Ecol Manage 258: 1166–1173.
38. Wolfe RE, Nishihama M, Fleig AJ, Kuyper JA, Roy DP, et al. (2002) Achieving sub-pixel geolocation accuracy in support of MODIS land science. Remote Sens Environ 83: 31–49.
39. Gamon JA, Field CB, Goulden ML, Griffin KL, Hartley AE, et al. (1995) Relationships between NDVI, canopy structure, and photosynthesis in three Californian vegetation types. Ecol Appl 5: 28–41.
40. Soudani K, Francois C (2014) Remote sensing: A green illusion. Nature 506: 165–166.
41. Morton DC, Nagol J, Carabajal CC, Rosette J, Palace M, et al. (2014) Amazon forests maintain consistent canopy structure and greenness during the dry season. Nature 506: 221–224.

10

Genetic Differentiation Revealed by Selective Loci of Drought-Responding EST-SSRs between Upland and Lowland Rice in China

Hui Xia[1], Xiaoguo Zheng[1], Liang Chen[1], Huan Gao[1], Hua Yang[1], Ping Long[1], Jun Rong[2], Baorong Lu[3], Jiajia Li[1], Lijun Luo[1]*

1 Shanghai Agrobiological Gene Center, Shanghai, China, **2** Center for Watershed Ecology, Institute of Life Science and Key Laboratory of Poyang Lake Environment and Resource Utilization, Ministry of Education, Nanchang University, Nanchang, China, **3** Ministry of Education Key Laboratory for Biodiversity and Ecological Engineering, Fudan University, Shanghai, China

Abstract

Upland and lowland rice (*Oryza sativa* L.) represent two of the most important rice ecotypes adapted to ago-ecosystems with contrasting soil-water conditions. Upland rice, domesticated in the water-limited environment, contains valuable drought-resistant characters that can be used in water-saving breeding. Knowledge about the divergence between upland and lowland rice will provide valuable cues for the evolution of drought-resistance in rice. Genetic differentiation between upland and lowland rice was explored by 47 Simple Sequence Repeats (SSRs) located in drought responding expressed sequence tags (ESTs) among 377 rice landraces. The morphological traits of drought-resistance were evaluated in the field experiments. Different outlier loci were detected in the *japonica* and *indica* subspecies, respectively. Considerable genetic differentiation between upland and lowland rice on these outlier loci was estimated in *japonica* ($Fst = 0.258$) and *indica* ($Fst = 0.127$). Furthermore, populations of the upland and lowland ecotypes were clustered separately on these outlier loci. A significant correlation between genetic distance matrices and the dissimilarity matrices of drought-resistant traits was determined, indicating a certain relationship between the upland-lowland rice differentiation and the drought-resistance. Divergent selections occur between upland and lowland rice on the drought-resistance as the *Qsts* of some drought-resistant traits are significantly higher than the neutral *Fst*. In addition, the upland- and lowland-preferable alleles responded differently among ecotypes or allelic types under osmotic stress. This shows the evolutionary signature of drought resistance at the gene expression level. The findings of this study can strengthen our understanding of the evolution of drought-resistance in rice with significant implications in the improvement of rice drought-resistance.

Editor: Chengdao Li, Department of Agriculture and Food Western Australia, Australia

Funding: The work was supported by the following: Natural Science Foundation of China (Grant no. 31200279); Postdoctoral Science Foundation, Shanghai (No. 12R21421300); National Basic Research Program (Grant no. 2012CB114305); Shanghai "Phosphor" Science Foundation (Grant no. 11QA1405900); the National High-Tech Research and Development Program of China (863 Plan)(Grant No. 2012AA101102; 2014AA10A603); Shanghai Seed projects (2014.01); and Talented Person Project (2010CI120). The funders had no role in study design, data collection and analysis, decision to publish, or preparation of the manuscript.

Competing Interests: The authors have declared that no competing interests exist.

* Email: lijun@sagc.org.cn

Introduction

How plants have developed resistance to biotic and abiotic stresses is a fundamental question in plant biology. To address this question, plant ecotypes grown in contrasting environments are studied as they are likely under divergent selections, resulting in adaptive divergence [1]. Although ecotypes of wild species adapted to different ecosystems are served as ideal systems for studying the evolution of resistance to biotic and abiotic stresses [2–5], literatures on different ecotypes of a crop to a specific stress are still rare.

The Asian cultivated rice (*Oryza sativa* L.) is one of the most important cereal crops in the world by providing stable food for > 50% of the total global population (ref). However, rice production has encountered severe challenges due to frequent droughts and water shortages [6]. The utilization of natural variations of water-saving and drought-resistant characters in upland rice is an effective solution to improve the drought-resistance for rice [7–8]. Upland rice has adapted to the water-limited and rain-fed rice ecosystems, facing the higher risk of drought [9]. On the contrary, lowland rice is commonly planted in fields with irrigation facilities with relatively lower risks of drought. Upland rice may accumulate more drought-resistant genetic variances than lowland rice, leading to potential adaptive divergence of this rice ecotype on drought-resistance [7]. Therefore, studying differentiation between upland and lowland rice provides ideal systems for understanding the evolution of drought resistance and exploring drought-resistant genetic resources in crops. Previous attempts using neutral markers to explore differentiation between upland and lowland rice were not very successful [10–11], while another study using the functional-based markers has detected some ecotype-specific genetic features between upland and lowland rice [12].

The simple sequence repeats (SSRs) located in the ESTs (expressed sequence tags) are developed and widely applied in population genetics and breeding since the last decade [13]. The expression of drought-responding EST-SSRs is up- or down- regulated by drought stress. Although these genic SSRs are not always selective among ecotypes, they have a higher probability of being associated with the drought-resistance. In this study, 47 drought-responding EST-SSR loci were selected to explore the genetic differentiation between upland and lowland rice ecotypes from China. The questions to be addressed are: 1) Does considerable differentiation occur on drought-responding EST-SSR loci between upland and lowland rice ecotypes? 2) Is the differentiation between upland and lowland rice related to drought-resistance? 3) How do the upland- and lowland-differential alleles of the selective loci respond to drought stress? Answering these questions will facilitate our understanding of the evolution of drought-resistance in rice and give cues in rice breeding for drought-resistance.

Materials and Methods

Plant materials

A total of 377 rice landraces from Yunnan, Guizhou, Guangxi, Jiangsu, and Hebei province in China were used to study the differentiation between upland and lowland rice ecotypes (Table 1). Based on the information provided by the National Core Germplasm Database (http://crop.agridata.cn/A010110.asp), the landraces of upland rice from the five provinces were accounted for ~80% of the total upland rice germplasm in China. The predefined subspecies and ecotypes of rice landraces were provided by the institutes that collected them. Thus, crop materials in this experiment can be divided into four groups: *jap*-upland, *jap*-lowland, *ind*-upland, and *ind*-lowland. Additionally, 22 common wild rice strains were also used as reference while studying the differentiation among these two rice ecotypes. Twenty-two drought-resistant traits of 56 *japonica* and 49 *indica* materials were further evaluated in the field experiments rice landraces.

Rice DNA extraction and SSR genotyping

Rice genomic DNA was extracted from the 10-day-old seedlings following the common cetyltrimethyl ammonium bromide (CTAB) protocol. Three individual seedlings of a landrace were mixed together to include the genetic variations within a material. To focus on the drought-related genetic features, 47 polymorphic drought-responding EST-SSRs (SSRs locating in drought-responding ESTs) across 12 rice chromosomes were selected for genotyping (Table S1). The three primers system was applied in the polymerase chain reaction (PCR), containing a pair of EST-SSR primers and a rice-universal M13 sequence. The M13 sequence (5′ to 3′: CACGACGTTGTAAAACGAC) was labeled with fluorescent dyes at the 5′ end. The forward primer of the EST-SSR was also joint with the rice common M13 sequence at the 5′ end. The 20 ul reaction system contains 1×buffer ($Mg2+$), 2 mM each of dNTP, 50 ng of genomic DNA, 1 U of Taq polymerase, 16 mM of fluorescent dye labeled M13 sequence, 4 mM of the SSR primer liked with M13, and 20 mM of the other SSR primer. The PCR program started with a denaturation period of 4 min at 94°C, followed by 34 cycles of 40 s at 94°C, 30 s at their respective Tm (Table S1), and 40 s at 72°C, and ended after 10 min final extension at 72°C. The PCR products were then analyzed on ABI 3130XL (Applied Biosystems, USA) using ROX500 as the internal standard. The resulting chromatograms were analyzed and scored by Peakscanner ver. 1.0.

Field drought-resistance evaluation

The field evaluation of the drought-resistant traits was conducted at the Zhuanghang experimental station of Shanghai Academy of Agriculture Science in 2012. Two treatments were designed as control (CK) and drought (DT). Plants in CK were grown in a paddy field following the normal irrigated rice cultivation, while plants in DT were grown in a similar field nearby. The surface level of the DT field was approximately 2 meters higher than the CK field to simulate the upland condition and drain away the water more quickly. The DT field had a drought-resistance screen facility, which equipped a moveable platform capable of stretching out to keep the rainfall away when necessary. The germinated seeds were directly seeded both in the CK and DT on May 30th. However, water in the DT field was drained away on July 15th and it was never irrigated until August 25th. Individuals of each rice landrace were planted in a plot with 3×4 grids of 20 cm intervals between two seedlings. Consequently, there were 105 plots in one experimental block. Three replicates were included and arranged following the randomized complete block design. The number of stomas per leaf area, the excised leaf water (EWR) loss rate in 4 hours, and the root-shoot ratio were measured from leaves of main tillers in CK 30 days after the beginning of drought treatment. Relative water content (RWC) of leaf samples was measured 16 days after drought both in CK and DT. The content of malonaldehyde (MDA) was measured from fresh leaf tissues 40 days after drought following the protocol of the test kit (Nanjing Jiancheng Bioengineer Institute). The flag leaf length and width were measured at the heading stage both in CK and DT. All morphological traits were measured from three individuals from each replicate. The yield related traits were measured from four individuals of each replicate after harvest. The general drought resistance of a rice landrace was quantified by the drought-resistant index (DRI) [14]. It was calculated as $P_d/P_C*(P_d/P_{ad})$, where P_d is the yield production in drought treatment, P_C is the

Table 1. Summary of rice landraces used in the genotyping experiment and their inferred subspecies, ecotypes, and sources.

Groups Regions	*Japonica*		*Indica*	
	Upland	**Lowland**	**Upland**	**Lowland**
Guangxi	15	13	66	33
Yunan	15	6	15	24
Guizhou	42	17	12	9
Jiangsu	14	24	20	11
Hebei	22	19	0	0
Total	108	79	113	77

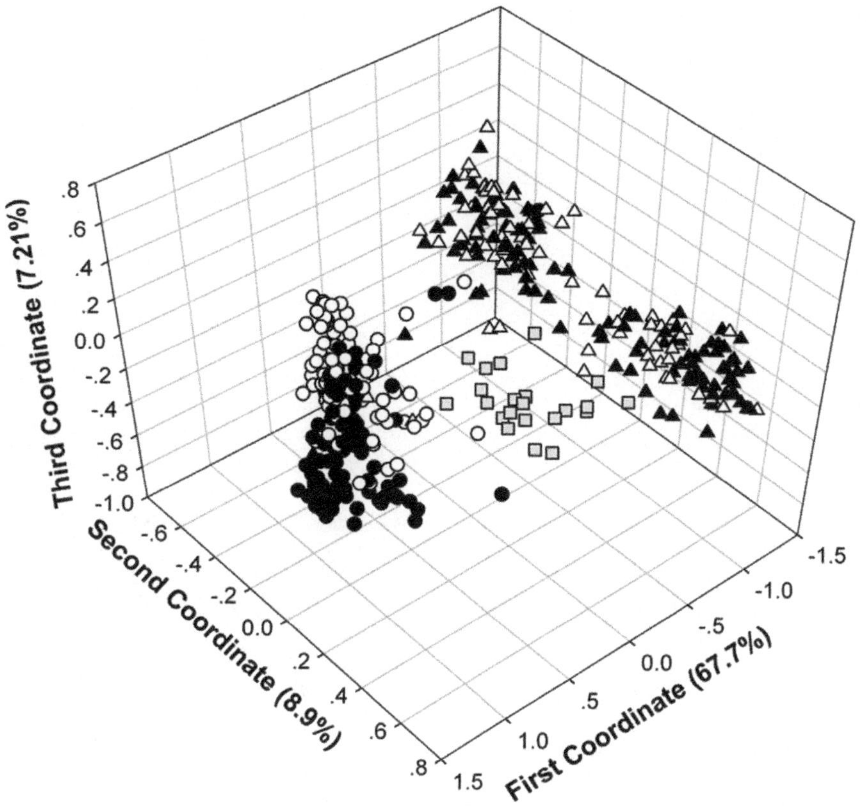

Figure 1. Population structure and differentiation investigated by Principal Coordinate Analysis. White color indicated lowland rice; dark color indicated upland rice; grey color and square indicated wild rice. Circles and triangles indicated *japonica* and *indica* subspecies, respectively.

yield production in control treatment, and P_{ad} is the average yield production of total materials in drought treatment.

Missing data and null alleles in data scoring

For genotyping, any null alleles were genotyped twice to avoid the manipulation errors. 337 putative null alleles were detected out of the total 18753 individual-loci combinations (1.8%), which was much lower than that expected by chance at the 5% level. The putative null alleles were treated as missing data in the further analyses.

Population structure and genetic diversity analysis

The model-based program STRUCTURE ver. 2.3.3 was used to infer population structure using the admixture model with a burn-in length of 50,000 and a run length of 50,000 for Markov Chain Monte Carlo iterations. Ten simulations were run for each K from K = 1 to K = 8 and the results were generated together using software CLUMPP_Windows.1.1.2. The Evanno's K was applied to determine the inferred K value [15]. The Principal Coordinate Analysis (PCoA) implemented in GenAlex ver. 6.43 was conducted to investigate the genetic distances between rice landraces and wild rice using the total 47 SSR loci. Genetic diversity was quantified by four estimators as gene diversity, number of alleles, heterozygosity, and polymorphism information

Table 2. Four genetic diversity estimators (mean ± standard deviations) analyzed in five groups using total loci set by PowerMarker ver. 3.25.

Group	Sample size	Allele No.	Gene Diversity	Heterozygosity	PIC
Jap-upland	108	6.88±0.59 b	0.435±0.030 a	0.036±0.008 b	0.399±0.032 a
Jap-lowland	79	6.49±0.46 a	0.428±0.040 a	0.024±0.004 a	0.402±0.033 a
Ind-upland	113	6.94±0.43 b	0.489±0.033 b	0.033±0.006 b	0.440±0.028 b
Ind-lowland	77	6.46±0.38 a	0.511±0.028 c	0.033±0.006 b	0.466±0.026 c
O. rufipogon	22	7.32±0.46 c	0.676±0.028d	0.336±0.023 c	0.636±0.025d
Overall	399	10.12±0.78	0.642±0.022	0.051±0.006	0.586±0.024

The different letters behind the values indicated significant differences ($p<0.05$) among groups by one-way ANOVA.

Table 3. Outlier loci detected under selection among upland and lowland ecotypes at 95% confidence by three outlier tests.

Method/Software	Code of locus	
	Japonica group	*Indica* group
Lositan	***E359, E647, E1238, E1899, E3735***	***E1177, E4208***
BayeScan	E214, E385, E674, E986, E1161, E1177, E1188, ***E1238***, E1615	***E647***, E1188
Detsel	***E359, E647, E1899***, E1941, ***E3735,***	E399, ***E647, E1177***, E1350, E1719, ***E4208***, E4632

Loci in **bold** and *italic* indicated they were detected at least twice.

content (PIC) using the software PowerMarker ver. 3.25 based on the total 47 EST-SSR loci. The standard deviations were calculated from 100 times bootstrap.

Neutrality tests for EST-SSR loci

Most of the genic markers used in this study were neutral among ecotypes as very low *Fsts* were recorded among rice ecotypes, similar with that using putatively neutral genomic SSRs [11]. Three *Fst*-based outlier tests were included to detect signs of selection among these genic SSRs respectively in the *indica* and *japonica* subspecies as indicated by the population structure. First, LOSITAN was used to detect loci under selection at the 95% confidence level [16]. An initial run with 100,000 simulations was conducted and followed by computing the distribution of neutral *Fst* using the putatively neutral loci derived from the initial run with 500,000 simulations. Second, the hierarchical-Bayesian method developed by Foll and Gaggiotti (2008) was applied to detect outlier loci using the software BAYESCAN (http://cmpg.unibe.ch/software/bayescan/) on the 95% posterior probabilities [17]. The parameters were set to 10 pilot runs of 5,000 iterations and additional burn-in of 50,000 iterations. Third, the software DETSEL 1.0 was applied to do pairwise comparisons to identify the outliers. For DETSEL, the coalescent simulation was conducted with the following parameters: mutation rate (infinite allele model) 0.005, 0.001, and 0.0001; ancestral population size $Ne = 500$, 1,000, and 10,000; population size before the split $N_0 = 100$; time since an assumed bottleneck $T_0 = 50$, 100, and 1,000 generations; and time since divergence $t = 100$ generations. Loci falling outside the specified "probability region (95% levels)" were considered as outliers [18]. Based on these outlier tests, two levels of loci sets could be defined from the total loci set as: (1) the neutral loci, which were not detected in any of the three tests, and (2) the decisive selective loci, which were detected at least twice by the three outlier tests.

Genetic differentiation among upland and lowland rice

The general level of genetic differentiation among different groups was quantified by pairewise *Fst* calculated in Arlequin ver. 3.1 using total loci set with 1,000 permutations. To estimate the level of differentiation on selective loci, *Fst* was calculated from the decisive selective loci set among upland and lowland ecotypes in *japonica* and *indica*, respectively.

Japonica landraces were from 5 provinces and *indica* landrace were from 4 provinces. Thus, there were 10 *japonica* and 8 *indica* populations included in this study. To explore the influence of the genetic differentiation on population structures, genetic distance (GD) matrices were calculated by GenAlex ver. 6.43 with 999 permutations using the corresponding neutral or decisive selective loci. The genetic distance matrix obtained form the calculation was used for the cluster analysis according to the un-weighted pair-group method with arithmetic averages (UPGMA) *via* the software NTSYS ver. 2.10e.

Testing whether the upland-lowland rice differentiation was related to drought-resistance

According to previous studies, the comparison of quantitative genetic divergence (*Qst*) and neutral genetic divergence (*Fst*) can be used to detect adaptive evolution. If the *Qst* is significantly higher than the *Fst* calculated by the neutral loci, it indicates that the directional selection drives phenotypic divergence and results in ecological adaptation [19]. In this study, the *Qst* of each trait was calculated as: $Qst = V_{pop}/(V_{pop}+2V_{ind})$, in which V_{pop} was the variance among-population and V_{ind} was the variance within-population. Although the SSRs located in the ESTs, the *Fst* generated among upland and lowland rice was as low as genomic SSRs [11]. This suggests most of these markers were neutral among the two ecotypes. The 95% confidence interval of *Qst* was calculated using R program with 1000 bootstraps. The *Fst* derived from neutral loci and its standard errors were calculated by Arlequin ver. 3.1 with 1000 permutations. Any significant differences between the *Qst* and *Fst* at $p = 0.05$ level was determined when $|Qst-Fst| > 2SQRT\ (SE_{Qst}^2+SE_{Fst}^2)$.

Table 4. Pairewise *Fst* among five experimental groups using total loci set.

Group	*Jap*-upland	*Jap*-lowland	*Ind*-upland	*Ind*-lowland	*O. rufipogon*
Jap-upland	0.0000	+	+	+	+
Jap-lowland	0.0714	0.0000	+	+	+
Ind-upland	0.4426	0.4649	0.0000	+	+
Ind-lowland	0.4255	0.4475	0.0302	0.0000	+
O. rufipogon	0.2683	0.2798	0.2405	0.2207	0.0000

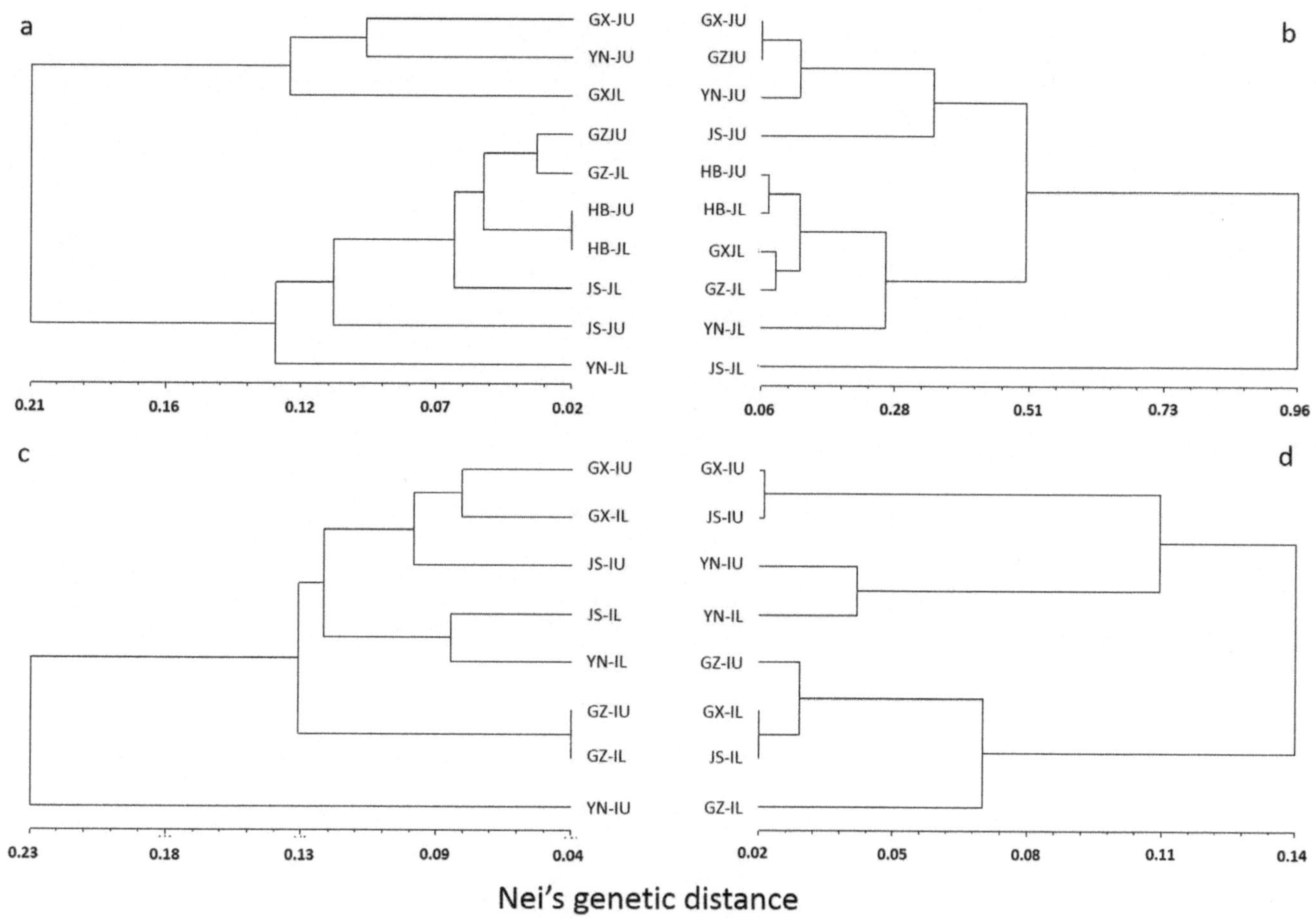

Figure 2. Regional populations cluster analysis using UPGMA method. a) *japonica* materials using neutral loci, b) *japonica* materials using decisive selective loci, c) *indica* materials using neutral loci, d) *indica* materials using decisive selective loci. GX, Guangxi; GZ, Guizhou; HB, Hebei; JS, Jiangsu; YN, Yunnan. U = upland; L = lowland.

To test whether the genetic differentiation between upland and lowland rice was resulting from divergent selections on drought-resistance, Mantel tests were conducted between the individual based GD matrix using the neutral or decisive selective loci and matrix constructed from differentiated morphological traits. The differentiated traits were defined as the traits having significant differences between upland and lowland rice by independent *t*-test using SPSS ver. 15.0. The individual based GD matrix was constructed by GenAlex ver. 6.43. The dissimilarity matrices of differentiated traits were constructed by Hierarchical Cluster Analysis on SPSS ver. 15.0.

Test the expression of ecotype-specific alleles of the selective loci

The decisive selective loci may play a role during upland rice adapting to water-limited upland conditions. To test this hypothesis, the expression levels of upland- and lowland-preferable alleles of the decisive selective loci (E647 and E1899 in *japonica*, E647 and E1177 in *indica*) were measured. Here, upland- or lowland-preferable allele was defined as a predominant allele in upland or lowland rice whose frequency was 20% higher than that in lowland or upland ecotype. The allele frequency was calculated as the number of counts of an allele/the total number of counts of all alleles in a given group *via* the software GenAlex ver. 6.43. Four landraces of each ecotype conferring the ecotype-preferable alleles were included. Three-week-old seedlings in the growth chamber were treated with 20% polyethylene glycol (PEG) 6000 to cause osmotic stress (DT). Some materials were kept in nutrition solutions as controls (CK). When most of the seedlings showed signs of leaf rolling 5 hours after treatment, leaves of three seedlings in DT and CK were collected and kept in liquid nitrogen until RNA extraction. The RNA extraction was followed by the kits of TRNzol A^+ (TianGen Biotech Co. Ltd.). The level of gene expressions were quantified by RealTime-PCR using $SYBR^R$ premix Ex Taq (Takara Bio. Inc.). *Actin* was used for reference. The expression change value was calculated as: the expression level in DT over that in CK. In addition, to test whether the ecotype-preferable alleles of the selective loci had any impacts on rice drought-resistance, the drought-resistant traits were further compared among upland and lowland rice containing differently expressed ecotype-preferable alleles (E647 and E1899 in *japonica*, and E1177 in *indica*).

Results

Population structure and general genetic diversity

A total of 477 alleles were detected in the 47 drought responding EST-SSR loci with number of alleles per locus ranging from 4–25. When running the STRUCTURE simulation using the total materials, there was a sharp peak in Evanno's ΔK at

Table 5. 22 drought-resistant traits measured in control treatment (CK) and drought treatment (DT) and their *Qst* calculated in *Jap*-upland/lowland and *Ind*-upland/lowland pairs (mean ± standard error).

Traits	Jap-upland	Jap-lowland	J-*Qst*	Ind-upland	Ind-lowland	I-*Qst*
RWC (CK)	**0.936±0.005***	**0.910±0.010***	0.052±0.0012	0.898±0.010	0.919±0.006	0.034±0.0007
RWC (DT)	0.871±0.006	0.853±0.011	0.032±0.0010	0.830±0.009	0.843±0.009	0.021±0.0007
No. of stomas per area	**13.9±0.3***	**15.1±0.5***	0.048±0.0012	14.8±0.3	14.7±0.4	0.011±0.0005
EWR	0.199±0.011	0.194±0.009	0.008±0.0004	0.219±0.011	0.219±0.011	0.010±0.0004
Root-shoot ratio	**0.215±0.006***	**0.259±0.014***	0.091±0.0016	0.225±0.009	0.239±0.014	0.017±0.0007
MDA (CK)	0.97±0.02	0.91±0.02	0.016±0.0006	**0.90±0.02***	**0.97±0.02***	**0.072±0.0016***
MDA (DT)	1.28±0.02	1.30±0.02	0.043±0.0010	1.27±0.02	1.26±0.02	0.010±0.0004
Flag leaf length (CK)	35.9±1.1	33.2±1.1	0.033±0.0010	36.0±1.5	38.0±1.6	0.020±0.0008
Flag leaf length (DT)	**32.1±1.2***	**24.5±1.1***	**0.171±0.0019***	32.1±1.6	28.5±2.2	0.032±0.0010
Flag leaf width (CK)	**1.62±0.05***	**1.33±0.05***	**0.131±0.0021***	1.45±0.05	1.51±0.05	0.016±0.0006
Flag leaf width (DT)	**1.70±0.06***	**1.31±0.07***	**0.151±0.0015***	1.33±0.07	1.35±0.06	0.009±0.0004
No. of panicles (CK)	**6.65±0.32***	**9.70±0.73***	**0.141±0.0021***	10.54±0.53	10.65±0.53	0.009±0.0004
No. of panicles (DT)	6.01±0.35	7.05±0.40	0.043±0.0011	11.12±0.67	10.24±0.85	0.018±0.0007
100-grain weight (CK)	2.36±0.06	2.50±0.07	0.030±0.0010	2.39±0.04	2.30±0.06	0.026±0.0009
100-grain weight (DT)	2.43±0.06	2.54±0.07	0.022±0.0008	**2.25±0.04***	**2.07±0.07***	**0.066±0.0015***
No. of seeds (CK)	**551.6±32.5***	**728.2±48.2***	**0.096±0.0017***	903.8±39.6	1079.7±70.2	**0.051±0.0013***
No. of seeds (DT)	480.0±29.3	445.7±29.5	0.014±0.0006	755.8±45.0	840.8±54.8	0.026±0.0010
Seed-set rate (CK)	**0.791±0.016***	**0.851±0.012***	0.086±0.0015	0.827±0.009	0.839±0.015	0.020±0.0008
Seed-set rate (DT)	0.822±0.009	0.835±0.020	0.018±0.0009	0.838±0.009	0.821±0.016	0.019±0.0006
Yield (CK)	**13.13±0.88***	**18.17±1.26***	**0.108±0.0018***	21.51±0.87	24.56±1.44	0.036±0.0010
Yield (DT)	11.25±0.53	11.21±0.80	0.011±0.0005	16.84±1.02	17.35±1.23	0.012±0.0006
DI	**0.89±0.13***	**0.57±0.07***	0.042±0.0009	1.02±0.11	0.93±0.12	0.017±0.0007

RWC: leaf relative water content, EWR: excised leaf water loss rate, MDA: malonaldehyde, DI: drought index.
The values in **bold** and with "*" indicated significant differences between upland and lowland ecotypes or significant difference between *Qst*s and the neutral *Fst*.

$K = 2$ ($\Delta K = 3718.5$) which divided all materials into *japonica* and *indica* groups (Figure S1a). In *japonica*, the sharp peak occurs when $K = 2$ ($\Delta K = 7.6$), which divided landraces into two groups as upland and lowland rice occupied the majority in each (Figure S1b). In *indica*, the sharp peak was also occurs when $K = 2$ ($\Delta K = 2760.1$), which separated rice landraces of Guangxi from other regions (Figure S1c). However, there was a lower peak when $K = 4$ ($\Delta K = 78.0$), in which upland rice (white) was distinguishable from lowland rice (thin gray) in Guangxi (Figure S1d). Similar to the results of STRUCTURE, PCoA indicated that *japonica* and *indica* landraces were separated along the first coordinate (x-axis) while common wild rice located at the center (Figure 1), suggesting that upland-lowland rice differentiation should be separately analyzed in *japonica* and *indica* subspecies. Differentiation between upland and lowland ecotypes was more apparent in *japonica* than that in *indica* (Figure 1). The level of genetic diversity was similar between upland and lowland rice ecotypes, while subspecies (*japonica vs. indica*) and species (*O. sativa vs. O. rufipogon*) showed considerable differences (Table 2).

Selective loci detected by neutrality tests

Among the 47 drought-induced EST-SSR loci, 5, 9, and 5 loci were detected to be selective by Lositan, BayeScan, and Detsel, respectively, in *japonica* subspecies. Five loci were detected to be the decisive selective loci (Table 3). In the *indica* subspecies, 2, 2 and 7 loci were detected to be selective by Lositan, BayeScan, and Detsel, respectively. Three loci were considered as the decisive selective loci in *indica* (Table 3).

Genetic differentiation among upland and lowland rice

The values of *Fst* were as low as 0.0714 in *japonica* or 0.0302 in *indica* types of upland and lowland rice. The value was much lower than that between *japonica* and *indica* varieties (Table 4). This result was similar to a previous study in which *Fst* among different ecotype was recorded as 0.068 in *japonica* using putative neutral SSR markers [11]. However, the values of *Fst* between upland and lowland rice became much higher in *japonica* (0.285) and *indica* (0.127) when using their respective decisive selective loci. In the UPGMA clusters, upland and lowland rice populations were grouped separately when using the decisive selective loci, while populations of landraces from the same regions were clustered together when the neutral loci were used (Figure 2). These results suggest differentiation occurred between upland and lowland rice generally on the selective loci.

Divergent selection on drought-resistant traits

Drought resistant traits were evaluated in the field experiment. Significant differences were detected on 11 traits between upland and lowland ecotypes in *japonica*, while significant differences were detected on only 2 traits in *indica* (Table 5). The comparison between *Qst* with neutral *Fst* was used to test any divergent selection on the drought-resistant traits. The values of the *Qst* ranged from 0.009~0.171 in *japonica* and from 0.009~0.072 in

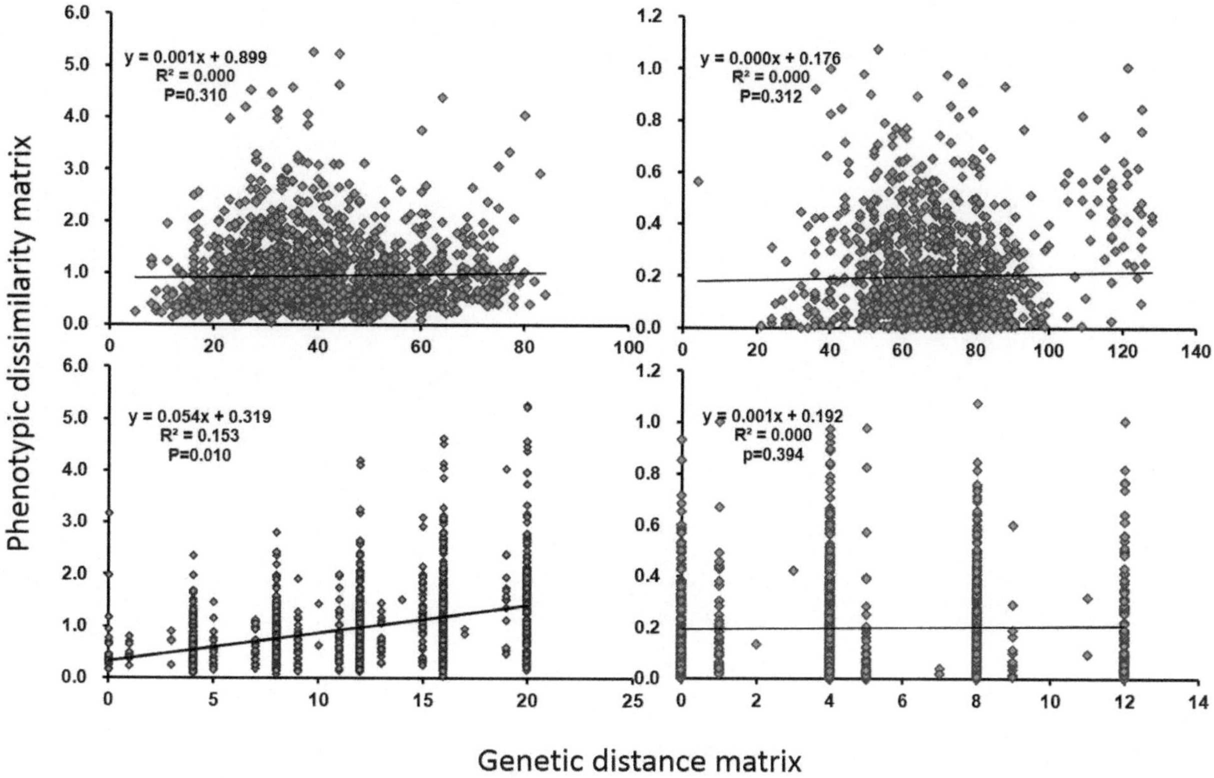

Figure 3. Correlations between matrices of genetic distance from different loci sets and the dissimilarity matrices from selective morphological traits. a) neutral loci in *japonica*; b) decisive selective loci in *japonica*; c) neutral loci in *indica*; d) decisive selective loci in in *indica*.

indica. The *Qsts* of the most differentiated traits were significantly higher than the neutral *Fsts* in *japonica* (0.0898±0.0024) and *indica* (0.0363±0.0017) materials (Table 5). This result suggested that some of the drought-resistant traits were under divergent selection during domestication.

The genetic differentiation associated with morphological differentiation on drought-resistant traits between upland and lowland rice

Given that some drought-resistant traits were differentially selected in upland and lowland rice, their genetic differentiation may be associated with the evolution of drought-resistance. To test this, Mantel test was conducted between the individual-based genetic distance (DG) matrix and dissimilarity matrix constructed from morphological traits. In *japonica*, the dissimilarity matrix constructed from the differentiated traits was significantly correlated with the GD matrix using decisive selective loci (Figure 3b), while it was not correlated with the neutral GD matrix (Figure 3a). These results suggested a strong association between the differentiations on the drought-resistance and on the selective loci in *japonica*. However, the morphological dissimilarity matrix was neither significantly correlated with GD matrix from the selective loci nor with that from the neutral loci in *indica* (Figure 3c, 3d).

Expression of ecotype-preferable alleles and their impacts on drought-resistant traits

In *japonica* subspecies, the allele-6 of locus E647 was upland-preferable (taking account for 87.9% in upland rice but only 50.6% in lowland rice), while allele-2 was lowland-preferable (taking account for 40.4% in lowland rice but 0 in upland rice). These two alleles expressed similarly in normal conditions (CK). However, the expression of allele-6 down-regulated largely (> 50%) while the expression of allele-2 remained at the same level in DT. Thus, the expression change values of the allele-2 (1.146±0.232) were marginally higher than the allele-6 (0.625±0.158, $p<0.10$) (Figure 4a). The allele-4 of E1899 was upland-preferable (taking account for 71.0% in upland rice but only 16.7% in lowland rice) and the allele-3 was lowland-preferable (taking account for 79.5% in lowland rice but 23.8% in upland rice). This EST was up-regulated in lowland rice ecotype while its expression was down-regulated in upland rice under DT. Thus, the expression change value of the allele-3 in lowland rice (1.832±0.583) was marginally higher than the allele-4 in upland rice (0.413±0.111, $p<0.10$) (Figure 4b).

In *indica* subspecies, the allele-6 of E647 was upland-preferable (taking account for 71.4% in upland rice but 50.7% in lowland rice), while the allele-7 was lowland-preferable (taking account for 49.3% in lowland rice but only 23.7% in upland rice). The allele-8 of E1177 was upland-preferable (taking account for 44.1% in upland rice but only 17.3% in lowland), while the allele-9 lowland-preferable (accounted for 74.7% in lowland rice but only 44.5% in upland). The expression change values of E647 were similar among ecotypes or allelic types (Figure 4c), while that of E1177 was much higher in lowland ecotype or lowland-preferable alleles than in upland rice or upland-preferable (Figure 4d).

As these ESTs differently expressed between allelic types or ecotypes, the drought-resistant traits were then compared among rice ecotypes conferring their ecotype-preferable alleles. As expected, many significant differences not previously detected among total upland and lowland rice were now detected (Table S2). For example, the upland rice conferring the allele-8 and

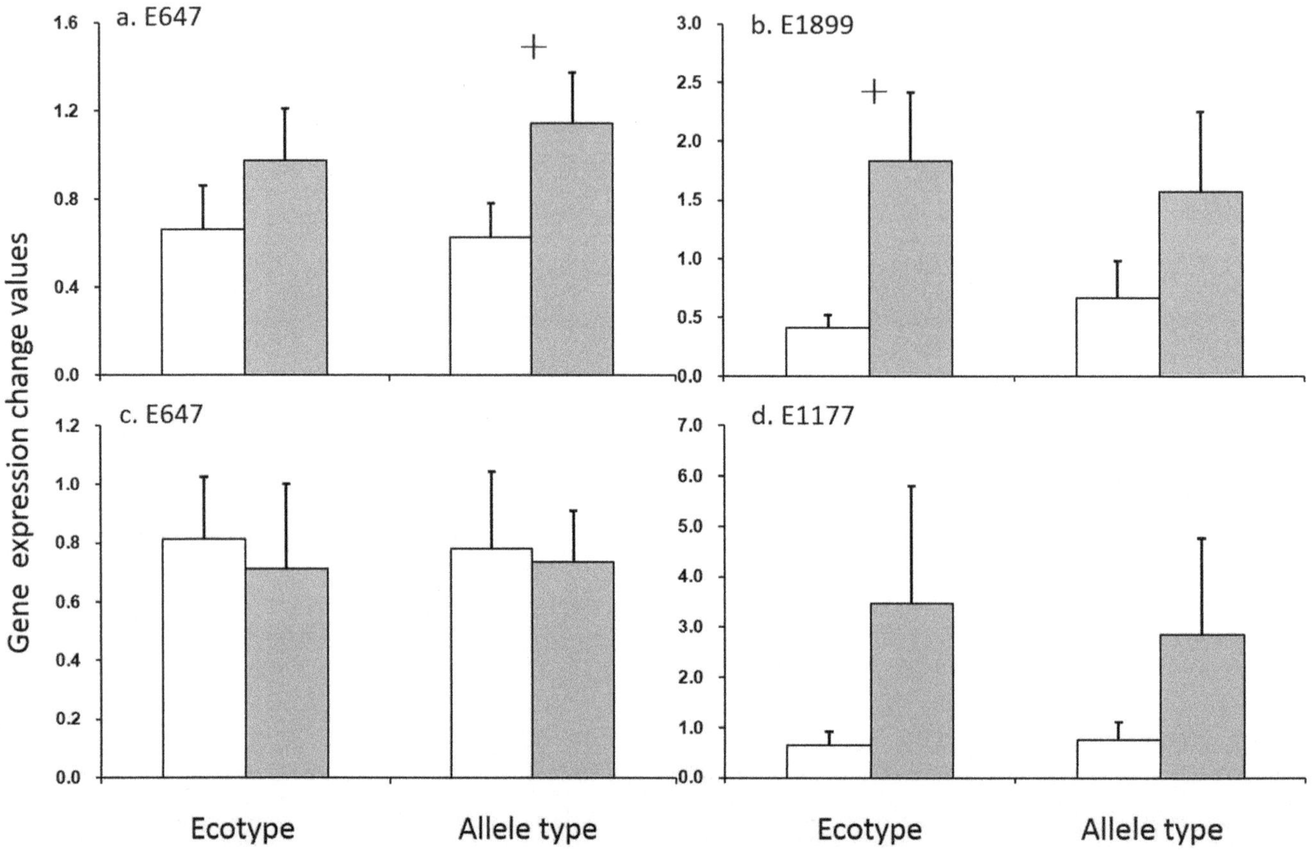

Figure 4. The expression change values (mean ± SE) between different ecotypes or alleles. The expression change values of E647 (a) and E1899 (b) between ecotypes or alleles in *japonica* and these of E647 (c) and E1177 (d) between ecotypes or alleles in *indica*. The expression change values were calculated as: gene expression in PEG/that in water. The white bar indicates upland rice or upland-specific allele, while the black bar indicates lowland rice or lowland-specific allele. "+" indicates differences at the significant level of $p<0.10$.

lowland rice conferring allele-9 of E1177 exhibited significant difference on the root-shoot ratio in *indica*. This was accordant to the annotated function of gene *Os06g0633300* (Table 6). Other genes containing the selective EST-SSRs were also considered to be related with resistance to abiotic stress given their annotations (Table 6), suggesting that these selective loci might play roles in rice drought-resistance.

Discussion

Considerable level of upland-lowland rice differentiation on selective loci

As in previously studies, obvious differentiation was reported on drought-resistant morphological traits between upland and lowland rice [9]. However, the general level of genetic differentiation between upland and lowland rice was very low in our study,

Table 6. Gene symbol, gene ID, and the annotated functions of the decisive selective loci detected in this study.

Locus	Gene symbol	Gene ID	Names	Predicted function
E647	Os01g0607400	4324222	hypothetical protein	Similar to STYLOSA protein
E359	Os06g0702600	4341978	hypothetical protein	Similar to Auxin response factor 7a
E1899	Os12g0563600	4352535	hypothetical protein	Protein of unknown function, DUF538 family protein
E3735	Os07g0260000	4342870	hypothetical protein	Protein prenyltransferase domain containing protein
E1238	Os10g0554200	4349339	hypothetical protein	TGF-beta receptor, type I/II extracellular region family protein
E1177	Os06g0633300	4341588	hypothetical protein	Phytosulfokines 1 precursor [Contains: Phyto sulfokine-alpha (PSK- alpha) (Phytosulfokine-a); Phytosulfokine-beta (PSK-beta) (Phytosulfokine-b)]
E4208	Os07g0546500	4343527	hypothetical protein	Conserved hypothetical protein

similar to the results in previous studies [10–11]. However, *Fst* calculated based on the outlier loci was considerably high between upland and lowland rice ecotypes, matching the level of *Fst* between *O. sativa* and *O. rufipogon* (0.252) in our study. These results suggested that adaptive divergence among upland and lowland rice occurred on these selective loci. Strong selections on the adaptive loci always affected population structures [20–21]. For example, wheat grown in different ecosystems was clustered separately based on the drought-responding gene TaSnRK2.7 [21]. Similar results were found in this study as revealed by the UPGMA cluster, providing evidences that these selective EST-SSRs received uniform selections among regions. It is noteworthy that morphological differentiation between upland and lowland ecotypes was also greater in *japonica* than in *indica*, consistent with the genetic data in these two subspecies.

Upland-lowland rice differentiation was driven by divergent selection on drought-resistance

Theoretically, drought-resistance alleles are more strongly selected in upland rice than that in lowland rice, due to its adaption to the water-limited environment. In this study, the general drought-resistance gained from field experiments was much higher in upland rice than that in lowland rice. The higher *Qst* than neutral *Fst* suggests some drought-resistant traits are under the divergent selections, leading genetic differentiation on the outlier loci between upland and lowland rice. The genetic differentiation between upland and lowland rice is likely associated with drought-resistance based on the Mantel tests between the genetic distance and morphological dissimilarity, especially in *japonica*. The formation of upland- or lowland-preferable alleles always result from divergent selections [22]. Different expression of the ecotype-preferable alleles under their favorable conditions could be considered as the signature of natural selection [23]. In this study, we found different expression patterns among upland- and lowland-preferable alleles encountering osmotic stress, adding further evidence of the adaptive divergence between upland and lowland rice. Based on these results, we conclude that genetic differentiation between upland and lowland rice is likely driven by divergent selection on the drought-resistance.

Candidate genes for drought-resistance gene from the selective loci

Recently studies disclosed that it was possible to indentify potential functional genes *via* studying adaptive divergence in natural populations [24], crops [5], and even in rice [25]. In this study, we found potential candidate genes that may play roles in rice drought-resistance. For example, *Os06g0633300* (E1177) encodes the rice phytosulfokine 1 precursor (*OsPSK1*). Its over-expression can promote rice cell division [26] and root development [27]. Interestingly, upland and lowland rice conferring different ecotype-referable alleles of E1177 exhibited significant differences on root-shoot ratio, a key character of drought-avoidance. Besides, genes with similar annotations as *Os01g0607400* (E647) and *Os12g0563600* (E1899) were also reported to be associated with plant stress responses by previous studies in other plant species [28–29]. These results provide strong evidences that these candidate genes should have played some roles in rice drought-resistance. However, their functions in rice need further investigation. In a word, we can explore more drought-resistant genes by studying the genetic divergence among the upland and lowland rice.

Conclusions

Crops adapted to different agro-ecosystems always promote the variation of agricultural important genes. In this study, several outlier loci were detected between upland and low rice ecotypes. A considerable degree of genetic differentiation between upland and lowland rice was detected both at the DNA and gene expression level on these outlier loci. Results from this study reveals that the genetic differentiation among the two rice ecotypes is most likely driven by divergent selection on drought-resistant traits. The findings of this study not only help us to understand the underlying molecular basis of adaptive divergence, but also provide valuable implications for rice domestication and breeding, especially on the drought-resistance in rice.

Supporting Information

Figure S1 Population structures inferred by STRUCTURE. a) *Japonica* and *indica* subspecies were separated when K = 2. b) Inferred population structures in *japonica* subspecies when K = 2. c) Inferred population structures in *indica* subspecies when K = 2. d) Inferred population structures in *indica* subspecies when K = 4, in which some upland rice and lowland rice were separated.

Table S1 Basic information of the used 47 drought-induced EST-SSR loci.

Table S2 22 drought-resistant traits measured in upland and lowland rice materials containing their ecotype-preferable alleles (indicated in parentheses) in control treatment (CK) and drought treatment (DT) (mean ± standard error). The values in bold and with "*" indicated significant differences between upland and lowland ecotypes.

Table S3 Original data for SSR scoring and field evaluated traits (xls).

Acknowledgments

Rice materials were kindly provided by Rice Research Institute of Guangxi Academy of Agriculture Science, Hebei Academy of Agriculture Science, Yunnan Academy of Agriculture Science, and Jiangsu Academy of Agriculture Science.

Author Contributions

Conceived and designed the experiments: HX LC BRL LJL. Performed the experiments: HX XGZ HG HY PL JR JJL. Analyzed the data: HX XGZ LC PL JR. Contributed reagents/materials/analysis tools: HY JR LJL. Wrote the paper: HX LC JR BRL LJL.

References

1. Nosil P, Funk DJ, Ortiz-Barrientos D (2009) Divergent selection and heterogeneous genomic divergence. Mol Ecol 18: 375–402.
2. Shikano T, Ramadevi J, Merilä J (2010) Identification of local- and habitat-dependent selection: scanning functionally important genes in nine-spined sticklebacks (*Pungitius pungitius*). Mol Biol Evol 27: 2775–2789.
3. Hübner S, Höffken M, Oren E, Haseneyer G, Stein N, et al. (2009) Strong correlation of wild barley (*Hordeum spontaneum*) population structure with temperature and precipitation variation. Mol Ecol 18: 1523–1536.

4. Whitehead A, Triant DA, Champlin D, Nacci D (2010) Comparative transcriptomics implicates mechanisms of evolved pollution tolerance in a killifish population. Mol Ecol 19: 5186–5203.
5. Xia H, Camus-Kulandaivelu L, Stephan W, Tellier A, Zhang ZW (2010) Nucleotide diversity patterns of local adaptation at drought-related candidate genes in wild tomatoes. Mol Ecol 19: 4144–4154.
6. Bernier J, Atlin GN, Serraj R, Kumar A, Spaner D (2008) Breeding upland rice for drought resistance. J Sci Food Agr 88: 927–939.
7. Farooq M, Wahid A, Lee DJ, Ito O, Siddique KH (2009) Advances in drought resistance of rice. Crit Rev Plant Sci 28: 199–217.
8. Luo LJ (2010) Breeding for water-saving and drought-resistance rice (WDR) in China. J Exp Bot 61: 3509–3517.
9. IRRI (International Rice Research Institute) (1986) Upland rice:A global perspective. Los Baiios, Philippines.
10. Yu LX, Nguyen HT (1994) Genetic variation detected with RAPD markers among upland and lowland rice cultivars (*Oryza sativa* L.) Theor Appl Genet 87: 668–672.
11. Zhang DL, Zhang HL, Wang MX, Sun JL, et al. (2009) Genetic structure and differentiation of *Oryza sativa* L. in China revealed by microsatellites. Theor Appl Genet 119: 1105–1117.
12. Ishikawa R, Maeda K, Harada T, Niizeki M, Saito K (1992) Genotypic variation for 17 isozyme genes among Japanese upland varieties in rice. Japan J Breeding 42: 737–746.
13. Varshney RK, Graner A, Sorrells ME (2005) Genic microsatellite markers in plants: features and applications. TRENDS Biotech 23: 48–55.
14. Lan JS, Hu FS, Zhang JR (1990) The concept and statistical method of drought resistance index in crops. Acta Agriculturae Boreali-Sinica 5: 20–25 (in Chinese with English abstract)
15. Evanno G, Regnaut S, Goudet J (2005) Detecting the number of clusters of individuals using the software STRUCTURE: a simulation study. Mol Ecol 14: 2611–2620.
16. Antao T, Lopes A, Lopes RJ, Beja-Pereira A, Luikart G (2008) LOSITAN: a workbench to detect molecular adaptation based on a F_{ST}-outlier method. BMC Bioinformatics 9: 323.
17. Foll M, Gaggiotti O (2008) A genome-scan method to identify selected loci appropriate for both dominant and codominant markers: a Bayesian perspective. Genetics 180: 977–993.
18. Vitalis R, Dawson K, Boursot P, Belkhir K (2003) DetSel 1.0: a computer program to detect markers responding to selection. J Heredity 94: 429–431.
19. Miller JR, Wood BP, Hamilton MB (2008) *FST* and *QST* Under Neutrality. Genetics, 180, 1023–1037.
20. Martin MA, Mattioni C, Cherubini M, Taurchini D, Villani F (2010) Genetic diversity in European chestnut populations by means of genomic and genic microsatellite markers. Tree Genetics & Genomes 6: 735–744.
21. Zhang H, Mao X, Wu X, Wang C, Jing R (2011) An abiotic stress response gene *TaSnRK2.7-B* in wheat accessions: genetic diversity analysis and gene mapping based on SNPs. Gene 478: 28–34.
22. Steele KA, Edwards G, Zhu J, Witcombe JR (2004) Marker-evaluated selection in rice: shifts in allele frequency among bulks selected in contrasting agricultural environments identify genomic regions of importance to rice adaptation and breeding. Theor Appl Genet 109: 1247–1260.
23. Knight CA, Vogel H, Kroymann J, Shumate A, Witsenboer H, Mitchell-Olds T (2006) Expression profiling and local adaptation of *Boechera holboellii* populations for water use efficiency across a naturally occurring water stress gradient. Mol Ecol 15: 1229–1237.
24. Ross-Ibarra J, Morrell PL, Gaut BS (2007) Plant domestication, a unique opportunity to identify the genetic basis of adaptation. Proc Natl Acad Sci, USA 104: 8641–8648.
25. Lyu J, Zhang SL, Dong Y, et al. (2013) Analysis of elite variety tag SNPs reveals an important allele in upland rice. Nature Communications, 4: 2138. doi: 10.1038/ncomms3138.
26. Yang HP, Matsubayashi Y, Nakamura K, Sakagami Y (1999) *Oryza sativa PSK* gene encodes a precursor pf phytosulfokine-α, a sulfated peptide growth factor found in plants. Proc Natl Acad Sci, USA 96: 13560–13565.
27. Huang JY, Wang YF, Yang JS (2010) Over-expression of OsPSK3 increases chlorophyll content of leaves in rice. Hereditas 32: 1281–1289.
28. Ichitani K, Namigoshi K, Sato K, Taura S, Aoki M, et al. (2007) Fine mapping and allelic dosage effect of *Hwc1*, a complementary hybrid weakness gene in rice. Theor Appl Genet 114: 1407–1415.
29. Gholizadeh A (2011) Heterologous expression of stress-responsive duf538 domain containing protein and its morpho-biochemical consequences. Prot J 30: 351–358.

11

Water Filtration Using Plant Xylem

Michael S. H. Boutilier[◑], Jongho Lee[◑], Valerie Chambers, Varsha Venkatesh, Rohit Karnik*

Department of Mechanical Engineering, Massachusetts Institute of Technology, Cambridge, Massachusetts, United States of America

Abstract

Effective point-of-use devices for providing safe drinking water are urgently needed to reduce the global burden of waterborne disease. Here we show that plant xylem from the sapwood of coniferous trees – a readily available, inexpensive, biodegradable, and disposable material – can remove bacteria from water by simple pressure-driven filtration. Approximately 3 cm^3 of sapwood can filter water at the rate of several liters per day, sufficient to meet the clean drinking water needs of one person. The results demonstrate the potential of plant xylem to address the need for pathogen-free drinking water in developing countries and resource-limited settings.

Editor: Zhi Zhou, National University of Singapore, Singapore

Funding: This work was supported by the James H. Ferry, Jr. Fund for Innovation in Research Education award to R.K. administered by the Massachusetts Institute of Technology. SEM imaging was performed at the Harvard Center for Nanoscale Systems, a member of the National Nanotechnology Infrastructure Network (NNIN), which is supported by the National Science Foundation under NSF award no. ECS-0335765. The funders had no role in study design, data collection and analysis, decision to publish, or preparation of the manuscript.

Competing Interests: The authors have declared that no competing interests exist.

* E-mail: karnik@mit.edu

◑ These authors contributed equally to this work.

Introduction

The scarcity of clean and safe drinking water is one of the major causes of human mortality in the developing world. Potable or drinking water is defined as having acceptable quality in terms of its physical, chemical, and bacteriological parameters so that it can be safely used for drinking and cooking [1]. Among the water pollutants, the most deadly ones are of biological origin: infectious diseases caused by pathogenic bacteria, viruses, protozoa, or parasites are the most common and widespread health risk associated with drinking water [1,2]. The most common water-borne pathogens are bacteria (e.g. *Escherichia coli*, *Salmonella typhi*, *Vibrio cholerae*), viruses (e.g. adenoviruses, enteroviruses, hepatitis, rotavirus), and protozoa (e.g. giardia) [1]. These pathogens cause child mortality and also contribute to malnutrition and stunted growth of children. The World Health Organization reports [3] that 1.6 million people die every year from diarrheal diseases attributable to lack of access to safe drinking water and basic sanitation. 90% of these are children under the age of 5, mostly in developing countries. Multiple barriers including prevention of contamination, sanitation, and disinfection are necessary to effectively prevent the spread of waterborne diseases [1]. However, if only one barrier is possible, it has to be disinfection unless evidence exists that chemical contaminants are more harmful than the risk from ingestion of microbial pathogens [1]. Furthermore, controlling water quality at the point-of-use is often most effective due to the issues of microbial regrowth, byproducts of disinfectants, pipeline corrosion, and contamination in the distribution system [2,4].

Common technologies for water disinfection include chlorination, filtration, UV-disinfection, pasteurization or boiling, and ozone treatment [1,2,5]. Chlorine treatment is effective on a large scale, but becomes expensive for smaller towns and villages. Boiling is an effective method to disinfect water; however, the amount of fuel required to disinfect water by boiling is several times more than what a typical family will use for cooking [1]. UV-disinfection is a promising point-of-use technology available [1], yet it does require access to electricity and some maintenance of the UV lamp, or sufficient sunlight. While small and inexpensive filtration devices can potentially address the issue of point-of-use disinfection, an ideal technology does not currently exist. Inexpensive household carbon-based filters are not effective at removing pathogens and can be used only when the water is already biologically safe [1]. Sand filters that can remove pathogens require large area and knowledge of how to maintain them [1], while membrane filters capable of removing pathogens [2,4] suffer from high costs, fouling, and often require pumping power due to low flow rates [6] that prevents their wide implementation in developing countries. In this context, new approaches that can improve upon current technologies are urgently needed. Specifically, membrane materials that are inexpensive, readily available, disposable, and effective at pathogen removal could greatly impact our ability to provide safe drinking water to the global population.

If we look to nature for inspiration, we find that a potential solution exists in the form of plant xylem – a porous material that conducts fluid in plants [7]. Plants have evolved specialized xylem tissues to conduct sap from their roots to their shoots. Xylem has evolved under the competing pressures of offering minimal resistance to the ascent of sap while maintaining small nanoscale pores to prevent cavitation. The size distribution of these pores – typically a few nanometers to a maximum of around 500 nm, depending on the plant species [8] – also happens to be ideal for filtering out pathogens, which raises the interesting question of whether plant xylem can be used to make inexpensive water filtration devices. Although scientists have extensively studied plant xylem and the ascent of sap, use of plant xylem in the context of water filtration remains to be explored. Measurements of sap flow in plants suggest that flow rates in the range of several liters per hour may be feasible with less than 10 cm-sized filters, using only gravitational pressure to drive the flow [7].

We therefore investigated whether plant xylem could be used to create water filtration devices. First, we reason which type of plant xylem tissue is most suitable for filtration. We then construct a simple water filter from plant xylem and study the resulting flow rates and filtration characteristics. Finally, we show that the xylem filter can effectively remove bacteria from water and discuss directions for further development of these filters.

Materials and Methods

Materials

Branches were excised from white pine growing on private property in Massachusetts, USA, with permission of the owner and placed in water. The pine was identified as *pinus strobus* based on the 5-fold grouping of its needles, the average needle length of 4.5 inches, and the cone shape. Deionized water (Millipore) was used throughout the experiments unless specified otherwise. Red pigment (pigment-based carmine drawing ink, Higgins Inks) was dissolved in deionized water. Nile-red coated 20 nm fluorescent polystyrene nanospheres were obtained from Life Technologies. Inactivated, Alexa 488 fluorescent dye labeled *Escherichia coli* were obtained from Life Technologies. Wood sections were inserted into the end of 3/8 inch internal diameter PVC tubing, sealed with 5 Minute Epoxy, secured with hose clamps, and allowed to cure for ten minutes before conducting flow rate experiments.

Construction of the Xylem Filter

1 inch-long sections were cut from a branch with approximately 1 cm diameter. The bark and cambium were peeled off, and the piece was mounted at the end of a tube and sealed with epoxy. The filters were flushed with 10 mL of deionized water before experiments. Care was taken to avoid drying of the filter.

Filtration and Flow Rate Experiments

Approximately 5 mL of deionized water or solution was placed in the tube. Pressure was supplied using a nitrogen tank with a pressure regulator. For filtration experiments, 5 psi (34.5 kPa) pressure was used. The filtrate was collected in glass vials. For dye filtration, size distribution of the pigments was measured for the input solution and the filtrate. Higgins pigment-based carmine drawing ink, diluted ~1000× in deionized water, was used as the input dye solution. For bacteria filtration, the feed solution was prepared by mixing 0.08 mg of inactivated *Escherichia coli* in 20 mL of deionized water ($\sim 1.6 \times 10^7$ mL^{-1}) after which the solution was sonicated for 1 min. The concentration of bacteria was measured in the feed solution and filtrate by enumeration with a hemacytometer (inCyto C-chip) mounted on a Nikon TE2000-U inverted epifluorescence microscope. Before measurement of concentration and filtration experiments, the feed solution was sonicated for 1 min and vigorously mixed.

Imaging

Xylem structure was visualized in a scanning electron microscope (SEM, Zeiss Supra55VP). Samples were coated with gold of 5 nm thickness before imaging. To visualize bacteria filtration, 5 mL of solution at a bacteria concentration of $\sim 1.6 \times 10^7$ mL^{-1} was flowed into the filter. The filter was then cut longitudinally with a sharp blade. One side of the sample was imaged using a Nikon TE2000-U inverted epifluorescence microscope and the other was coated with gold and imaged with the SEM. Optical images were acquired of the cross section of a filter following filtration of 5 mL of the dye at a dilution of ~100×.

Particle Sizing

Dynamic light scattering measurements of particle size distributions were performed using a Malvern Zetasizer Nano-ZS.

Results

Xylem Structure and Rationale for use of Conifer Xylem

The flow of sap in plants is driven primarily by transpiration from the leaves to the atmosphere, which creates negative pressure in the xylem. Therefore, xylem evolution has occurred under competing pressures of providing minimal resistance to the flow of sap, while protecting against cavitation (i.e. nucleation) and growth of bubbles that could stop the flow of sap and kill the plant, and to do this while maintaining mechanical strength [7]. The xylem structure comprises many small conduits that work in parallel and operate in a manner that is robust to cavitation [7,8] (Figure 1). In woody plants, the xylem tissue is called the sapwood, which often surrounds the heartwood (i.e. inactive, non-conducting lignified tissue found in older branches and trunks) and is in turn surrounded by the bark (Figure 1b,c). The xylem conduits in gymnosperms (conifers) are formed from single dead cells and are called tracheids (Figure 1c), with the largest tracheids reaching diameters up to 80 μm and lengths up to 10 mm [7]. Angiosperms (flowering plants) have xylem conduits called vessels that are derived from several cells arranged in a single file, having diameters up to 0.5 mm and lengths ranging from a few millimeters to several meters [7]. These parallel conduits have closed ends and are connected to adjacent conduits via "pits" [8] (Figure 1d,e). The pits have membranes with nanoscale pores that perform the critical function of preventing bubbles from crossing over from one conduit to another. Pits occur in a variety of configurations; Figure 1d,e shows torus-margo pit membranes that consist of a highly porous part shaped like a donut (margo) and an impermeable part in the center called torus, occurring in conifers [8]. The porosity of the pit membranes ranges in size from a few nanometers to a few hundred nanometers, with pore sizes in the case of angiosperms tending to be smaller than those in gymnosperms [8,9]. Pit membrane pore sizes have been estimated by examining whether gold colloids or particles of different sizes can flow through [8,10]. Remarkably, it was observed that 20 nm gold colloids could not pass through inter-vessel pit membranes of some deciduous tree species [10], indicating an adequate size rejection to remove viruses from water. Furthermore, inter-tracheid pit membranes were found to exclude particles in the 200 nm range [8], as required for removal of bacteria and protozoa.

Since angiosperms (flowering plants, including hardwood trees) have larger xylem vessels that are more effective at conducting sap, xylem tissue constitutes a smaller fraction of the cross-section area of their trunks or branches, which is not ideal in the context of filtration. The long length of their xylem vessels also implies that a large thickness (centimeters to meters) of xylem tissue will be required to achieve any filtration effect at all – filters that are thinner than the average vessel length will just allow water to flow through the vessels without filtering it through pit membranes (Figure 1a). In contrast, gymnosperms (conifers, including softwood trees) have short tracheids that would force water to flow through pit membranes even for small thicknesses (<1 cm) of xylem tissue (Figure 1a). Since tracheids have smaller diameters and are shorter, they offer higher resistance to flow, but typically a greater fraction of the stem cross-section area is devoted to

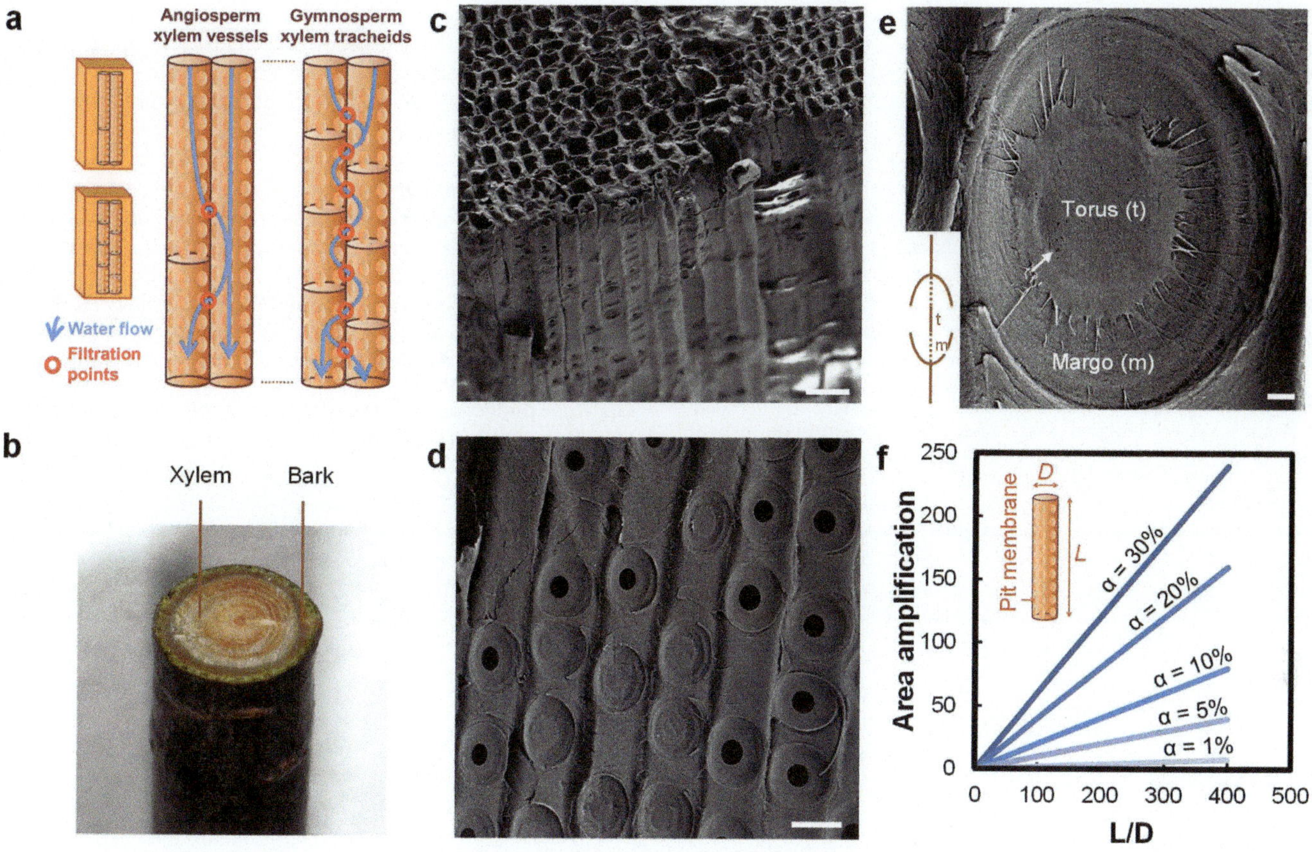

Figure 1. Xylem structure. a) Structure of xylem vessels in flowering plants and tracheids in conifers. Longer length of the vessels can provide pathways that can bypass filtration through pit membranes that decorate their circumference. b) Photograph of ~1 cm diameter pine (*pinus strobus*) branch used in the present study. c) Scanning electron microscope (SEM) image of cut section showing tracheid cross section and lengthwise profile. Scale bar is 40 μm. d) SEM image showing pits and pit membranes. Scale bar is 20 μm. e) Pit membrane with inset showing a cartoon of the pit cross-section. The pit cover has been sliced away to reveal the permeable margo surrounding the impermeable torus. Arrow indicates observed hole-like structures that may be defects. The margo comprises radial spoke-like structures that suspend the torus, which are only barely visible overlaying the cell wall in the background. Scale bar is 1 μm. f) Dependence of area amplification, defined as the pit membrane area divided by the nominal filter area, on the tracheid aspect ratio *L/D* and fractional area α occupied by pit membranes.

conducting xylem tissue. For example, in the pine branch shown in Figure 1b used in this study, fluid-conducting xylem constitutes the majority of the cross-section. This reasoning leads us to the conclusion that in general the xylem tissue of coniferous trees – i.e. the sapwood – is likely to be the most suitable xylem tissue for construction of a water filtration device, at least for filtration of bacteria, protozoa, and other pathogens on the micron or larger scale.

The resistance to fluid flow is an important consideration for filtration. Pits can contribute a significant fraction (as much as 30–80%) [7,8] of the resistance to sap flow, but this is remarkably small considering that pit membrane pore sizes are several orders of magnitude smaller than the tracheid or vessel diameter. The pits and pit membranes form a hierarchical structure where the thin, highly-permeable pit membranes are supported across the microscale pits that are arranged around the circumference of the tracheids (Figure 1a). This arrangement permits the pit membranes to be thin, offering low resistance to fluid flow. Furthermore, the parallel arrangement of tracheids with pits around their circumference provides a high packing density for the pit membranes. For a given tracheid with diameter D and length L, where pit membranes occupy a fraction α of the tracheid wall area, each tracheid effectively contributes a pit membrane area of $\pi DL\alpha/2$, where the factor of 2 arises as each membrane is shared by two tracheids. However, the nominal area of the tracheid is only $\pi D^2/4$, and therefore, the structure effectively amplifies the nominal filter area by a factor of $2\alpha(L/D)$ (Figure 1f). The images in Figure 1c indicate that typical $D \sim 10$–15 μm and $\alpha \sim 0.2$ yield an effective area amplification of ~20 for tracheid lengths of 1–2 mm. Therefore, for a filter made by cutting a slice of thickness $\sim L$ of the xylem, the actual membrane area is greater by a large factor due to vertical packing of the pit membranes. Larger filter thicknesses further increase the total membrane area, but the additional area of the membrane is positioned in series rather than in parallel and therefore reduces the flow rate, but potentially improves the rejection performance of the filter due to multiple filtration steps as shown in Figure 1a.

Construction of the Xylem Filter and Measurement of Flow Rate

The xylem filter device was constructed by simply peeling off the bark and cambium from a section of the pine branch and inserting it into a tube (Figure 2a). Although a simple tube fastener could provide a leak-tight seal between the tube and the xylem, we used epoxy to ensure that there was no inadvertent leakage. When deionized water was loaded into the tube above the xylem and subjected to pressure in the 0.5–5 psi (3.45 to 34.5 kPa) range, we found that water readily flowed through the xylem. The flow rate was proportional to applied pressure (Figure 2b), which allowed for the extraction of the hydrodynamic conductivity K ($m^2\ Pa^{-1}\ s^{-1}$) of the filter, defined by

$$Q = KA\frac{\Delta P}{l} \qquad (1)$$

where Q is the volumetric flow rate (in $m^3\ s^{-1}$) under pressure difference ΔP across the filter, while l and A are the thickness and the cross-section area of the filter, respectively. The observed conductivities for three different filters were in the range of ~5–$6\times10^{-10}\ m^2\ Pa^{-1}\ s^{-1}$ (Figure 2c), or equivalently, ~0.5–0.6 kg $s^{-1}\ m^{-1}\ MPa^{-1}$ when defined with respect to mass flow rate of water.

Biologists have performed similar flow rate measurements by cutting a section of a plant stem under water, flushing to remove any bubbles, and applying a pressure difference to measure the flow rate [11,12]. Xylem conductivities of conifers [7] typically range from 1–4 kg $s^{-1}\ m^{-1}\ MPa^{-1}$, which compares very well with the conductivities measured in our experiments. Lower conductivities can easily result from introduction of bubbles [11] or the presence of some non-conducting heartwood. We can therefore conclude that the flow rate measurements in our devices are consistent with those expected from prior literature on conductivity of conifer xylem.

Filtration of Pigment Dye

After construction of the filter, we tested its ability to filter a pigment dye with a broad particle size distribution. The red color of the feed solution disappeared upon filtration (Figure 3a) indicating that the xylem filter could effectively filter out the dye.

Since the dye had a broad pigment size distribution, we investigated the size-dependence of filtration by quantifying the pigment size distribution before and after filtration using dynamic light scattering. We found that the feed solution comprised particles ranging in size from ~70 nm to ~500 nm, with some larger aggregates (Figure 3b). In contrast, the filtrate particle size distribution peaked at ~80 nm, indicating that larger particles were filtered out. In a separate experiment, we observed that 20 nm fluorescent polystyrene nanoparticles could not be filtered by the xylem filter, confirming this size dependence of filtration.

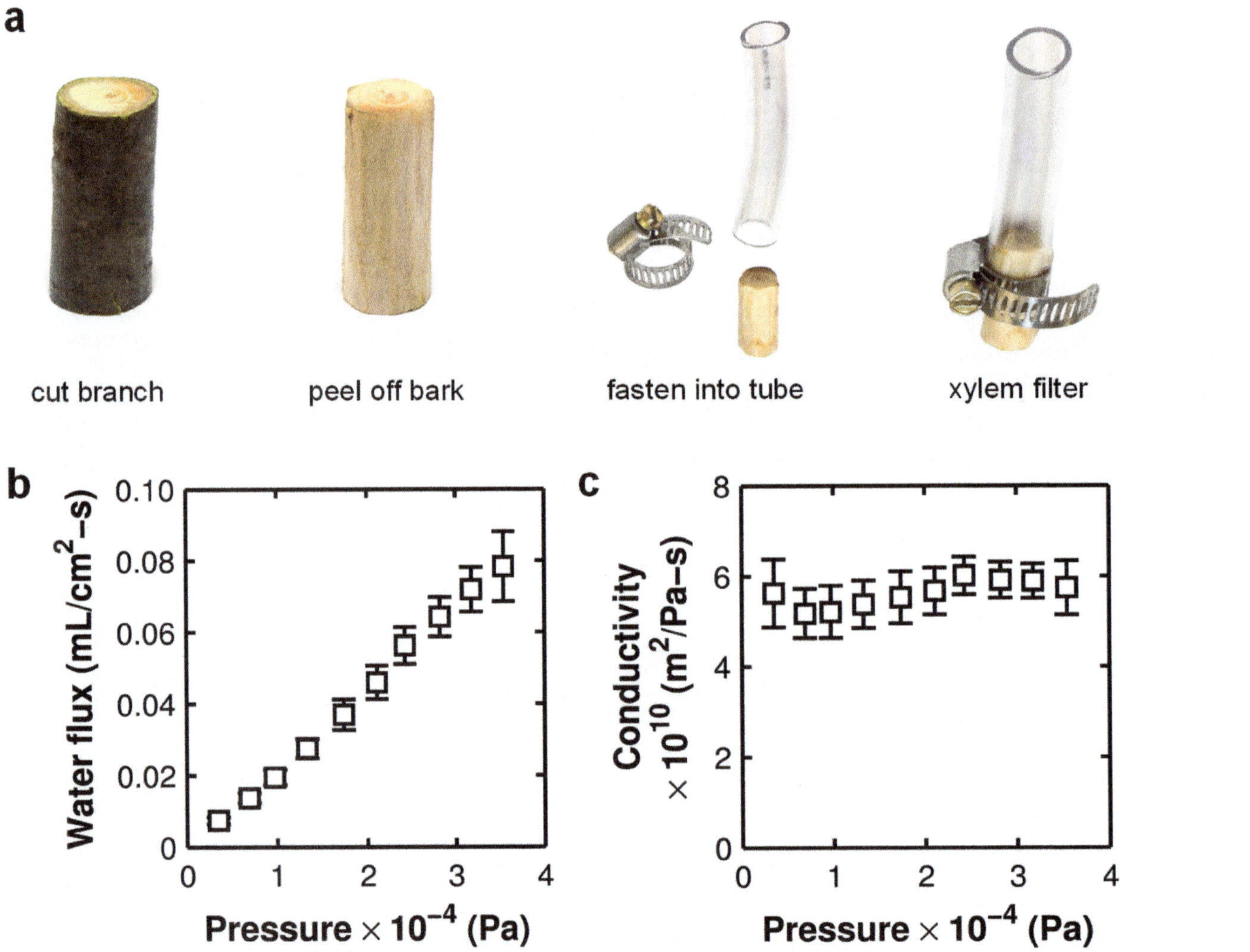

Figure 2. Xylem filter. a) Construction of xylem filter. b) Effect of applied pressure on the water flux through the xylem filter. c) Hydrodynamic conductivity of the filter extracted at each measured pressure using the total filter cross-section area and thickness as defined by Equation 1. Error bars indicate ±S.D. for measurements on three different xylem filters.

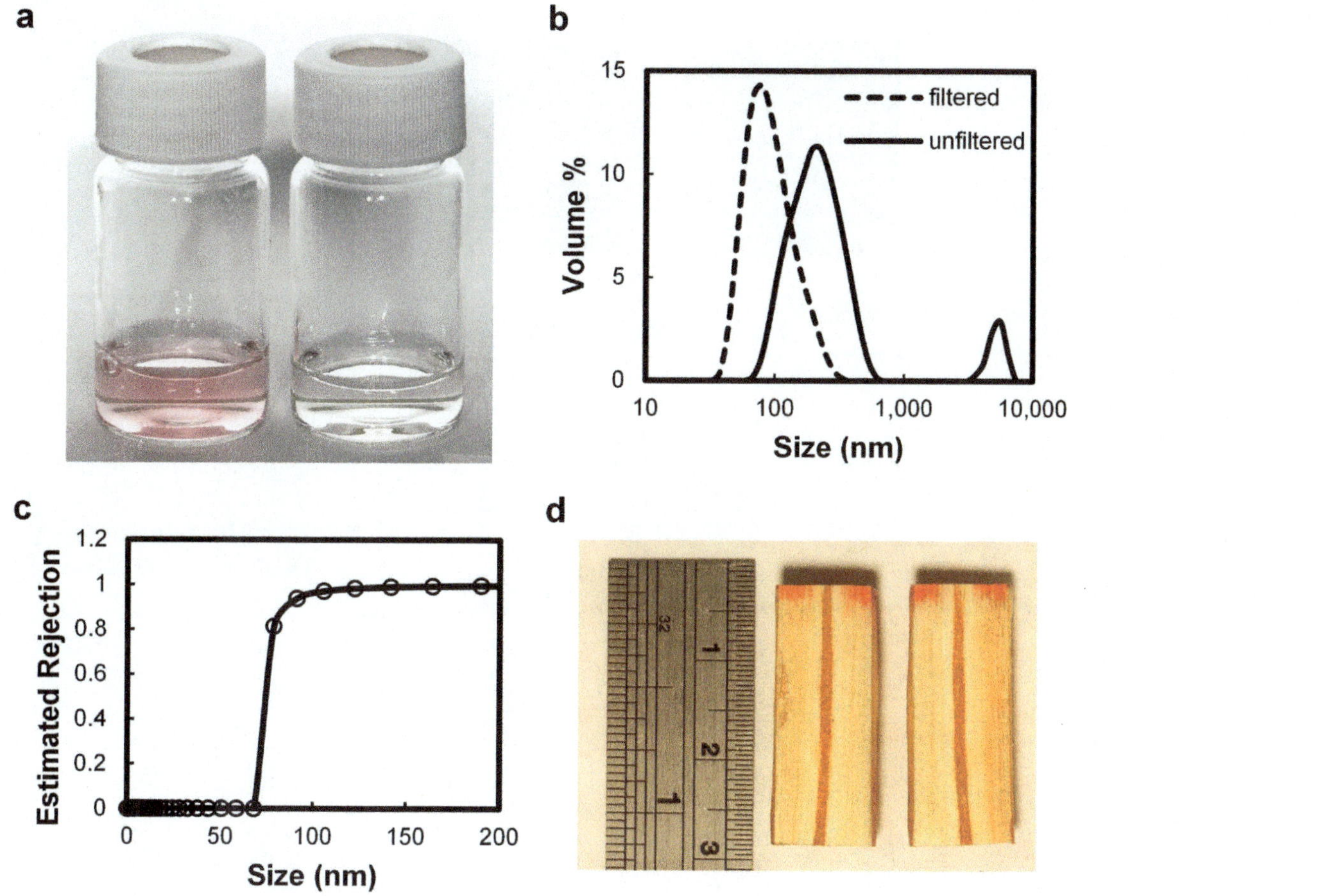

Figure 3. Filtration performance of the xylem filter. a) Feed solution of a pigment dye before filtration (left), compared to the filtrate (right). b) Size distribution of the pigment particles in the feed and filtrate solutions measured by dynamic light scattering. c) Dependence of the rejection on the particle size estimated from the data in (b). d) Cross-section of the xylem filter after filtration. Scale is in centimeters and inches.

Assuming that pigment particles 70 nm or less in size were not rejected, the size distributions before and after filtration enable calculation of the rejection performance of the xylem filter as a function of particle size (Figure 3c). We find that the xylem filter exhibits excellent rejection for particles with diameters exceeding 100 nm, with the estimated rejection exceeding 99% for particles over 150 nm. Smaller particles are expected to pass through the larger porosity of the pit membrane: SEM images in Figure 1e indicate sub-micron spacing between the radial spoke-like margo membrane through which the pigment particles can pass, although the porosity is difficult to resolve in the SEM image.

After filtration, we cut the xylem filter parallel to the direction of flow to investigate the distribution of dye in the filter. We observed that the dye was confined to the top 2–3 millimeters of the xylem filter (Figure 3d), which compares well with the tracheid lengths on the millimeter scale expected for coniferous trees [7]. These results show that the majority of the filtration occurred within this length scale, and suggests that the thickness of the xylem filter may be reduced to below 1 cm while still rejecting the majority of the dye.

Filtration of Bacteria from Water

Finally, we investigated the ability of the xylem filter to remove bacteria from water. As a model bacterium, we used fluorescently labeled and inactivated *Escherichia coli* bacteria that are cylindrical in shape with a diameter of ~1 μm. Use of fluorescently labeled bacteria enabled easy enumeration of their concentrations, and also allowed us to track the location in the xylem filter where they were trapped. Since filtration is dominated by size-exclusion at this length scale, we do not expect modification with the dye to significantly affect filtration characteristics. Filtration using three different xylem filters showed nearly complete rejection of the bacteria (Figure 4a). Using a hemacytometer to count the bacteria, we estimate that the rejection was at least 99.9%.

To investigate the mechanism of filtration, the xylem filter was cut parallel to the direction of flow after filtration. When examined under a fluorescence microscope, we observed that the bacteria accumulated over the donut-shaped margo pit membranes (Figure 4b). This distribution is consistent with the expectation that the bacteria are filtered by the porous margo of the pit membranes. The distribution of trapped bacteria was not uniform across the cross section of the filter. Similar to the case of the dye, bacteria were observed only within the first few millimeters from the end through which the solution was infused (indicated by the white arrow in Figure 4c). In addition, the low-magnification fluorescence image shows that the bacteria had accumulated primarily over pit membranes at the bottom of the tracheids, which is again not unexpected. Further investigation by SEM clearly showed individual bacterial cells accumulated on the pit membranes over the porous margo (Figure 4d,e). These results confirm the pit membranes as the functional units that provide the filtration effect in the xylem filter.

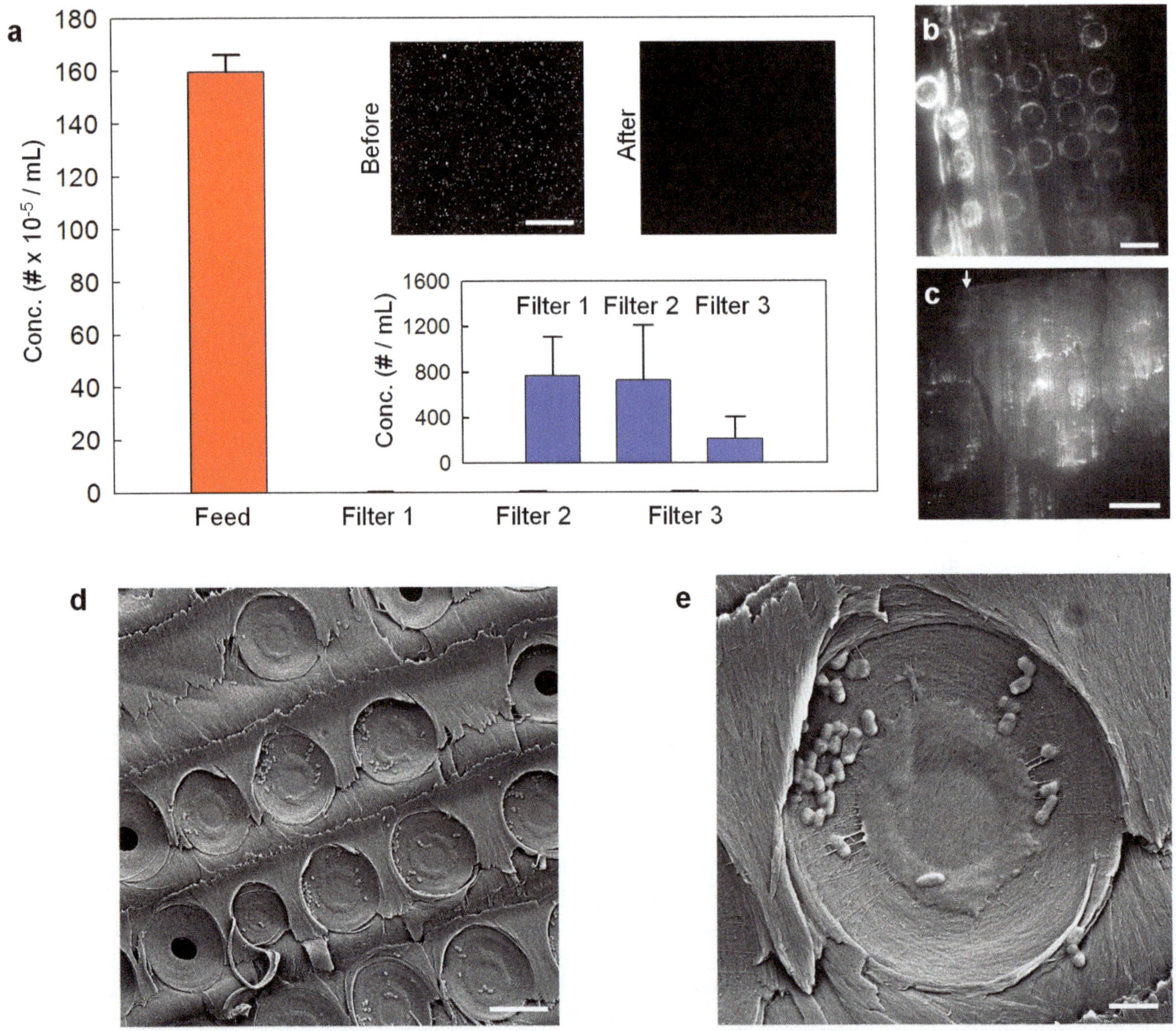

Figure 4. Filtration of model bacteria by the xylem filter. a) Concentrations of bacteria in the feed and filtrate solutions. Inset shows fluorescence images of the two solutions. Scale bar is 200 μm. Error bars indicate ±S.D. for experiments performed on three different xylem filters. b) Fluorescence image of xylem filter cross-section showing accumulation of bacteria over the margo pit membranes. Scale bar is 20 μm. c) Low-magnification fluorescence image shows that bacteria are trapped at the bottoms of tracheids within the first few millimeters of the top surface. Scale bar is 400 μm. Arrow indicates top surface of the xylem filter and also the direction of flow during filtration. Autofluorescence of the xylem tissue also contributes to the fluorescence signal in (b) and (c). d), e) SEM images showing bacteria accumulated on the margo pit membranes after filtration. Scale bars are 10 μm and 2 μm, respectively.

Discussion

Wood has been investigated recently as a potential filtration material [13], showing moderate improvement of turbidity. While we showed filtration using freshly cut xylem, we found that the flow rate dropped irreversibly by over a factor of 100 if the xylem was dried, even when the xylem was flushed with water before drying. We also examined flow through commercially available kiln-dried wood samples cut to similar dimensions. Wood samples that exhibited filtration showed two orders of magnitude smaller flow rates than in the fresh xylem filter, while those that had high flow rates did not exhibit much filtration effect and seemed to have ruptured tracheids and membranes when observed under SEM. Wetting with ethanol or vacuuming to remove air did not significantly increase the flow rate in the wood samples that exhibited the filtration effect, suggesting that the pit membranes may have a tendency to become clogged during drying. These results are consistent with literature showing that the pit membranes can become irreversibly aspirated against the cell wall, blocking the flow [14]. In fact, the pit membranes in the SEM images (Figure 1d,e and Figure 4d,e), which were acquired after drying the samples, appear to be stuck to the walls. Regardless, our results demonstrate that excellent rejection (> 99.9%) of bacteria is possible using the pit membranes of fresh plant xylem, and also provide insight into the mechanism of filtration as well as guidelines for selection of the xylem material.

Peter-Varbanets et al. [2] have outlined the key requirements for point-of-use devices for water disinfection: a) performance (ability to effectively remove pathogens), b) ease of use (no time-consuming maintenance or operation steps), c) sustainability (produced locally with limited use of chemicals and non-renewable energy), and d) social acceptability. Meeting all of these requirements has proved to be challenging, but point-of-use methods that have been successfully used for low-cost water treatment in developing countries include free-chlorine/solar disinfections, combined coagulant-chlorine disinfection, and biosand/ceramic filtrations [5]. While chlorine is a very effective biocide, its reaction with organic matter can produce carcinogenic by-products [15] and some waterborne pathogens such as *Cryptosporidium parvum* and *Mycobacterium avium* are resistant to the chlorine [16]. Solar disinfection based on ultraviolet irradiation can effectively inactivate *C. parvum*, but this requires low turbidity of source water [17] and is not effective for control of viruses [16]. Filtration based on biosand and ceramic filters is also effective at removing pathogens, but the effectiveness against viruses is low or unknown [18]. Coagulation combined with chlorine disinfection removes or inactivates viruses and pathogens effectively. However, necessity of an additional filtration step and relatively high cost are potential barriers for practical use [18]. Among these methods, a review on field studies by Sobsey et al. [5] suggested that biosand and ceramic filtration are the most effective methods in practice, because once the apparatus is installed, the effort for use and dosage is significantly reduced and therefore promotes persistent use compared to disinfection approaches. Although membrane-based filtration is the most widely used for water treatment in industrialized nations and the cost of membranes has significantly decreased, membranes are still unaffordable to poor communities in the developing world [2]. Ultrafiltration systems run by hydrostatic pressure [19] and some recently invented point-of-use devices using ultrafiltration membranes may provide water to developing regions at reasonable cost [2]. However, membranes still require specialized chemicals and processes for manufacture, and need cleaning or replacement.

Xylem filter technology could be an attractive option for low-cost and highly efficient point-of-use water treatment by filtration, overcoming some of the challenges associated with conventional membranes. Xylem filters could provide the advantage of reduced human effort compared to existing point-of-use water treatment options, requiring only simple periodic filter replacement. In addition, the pressures of 1–5 psi used here are easily achievable using a gravitational pressure head of 0.7–3.5 m, implying that no pumps are necessary for filtration. The measured flow rates of about 0.05 mL/s using only ~1 cm^2 filter area correspond to a flow rate of over 4 L/d, sufficient to meet the drinking water requirements of one person [20]. This is comparable to chlorination and biosand filtration, which have the highest production rates of prevalent point-of-use water treatment methods, and far exceeds typical production rates for solar disinfection. Xylem filters could potentially be produced locally and inexpensively, and disposed of easily owing to their biodegradability. The high flow rates and low cost would certainly help address the issues of maintenance and replacement. For example, 200 filters of 10 cm^2 area and 0.5 cm thickness could be packaged into a volume of about 1 L, which will be inexpensive and last a few years even with weekly replacement. Furthermore, as suggested by the dye filtration experiment, xylem filters should be able to significantly reduce water turbidity, enhancing the aesthetic qualities of the drinking water, which is hardly achieved by chlorination and solar disinfection.

Wood is an easily available material. While use of fresh xylem does not preclude its use as a filter material, dried membranes have definite practical advantages. Therefore, the process of wood drying and its influence on xylem conductivity needs further study. In particular, processes that yield intact yet permeable xylem tissues that can be stored long-term will be helpful for improving the supply chain if these filters are to be widely adopted. In addition, flow through xylem of different plants needs to be studied to identify locally available sources of xylem, which will truly enable construction of filters from locally available materials. In the present study, we report results using xylem derived from only one species. These xylem filters could not filter out 20 nm nanoparticles, which is a size comparable to that of the smallest viruses. It will be interesting to explore whether there exist any coniferous species that have pit membranes with smaller pore sizes that can filter out viruses, or whether conifer xylem can be impregnated with particles such as carbon black to improve rejection of viruses. In their absence, angiosperms with short vessels that yield practical filter lengths may be the best alternative due to their smaller pit membrane pore sizes [8]. Further exploration of xylem tissues from different plants with an engineering perspective is needed to construct xylem filters that can effectively reject viruses, exhibit improved flow rates, or that are amenable to easy storage. It is also conceivable that plants could be selected or developed for enhanced filtration characteristics, as has been the norm in agriculture for enhancement of many desirable characteristics including resistance to pests, flavor, or productivity.

Conclusions

Plant xylem is a porous material with membranes comprising nanoscale pores. We have reasoned that xylem from the sapwood of coniferous trees is suitable for disinfection by filtration of water. The hierarchical arrangement of the membranes in the xylem tissue effectively amplifies the available membrane area for filtration, providing high flow rates. Xylem filters were prepared by simply removing the bark of pine tree branches and inserting the xylem tissue into a tube. Pigment filtration experiments revealed a size cutoff of about 100 nm, with most of the filtration occurring within the first 2–3 mm of the xylem filter. The xylem filter could effectively filter out bacteria from water with rejection exceeding 99.9%. Pit membranes were identified as the functional unit where actual filtration of the bacteria occurred. Flow rates of about 4 L/d were obtained through ~1 cm^2 filter areas at applied pressures of about 5 psi, which is sufficient to meet the drinking water needs of one person. The simple construction of xylem filters, combined with their fabrication from an inexpensive, biodegradable, and disposable material suggests that further research and development of xylem filters could potentially lead to their widespread use and greatly reduce the incidence of waterborne infectious disease in the world.

Acknowledgments

The authors thank Yukiko Oka for assistance with preparation of illustrations and Sunandini Chopra for help with dynamic light scattering measurements.

Author Contributions

Conceived and designed the experiments: MSHB JL VV VC RK. Performed the experiments: MSHB JL VV VC. Analyzed the data: MSHB JL RK. Contributed reagents/materials/analysis tools: VC. Wrote the paper: MSHB JL RK.

References

1. Gadgil A (1998) Drinking water in developing countries. Annual Review of Energy and the Environment 23: 253–286.
2. Peter-Varbanets M, Zurbrügg C, Swartz C, Wouter Pronk W (2009) Decentralized systems for potable water and the potential of membrane technology. Water Research 43: 245–265.
3. World Health Organization. Available: http://www.who.int/water_sanitation_health/mdg1/en/. Accessed 2014 Jan 14.
4. Madaeni SS (1999) The application of membrane technology for water disinfection. Water Research 33: 301–308.
5. Sobsey MD, Stauber CE, Casanova LM, Brown JM, Elliott MA (2008) Point of use household drinking water filtration: A practical, effective solution for providing sustained access to safe drinking water in the developing world. Environmental Science and Technology 42: 4261–4267.
6. Loo SL, Fane AG, Krantz WB, Lim TT (2012) Emergency water supply: A review of potential technologies and selection criteria. Water Research 46: 3125–3151.
7. Sperry JS (2003) Evolution of water transport and xylem structure. International Journal of Plant Sciences 164: S115–S127.
8. Choat B, Cobb AR, Jansen S (2008) Structure and function of bordered pits: new discoveries and impacts on whole-plant hydraulic function. New Phytologist 177: 608–625.
9. Jansen S, Choat B, Pletsers A (2009) Morphological variation of intervessel pit membranes and implications to xylem function in angiosperms. American Journal of Botany 96: 409–419.
10. Choat B, Ball M, Luly J, Holtum J (2003) Pit membrane porosity and water stress-induced cavitation in four co-existing dry rainforest tree species. Plant Physiology 131: 41–48.
11. Espino S, Schenk HJ (2011) Mind the bubbles: achieving stable measurements of maximum hydraulic conductivity through woody plant samples. Journal of Experimental Botany 62: 1119–1132.
12. van Ieperen W, van Meeteren U, van Gelder H (2000) Fluid ionic composition influences hydraulic conductance of xylem conduits. Journal of Experimental Botany 51: 769–776.
13. Sens ML, Emmendoerfer ML, Muller LC (2013) Water filtration through wood with helical cross-flow. Desalination and Water Treatment.
14. Petty JA (1972) Aspiration of bordered pits in conifer wood. Proceedings of the Royal Society Series B-Biological Sciences 181: 395-+.
15. Dunnick J, Melnick R (1993) Assessment of the carcinogenic potential of chlorinated water - Experimental studies of chlorine, chloramine, and trihalomethanes. Journal of the National Cancer Institute 85: 817–822.
16. Shannon M, Bohn P, Elimelech M, Georgiadis J, Marinas B, et al. (2008) Science and technology for water purification in the coming decades. Nature 452: 301–310.
17. Mintz E, Bartram J, Lochery P, Wegelin M (2001) Not just a drop in the bucket: Expanding access to point-of-use water treatment systems. American Journal of Public Health 91: 1565–1570.
18. Lantagne DS, Quick R, Mintz ED (2006) Household water treatment and safe storage options in developing countries: A review of current implementation practices. Woodrow Wilson International Center for Scholars. Environmental Change and Security Program.
19. Arnal J, Fernandez M, Verdu G, Garcia J (2001) Design of a membrane facility for water potabilization and its application to Third World countries. Desalination 137: 63–69.
20. Reed B, Reed B (2011) How much water is needed in emergencies. World Health Organization. Available: http://www.who.int/water_sanitation_health/publications/2011/tn9_how_much_water_en.pdf. Accessed 2014 Jan 14.

12

Biodegradation of Microcystins during Gravity-Driven Membrane (GDM) Ultrafiltration

Esther Kohler[1], Jörg Villiger[1], Thomas Posch[1], Nicolas Derlon[2], Tanja Shabarova[1], Eberhard Morgenroth[2,3], Jakob Pernthaler[1], Judith F. Blom[1]*

1 Limnological Station, Institute of Plant Biology, University of Zurich, Kilchberg, Switzerland, **2** Eawag: Swiss Federal Institute of Aquatic Science and Technology, Dübendorf, Switzerland, **3** Institute of Environmental Engineering, ETH Zurich, Zurich, Switzerland

Abstract

Gravity-driven membrane (GDM) ultrafiltration systems require little maintenance: they operate without electricity at ultra-low pressure in dead-end mode and without control of the biofilm formation. These systems are already in use for water purification in some regions of the world where adequate treatment and distribution of drinking water is not readily available. However, many water bodies worldwide exhibit harmful blooms of cyanobacteria that severely lower the water quality due to the production of toxic microcystins (MCs). We studied the performance of a GDM system during an artificial *Microcystis aeruginosa* bloom in lake water and its simulated collapse (i.e., the massive release of microcystins) over a period of 21 days. Presence of live or destroyed cyanobacterial cells in the feed water decreased the permeate flux in the *Microcystis* treatments considerably. At the same time, the microbial biofilms on the filter membranes could successfully reduce the amount of microcystins in the filtrate below the critical threshold concentration of 1 µg L^{-1} MC for human consumption in three out of four replicates after 15 days. We found pronounced differences in the composition of bacterial communities of the biofilms on the filter membranes. Bacterial genera that could be related to microcystin degradation substantially enriched in the biofilms amended with microcystin-containing cyanobacteria. In addition to bacteria previously characterized as microcystin degraders, members of other bacterial clades potentially involved in MC degradation could be identified.

Editor: Rajeev Misra, Arizona State University, United States of America

Funding: This research was supported by the ProDoc program "Predictive Toxicology" (PDFMP3_132466), funded by the Swiss National Science Foundation. J. Villiger was supported by the Swiss Federal Department of Foreign Affairs through Polish-Swiss Research Program, project PSPB-036/2010: Diversity and Ecology of Mixotrophic Nanoflagellates in the Gulf of Gdańsk (DEMONA). The funders had no role in study design, data collection and analysis, decision to publish, or preparation of the manuscript.

Competing Interests: The authors have declared that no competing interests exist.

* Email: blom@limnol.uzh.ch

Introduction

During the last century, the anthropogenic input of nutrients into freshwaters has resulted in a distinct increase of cyanobacterial biomass in many water bodies worldwide [1]. Climate change and global warming may even increase the frequency and intensity of cyanobacterial blooms in the future [2]. Some cyanobacteria represent a major challenge for drinking water usage due to their production of microcystins (MCs), toxic secondary metabolites that affect a wide range of animals and humans [3]. The major route of human exposure to MCs is via oral ingestion, mainly due to the consumption of drinking water [4]. Even a subchronic dose of MCs in drinking water may elevate the rate of liver cancer, as was shown in China, where prevalent incidences of liver cancer correlated with MC-contaminations in drinking water [5]. Consequently, the WHO has developed a guideline for MCs in drinking water stating that an average exposure generally should be below the level of 1 µg L^{-1} [4]. *Microcystis aeruginosa* is known to form massive blooms in many lakes worldwide, and it produces MCs in high amounts. The concentrations of the intracellular MCs range between 0.3 to 15 µg L^{-1} [6] and up to 400 µg L^{-1} [7] in cyanobacterial blooms, however, occasionally high concentrations of up to 1400 µg L^{-1} were found [8]. Elimination of the MCs from drinking water is therefore highly desirable.

The largely cell-bound MCs are eliminated by removing the intact cyanobacterial cells by conventional water treatment procedures such as coagulation or flocculation [4]. However, cell damage (e.g. during the collapse of a bloom) will release toxins into water, and the above mentioned procedures will not sufficiently remove MCs from drinking water. Strategies such as powdered activated carbon [9], sediment sorption [10], or ozonisation [11] have been suggested to effectively eliminate dissolved MCs. However, these treatments are costly in terms of development and management (energy, need of chemicals) and thus not suitable for the application in developing and transient countries.

Gravity driven membrane (GDM) ultrafiltration is considered for drinking water production as a relevant alternative to common appliances [12]. GDM uses a simple set-up [13], which is inexpensive, electricity-free, easy to use, and it is already known to provide an effective barrier against pathogens, disease vectors and suspended solids [14]. Microbial activity as well as the total organic carbon content in the feed water have been shown to affect the performance of the GDM ultrafiltration without control of the biofilm formation (no backwashing or chemical cleaning) [13]. However, nothing is currently known about the possible degradation processes of intact cyanobacterial cells or of toxins such as MCs in these point-of-use membrane systems. In recent years, biodegradation by heterotrophic bacteria has been recognized as an alternative way to eliminate MCs [15]. A few bacterial isolates capable of MC degradation have been already characterized [16]. The best studied MC degrading bacteria are belonging to the *Alphaproteobacteria* such as *Sphingomonas* sp. [17], *Sphingopyxis* sp. [18], or *Novosphingobium* sp. [19]. However, only a few studies have tried to link the composition of bacterial communities in plankton [20] or in biofilms of biological drinking water treatment facilities [21] with the ability of these systems to degrade microcystins.

The objective of our study was to examine possible degradation of MCs by the microbial biofilm of a GDM ultrafiltration system. We simulated cyanobacterial blooms and their collapse (and thus the release of the cell-bound MCs into the surrounding water) and determined the ability of the microbial biofilm to remove MCs during drinking water production. We also analysed the composition of microbial assemblages of the biofilms of the GDM ultrafiltration systems by next generation sequence analyses in order to obtain information about the microorganisms that might potentially be involved in this process.

Materials and Methods

Cyanobacterial cultures and quantification of microcystins

Axenic cultures of Microcystis aeruginosa PCC 7806 were kept at 20°C in Cyano-medium in several Erlenmeyer flasks under constant light at 5 µmol quanta m^{-2} s^{-1} from fluorescent tubes. Fresh Microcystis aeruginosa PCC 7806 cultures were taken every three to four days from the cultivation for the ongoing experiment. The cell number of the cyanobacterial culture was determined by flow cytometry (described below), and the MC concentration was quantified by high-performance liquid-chromatography (HPLC) as followed: A volume of 5 mL of the culture was frozen at −23°C for three hours. After thawing, 7.5 mL of 100% methanol (MeOH) were added to achieve a 60% aqueous methanolic solution. The extract was centrifuged for 15 min at 25′700 g. HPLC analysis was performed on a Shimadzu 10 AVP system with photodiode array detector (PDA) and a Hydrosphere C18 column (YMC, 4.6×250 mm, Stagroma, Switzerland), using solvent A: UV-treated H_2O containing 0.05% trifluoroacetic acid (TFA, Merck) and solvent B: acetonitrile and 0.05% TFA. A gradient was achieved by applying linear increases in two steps (solvent B from 35% to 70% in 30 min, 70% to 100% in 2 min). For the quantification procedure, calibration curves for MC-LR and [D-Asp^3] MC-LR, the two MCs of M. aeruginosa PCC 7806 had to be established: The two MCs were isolated in high purity (>99%, HPLC) from Microcystis aeruginosa PCC 7806, and their specific molar absorption coefficient was used to prepare accurate standard solutions between 1 and 10 µg mL^{-1}. The calibration curves were based on the peak area recorded at a wavelength of 239 nm. The microcystin quantification was done in duplicate and is referred to as the sum of the concentrations of MC-LR and [D-Asp^3] MC-LR.

Experimental setup of the Gravity-Driven-Membrane (GDM) system

Water from a depth of 5 m of Lake Zurich was continuously pumped by a fountain pump (Nautilus 450, Oase GmbH, Hörstel, Germany) to a storage tank (6 l volume, kept in the dark at room temperature). This storage tank was connected by silicon tubes (Saint-Gobin) to six parallel membrane modules consisting of filter holders of 48 mm inner diameter (Whatman, Maidstone, Kent, UK) and polyethersulfone ultrafiltration membranes with a 150 kDa nominal cut-off (PBHK, Biomax Millipore, Billerica, MA, USA). A hydrostatic pressure of 0.65 mbar was received by keeping the storage tank 0.65 m above the membrane surface. Overflow conditions at the storage tank guaranteed constant transmembrane pressure. Filter holders, silicon tubes and glass bottles for collection of filtrate water were autoclaved prior to experiment. Ultrafiltration membranes have been soaked in nanopure water (Bearnstead, Thermo Scientific, Basel, Switzerland) for 24 h before starting the experiment. The six membrane systems were split into three different treatments, each with two replicates (Figure 1): the control treatment (referred to as CON) received lake water only. A Microcystis bloom was simulated in two replicates: the lake water was enriched with the cyanobacterium Microcystis aeruginosa PCC 7806 (about 2×10^8 cells) once every 24 h (treatment subsequently referred to as LMA: living Microcystis aeruginosa). A collapsing cyanobacterial bloom was simulated in the last treatment. A culture of Microcystis aeruginosa PCC 7806 was first frozen for 3 h at −20°C, and thawed before adding to the membrane system once every 24 h (referred to as DMA: dead Microcystis aeruginosa). The cyanobacterial cells were directly added above the filtration membrane module to avoid MC degradation processes in the storage tank or during tubing passages. All filter systems were kept in the dark. The filtrate water of all treatments was collected every 24 h, and the volume was determined to quantify the permeate flux (as L m^{-2} h^{-1}). A subsample of 1 mL of each filtrate was fixed with 50 µl glutaraldehyde (2.5% final concentration) for cell enumeration at the flow cytometer. The rest was stored at 4°C for microcystin (MC) quantification (usually done within 24 to 48 h). The experiment was running for 21 days. At the end of the experiment, the ultrafiltration membranes were cut into three equal parts that were subjected to the following analyses: (i) One part of the filter was used to quantify MCs by HPLC that possibly remained in or attached to the biofilm on the filters. (ii) Phylogenetic analyses (454 tag pyrosequencing) were performed with the biofilm on another filter part, and (iii) the last part was used to take a closer look on biofilm structures by non-invasive methods such as Optical Coherence Tomography (OCT) (model 930 nm Spectral Domain, Thorlabs GmbH, Dachau, Germany). OCT images were analysed for average thickness and relative roughness using image analysis software developed under Matlab (MathWorks, Natick, US) [12]. To determine the exact membrane area of the pieces image analysis (Zeiss, AxioVision 4.7) was applied using a microscope (AxioImager.Z1, 1 x EC Plan-Neofluar, Zeiss) and a CCD camera (AxioCam MRm, 12 bit grayscale, 1388×1040 px, Zeiss).

Flow cytometric enumeration of cells in the filtrate water

Samples for flow cytometry were stained with DAPI (4′,6-diamidino-2-phenylindole, 1 µg mL^{-1} final concentration) for 15 min in the dark. Subsequently, samples were analysed using an Influx V-GS cell sorter (Becton Dickinson, Inc., San Jose, CA) equipped with a UV laser (60 mW, 355 nm; CY-PS; Lightwave Electronics) for detection of DAPI fluorescence, and a blue laser (200 mW, 488 nm; Sapphire; Coherent Inc.) for scattered light

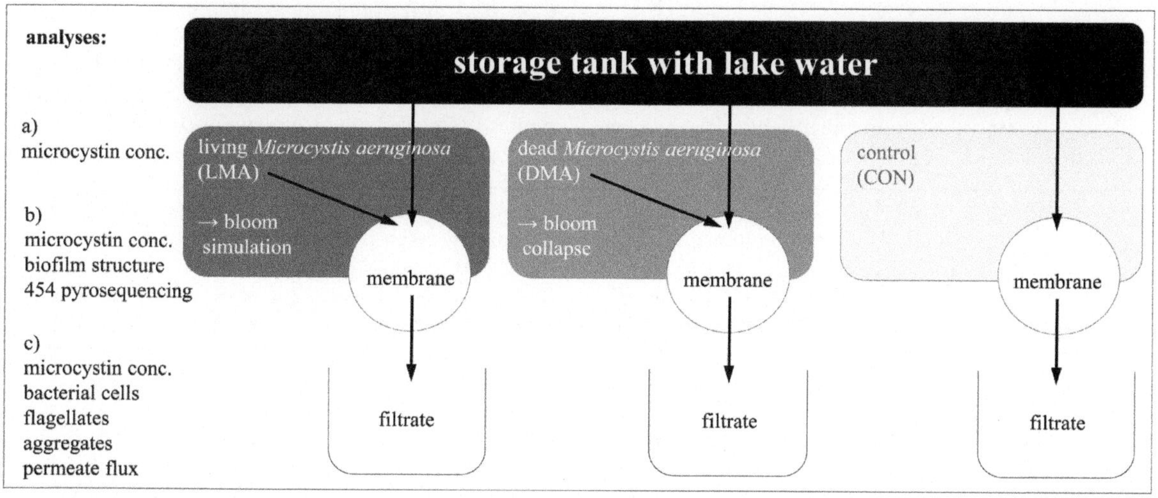

Figure 1. Schematic view of the gravity-driven membrane (GDM) system. Depicted are the three different operating treatments (LMA, DMA, and CON), the ultrafiltration membrane (150 kDa) and the filtrate collection. Analyses that were carried out during this study a) in the feed water, b) in/on the biofilm, and c) in the filtrate are listed on the left side of this overview.

and autofluorescence of flagellates. If necessary, and to avoid particle coincidence, samples were diluted with sheath fluid (2.5 g L^{-1} NaCl; filtered by 0.2 µm-pore-size). Sample volume was calculated from the analysed sample weight. Data obtained by flow cytometry were analysed with the custom-made software ViiGate 1.0a. Bacterial cells were identified using side scattered light (SSC) versus DAPI fluorescence (431 nm), flagellates were determined on the basis of SSC versus green fluorescence (531 nm). Bacterial aggregates were operationally defined by their DAPI fluorescence and scatter properties equal to or higher than that of flagellates [22]. Cyanobacterial cells of the *Microcystis aeruginosa* PCC 7806 culture were identified using SSC versus their auto-fluorescence at 692 nm (without prior staining).

Extraction of microcystins

The entire filtrate was collected for solid phase extraction of MCs. The C18 cartridges (1 g, 60 mL, Mega Bond Elute, Varian, Agilent Technologies, Basel, Switzerland) were first equilibrated with 10% MeOH before adding the filtrate water. Afterwards, the MCs were eluted with 100% methanol. The samples were dried in a vacuum rotary evaporator at 40°C and 35 mbar. The residues were re-suspended in 1 ml 60% MeOH, the microcystin quantification was performed by HPLC in duplet. The MC concentrations in the filtrate were always measured 24 h after injection and were expressed as percentage of the MCs that were removed from the system (removal efficiency) or as MC removal rates (ug L^{-1} d^{-1}).

At the end of the experiment, GDM polyethersulfone membranes were first frozen at −23°C for 3 h to isolate the MCs from the biofilm. After thawing, the biofilms were extracted twice with 5 mL 60% MeOH for 1 h, and both extracts were combined. Accordingly, the solvent was evaporated (40°C and 35 mbar), and the samples were prepared for HPLC analysis. The residues were re-suspended in 1 ml 60% MeOH, and centrifuged for 5 min at 10′000 rpm. Afterwards, the supernatants were taken for microcystin quantification by HPLC (as described above).

454 tag pyrosequencing analysis

Prior to the analysis, the filter parts of both replicates were pooled. The DNA extraction of the biofilm bacteria was performed using the UltraClean Water DNA isolation kit (MO BIO Laboratories, Inc.). Subsamples of 300 µl of DNA suspension (final concentration 4–10 ng μL^{-1}) of all three DNA extractions (CON, LMA, and DMA) were sent to Research and Testing Laboratory, Inc. (Lubbock, TX, USA) for further processing. Partial 16S rRNA gene encoding sequences were obtained from 454 pyrosequencing (Roche FLX platform) following Assay b.9 by using the primer pair 799F and 1115R that exclusively amplify DNA of heterotrophic bacteria and exclude cyanobacteria from the process [23]. Raw data (68′981 Reads; mean raw read length 378.4 base pairs) were processed by a custom-made pipeline on a local computer cluster consisting of 16 units (each equipped with an 8 core AMD FX-8150 CPU, 16 GB RAM and a 128 GB SSD hard disk) and a separate control workstation. The program was developed in DELPHI and run under Windows 7. The processing of the raw data is extensively described elsewhere [24]. In brief, reads were denoised at the level of flowgrams according to Quince and co-workers [25]. Afterwards, quality filtering strategies were applied to finally end up with the number of 27′391 sequences (raw reads reduced by 60%) corresponding to 956 operational taxonomic units (OTUs, 3% similarity). A distance matrix was calculated and the OTUs were produced after the pairwise alignment (Needleman-Wunsch algorithm) by average linkage at similarity levels of 97%. OTUs were assigned to taxonomic entities on the level of similarity of the OTU to the most closely related sequence in the SILVA reference data base (release 109) [26]. OTUs were grouped into different levels of sequence identity: ≤ 3% divergence in 16S rRNA gene sequence corresponds to species level, ≤5% to genus level, and ≤10% to family level. Finally, the OTUs of the both treatments were compared with each other. Only OTUs were included that were >0.5% of sequences per sample (CON or LMA+DMA) and that occurred at least ten times more frequently in one treatment than the other. Shannon's diversity index was estimated using the formula $H' = -\Sigma(P_i * \ln P_i)$, where P_i is the relative abundance of the sequences per sample. The index H′ is used to characterize the diversity of species or species-like units (OTUs) in a community.

Results

Physical parameters: Permeate Flux and structure of the biofilms

Flux stabilization was observed approximately after eight to ten days of the experiment in all three treatments albeit great differences between the control and both Microcystis treatments (Figure 2). A mean flux of 4.7 L m^{-2} h^{-1} was measured after 12 days in the CON treatment. In one of the two replicates, the flux stayed constant until the end of the experiment. In the second replicate, the flux increased slowly to 6.9 L m^{-2} h^{-1} on day 21. Accordingly, the mean thickness (as assessed by OTC measurements) of the biofilms of both control replicates at the end of the experiment were slightly different. The biofilm of the first replicate had a thickness of about 125 (±23) µm; the biofilm of the second replicate was 96 (±17) µm thick. The second replicate was less heterogeneous than the first replicate, but both exhibited low relative roughness values of 0.49 and 0.35, respectively. The permeate flux in both Microcystis treatments showed a similar trend. Stabilization could be observed at a mean flux of 1.6 L m^{-2} h^{-1} in the LMA and of 2.0 L m^{-2} h^{-1} in the DMA treatment. In both treatments, mean permeate flux decreased further to 1.0 L m^{-2} h^{-1} and 1.36 L m^{-2} h^{-1}, as measured at the end of the experiment. Thus, the mean flux in the Microcystis replicates was about 80% lower than in the CON treatment. Biofilms in the DMA treatment were about six to seven times thicker as in the CON treatment with values of 625 (±33) µm and 796 (±29) µm for both replicates. Unfortunately, a quantification of the biofilm thickness in the LMA treatment could not be carried out.

Microcystin removal

The microcystin removal efficiency of the biofilms was calculated from the amount of MCs that was injected via living or dead Microcystis cells into the systems and the amount that was measured 24 hours later in the filtrates (Figure 3, upper panel). In the LMA treatment, both replicates were working similarly, showing already high removal efficiency of almost 70% at the beginning. After 10 days of the experiment, the biofilms of both replicates showed nearly 100% removal efficiency; complete removal was achieved after 15 days and remained constant until the end of the experiment. Both replicates of the DMA treatment started with a low removal efficiency of about 10% during the first three days of the experiment. One of the replicates developed a biofilm that was able to remove the MCs completely after 15 days. The MC removal efficiency of the biofilm of the second replicate decreased again to less than 80% (Figure 3, upper panel). The MC removal rates in both treatments increased during the course of the experiment up to 440, respectively 300 µg L^{-1} d^{-1} (Figure 3, lower panel).

Altogether, 4.6×10^9 living or dead cells of M. aeruginosa (containing 262.5 µg MCs) were added to each MC-replicate of the GDM systems throughout the duration of the experiment (Table 1). About 10% (27 µg) of the added MCs were found again in the filtrates of each replicate of the LMA treatments during the entire experiment, 96.5 µg (36.8%) were found on the LMA filters at the end of the experiment. Thus, 140 µg of MCs were removed in both replicates during the course of the experiment. Only 0.5 µg MC was found on the filter in the DMA treatments, 99 µg, respectively 121 µg have been collected in the DMA filtrates of both replicates; 141–163 µg of the MCs were degraded during the course of the experiment.

Cell numbers

Bacterial single cell numbers in the CON filtrates constantly increased during the course of the experiment (Figure 4, upper panel) up to 0.1 and 0.3×10^6 cells mL^{-1}, respectively. Similarly, flagellate and bacterial aggregate numbers increased slowly, but stayed on comparable low mean levels of 280 flagellates mL^{-1} and 78 aggregates mL^{-1} on day 21 of the experiment (Figure 4). In contrast, both MC treatments showed higher cell numbers as

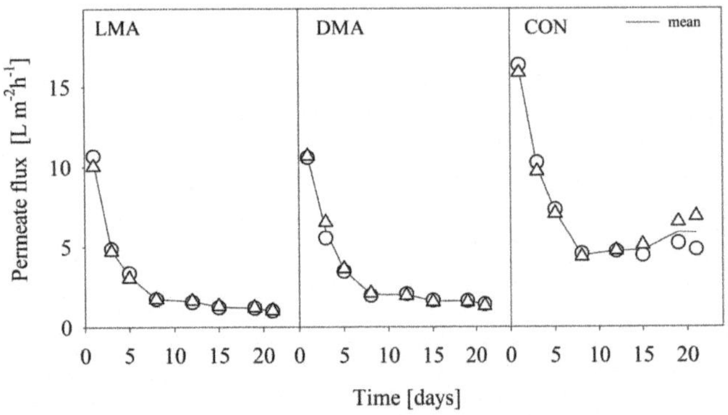

Figure 2. Evolution of the permeate flux. The flux is shown in L m^{-2} h^{-1} for the filtration of differently treated feed water sources (LMA, DMA, and CON according to Figure 1). The two replicates per system are shown as circles and triangles, and are connected by the mean.

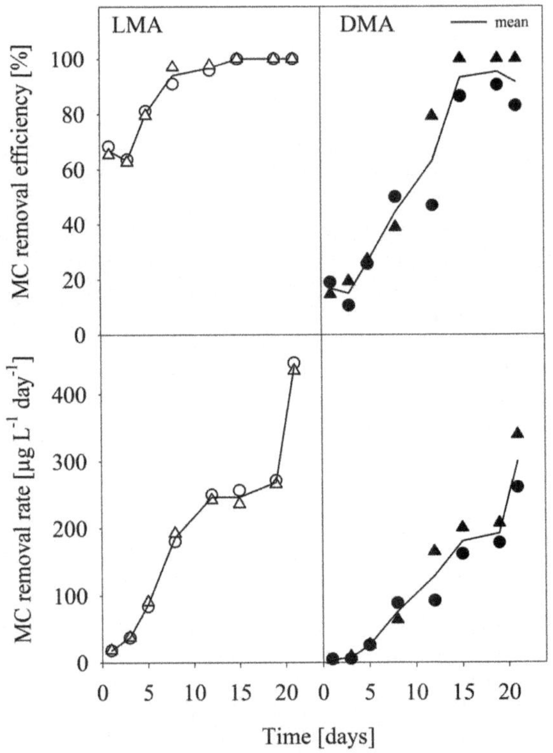

Figure 3. Microcystin removal and removal rates. Time course of the microcystins (MCs) removal efficiency of the GDM system (upper panel) and the MC removal rate (lower panel) in the two *Microcystis* treatments (LMA and DMA according to Figure 1). The two replicates per system are shown as circles and triangles, and are connected by the mean.

Table 1. Microcystin concentrations.

	LMA		DMA	
	Replicate 1	**Replicate 2**	**Replicate 1**	**Replicate 2**
Total amount of MCs injected over 21 days	262.5 (100)	262.5 (100)	262.5 (100)	262.5 (100)
Amount of MCs in the filtrate on day 21	0 (0)	0 (0)	0 (0)	2.9 (1.11)
MCs on the filter at the end of the experiment (on day 21)	96.5 (36.8)	96.5 (36.8)	0.5 (0.2)	0.5 (0.2)
Total amount of MCs in the filtrates, collected for 21 days	28 (10.6)	26 (9.9)	99 (37.7)	121 (46.1)
MCs loss	138 (52.6)	140 (53.3)	163 (62.1)	141 (53.7)
Threshold concentration of less than 1 µg L^{-1} was reached on day	15	15	15	Not reached

Overview about the amount of MCs [µg; (%)] totally injected, found on the biofilm or in the filtrate, and the estimated loss of MCs during the experiment. Last row shows the day, at which the threshold concentration of less than 1 µg L^{-1} was reached (LMA and DMA according to Figure 1).

compared to the CON treatment. Between 0.8 and 0.5×10^6 bacterial single cells mL^{-1} were found in both replicates of the LMA treatment at the end of the experiment. Comparable numbers of bacterial single cell numbers were determined also in the replicate 1 of the DMA treatment (0.4×10^6 mL^{-1}). The bacterial single cell numbers in the corresponding second replicate of the DMA treatment were about ten times higher (3.9×10^6 mL^{-1}). Flagellate numbers in the filtrates of the LMA treatment were comparable high at the end of the experiment (mean value 0.67×10^4 mL^{-1}), as well as bacterial aggregates (mean value 0.8×10^3 mL^{-1}). Comparable numbers of flagellates were found in the filtrate of the first replicate of the DMA treatment (0.54×10^4 mL^{-1}) as well as the highest amount of aggregates (4.0×10^3 mL^{-1}) at the end of the experiment. However, rather low numbers of flagellates and aggregates were found in the second replicate of the DMA treatment, 0.15×10^4 mL^{-1} and 0.1×10^3 mL^{-1}, respectively (Figure 4).

Phylogenetic analyses

A total of 27′391 sequences, assigned to 956 OTUs, were evaluated after removing low quality reads and chimeric sequences. For the bacterial communities of the MC-treated biofilms, about 452 OTUs (8′064 sequences; LMA) and 378 OTUs (8′048 sequences; DMA) were determined (Figure 5A); slightly more sequences were obtained for the CON bacterial community (11′279 sequences; 551 OTUs). The bacterial communities of the three biofilms shared only 11% of all OTUs (105 OTUs), but comprised 54% of all sequences (Figure 5A and 5B). The OTUs therein were large, and consisted of 140 sequences on average. This core community consisted predominantly of Sphingobacteriales (Bacteroidetes; 48% of the shared OTUs) and Comamonadaceae (Betaproteobacteria, 45% of the shared OTUs). The CON treatment consisted of the largest amount of OTUs (35% of all OTUs) that were unique to this special treatment, but comprised only 1764 sequences (5.2 sequences per OTU). LMA and DMA shared 108 OTUs, the average sizes of these OTUs were about 50 sequences per OTU (Figure 5A and 5B).

Most of the sequences in the combined LMA and DMA (MA) communities received taxonomic assignments at least to the genus level (93% in the LMA treatment, 96% in the DMA treatment), affiliation to families could be determined for 97%, respectively 99% of these sequences. However, only 44% of all sequences in the CON communities were assigned to the genus level, and 52% to the family level. With a high divergence in similarity of more than 12% to the closest known relative, two large OTUs were assigned to either Myxococcales (1′990 sequences) or to Fibrobacteres (2′357 sequences).

Bacterial taxa favoured by Microcystis addition

Addition of Microcystis cells led to compositional differentiation between the communities of both MC-treated and the CON biofilms. Over one third (38.3%) of all sequences in the CON assemblage was affiliated with Deltaproteobacteria, a class that was significantly less abundant in the combined LMA and DMA (MA) communities (5.0%) (Figure 6). The Fibrobacteres (36.6%) and the Alphaproteobacteria (19.4%) were the second and third most abundant taxa affiliated with the CON assemblage, both taxa were underrepresented in MA assemblage as well. These three classes comprised almost 95% of all sequences in the CON assemblage. However, more than one third (40.4%) of all sequences in the MA assemblage was affiliated with Betaproteobacteria that were found only marginally in the CON communities. Firmicutes (22.4%) and Gammaproteobacteria (12.8%) were the second and third most abundant taxa in the MA assemblage, but were not found in the communities of the CON treatment. These three classes comprised 75.6% of all sequences. Additionally, minor fractions of Candidate division TM7 and Bacteroidetes were present in both assemblages, Spirochaetes only to a minor extent in the MA assemblage (1.3%).

Most of the bacterial genera found in the CON assemblage were typically isolated from microbial biofilms in drinking or freshwater reservoirs or pipelines, such as Hirschia, Phenylobacterium, or Comamonas. Haliangium or the Myxococcales were present in anaerobic filter sediments or suboxic freshwater ponds (Table S1 in File S1). The Fibrobacteres were isolated before from upper sediment layers and biofilm samples. Bacterial sequences affiliated with Sphingomonas containing MC degradation proteins were only found in the CON assemblage, where they made up 12% of all sequences (Table S1 in File S1). Other genera of the Alphaproteobacteria that contain MC degradation proteins or MC dependent proteins have been found only in the MA assemblage, such as Azospirillum or Magnetospirillum, as well as the Rheinheimera (belonging to the Gammaproteobacteria) or Spirochaetales (Table S2 in File S1). Some other bacterial genera have been repeatedly found in the Planktothrix layer of Lake Zurich, such as Variovorax (5.4% of all sequences in the MA assemblage) or unknown genera belonging to the Sphingobacteriales (6.5%). Paucibacter known to degrade MCs has been only found in the MA assemblage (3.2%). The Firmicutes (22.4% of all sequences of the MA assemblage) comprised mainly of Acidaminobacter that are typical inhabitants of suboxic freshwater ponds.

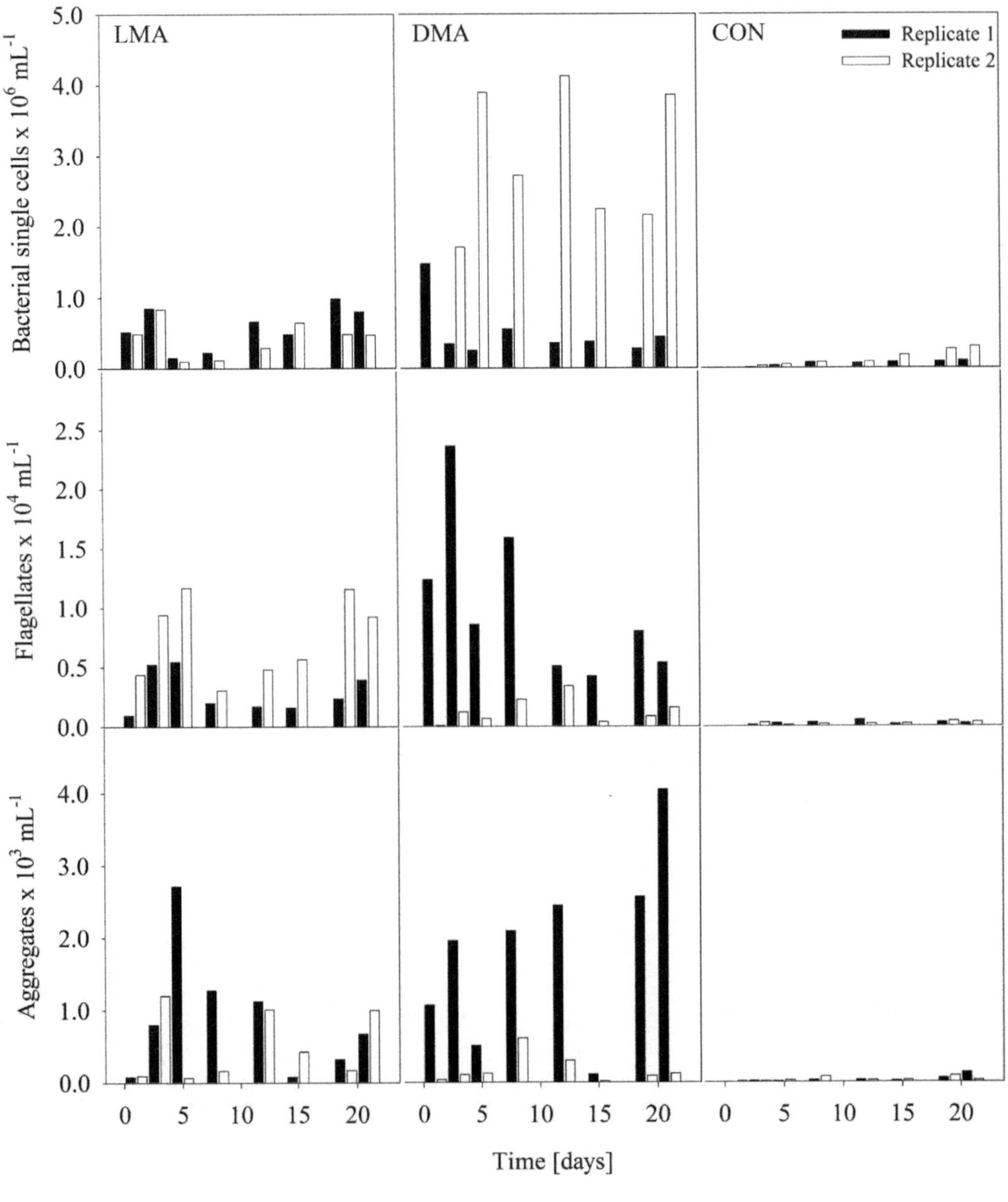

Figure 4. Cell abundances in the filtrate. Regrowth of bacterial single cells (upper panel), flagellates (middle panel) and flagellate-inedible bacterial aggregates (lower panel) in the filtrate of the three different treatments (LMA, DMA, and CON according to Figure 1).

Discussion

Performance of the GDM system and fate of microcystins

Two factors have a strong impact on the performance of GDM systems: the composition of the microbial community (bacteria and predators) and the organic carbon content of the feed water [12]. In the latter study, the almost complete absence of metazoan organisms resulted in smooth and homogeneous biofilm structures. In our study, this was indicated by the comparably low values of relative roughness of both CON replicates. In the absence of predation, the total organic carbon (TOC) content governs the permeability of the biofilms. The decrease of the permeate flux with increasing TOC content is due to, both, a higher accumulation of particulate matter (originated from the influent) and a higher bacterial growth because of the higher nutrient loads upon cell lysis [12]. Cyanobacterial blooms are such a source of high TOC. The enormous load of cyanobacterial biomass in both Microcystis treatments resulted in the massive increase in biofilm thickness. Thus, both Microcystis treatments lost 80% of their performance, which was shown by a much lower flux and permeate, respectively.

At the same time, these biofilms were able to degrade MCs. A complete removal of the MCs by the biofilms in our study could be achieved after 15 days, which was one week after the stabilization of the flux, i.e., after the development of a stable biofilm. It is conceivable that after that time bacteria capable of the degradation of MCs had accumulated on the biofilm. The key role of the bacterial biofilm in MC degradation can be confirmed by comparing the performance of the GDM system during the initial phase of the experiment and after flux stabilization. A reduction of 10% during the first three days of the experiment (Figure 3) cannot be attributed to the not yet developed biofilm but rather to

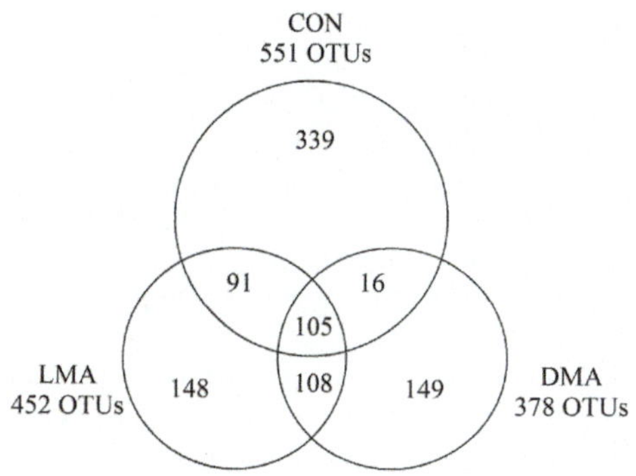

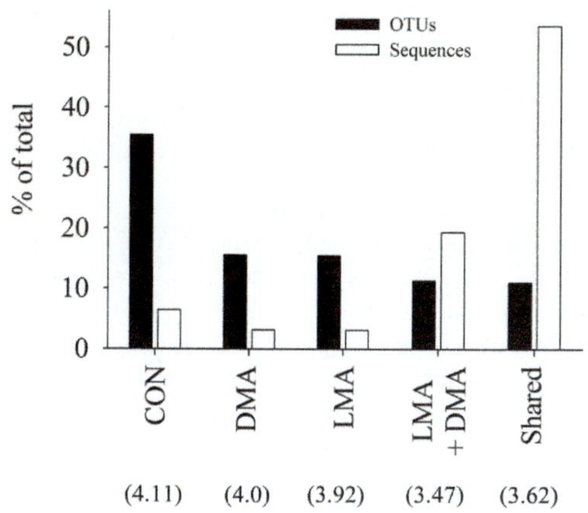

Figure 5. Overview of microbial diversity analysis. Upper panel: Venn diagram of shared OTUs among the biofilms of the three treatments. Lower Panel: percentage of operational taxonomic units (OTUs) and sequences found in the different treatments. Number in brackets marks the Shannon's diversity index (*H'*) for each treatment (LMA, DMA, and CON according to Figure 1).

adsorption processes, either to the filtration units or to cyanobacterial cells that retained on the filter membrane.

However, the potential of biofilms to remove MCs does not only depend on the presence and fast accumulation of possible MC degrading bacteria, but also on their efficiencies to biodegrade these MCs. Degradation rates of various bacterial strains might range between 1.5 up to several 10′000 μg L^{-1} d^{-1} [16] Already after 15 days, the added MCs were completely removed. In our study, MC removal rates increased constantly after a short lag phase, but did not reach a plateau. Therefore, the degradation capability of the aged biofilm might have further increased and even higher concentrations of MCs might be successfully degraded. The previous natural exposure of the feed water to cyanobacteria and MCs (as is the case for water from Lake Zurich) seems to be a significant driver for a fast performance of a MC degrading biofilm and thus its accelerated maturation [27]. Lake Zurich contains large populations of Planktothrix rubescens, a filamentous microcystin-producing cyanobacterium, which is frequently found in natural pre-alpine lakes [2]. It should be noted that the biofilm in our study was selectively grown under Microcystis bloom conditions. It remains to be investigated, how a bacterial biofilm without prior presence of MCs would respond to Microcystis addition. Another interesting aspect, which was beyond the scope of our experiment, would be the comparison of the bacterial composition in the biofilms by comparing MC-producing with MC-non-producing strains. This would allow for a more specific assessment, which shifts in microbial community compositions were caused by other substrates provided by the addition of cyanobacterial biomass.

Drinking water quality

Pro- and eukaryotic microbes can regrow during drinking water treatment and the distribution of non-chlorinated potable water [28]. The total bacterial cell counts in tap and mineral water may reach values between 1.5×10^5 to 5×10^5 cells mL^{-1} [29,30]. In our study, bacterial abundances in the CON treatment were within this range. Bacteria most likely regrew in biofilms underneath the filter membranes or on materials downstream the filtration process and dropped into the filtrate [29]. Inadequate disinfection, hydraulic retention time, flow regime, pipe material, temperature, source of water or corrosion could lead to the development of a microbial biofilm. However, excessive regrowth

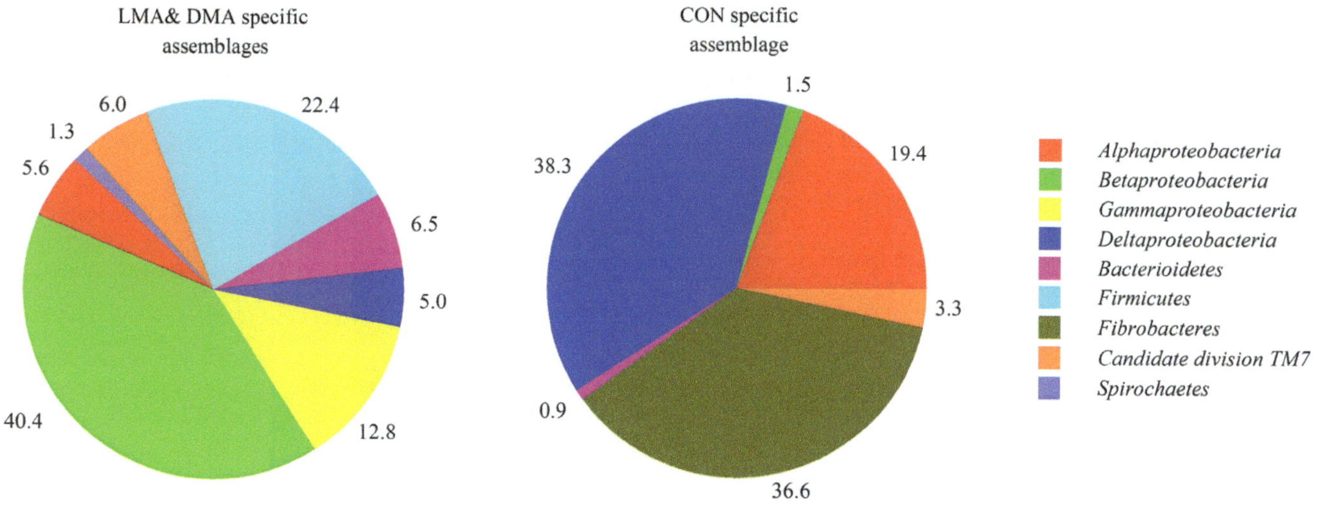

Figure 6. Phylogenetic composition of bacterial biofilms. Phylogenetic composition [%] of bacteria specific for the (combined) LMA and DMA assemblage (left) and for the CON (right) assemblage. Specificity for either assemblage was defined by the size of the OTU (>0.5% of sequences per sample) and by a ten times higher occurrence in one category than the other (LMA, DMA, and CON according to Figure 1).

in drinking water supply systems can be also triggered by nutrient introduction. It is conceivable that the massive load of cyanobacterial biomass in our study led to higher bacterial cell numbers, not only under the membrane filter but also in the filtrate. The increased availability of various substrates released from broken cells in the DMA treatment resulted in even higher cell numbers in the corresponding filtrates. As the drinking water quality may be deteriorated by the presence of pathogenic bacteria, an adequate storage of the drinking water is important to avoid re-contamination [31]. The focus in our study was on the removal of cyanobacterial toxins. However, the effect of other components of cyanobacterial cells that enhance regrowth remains to be investigated.

Drinking water storage or distribution systems represent functional ecosystems with well-established and structured microbial communities [32]. The increasing numbers of bacteria support the succession of protists: on average 10^5 cell L^{-1} were found in the water phase and 10^3 cells cm^{-2} in different biofilms [32]. As a result, some bacteria may aggregate to overcome the predation pressure by protists as it was shown in field and laboratory studies before [15]. The regrowth of bacteria underneath the filter membranes or on materials downstream the filtration process of the LMA treatments in our study stimulated the growth of bacterivorous flagellates, followed by an increase in bacterial aggregates (Figure 4). Interestingly, the DMA replicates showed two different responses: high numbers of free-living planktonic bacteria were established in the absence of flagellates in one replicate of the DMA treatment, whereas the numbers of free-living bacteria in the second replicate were low, as bacteria either were grazed by flagellates or formed aggregates in order to resist predation. This illustrates that there might be variability in the primary microbial assemblages of such filtrates that may have consequences for the development of drinking water quality.

Differences in microbial biofilms

Comparative phylogenetic analyses of 16S rDNA has increased our understanding of microbial diversity in environmental samples, since only few of the identifiable major phyla within the domain Bacteria have cultivable representatives [33]. Many of these uncultivated bacteria are found in diverse habitats in extraordinarily high abundances and might be at best only remotely related to strains that have been characterized by phenotype or by genome sequencing [33]. This in turn implies that there is very limited understanding of their respective physiologies, e.g., their substrate degradation potential. Calculating distances to sequences in a well-curated database [26] allowed us to classify OTUs to the closest taxonomic level. The CON treatment was the most diverse of all treatments, as reflected by the high number of small OTUs and an H′ index of 4.11. With the exception of river and stream habitats [34] biofilms that occur under more oligotrophic conditions (as in the CON treatment) seem to be rather understudied. As a consequence, more "exotic" species were found that had a high distance to the closest known relatives (CKR), such as uncultured Myxococcales (11.3–12.1% distance to CKR) and Fibrobacteres (12.7% distance to CKR) (Table S1 in File S1).

The massive load of cyanobacterial biomass selected for a few large OTUs (H′ = 3.47 of the combined MA) and the bacterial taxa represented by these OTUs were from a more "known" bacterial diversity (Table S2 in File S1). An intriguing example for typical freshwater bacteria that were present in our MA samples is a set of OTUs with <3% distance to known culturable genera, such as Azospirillum, Pelomonas, or Undibacterium (Table S2 in File S1). Moreover, oxygen subsaturation in the thick biofilms of the MA treatment was suggested by the presence of obligate anaerobic bacteria such as Desulfovibrio, Acidaminobacter, Fusibacter, or Spirochaetales. The ability to remove MCs appears to be not only common for aerobic bacteria, but can also be a feature of anaerobic microorganisms (not otherwise specified) that were found in lake sediments or sediments of water recharge facilities [35]. It seems that several Spirochaeta contain the MC-LR degradation protein MlrC (Table S2 in File S1). However, further studies are needed to determine possible MC degrading bacteria in anoxic environments. Interestingly, Bacteroidetes such as Sphingobacteriales that are common in the Planktothrix layer of Lake Zurich [36] were also present in high proportions in the MA treatments only. The natural co-occurrence of these bacteria with MC-producing cyanobacteria might indicate a possible and so far unknown role in the MC-degradation process.

Previous investigations have suggested the Sphingomonadales (Alphaproteobacteria) being the major MC degraders in aquatic environments: genetic studies on Sphingomonadaceae revealed the distinct gene cluster mlrABCD to be involved in MC removal [37], because it encodes for an enzymatic ring cleavage and thus a linearization of the MCs. However, during a massive cyanobacterial bloom in Lake Erie only ~1% of the total bacterial community could be attributed to Sphingomonadales [38] and also the metagenomic identification of bacterioplankton taxa involved in MC degradation revealed only a minor importance of these bacteria [20]. Our data also suggest that Sphingomonadales may not necessarily be relevant in MC degradation, as they only enriched in the CON treatment (Figure 6, Table S1 in File S1).

Our data indicate that Betaproteobacteria may be more important amongst the major MC degradation bacteria as they constituted the major fraction (>40%) in the MA biofilms but were hardly found in the CON assemblages. OTUs were found in high quantities that were closely related to Paucibacter, capable of degrading MCs [39] and Variovorax, containing MC-LR degradation proteins, (Table S2 in File S1). The importance of Betaproteobacteria (mainly Burkholderiales and Methylophilales) has already been suggested before based on laboratory microcosms experiments amended with MCs [38]. Interestingly, recent studies revealed that Betaproteobacteria such as Methylophilales were capable of degrading MCs but lacked the mlr cluster, thereby providing an alternative and so far unknown means of MC removal [20]. Studies focussing on the detection of the mlrA gene as the only marker for MC degradation bacteria might therefore underestimated the possible presence of other MC degrading bacterial taxa [16,20,21].

Conclusion

i. We demonstrated that GDM ultrafiltration systems provide a fast and efficient way to remove MCs from drinking water. However, it should be noted that complete MC degradation only took place one week after establishment of a stable biofilm.
ii. Addition of live or dead Microcystis cells led to remarkable differences between the bacterial communities of both MC-treated and the CON biofilms.
iii. Betaproteobacteria were identified as potentially important taxa for MC degradation in the MA biofilms. Additionally, Spirochaeta and Bacteroidetes such as Sphingobacteriales were enriched in these biofilms, and might indicate their so far unknown role in the MC degradation process.

Supporting Information

File S1 File S1 contains two supplemental tables: **Table S1, Phylogenetic composition.** Affiliation of bacteria in the CON assemblage (OTUs specific for CON treatment), number of OTUs and sequences, and phylogenetic distances of OTUs and associated sequences in the CON assemblage to the most closely related genotype in the SILVA reference database. **Table S2, Phylogenetic composition.** Affiliation of bacteria in the LMA & DMA assemblage (OTUs specific for the *Microcystis* treatment), number of OTUs and sequences, and phylogenetic distances of OTUs and associated sequences in the LMA & DMA assemblage to the most closely related genotype in the SILVA reference database.

Acknowledgments

We thank Eugen Loher for his valuable help with the experimental setup, and M. M. Salcher for searching the worldwide ARB for close relatives. C. Ewert is being acknowledged for her valuable help in the accomplishment of the experiment and in the microcystin measurements.

Author Contributions

Conceived and designed the experiments: EK TP ND EM JP JFB. Performed the experiments: EK JV TP ND TS. Analyzed the data: EK JV TP ND EM JP JFB. Contributed reagents/materials/analysis tools: EM JP. Contributed to the writing of the manuscript: EK TP ND EM JP JFB.

References

1. Paerl HW, Huisman J (2008) Climate - Blooms like it hot. Science 320: 57–58.
2. Posch T, Köster O, Salcher MM, Pernthaler J (2012) Harmful filamentous cyanobacteria favoured by reduced water turnover with lake warming. Nat Clim Chang 2: 809–813.
3. Carmichael WW, Azevedo S, An JS, Molica RJR, Jochimsen EM, et al. (2001) Human fatalities from cyanobacteria: Chemical and biological evidence for cyanotoxins. Environ Health Perspect 109: 663–668.
4. WHO (2011) Guidelines for Drinking-water quality. Fourth edition. Geneva.
5. Ueno Y, Nagata S, Tsutsumi T, Hasegawa A, Watanabe MF, et al. (1996) Detection of microcystins, a blue-green algal hepatotoxin, in drinking water sampled in Haimen and Fusui, endemic areas of primary liver cancer in China, by highly sensitive immunoassay. Carcinogenesis 17: 1317–1321.
6. Sabart M, Pobel D, Briand E, Combourieu B, Salencon MJ, et al. (2010) Spatiotemporal variations in microcystin concentrations and in the proportions of microcystin-producing cells in several *Microcystis aeruginosa* populations. Appl Environ Microbiol 76: 4750–4759.
7. Dyble J, Fahnenstiel GL, Litaker RW, Millie DF, Tester PA (2008) Microcystin concentrations and genetic diversity of *Microcystis* in the lower Great Lakes. Environ Toxicol 23: 507–516.
8. Jones GJ, Orr PT (1994) Release and degradation of microcystin followin algicide treatment of a *Microcystis aeruginosa* bloom in a recreational lake, as determined by HPLC and protein phosphatase inhibition assay. Water Res 28: 871–876.
9. Campinas M, Rosa MJ (2010) Removal of microcystins by PAC/UF. Sep Purif Technol 71: 114–120.
10. Grutzmacher G, Wessel G, Klitzke S, Chorus I (2010) Microcystin elimination during sediment contact. Environ Sci Technol 44: 657–662.
11. Hoeger SJ, Dietrich DR, Hitzfeld BC (2002) Effect of ozonation on the removal of cyanobacterial toxins during drinking water treatment. Environ Health Perspect 110: 1127–1132.
12. Derlon N, Koch N, Eugster B, Posch T, Pernthaler J, et al. (2013) Activity of metazoa governs biofilm structure formation and enhances permeate flux during Gravity-Driven Membrane (GDM) filtration. Water Res 47: 2085–2095.
13. Peter-Varbanets M, Hammes F, Vital M, Pronk W (2010) Stabilization of flux during dead-end ultra-low pressure ultrafiltration. Water Res 44: 3607–3616.
14. Peter-Varbanets M, Zurbrugg C, Swartz C, Pronk W (2009) Decentralized systems for potable water and the potential of membrane technology. Water Res 43: 245–265.
15. Christoffersen K, Lyck S, Winding A (2002) Microbial activity and bacterial community structure during degradation of microcystins. Aquat Microb Ecol 27: 125–136.
16. Dziga D, Wasylewski M, Wladyka B, Nybom S, Meriluoto J (2013) Microbial degradation of microcystins. Chem Res Toxicol 26: 841–852.
17. Park HD, Sasaki Y, Maruyama T, Yanagisawa E, Hiraishi A, et al. (2001) Degradation of the cyanobacterial hepatotoxin microcystin by a new bacterium isolated from a hypertrophic lake. Environ Toxicol 16: 337–343.
18. Ho LN, Gaudieux AL, Fanok S, Newcombe G, Humpage AR (2007) Bacterial degradation of microcystin toxins in drinking water eliminates their toxicity. Toxicon 50: 438–441.
19. Jiang YG, Shao JH, Wu XQ, Xu Y, Li RH (2011) Active and silent members in the *mlr* gene cluster of a microcystin-degrading bacterium isolated from Lake Taihu, China. FEMS Microbiol Lett 322: 108–114.
20. Mou XZ, Lu XX, Jacob J, Sun SL, Heath R (2013) Metagenomic identification of bacterioplankton taxa and pathways iInvolved in microcystin degradation in Lake Erie. Plos One 8.
21. Shimizu K, Maseda H, Okano K, Hiratsuka T, Jimbo Y, et al. (2013) Determination of microcystin-LR degrading gene *mlrA* in biofilms at a biological drinking water treatment facility. Maejo Int J Sci Technol 7 (Special Issue): 22–35.
22. Blom JF, Zimmermann YS, Ammann T, Pernthaler J (2010) Scent of Danger: Floc formation by a freshwater bacterium is induced by supernatants from a predator-prey coculture. Appl Environ Microbiol 76: 6156–6163.
23. Chelius MK, Triplett EW (2001) The diversity of archaea and bacteria in association with the roots of *Zea mays* L. Microb Ecol 41: 252–263.
24. Shabarova T, Villiger J, Morenkov O, Niggemann J, Dittmar T, et al. (2014) Bacterial community structure and dissolved organic matter in repeatedly flooded subsurface karst water pools. FEMS Microbiol Ecol 89: 111–126.
25. Quince C, Lanzen A, Davenport RJ, Turnbaugh PJ (2011) Removing noise from pyrosequenced amplicons. Bmc Bioinformatics 12: Article Number 38.
26. Pruesse E, Quast C, Knittel K, Fuchs BM, Ludwig WG, et al. (2007) SILVA: a comprehensive online resource for quality checked and aligned ribosomal RNA sequence data compatible with ARB. Nucleic Acids Res 35: 7188–7196.
27. Li JM, Shimizu K, Maseda H, Lu ZJ, Utsumi M, et al. (2012) Investigations into the biodegradation of microcystin-LR mediated by the biofilm in wintertime from a biological treatment facility in a drinking-water treatment plant. Bioresour Technol 106: 27–35.
28. Leclerc H, Moreau A (2002) Microbiological safety of natural mineral water. Fems Microbiol Rev 26: 207–222.
29. Hammes F, Berney M, Wang YY, Vital M, Köster O, et al. (2008) Flow-cytometric total bacterial cell counts as a descriptive microbiological parameter for drinking water treatment processes. Water Res 42: 269–277.
30. Yamaguchi N, Torii M, Uebayashi Y, Nasu M (2011) Rapid, semiautomated quantification of bacterial cells in freshwater by using a microfluidic device for on-chip staining and counting. Appl Environ Microbiol 77: 1536–1539.
31. Roberts L, Chartier Y, Chartier O, Malenga G, Toole M, et al. (2001) Keeping clean water clean in a Malawi refugee camp: a randomized intervention trial. Bull World Health Organ 79: 280–287.
32. Sibille I, Sime-Ngando T, Mathieu L, Block JC (1998) Protozoan bacterivory and *Escherichia coli* survival in drinking water distribution systems. Appl Environ Microbiol 64: 197–202.
33. Rappe MS, Giovannoni SJ (2003) The uncultured microbial majority. Annu Rev Microbiol 57: 369–394.
34. Battin TJ, Kaplan LA, Newbold JD, Hansen CME (2003) Contributions of microbial biofilms to ecosystem processes in stream mesocosms. Nature 426: 439–442.
35. Holst T, Jørgensen NOG, Jørgensen C, Johansen A (2003) Degradation of microcystin in sediments at oxic and anoxic, denitrifying conditions. Water Res 37: 4748–4760.
36. Van den Wyngaert S, Salcher MM, Pernthaler J, Zeder M, Posch T (2011) Quantitative dominance of seasonally persistent filamentous cyanobacteria (*Planktothrix rubescens*) in the microbial assemblages of a temperate lake. Limnol Oceanogr 56: 97–109.
37. Shimizu K, Maseda H, Okano K, Kurashima T, Kawauchi Y, et al. (2012) Enzymatic pathway for biodegrading microcystin LR in *Sphingopyxis* sp C-1. J Biosci Bioeng 114: 630–634.
38. Mou XZ, Jacob J, Lu XX, Robbins S, Sun SL, et al. (2013) Diversity and distribution of free-living and particle-associated bacterioplankton in Sandusky Bay and adjacent waters of Lake Erie Western Basin. J Gt Lakes Res 39: 352–357.
39. Rapala J, Berg KA, Lyra C, Niemi RM, Manz W, et al. (2005) *Paucibacter toxinivorans* gen. nov., sp nov., a bacterium that degrades cyclic cyanobacterial hepatotoxins microcystins and nodularin. Int J Syst Evol Microbiol 55: 1563–1568.

13

Long-Term Corticosterone Exposure Decreases Insulin Sensitivity and Induces Depressive-Like Behaviour in the C57BL/6NCrl Mouse

Eva L. van Donkelaar[1]*, Koen R. D. Vaessen[1,2], Jodi L. Pawluski[1,3,4], Annerieke S. Sierksma[1,5,6], Arjan Blokland[7], Ramón Cañete[8], Harry W. M. Steinbusch[1]

1 School for Mental Health and Neuroscience, Faculty of Health, Medicine and Life Sciences, Maastricht University, Maastricht, The Netherlands, **2** Central Laboratory Animal Research Facility, Faculty of Veterinary Medicine, University of Utrecht, Utrecht, The Netherlands, **3** University of Liege, GIGA-Neurosciences, Liège, Belgium,, **4** Department of Biological Sciences, Irvine Hall, Ohio University, Athens, Ohio, United States of America, **5** VIB Center for the Biology of Disease, Leuven, Belgium, **6** Center for Human Genetics, KU Leuven, Leuven, Belgium, **7** Department of Neuropsychology and Psychopharmacology, Faculty of Psychology and Neuroscience, Maastricht University, Maastricht, The Netherlands, **8** Unidad de Endocrinología Pediátrica del Hospital Universitario Reina Sofía, Instituto Maimónides de Investigación Biomédica de Córdoba (IMIBIC), Córdoba, Spain

Abstract

Chronic stress or long-term administration of glucocorticoids disrupts the hypothalamus-pituitary-adrenal system leading to continuous high levels of glucocorticoids and insulin resistance (IR). This pre-diabetic state can eventually develop into type 2 diabetes mellitus and has been associated with a higher risk to develop depressive disorders. The mechanisms underlying the link between chronic stress, IR and depression remains unclear. The present study aimed to establish a stress-depression model in mice to further study the effects of stress-induced changes upon insulin sensitivity and behavioural consequences. A pilot study was conducted to establish the optimal administration route and a pragmatic measurement of IR. Subsequently, 6-month-old C57BL/6NCrl mice were exposed to long-term oral corticosterone treatment via the drinking water. To evaluate insulin sensitivity changes, blood glucose and plasma insulin levels were measured at different time-points throughout treatment and mice were behaviourally assessed in the elevated zero maze (EZM), forced swimming test (FST) and open field test to reveal behavioural changes. Long-term corticosterone treatment increased body weight and decreased insulin sensitivity. The latter was revealed by a higher IR index and increased insulin in the plasma, whereas blood glucose levels remained unchanged. Corticosterone treatment induced longer immobility times in the FST, reflecting depressive-like behaviour. No effects were observed upon anxiety as measured in the EZM. The effect of the higher body weight of the CORT treated animals at time of testing did not influence behaviour in the EZM or FST, as no differences were found in general locomotor activity. Long-term corticosterone treatment via the drinking water reduces insulin sensitivity and induces depressive-like behaviour in the C57BL/6 mouse. This mouse model could thus be used to further explore the underlying mechanisms of chronic stress-induced T2DM and its association with increased prevalence of major depressive disorder on the short-term and other behavioural adaptations on the longer term.

Editor: Christian Holscher, University of Lancaster, United Kingdom

Funding: EVD was financially supported by the Internationale Stichting Alzheimer Onderzoek (ISAO, grant number 09551; https://www.alzheimer.nl/) and ZonNW (grant number 114024017; http://www.zonmw.nl/nl/). The funders had no role in study design, data collection and analysis, decision to publish, or preparation of the manuscript.

Competing Interests: The authors have declared that no competing interests exist.

* Email: eva.vandonkelaar@maastrichtuniversity.nl

Introduction

Chronic stress or long-term glucocorticoid administration, as well as (genetic) impairments in HPA-axis or corticoid receptor function are proposed to participate in the aetiology and progression of many health problems, including psychiatric disorders such as major depression and anxiety. Numerous studies have documented hypercortisolaemia, HPA-axis hyperactivity and reduced stress responsivity in patients with depression [1]. Moreover, both environmental and genetic risk factors for depression appear to correlate with increased HPA-axis activity in adulthood [2] and a direct relationship has been observed between peripheral cortisol levels and the severity of depressive symptoms [3].

One of the major metabolic effects of chronic hypercortisolaemia and a key feature in the development of the metabolic syndrome is insulin resistance (IR), a condition characterized by persistent high levels of peripheral glucose (hyperglycaemia) and insulin (hyperinsulinaemia) due to a reduced ability of insulin to lower blood glucose levels by uptake into muscle tissue (insulin insensitivity) [4]. Chronic peripheral hyperinsulinaemia, the so-called pre-diabetic state, can eventually develop into Type 2 diabetes mellitus (T2DM). In addition, glucocorticoids can induce IR in the hippocampus, which explains why the neurological consequences observed in experimental models of T2DM are similar to those observed following chronic stress. Moreover, longitudinal and cross-sectional associations exist between mood disorders and T2DM [5] and reveal that the prevalence of T2DM is increased by two-fold in persons with mood disorders and vice-versa [6]. Glucocorticoids may thus be common mechanistic mediators in the pathophysiological consequences of both T2DM and stress-related disorders [7]. In addition, stress-induced insulin dysregulation and affective disturbances are risk factors for the development of Alzheimer's disease (AD) [8–14]. Recent evidence suggests that cognitive impairment in diet-induced insulin resistant rats may result from glucocorticoid-mediated deficits in neurogenesis and synaptic plasticity [15]. Additionally, glucocorticoids have been shown to aggravate tangle-formation in the hippocampus [16] and affect aβ-processing [17] in rats. Moreover, insulin resistance and the subsequent impairment of insulin signalling in the brain affects the expression and metabolism of Aβ and tau protein [8,18–20]. However, the exact link between chronic stress, major depression and AD has not been established yet and the underlying mechanisms remain unclear [21–23].

In mice, long-term daily injections with corticosterone induces depressive-like behaviour and has thus been proposed as a reliable mouse model for studying the implication of stress and glucocorticoids in depression [24]. Insulin resistance has been studied in rats by long-term treatment with low doses of the synthetic glucocorticoid dexamethasone [25] and in mice, similar changes as those observed in individuals suffering from the metabolic syndrome were observed after a long-term, high dose of corticosterone treatment via the drinking water [26]. However, the combined effect of chronic elevated levels of glucocorticoids upon glucose-insulin regulation and depressive-like behavioural changes in rodents remains unclear. A mouse model that mimics chronic stress-induced glucose-insulin dysregulation not only allows for the exploration of acute behavioural changes, but additionally provides the opportunity to link the short-term system adaptations to potential long-term physiological and behavioural consequences of peripheral and central metabolic changes. The present study, therefore, aimed to establish a mouse model of chronic stress-induced insulin resistance and study its effects upon affective behaviour. The proposed model seems suitable, also for potential future application in specific transgenic mouse models.

Materials and Methods

Animals

Virgin female C57BL/6NCrl mice (Charles River, L'Arbresle, France) were crossed with male APPswe/PS1ΔE9 mice (Jackson Laboratory, strain #005864, Bar Harbor, ME, USA) at age 8–12 weeks as described by Sierksma *et al.* [27]. A total of 52 male wild type offspring was used of which 12 were handled at age 4-months-old and included for an initial study into the potential of a chronic stress model to induce glucose/insulin dysregulation. The pilot study aimed to establish the optimal administration route for CORT treatment, i.e., orally via drinking water compared to daily subcutaneous (s.c.) injection. In addition, treatment outcome was used to establish a pragmatic measurement of IR that could easily distinguish between normal and insulin resistant animals. Forty mice, age 6-month-old, were used for the main study of which experimental procedures were based upon the treatment results of the pilot study.

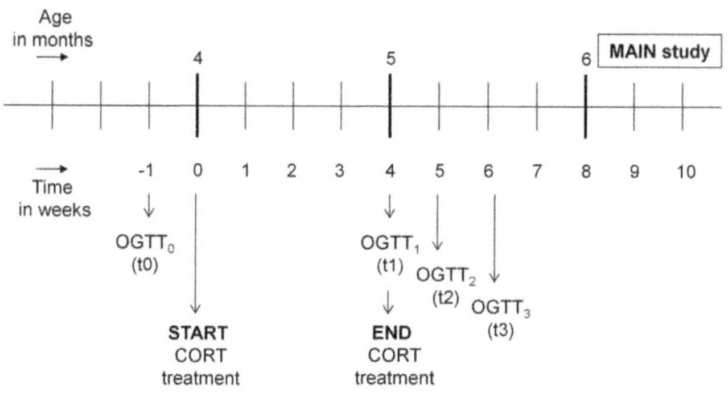

Figure 1. Experimental overview of the pilot study. CORT: corticosterone; OGTT: oral glucose tolerance test.

Experimental protocols were reviewed and approved by the local ethical Committee on Animal Experimentation of the Maastricht University (DEC2010-190, DEC 2010-010) and met Dutch governmental guidelines. Throughout all experiments the principles of laboratory animal care were carefully followed in accordance with the official codes of practice guidelines (Experiments on Animals Act, The Netherlands, revised, 1996) and all efforts were made to minimize the number of mice used and their suffering.

Mice were individually housed on sawdust bedding in standard Macrolon cages and placed in an individually ventilated cage (IVC) unit, regulating temperature (±20°C) and humidity (40–60%). They were further maintained on a reversed 12:12-hr light-dark schedule (lights on from 18:00 h−06:00 h) and continuous background noise was provided by a softly playing radio. Food (Ssniff, Germany) and tap water were available *ad libitum* throughout the experiments, unless otherwise stated. Animal cages were supplied with paper rolls and tissues to enrich environmental conditions. Clean cages were provided on a weekly basis and mice were weighed weekly and prior to oral glucose administration (*pilot study*). Drinking water consumption was measured by weighing the bottles before and after refreshing them.

To facilitate environmental adaptation, mice were not treated or handled until several weeks after single housing on a reversed day-night cycle. During this initial adaptation period, mice were additionally habituated to general handling procedures, including weighing. For the pilot study, mice were also familiarized to oral (saline) administrations by gavage for subsequent oral glucose administration procedures. At all times, blood samples were taken by means of saphenous vein puncture, as this method has been proven to be appropriate for repeated sampling in the same animal in a quick and painless manner and without the requirement of anesthesia [28]. Mice had both hind paws shaved over the lateral saphenous vein, using a small and almost soundless hair cutting system (Kyone, model GT 100, The Netherlands). For shaving and blood sampling, the animals were placed in a thick cotton cloth, in such a way that the hind paws were freely accessible for the particular procedures. Habituation to the specific restraint procedures and hind paw shaving was included in the initial habituation period as mentioned above.

Behavioural testing for the main study did not start until 6 weeks into CORT treatment, at approximately 7.5 months of age. This was done deliberately, to make potential future comparisons with the behavioural and cognitive effects in the APPswe/PS1ΔE9 possible. Testing was always done in the (active) dark phase of the rodent day-night cycle. Mice were initially exposed to the elevated zero maze (EZM), followed by the forced swim test (FST) and open field test (OF), with a three-week and one-week interval, respectively (see also Figure 1 and 2 for experimental overviews).

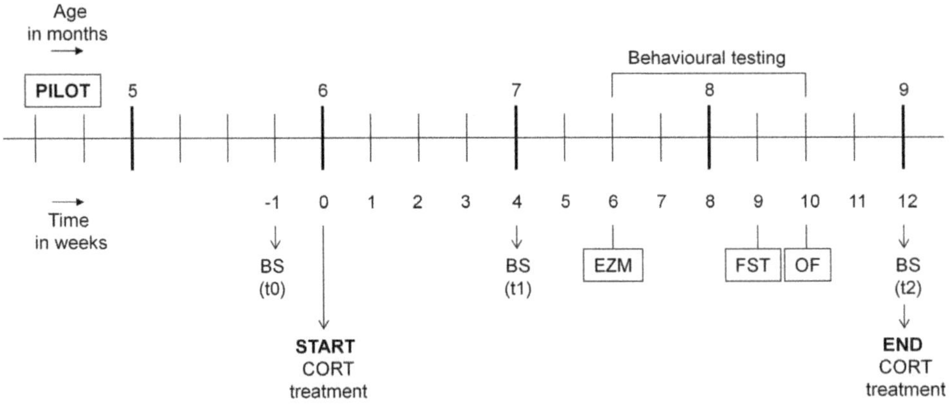

Figure 2. Schematic diagram of the different stages of the main study. BS: blood sample; CORT: corticosterone; EZM: elevated zero maze; FST: forced swimming test; OF: open field test.

Corticosterone treatment

Pure (≥92%) corticosterone (CORT) was used, obtained from SIGMA-ALDRICH (Zwijndrecht, The Netherlands). For the pilot study, mice were administered with either CORT (20 mg/kg; n = 3) or vehicle (50% propylene glycol; n = 3) via daily subcutaneous (s.c.) injection or received CORT (100 μg/ml; n = 3) or vehicle (1% ethanol; n = 3) in the drinking water for a total amount of 4 weeks. As CORT is hydrophobic, it was dissolved in ethanol and added to the drinking water to make a 1% ethanol solution. As ethanol is not an appropriate solvent for injectables, CORT was dissolved in propylene glycol and further diluted with saline for the s.c. injections. The vehicle injection group was administered with a 50% propylene glycol solution in saline (5 ml/kg). Injections were alternated daily between left or right flank or over the shoulders (into the loose skin over the neck) to avoid discomfort produced by the procedure or injection-site infection/local skin reaction. Potential interfering ethanol effects in the vehicle drinking water group compared to the other 3 experimental groups were avoided by providing all mice with 1% ethanol in the drinking water.

In the main study, CORT (100 μg/ml; n = 20) or vehicle (1% ethanol; n = 20) was administered via the drinking water for a total time of 12 weeks. Water bottles containing CORT were covered with tinfoil to avoid light-induced degradation of the compound and water quantity was monitored several times throughout the week. When calculating weekly water intake, any water spillage

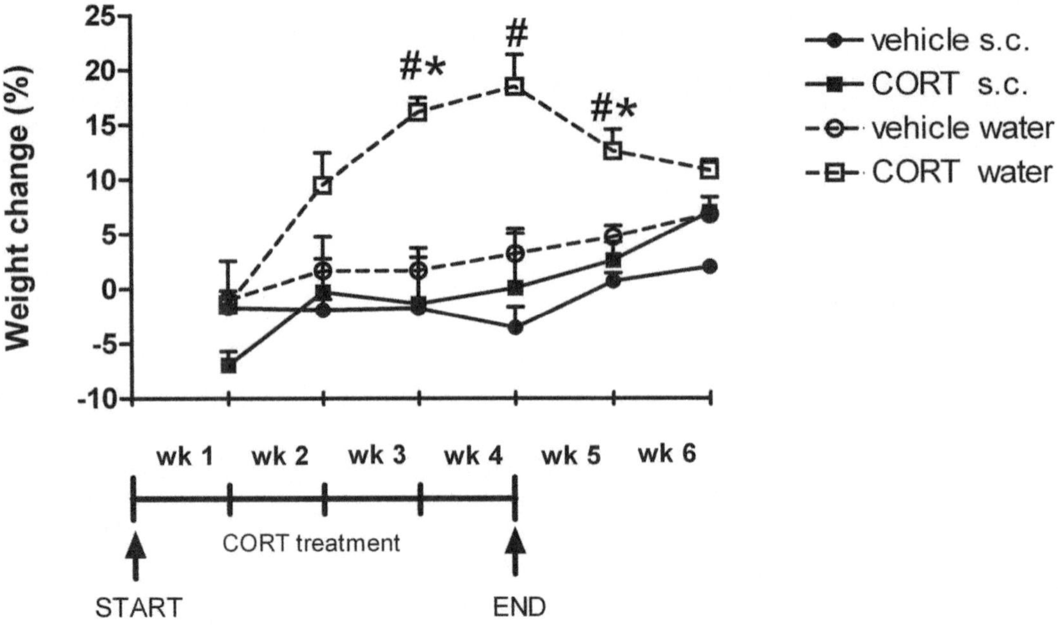

Figure 3. Effects of CORT treatment orally via the drinking water or by daily s.c. injection upon body weight (mean % change compared to body weight prior to start of treatment+S.E.M.) during the 4 week treatment period and 1 and 2 weeks after treatment cessation (week 5 and 6). The figure shows significant increased body weights in the group receiving CORT via the drinking water compared to the vehicle drinking water group (#) or CORT injection group (*). #/*$P<0.05$.

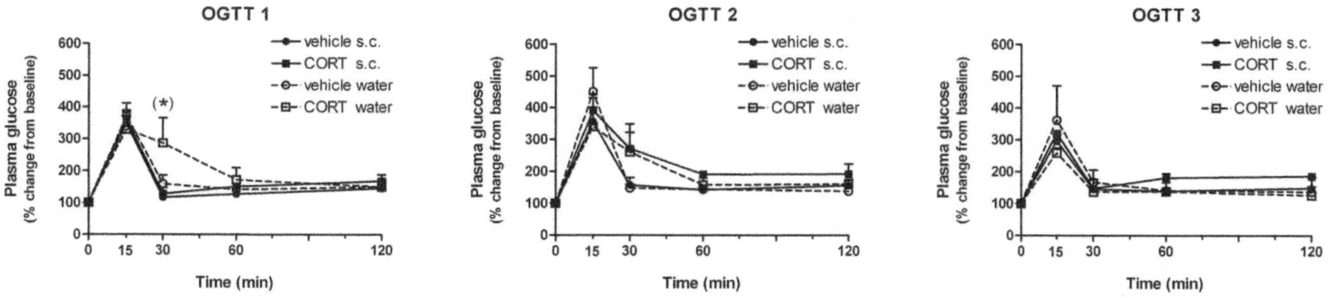

Figure 4. Plasma glucose levels (mean % change from baseline+S.E.M.) after acute glucose challenge in fasted animals receiving corticosterone (CORT) or vehicle either orally via the drinking water or by daily s.c. injection as measured by oral glucose tolerance tests (OGGT). Within each OGTT, plasma glucose levels are measured at baseline (0) and 15, 30, 60 and 120 min after an oral administration of glucose. Treatment effects upon glucose tolerance were measured at 3 time-points throughout the experiment, namely directly after 4 weeks treatment (OGTT1) and 1 and 2 weeks after treatment cessation (OGTT2, OGTT3, respectively). (*) *trend*; $P = 0.094$.

when removing bottles from cages was systematically taken into account by applying a correction factor. This factor was calculated by taken the mean spillage of water after taking the bottle out of the cage for 7 times in a row.

Insulin sensitivity assessment

Glucose tolerance test. The effects of CORT treatment (daily s.c. injection versus drinking water) on insulin sensitivity were evaluated by means of oral glucose tolerance tests (OGTT). An OGGT consists of assessing the efficacy of glucose clearing by insulin in the blood by measuring blood glucose and/or plasma insulin levels at different time points after an oral (body weight-matched) dose of glucose. Mice were fasted for 6 h after which a glucose solution (D-(+)-Glucose, SIGMA-ALDRICH, Zwijndrecht, The Netherlands) of 250 mg/ml was administered orally by gavage (2 g/kg) as derived from Adrikopoulos et al. [29]. Blood samples were taken just prior to glucose administration (baseline) and 15, 30, 60 and 120 minutes after glucose administration, as adapted from Ferrannini and Mari [30] and described by Gounarides et al. [31]. The first drop of blood was removed, the second drop was placed on a test strip (GLUCOCARD X-SENSOR Test Strips) for immediate glucose measurement with a hand-held glucose meter (GLUCOCARD X-meter, A. MENARINI Diagnostics, Benelux, N.V., Valkenswaard, The Netherlands). A subsequent blood sample (±20 µl) was collected in a heparinized microcentrifugation tube for plasma insulin measurement. In order to separate the plasma, blood samples were kept on ice, then centrifuged at 2000 rpm at 4°C for 20 min and finally stored at −80°C until further analysis. Plasma insulin levels were measured using an ultra sensitive mouse insulin ELISA Kit (Crystal Chem, inc., Downers Grove, USA). OGTT was performed prior to the start of the CORT treatment (OGTT0; baseline) and immediately after 4 weeks CORT treatment (OGTT1). To measure the persistence of treatment effects upon glucose tolerance, OGTTs were repeated 1 (OGTT2) and 2 weeks (OGTT3) after treatment cessation.

Homeostasis model assessment of insulin resistance index. The homeostasis model assessment of insulin resistance index (HOMA-IR), as described by Matthews *et al.* [32], is the most easily obtained measurement of insulin resistance (IR) and was used as a surrogate measure of *in vivo* insulin sensitivity. Based upon fasting blood glucose and plasma insulin levels (baseline, i.e. the blood sample prior to glucose administration), the following formula is used: HOMA-IR = fasting [insulin(mIU/l)]*fasting [glucose(mmol/l)]/22.5. Lee et al. [33] found that the HOMA-IR index correctly differentiated between insulin sensitivity and IR. In the drinking water condition, in addition to the OGTT measurements for insulin sensitivity assessment, HOMA-IR indices were calculated and compared for CORT effects directly after treatment (OGTT1; t1) and 1 (OGTT2; t2) and 2 weeks (OGTT3; t3) after treatment cessation. Blood samples for fasting blood glucose and plasma insulin samples were the same samples as those obtained from OGTT1, 2 and 3 at time 0 (baseline; prior to oral glucose administration).

Based upon the results of the effects of CORT treatment via daily s.c. injection versus drinking water upon body weight, glucose tolerance and insulin sensitivity, it was decided to rely on the HOMA-IR index for insulin sensitivity measurements in the main study. Baseline blood glucose and plasma insulin levels were

Table 1. Mean (± S.E.M.) fasting blood glucose (mmol/l), plasma insulin (ng/ml) values and HOMA-IR index directly after 4 weeks treatment via drinking water (t1) and 1 (t2) and 2 weeks (t3) after treatment cessation.

Time	Treatment	Glucose	Insulin	HOMA-IR
t1	vehicle	6.5±0.8	1.0±0.5	6.9±2.8
	CORT	4.9±1.0	8.7±3.9	38.7±10*
t2	vehicle	6.2±0.4	0.7±0.1	4.5±0.2
	CORT	6.3±0.5	0.7±0.1	5.0±0.9
t3	vehicle	8.0±1.4	1.7±0.3	14.5±3.4
	CORT	8.3±0.6	1.8±0.4	16.3±2.7

CORT: corticosterone; HOMA-IR: homeostasis model assessment of insulin resistance index *$P < 0.05$.

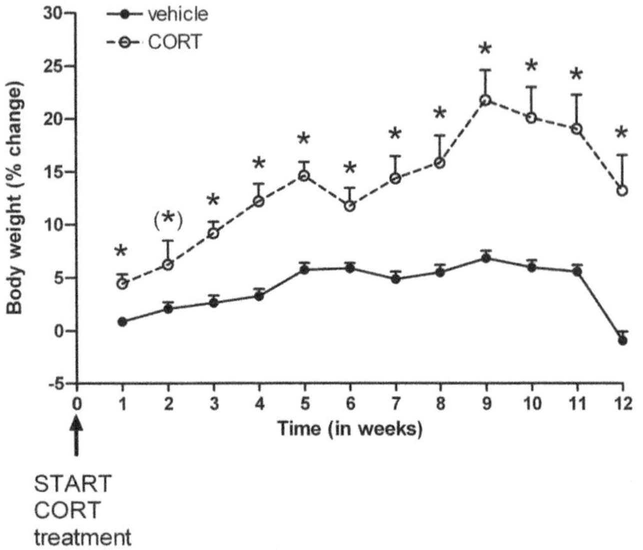

Figure 5. Effects of corticosterone (CORT) administration via the drinking water upon body weight (mean % change compared to body weight prior to start of treatment+S.E.M.) during the 12 week treatment period. Significant increased body weights in the CORT group are denoted with * (P<0.01); (*) *trend.*

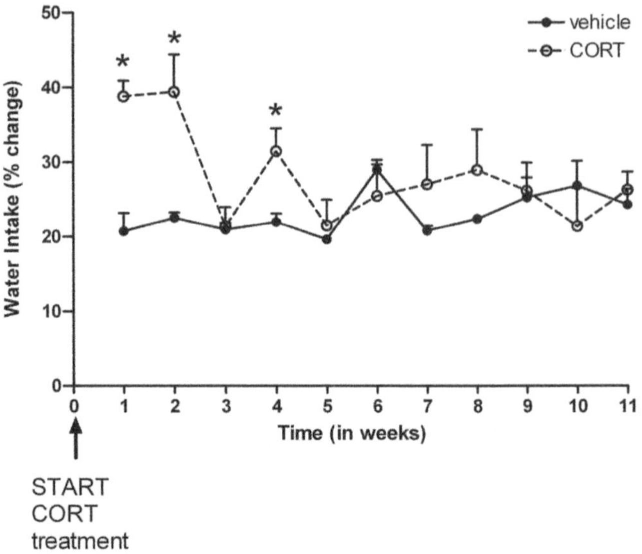

Figure 6. Effects of corticosterone (CORT) administration via the drinking water upon water intake (mean % change+S.E.M.) during the 12 week treatment period. Significant increased amounts of water intake in the CORT group are denoted with *(*P*<0.01).

measured at 3 different time-points throughout the main study, namely prior to the start of CORT treatment (baseline; t0) and 4 (t1) and 12 weeks (t2) into CORT treatment. Blood glucose and plasma insulin levels were used to calculate the HOMA-IR index as described earlier (see Figure 2 for an overview of the experimental procedures during the main study).

Behavioural testing

Elevated zero-maze test. The elevated zero-maze test (EZM) is a modification of the elevated plus-maze and models anxiety-like behaviour in rodents [34]. The maze consisted of a circular platform (50 cm in diameter), elevated 20 cm above floor level, with two opposite enclosed parts (50 cm high side walls) and two open parts equally divided along the circular runway (5 cm). This modified design allows the animal uninterrupted exploration and avoids undefined time spent in the central square as is the case in the traditional elevated plus-maze design [34]. Falls from the open parts were prevented by a 5 mm high edge. Both the side walls and the maze itself were made of black plastic, transparent for infrared light, and connected via an infrared video camera to a video tracking system (Ethovision Pro, Noldus Wageningen, The Netherlands). For testing, mice were placed in the middle of one of the open parts, facing a closed part, and allowed to explore the maze for a total time of 5 min. In between trials, the maze was thoroughly cleaned with water and 70% alcohol. Total time spent in closed and open parts and total distance travelled was video tracked and the percentage time spent in the open arms was calculated and corrected for the first latency to enter a closed arm.

Forced swimming test. A modified forced swimming test (FST) consisted of exposing the animals to an inescapable water stress, using the total time spent in an immobile position as a measure of their level of behavioural despair [35,36]. For the swim sessions, transparent Plexiglas cylinders (50 cm in height×19 cm in diameter), those normally used for rats, were filled with warm water (32±2°C) up to 20 cm. Water height was specifically chosen to prevent the mice from being able to touch the cylinder bottom with their hind paws or tail [37,38]. Four mice were tested in parallel and videotaped from above. Mice were not able to see one another due to grey separation panels placed between the four cylinders. The 6 min swim sessions were initiated by transferring the mice from their home cage to one of the cylinders. After each session, mice were removed from the water and transferred to a cage filled with paper towels for an initial 2 min recovery. Subsequently, they were dried with paper towels and returned to their home cages. In between trials, cylinders were cleaned with 70% alcohol and refilled with clean water.

The cylinders were positioned on top of a light box to create optimal illumination conditions for video tracking. A camera was placed 75 cm above the light box and connected to a video tracking system (Ethovision Pro, Noldus Wageningen, The Netherlands), which allowed automated recording of the total distance moved. Total immobility time for each mouse was scored manually by an independent researcher blind for experimental conditions.

Open field test. The open field (OF) was originally designed to simultaneously measure locomotion, exploration and anxiety by exposing mice to a novel and open space [39]. For the present study, only general locomotor activity was scored to be able to control for potential interfering effects due to differences in body weight. Due to the relatively small dimensions of the separate arenas, the test was not found appropriate for assessing additional anxiety-like behaviour in a valid manner [40]. Therefore, total time spent in corners, periphery or centre of the arena was not taken into account. However, defecation was used as indicator of emotionality/anxiety and was measured by the number of faecal boli produced during the test session.

The equipment was made of acrylic and consisted of a grey square arena (50×50×25 cm), subdivided by grey walls (height: 25 cm) into 4 separate arenas of 25×25×25 cm. Four mice were tested in parallel by placing them in the centre of the arenas, allowing them to move around freely for a total time of 20 min. After each 20 min session, faecal boli were counted per arena/mouse and arenas were subsequently thoroughly cleaned with 70% alcohol. Sessions were run under dimmed lightning conditions and videotaped from above by a camera connected

Table 2. Mean (± S.E.M.) fasting blood glucose (mmol/l), plasma insulin (ng/ml) values and HOMA-IR index at baseline (t0), after 4 weeks (t1) and after 12 weeks treatment (t2).

Time	Treatment	Glucose	Insulin	HOMA-IR
t0	vehicle	5.0±0.2		
	CORT	5.3±0.2		
t1	vehicle	5.5±0.2	1.1±0.2	6.8±1.1
	CORT	5.4±0.2	6.3±0.8*	37.1±4.5*
t2	vehicle	7.8±0.3	1.0±0.1	8.7±1.2
	CORT	6.1±0.4	3.7±0.9*	24.6±5.3*

CORT: corticosterone; HOMA-IR: homeostasis model assessment of insulin resistance index *$P<0.01$.

to a video tracking system (Ethovision Pro, Noldus Wageningen, The Netherlands) which allowed automated tracking of the total distance moved for each mouse separately.

Statistics

In the pilot study, changes in body weight and glucose tolerance over time were analyzed by means of a three-way analysis of variance (ANOVA) with Treatment and Administration Route as between-subject factors and Time as within-subject variable. In all other cases, changes over time were analyzed by means of two-way ANOVA with Treatment as between-subject factor and Time as within-subject variable. Separate one-way ANOVAs or independent Student's *t*-tests were performed to analyze differences between treatment conditions per time-point and the effects upon behaviour. Where appropriate, a post hoc Bonferroni test was used to further characterize the treatment effects.

In all cases the acceptable level of significance was set at $P<0.05$. All data are represented as mean ± standard error of the mean (S.E.M.). Analyses were performed using the Statistical Package for Social Science (SPSS) for Windows (IBM SPSS Statistics version 20). GraphPad Prism, version 4.02 for Windows was used for preparing the graphs.

Results

Pilot study

Body weight. Significantly increased body weights were found over the course of 6 weeks [$F(5, 40) = 18.19$; $P<0.001$]. However, changes over time depended on treatment [$F(5, 40) = 7.15$; $P<0.001$] and route of administration [$F(5, 40) = 5.35$; $P<0.01$]. Moreover, a significant time*treatment*route interaction was found [$F(5, 40) = 2.97$; $P<0.05$]. An overall treatment effect [$F(1, 8) = 5.85$; $P<0.05$] and effect of administration route [$F(1, 8) = 15.37$; $P<0.01$] was revealed, however, no significant treatment*route interaction was observed [$F(1, 8) = 3.21$; n.s.].

Separate one-way ANOVAs for treatment effects per week revealed no differences in body weight change at 1 [$F(3, 11) = 1.89$; n.s.] and 2 weeks [$F(3, 11) = 2.80$; n.s.] into CORT treatment. Significantly increased body weights were found after 3 weeks CORT treatment body weights were significantly increased [$F(3, 11) = 9.73$; $P<0.01$] with higher body weights in the CORT drinking water condition compared to the vehicle drinking water group and the group receiving CORT via injection as revealed by post hoc analyses. The same was true at the end (after 4 weeks) of CORT treatment [$F(3, 11) = 8.34$; $P<0.01$] with post hoc analyses showing significantly higher body weights in mice receiving CORT via the drinking water compared to mice receiving daily CORT injections. Only a trend was found toward increased body weights in the mice receiving CORT through drinking water compared to their vehicle counterparts ($P = 0.07$). One week after treatment cessation (week 5) body weights of mice receiving CORT via drinking water were still significantly higher compared to all other groups [$F(3, 11) = 13.15$; $P<0.01$]. Two weeks after treatment cessation body weights were still significantly increased [$F(3, 11) = 8.34$; $P<0.01$], however, post hoc analyses revealed increased body weights only in the CORT drinking water group compared to the vehicle injection group. At this time point, no differences were found between the CORT drinking water and

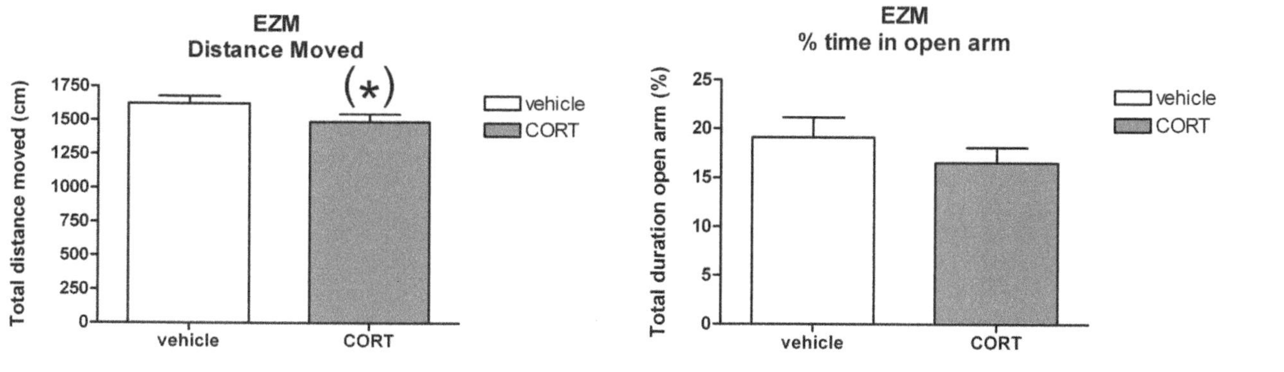

Figure 7. Effect of CORT administration via the drinking water (mean+S.E.M.) upon total distance moved and % time spent in the open arms of the elevated zero maze (EZM). (*) Trend towards difference between CORT and vehicle treatment ($P = 0.09$).

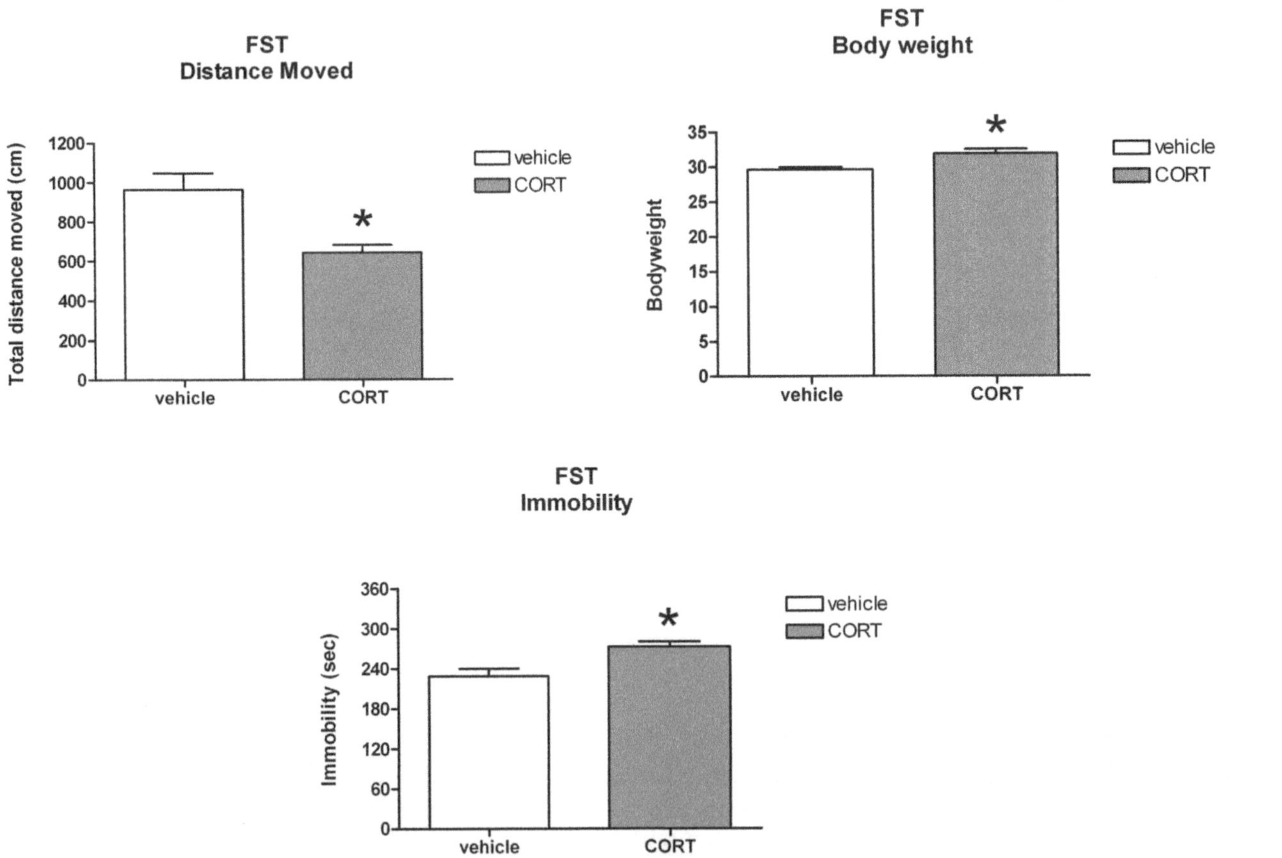

Figure 8. Effect of corticosterone (CORT) administration via the drinking water (mean+S.E.M.) upon total distance moved, body weight at time of testing and total immobility time in the forced swim test (FST). *Significant difference between CORT and vehicle treatment ($P<0.01$).

CORT injection condition or between the CORT drinking water and vehicle drinking water group (see Figure 3).

Glucose tolerance test. Glucose tolerance prior to the start of CORT treatment (OGGT0) is not depicted in a graph as it involves one baseline measurement of all mice together. Directly after 4 weeks CORT treatment (OGGT1), glucose tolerance differed over time, depending on route of administration [$F(4, 32)=4.77$; $P<0.01$]. In addition, a tendency towards a time*-treatment interaction effect [$F(4, 32)=2.40$; $P=0.070$] and time*treatment*route interaction effect [$F(4, 32)=2.46$; $P=0.065$] was observed. Subsequent one-way ANOVA for detailed analysis of treatment effects per time-point reveals a trend towards impaired glucose clearance 30 min after oral glucose administration in the CORT via drinking water condition compared to the other treatment groups ($P=0.094$). During OGTT2, only a significant time*treatment interaction [$F(2.73)=2.40$; $P=0.047$] was revealed. Overall treatment effects upon glucose tolerance were not found during any of the OGTTs (see Figure 4).

Blood glucose, plasma insulin and HOMA-IR. Although no effects were found of CORT treatment upon fasting blood glucose and plasma insulin levels (independent of administration route), a significant higher HOMA-IR index in the CORT drinking water group compared to the vehicle drinking water group directly after the 4 weeks treatment period (t1) was found [$t(4)=-3.06$; $P<0.05$]. Treatment cessation led to a normalisation as no differences were found in HOMA-IR indices 1 (t2) [$t(4)=-0.47$; n.s.] and 2 weeks after the end of CORT treatment (t3) [$t(4)=-0.40$; n.s.] (Table 1).

Main study

Body weight. Body weight (see Figure 5) significantly increased during the 12 week experimental period [$F(11, 341)=21.40$; $P<0.001$], but not for both groups (CORT versus vehicle) similarly [$F(11, 341)=5.44$; $P<0.001$]. An overall treatment effect revealed that mice receiving CORT via the drinking water gained significantly more weight compared to the control vehicle group [$F(1, 31)=37.02$; $P<0.001$]. Separate t-tests per week showed increased body weights in the CORT treated mice at all time points (all $P<0.01$), except for week 2, which resulted in a trend towards increased body weight [$t(37)=-1.75$; $P=0.09$].

Water intake. The amount of water intake changed over time [$F(10, 280)=4.06$; $P<0.001$] and differently depending on treatment condition [$F(10, 280)=6.08$; $P<0.001$]. However, no overall treatment effect was found [$F(1, 28)=1.15$; n.s.]. When analyzing differences between the treatment groups per time point, separate t-tests revealed significant higher amounts of water intake only at week 1 [$t(35)=-5.54$; $P<0.001$], 2 [$t(35)=-3.40$; $P<0.01$] and 4 [$t(35)=-2.92$; $P<0.01$] in the CORT treated animals (see Figure 6).

Blood glucose, plasma insulin and HOMA-IR. Table 2 shows the effects of CORT treatment upon blood glucose, plasma insulin and HOMA-IR index throughout the experiment. Blood glucose levels significantly increased over time [$F(2, 62)=30.84$;

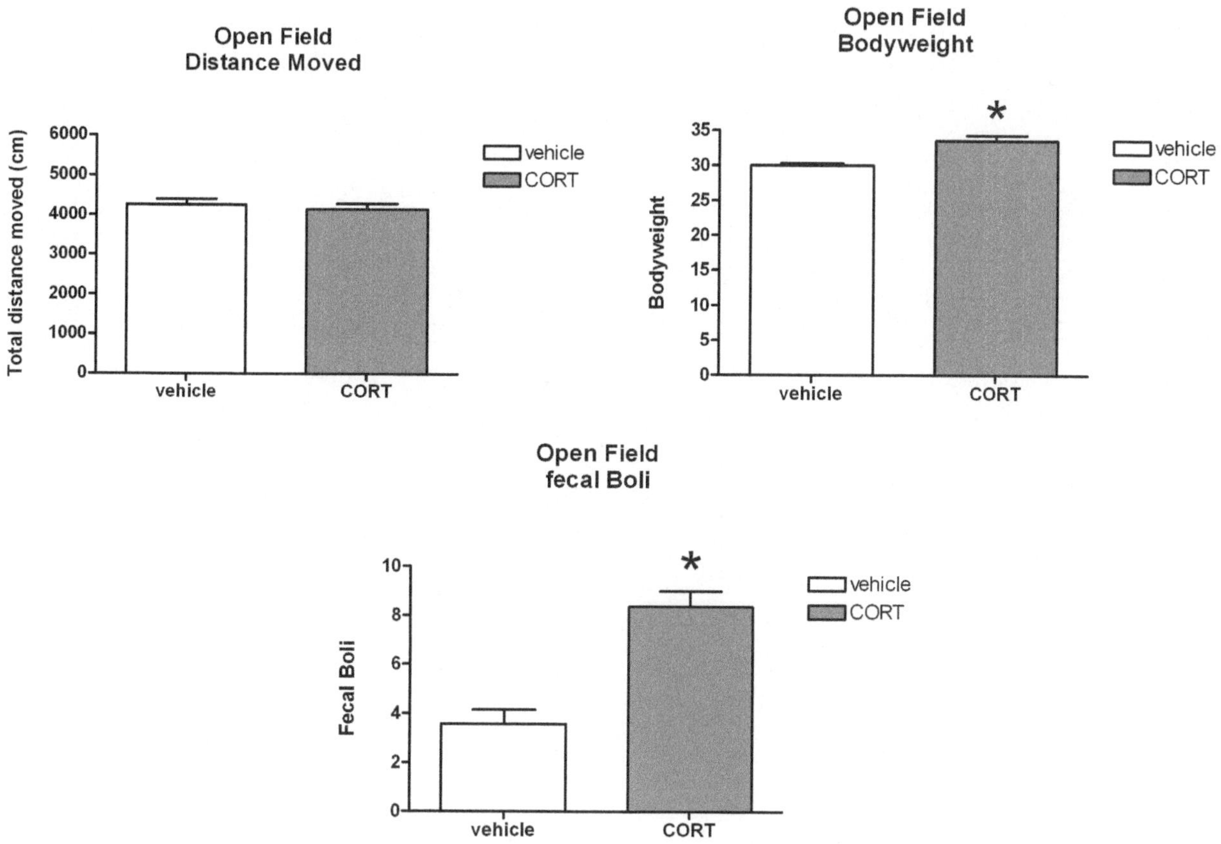

Figure 9. Effect of corticosterone (CORT) administration via the drinking water (mean+S.E.M.) upon total distance moved, body weight at time of testing and fecal boli production in the open field test (OF). *Significant difference between CORT and vehicle treatment ($P<0.01$).

$P<0.001$], but differently depending on treatment condition [$F(2, 62)=12.05$; $P<0.001$]. However, an overall treatment effect was not found, but did result in a trend towards increased blood glucose levels in the vehicle treated animals at 12 weeks [$F(1, 31)=3.97$; $P=0.06$].

Plasma insulin levels [$F(1, 19)=3.06$; n.s.] and HOMA-IR index [$F(1, 19)=1.44$; n.s.] did not change over time, but overall treatment effects were found ([$F(1, 19)=39.19$; $P<0.001$] and [$F(1, 19)=33.17$; $P<0.001$], respectively). Separate t-tests for treatment effects per time-point revealed significantly increased plasma insulin levels in CORT treated mice compared to the vehicle group after both 4 [$t(36)=-6.25$; $P<0.001$] and 12 weeks [$t(19)=-3.07$; $P<0.01$] of CORT treatment and similar effects were found for the HOMA-IR index after both 4 [$t(36)=-6.25$; $P<0.001$] and 12 weeks [$t(19)=-3.07$; $P<0.01$] of CORT treatment.

Behaviour. A trend was revealed towards less distance moved by CORT treated mice compared to the vehicle group [$t(34)=0.67$; $P=0.09$] in the EZM. However, the percentage of time spent in the open arms was not different between treatment conditions [$t(34)=0.60$; n.s.] (see Figure 7).

In the FST, the total distance moved was significantly lower in CORT treated mice [$t(34)=3.37$; $P<0.01$], whereas body weight at time of testing was significantly higher [$t(34)=-3.16$; $P<0.01$]. In addition, CORT treated mice showed more depressive-like behaviour as was reflected upon by the significantly increased immobility times in this group [$t(34)=-3.15$; $P<0.01$] (see Figure 8).

For the OF, no effects of CORT treatment upon total distance moved were found [$t(33)=0.55$; n.s.], whereas mean body weight of the CORT treated mice, like during the FST, was significantly higher compared to the vehicle group [$t(34)=-4.57$; $P<0.001$]. Moreover, the amount of faecal boli was significantly higher in the CORT treated animals [$t(34)=-5.63$; $P<0.001$] (see Figure 9).

Discussion

The present study aimed to establish a stress-IR-depression model in the mouse and specifically in the C57BL/6NCrl mouse since it is the background mouse strain used for all behavioural experiments in transgenic and knockout research [41]. To the best of our knowledge, this is the first study to explore the effects of chronic glucocorticoid exposure upon both insulin-glucose regulatory systems and affective behavioural changes in the mouse.

Prior to the main study, 12 mice were used to clarify the difference between continuous CORT exposure via the drinking water or by daily s.c. injection, the reversibility of physiological effects after treatment cessation and the reliability of oral glucose tolerance tests to indicate between-group changes in insulin sensitivity. No effect of 4 weeks CORT treatment via the drinking water or daily injection on glucose tolerance could be revealed. This is not in line with the severely compromised glucose tolerance found by Karatsoreos *et al.* [26] after 4 weeks CORT treatment via the drinking water. With the same dose and mouse strain, Karatsoreos *et al.* [26] observed significantly high plasma glucose levels even 120 min after glucose administration. However, they administered glucose via intraperitoneal (i.p.) injection which avoids the incretin response that potentiates the glucose-mediated insulin response [29] which could explain their higher glucose levels due to a lack of initial insulin secretion. Based upon the observation that a slight tendency was observed towards higher glucose levels 30 min after glucose administration only in the mice that received CORT via the drinking water (present study), together with the results of CORT administration via the drinking water as showed by Karatsoreos *et al.* [26], it was decided to continue further investigation for the drinking water group only. The food-deprivation state prior to glucose administration allowed us to collect fasting blood samples from which, besides blood glucose levels, also plasma insulin levels could be determined. From these parameters the HOMA-IR index was calculated which is a surrogate marker of IR and a useful non-invasive way of determining insulin sensitivity in mice [29]. Despite unchanged glucose and insulin levels, the HOMA-IR index was significantly higher in the CORT treated mice immediately after 4 weeks treatment, indicating reduced insulin sensitivity. These data also showed that treatment cessation reverses the treatment effect, as HOMA-IR index was back to normal levels 1 and 2 weeks after the end of treatment. Although it should be taken into account that the above mentioned pilot data is derived from only 3 animals per experimental group, the data served to provide information concerning the experimental protocol of the main study. CORT treatment was administered via drinking water and continuously throughout the entire experiment for valid behavioural assessment. Moreover, HOMA-IR index could be used as non-invasive measure to assess changes in insulin sensitivity instead of the relatively invasive OGTT procedure.

Daily s.c. or i.p. injections of glucocorticoids solutions might be more precise and standardized with respect of the amount of drug delivered at a specific time point thereby avoiding fluctuations of circulating drug concentrations. However, the acute stress effect of the injections can easily and seriously confound the long-term effects of continuous high levels of glucocorticoids [42]. Moreover, administering CORT via the drinking water is not only a technically low demanding and non-invasive method, it additionally allows for the maintenance of the naturally occurring diurnal rhythm of the hormone secretion [26]. Despite some inexplicable fluctuations in water intake at the very beginning of the treatment period, no difference in amount of intake was found between the treatment groups overall. All animals thus received the same amount of water and CORT over a one week period, i.e. until the weighing and refreshing of the drinking bottles. Throughout the entire 12 week treatment period, CORT treated mice significantly increased their body weight, which is in line with the observation that chronic stress in humans has been linked to obesity [43]. Unfortunately, the amount of food consumed was not taken into account for the present study. However, Karatsoreos *at al.*, [26] found similar increases in mouse body weight after 4 week continuous CORT treatment via the drinking water which was clearly attributable to CORT-induced hyperphagia with a significantly increased proportion of food consumed by body weight. They additionally report increased levels of plasma leptin which indicates a degree of resistance to the action of leptin which normally has an important role in the regulation of food intake.

In addition, the present study shows that CORT treatment significantly increased plasma insulin concentrations, whereas blood glucose levels remained unchanged. Insulin works as a key signal of blood glucose levels and mean plasma insulin levels can therefore be used to indicate the relative degree of insulin sensitivity with a higher plasma insulin concentration reflecting more insulin resistance [30]. Thus, the higher levels of insulin as observed in the present study, together with the unchanged levels of glucose in the blood suggest the development of resistance to the action of insulin as is also seen in T2DM. This was confirmed by the significantly higher HOMA-IR index in the CORT treated mice. However, although peripheral insulin resistance is one of the hallmarks of T2DM, development of diabetes additionally requires the onset of β-cell dysfunction [44]. Decreased insulin sensitivity together with increased body weight are features that nicely resemble some of the key physiological effects of hypercortisolaemia as observed in people suffering from Cushing's syndrome (excess of cortisol secretion) or the metabolic syndrome. Besides glucose intolerance and insulin resistance, the metabolic syndrome is further characterized by (abdominal) obesity, dyslipidemia and hypertension and increases the risk to develop T2DM or cardiovascular disease [45]. The results of the present study thus confirm the observation of Karatsoreos *et al.* [26], that treating mice with a relatively high dose of CORT (100 μg/ml) via the drinking water continuously for at least 4 weeks serves as a valid model of key features of the metabolic syndrome in the mouse, which precedes T2DM.

Chronic stress-induced HPA-axis dysfunction is well known to increase the risk to developing major depression [1] and the prevalence of T2DM appears to be increased by two-fold in persons with mood disorders and vice-versa [6]. The present study, therefore, assessed the effect of long-term CORT treatment upon both glucose-insulin dysregulation and affective behavioural changes. Whereas prolonged CORT treatment did not induce anxiety, significant increased immobility times were observed in the FST, reflecting depressive-like behaviour [46]. Similar increases in FST immobility times after prolonged glucocorticoid exposure in C57BL/6 mice have also been found by others, both after sub-chronic treatment (3 and 5 weeks) with daily s.c. injections [24] and after continuous exposure via the drinking water for either 2 weeks [47] or 4 weeks [48]. In line with this, decreased grooming behaviour [48] and a state of anhedonia, as measured by sucrose preference, has been reported [47,48]. Moreover, significant increased defecation (fecal boli) was observed during a 20 min open field session, indicative of disturbed emotionality rather than signs of fearfulness or anxiety [49]. Although at time of behavioural assessment, CORT treated mice had significantly higher body weights, behavioural effects cannot be attributed to differences in body weight as no differences in general locomotor activity (total distance moved) were observed in the OF.

General conclusion

Long-term continuous treatment with exogenous CORT via the drinking water in the mouse seems a valid and reversible model of decreased insulin sensitivity mimicking both the physiological and the behavioural effects of hypercortisolaemia-induced IR which can ultimately develop in T2DM. The model therefore serves to further explore the causal direction and underlying mechanism of the strong association between IR and major depression, which up until today remains unresolved. As the model was established in C57BL/6NCrl mice, which is the genetic background mouse of many genetically manipulated mouse strains, including the APP/PS1 Alzheimer mouse, it additionally allows further investigation into the long-term effects of stress-induced IR and depression as risk factors for developing Alzheimer's disease later in life.

Acknowledgments

The authors would like to thank Irene Vroegrijk and Ilia Karatsoreos for all the essential information they have provided to get to an optimal experimental protocol and Borja Roses Cañete for his assistance with the data analysis.

Author Contributions

Conceived and designed the experiments: EVD KV JP AS RC. Performed the experiments: EVD KV. Analyzed the data: EVD KV. Contributed reagents/materials/analysis tools: HS. Contributed to the writing of the manuscript: EVD KV JP AS AB.

References

1. Gillespie CF, Nemeroff CB (2005) Hypercortisolemia and depression. Psychosomatic medicine 67 Suppl 1: S26–28.
2. Bao AM, Meynen G, Swaab DF (2008) The stress system in depression and neurodegeneration: focus on the human hypothalamus. Brain Res Rev 57: 531–553.
3. Reus VI, Wolkowitz OM (2001) Antiglucocorticoid drugs in the treatment of depression. Expert opinion on investigational drugs 10: 1789–1796.
4. Haas JT, Biddinger SB (2009) Dissecting the role of insulin resistance in the metabolic syndrome. Curr Opin Lipidol 20: 206–210.
5. Anderson RJ, Freedland KE, Clouse RE, Lustman PJ (2001) The prevalence of comorbid depression in adults with diabetes: a meta-analysis. Diabetes Care 24: 1069–1078.
6. Gavard JA, Lustman PJ, Clouse RE (1993) Prevalence of depression in adults with diabetes. An epidemiological evaluation. Diabetes Care 16: 1167–1178.
7. Reagan LP, Grillo CA, Piroli GG (2008) The As and Ds of stress: metabolic, morphological and behavioral consequences. European journal of pharmacology 585: 64–75.
8. Craft S (2005) Insulin resistance syndrome and Alzheimer's disease: age- and obesity-related effects on memory, amyloid, and inflammation. Neurobiology of aging 26 Suppl 1: 65–69.
9. Haan MN (2006) Therapy Insight: type 2 diabetes mellitus and the risk of late-onset Alzheimer's disease. Nature clinical practice Neurology 2: 159–166.
10. Caraci F, Copani A, Nicoletti F, Drago F (2010) Depression and Alzheimer's disease: neurobiological links and common pharmacological targets. Eur - J Pharmacol 626: 64–71.
11. Biessels GJ, Kappelle LJ (2005) Increased risk of Alzheimer's disease in Type II diabetes: insulin resistance of the brain or insulin-induced amyloid pathology? Biochem Soc Trans 33: 1041–1044.
12. Briones A, Gagno S, Martisova E, Dobarro M, Aisa B, et al. (2011) Stress-induced anhedonia is associated with an increase in Alzheimer's disease-related markers. Br J Pharmacol 165: 897–907.
13. Solas M, Aisa B, Mugueta MC, Del Rio J, Tordera RM, et al. (2010) Interactions between age, stress and insulin on cognition: implications for Alzheimer's disease. Neuropsychopharmacology 35: 1664–1673.
14. Solas M, Aisa B, Tordera RM, Mugueta MC, Ramirez MJ (2013) Stress contributes to the development of central insulin resistance during aging: Implications for Alzheimer's disease. Biochim Biophys Acta 1832: 2332–2339.
15. Stranahan AM, Norman ED, Lee K, Cutler RG, Telljohann RS, et al. (2008) Diet-induced insulin resistance impairs hippocampal synaptic plasticity and cognition in middle-aged rats. Hippocampus 18: 1085–1088.
16. Elliott EM, Mattson MP, Vanderklish P, Lynch G, Chang I, et al. (1993) Corticosterone exacerbates kainate-induced alterations in hippocampal tau immunoreactivity and spectrin proteolysis in vivo. Journal of neurochemistry 61: 57–67.
17. Budas G, Coughlan CM, Seckl JR, Breen KC (1999) The effect of corticosteroids on amyloid beta precursor protein/amyloid precursor-like protein expression and processing in vivo. Neuroscience letters 276: 61–64.
18. Gasparini L, Gouras GK, Wang R, Gross RS, Beal MF, et al. (2001) Stimulation of beta-amyloid precursor protein trafficking by insulin reduces intraneuronal beta-amyloid and requires mitogen-activated protein kinase signaling. The Journal of neuroscience 21: 2561–2570.
19. Zhao L, Teter B, Morihara T, Lim GP, Ambegaokar SS, et al. (2004) Insulin-degrading enzyme as a downstream target of insulin receptor signaling cascade: implications for Alzheimer's disease intervention. The Journal of Neuroscience 24: 11120–11126.
20. Stoothoff WH, Johnson GV (2005) Tau phosphorylation: physiological and pathological consequences. Biochimica et biophysica acta 1739: 280–297.
21. Aznar S, Knudsen GM (2011) Depression and Alzheimer's disease: is stress the initiating factor in a common neuropathological cascade? J Alzheimers Dis 23: 177–193.
22. Riederer P, Bartl J, Laux G, Grunblatt E (2011) Diabetes type II: a risk factor for depression-Parkinson-Alzheimer? Neurotox Res 19: 253–265.
23. Sierksma AS, van den Hove DL, Steinbusch HW, Prickaerts J (2010) Major depression, cognitive dysfunction and Alzheimer's disease: is there a link? Eur J Pharmacol 626: 72–82.
24. Zhao Y, Ma R, Shen J, Su H, Xing D, et al. (2008) A mouse model of depression induced by repeated corticosterone injections. European journal of pharmacology 581: 113–120.
25. Severino C, Brizzi P, Solinas A, Secchi G, Maioli M, et al. (2002) Low-dose dexamethasone in the rat: a model to study insulin resistance. American journal of physiology Endocrinology and metabolism 283: E367–373.
26. Karatsoreos IN, Bhagat SM, Bowles NP, Weil ZM, Pfaff DW, et al. (2010) Endocrine and physiological changes in response to chronic corticosterone: a potential model of the metabolic syndrome in mouse. Endocrinology 151: 2117–2127.
27. Sierksma AS, Prickaerts J, Chouliaras L, Rostamian S, Delbroek L, et al. (2013) Behavioral and neurobiological effects of prenatal stress exposure in male and female APPswe/PS1dE9 mice. Neurobiol Aging 34: 319–337.
28. Hem A, Smith AJ, Solberg P (1998) Saphenous vein puncture for blood sampling of the mouse, rat, hamster, gerbil, guinea pig, ferret and mink. Laboratory animals 32: 364–368.
29. Andrikopoulos S, Blair AR, Deluca N, Fam BC, Proietto J (2008) Evaluating the glucose tolerance test in mice. Am J Physiol Endocrinol Metab 295: E1323–1332.
30. Ferrannini E, Mari A (1998) How to measure insulin sensitivity. Journal of hypertension: 895–906.
31. Gounarides JS, Korach-Andre M, Killary K, Argentieri G, Turner O, et al. (2008) Effect of dexamethasone on glucose tolerance and fat metabolism in a diet-induced obesity mouse model. Endocrinology 149: 758–766.
32. Matthews DR, Hosker JP, Rudenski AS, Naylor BA, Treacher DF, et al. (1985) Homeostasis model assessment: insulin resistance and beta-cell function from fasting plasma glucose and insulin concentrations in man. Diabetologia 28: 412–419.
33. Lee S, Muniyappa R, Yan X, Chen H, Yue LQ, et al. (2008) Comparison between surrogate indexes of insulin sensitivity and resistance and hyperinsulinemic euglycemic clamp estimates in mice. Am J Physiol Endocrinol Metab 294: E261–270.
34. Shepherd JK, Grewal SS, Fletcher A, Bill DJ, Dourish CT (1994) Behavioural and pharmacological characterisation of the elevated "zero-maze" as an animal model of anxiety. Psychopharmacology (Berl) 116: 56–64.
35. Porsolt RD, Bertin A, Jalfre M (1977) Behavioral despair in mice: a primary screening test for antidepressants. Archives internationales de pharmacodynamie et de thérapie 229: 327–336.
36. Porsolt RD, Le Pichon M, Jalfre M (1977) Depression: a new animal model sensitive to antidepressant treatments. Nature 266: 730–732.
37. Detke MJ, Lucki I (1996) Detection of serotonergic and noradrenergic antidepressants in the rat forced swimming test: the effects of water depth. Behavioural Brain Research 73: 43–46.
38. Lucki I (1997) The forced swimming test as a model for core and component behavioral effects of antidepressant drugs. Behavioural pharmacology 8: 523–532.
39. Walsh RN, Cummins RA (1976) The Open-Field Test: a critical review. Psychol Bull 83: 482–504.
40. Ennaceur A, Michalikova S, Chazot PL (2006) Models of anxiety: responses of rats to novelty in an open space and an enclosed space. Behav Brain Res 171: 26–49.
41. Gardier AM, Bourin M (2001) Appropriate use of "knockout" mice as models of depression or models of testing the efficacy of antidepressants. Psychopharmacology (Berl) 153: 393–394.
42. Herrmann M, Henneicke H, Street J, Modzelewski J, Kalak R, et al. (2009) The challenge of continuous exogenous glucocorticoid administration in mice. Steroids 74: 245–249.
43. Dallman MF, Pecoraro N, Akana SF, La Fleur SE, Gomez F, et al. (2003) Chronic stress and obesity: a new view of "comfort food". Proc Natl Acad Sci U S A 100: 11696–11701.
44. Nandi A, Kitamura Y, Kahn CR, Accili D (2004) Mouse models of insulin resistance. Physiol Rev 84: 623–647.
45. Grundy SM (2004) Metabolic syndrome: part I. Endocrinol Metab Clin North Am 33: ix-xi.
46. Cryan JF, Slattery DA (2007) Animal models of mood disorders: Recent developments. Current opinion in psychiatry 20: 1–7.
47. Gourley SL, Kiraly DD, Howell JL, Olausson P, Taylor JR (2008) Acute hippocampal brain-derived neurotrophic factor restores motivational and forced swim performance after corticosterone. Biol Psychiatry 64: 884–890.
48. David DJ, Samuels BA, Rainer Q, Wang JW, Marsteller D, et al. (2009) Neurogenesis-dependent and -independent effects of fluoxetine in an animal model of anxiety/depression. Neuron 62: 479–493.
49. Bindra D, Thompson WR (1953) An evaluation of defecation and urination as measures of fearfulness. J Comp Physiol Psychol 46: 43–45.

14

Virtuous and Vicious Virtual Water Trade with Application to Italy

Julia Anna Winter*, Paola Allamano, Pierluigi Claps

Department of Environment, Land and Infrastructure Engineering, Politecnico di Torino, Torino, Italy

Abstract

The current trade of agricultural goods, with connections involving all continents, entails for global exchanges of "virtual" water, i.e. water used in the production process of alimentary products, but not contained within. Each trade link translates into a corresponding virtual water trade, allowing quantification of import and export fluxes of virtual water. The assessment of the virtual water import for a given nation, compared to the national consumption, could give an approximate idea of the country's reliance on external resources from the food and the water resources point of view. A descriptive approach to the understanding of a nation's degree of dependency from overseas food and water resources is first proposed, and indices of water trade virtuosity, as opposed to inefficiency, are devised. Such indices are based on the concepts of self-sufficiency and relative export, computed systematically on all products from the FAOSTAT database, taking Italy as the first case study. Analysis of time series of the self-sufficiency and relative export can demonstrate effects of market tendencies and influence water-related policies at the international level. The goal of this approach is highlighting incongruent terms in the virtual water balances by the viewpoint of single products. Specific products, which are here referred to as "swap products", are in fact identified as those that lead to inefficiencies in the virtual water balance due to their contemporaneously high import and export. The inefficiencies due to the exchanges of the same products between two nations are calculated in terms of virtual water volumes. Furthermore, the cases of swap products are investigated by computing two further indexes denoting the ratio of virtual water exchanged in the swap and the ratio of the economic values of the swapped products. The analysis of these figures can help examine the reasons behind the swap phenomenon in trade.

Editor: Alejandro Raul Hernandez Montoya, Universidad Veracruzana, Mexico

Funding: The first author received support from the EU Project WARECALC (2008-230845). The funders had no role in study design, data collection and analysis, decision to publish, or preparation of the manuscript.

Competing Interests: The authors have declared that no competing interests exist.

* E-mail: julia.winter@polito.it

Introduction

The virtual water content of a product is the volume of water used to produce it, measured at the place where it is actually grown or manufactured. The adjective "virtual" refers to the fact that most of the water used in the production is ultimately not contained within the product. The real water content of products is generally negligible compared with their virtual water content [1].

The virtual water content of a given product varies greatly from place to place, depending mainly on the climate and on the technology adopted for irrigation and farming [2]. The variability of virtual water contents among different products and different locations has led to a diversification of case studies and research topics. Studies have been conducted at a global scale (e.g. [3], [4], [5]) but also in depth on single countries or areas (e.g. [6], [7], [8]) or on single products (e.g. [9], [10], [11]).

The global trade of all goods can also be translated into a corresponding network of virtual water trade, allowing quantification of the import and export fluxes of virtual water (e.g. [12], [13], [14]). Nations have been classified as net importers or exporters of virtual water according to the balance of trade fluxes entering and leaving the country (e.g. [15]).

It is therefore necessary to address the problems concerning water resources, such as water scarcity, water excess and deterioration of quality, not only at the agricultural district level and at the river basin level, but also at the nation and, ultimately, at a global level [15].

A nation can be considered "wise" when it exports products that are obtained from abundantly available resources within the country and imports products that are produced with resources that are scarcely available within the country. One of these resources could clearly be water; a water-scarce country might thus aim at behaving "wisely" by importing products that require a lot of water in their production and exporting products or services that require less water. However, this "virtuous" kind of trade does not (always) correspond to what trade figures show when examined on a long term and at the nation level, most likely because other factors, such as profit for instance, have a stronger influence on trade patterns.

To better examine this concept, it is important to determine the significance of the virtual water import for a given nation compared to what is nationally consumed; this could give an approximate idea of the country's reliance and dependency on external resources. The resources taken into consideration for this study are water resources destined for food production. Using the virtual water concept, a descriptive approach of the degree of dependency of a nation from overseas water resources is expanded in the next section in which indices are created, that could suggest a degree of trade "virtuosity" or "inefficiency". This requires establishing virtual water balances between a country and the rest of the world, in terms of exchanges of crops and livestock.

The immediate goal of this approach is to highlight incongruent terms in the virtual water balances, at a nation's scale and by the viewpoint of single products. In the subsequent sections, specific incongruences emerging in the trade of single products are highlighted to allow a clearer connection between economic driving forces and water resources. Italy is the nation considered in all examples shown in the paper, because it is among the leading OECD (Organisation for Economic Co-operation and Development) countries in the list of the greatest virtual water importers [8].

Materials and Methods

Descriptive Analysis of Production and Overall Trade

Data necessary for the investigations on production and trade presented in this paper are obtained from the database FAOSTAT (www.faostat.fao.org), that provides time-series of data relating to food and agriculture for some 200 countries. For this analysis primary and processed crops and primary and processed livestock are taken into consideration: 397 items in total for Italy - from year 1961 to year 2009. Exchanges are considered in terms of quantities imported (I), exported (E) and produced (P), in tonnes. Production data are paired with data on import and export and the balance term 'Consumption' (C) is defined as $C = P - E + I$ (considering constant storage stocks).

In order to have an overview of the overall trade of the country of interest, two ratios are computed for each product throughout the available time span:

i) the ratio between 'Import' and 'Consumption' (I/C), that shows the countries' dependency on overseas products; this term, named DP after *dependency*, indicates how much of the national request for crops or livestock is satisfied by import and not by internal production (the closer DP is to 0 the more is the commodity supply provided by the country itself);
ii) the ratio between 'Export' and 'Production' (E/P), named ER for *export ratio*, that shows the relative export attitude of the country; the index ER explains how much of the internal production is destined to be exported.

Items which have a high DP are items whose national demand is not satisfied by internal production; those which have a high ER are items whose production is destined mostly to be exported instead of being nationally consumed. When combining these two metrics, products can be found which show both a high DP and a high ER. These are products which behave very particularly: whilst there is a significant import, seemingly because of an impossibility of self sufficiency of the nation, there is for the same commodity a strong export of its production. Acknowledging that further internal processing of products is impossible to track, and assuming that the country does not function as a mere transit, i.e. imports are not consumed nationally but further exported, this feature configures a "vicious" kind of trade. The interest in computing DP and ER systematically on all products and on all the available years is in demonstrating possible "vicious" and incongruent trade figures for a given nation, at least in terms of virtual water budget.

Considering several products one may want to visualize the above indices on a bubble graph where the horizontal axis shows the dependency DP and the vertical axis the export ratio ER (Fig. 1). The production amount is reflected in the size of the bubbles, that also varies among the years. The radius of the bubble is assumed to scale with production amounts according to a logarithmic growth, i.e. when the radius doubles, a factor of 10 is applied. Logarithmic scaling allows to appreciate figures for the smallest productions. Everything is then scaled by the numeric amount of the highest production (that is reported in the caption of the figure).

The main information content of fig. 1 is the position of a commodity according to the quadrant where the two indices fall. The items located in the lower left quadrant are those which have very simple behaviour: they are produced locally mainly for the local consumption, keeping export and import to the minimum. This otherwise means that the production of such items seems to be calibrated on the request of the population and does not exceed it. The commodities located towards the upper right quadrant of the (DP, ER) diagram, instead, have at the same time high import and high export amounts compared respectively to consumption and production. These positions hence exemplify the incongruent type of trade discussed above. Figure 1 shows the ER-DP representation for the totality of the products analysed for Italy in three different years (1961, 1985 and 2009) which were taken as indicative of the complete time series. The colour green indicates primary and processed crops whereas light blue stands for primary and processed livestock. The products located in the upper right quadrant have evidently increased in number from 1961 to the year 2009.

A different way of examining information on production and trade is to visualize the time trend of a single product on the ER-DP diagram. The examples chosen for this time-series representation are Beans and Tobacco, whose location in the ER-DP diagram is highlighted by an orange star and that -among the products with high DP, high ER or both- presented the most significant trends. DP and ER data have been plotted versus time in figure 2, that shows for these commodities an increase in time in both DP and ER. The radius of the bubble is again proportional to the production but is not scaled by logarithm; the numeric amount of the highest production is reported in the caption of the figure. Economic considerations are likely to justify this peculiar evolution, as will be discussed hereinafter.

Results and Discussion

Virtual Water Inefficiency Assessment: The Swap Phenomenon

The previously shown behaviour for which a commodity is at the same time highly imported and highly exported will be referred to as the "swap" attitude. Between two trading countries a product can therefore be considered a swap product when being "unreasonably" exchanged both ways.

In order for a country to quantify the actual swap, information is necessary regarding import and export considering all fluxes entering (leaving) Country #1 from (to) the individual country of origin (destination). Such information can be obtained from a further section of the FAO database (i.e. the FAO Statistics TradeSTAT). A specific evaluation of the entity and evolution of this attitude is conducted on Italy as an initial case study.

The quantification of swap for the various products is done by comparing, year by year, the products' amounts traded by Italy with each different country: the minimum between import and export of each trade connection is selected as the quantity of product which is unreasonably exchanged, as expressed in the following equation:

$$swap_{i,j} = \min(import_{i,j}, export_{i,j})$$

where i indicates the traded commodity and j the country with which Italy is trading it. This is because a product should (ideally)

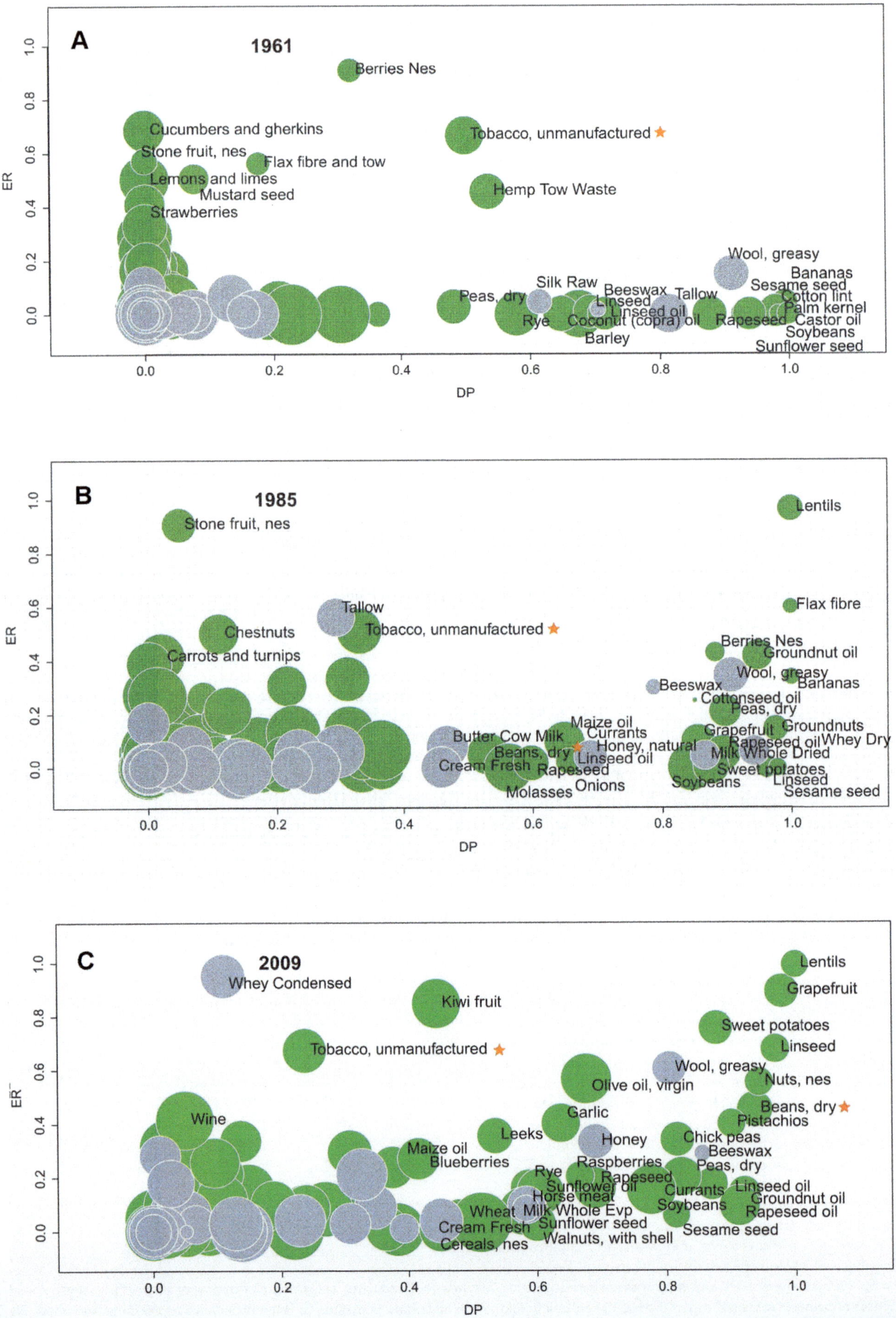
A
1961
Berries Nes
Cucumbers and gherkins
Tobacco, unmanufactured
Stone fruit, nes
Flax fibre and tow
Lemons and limes
Mustard seed
Hemp Tow Waste
Strawberries
Wool, greasy
Bananas
Sesame seed
Cotton lint
Palm kernel
Castor oil
Soybeans
Sunflower seed
Silk Raw
Peas, dry
Beeswax
Tallow
Linseed
Linseed oil
Rye
Coconut (copra) oil
Rapeseed
Barley
ER
DP
B
1985
Stone fruit, nes
Lentils
Tallow
Flax fibre
Chestnuts
Tobacco, unmanufactured
Carrots and turnips
Berries Nes
Groundnut oil
Wool, greasy
Bananas
Beeswax
Cottonseed oil
Peas, dry
Maize oil
Butter Cow Milk
Currants
Grapefruit
Groundnuts
Rapeseed oil
Whey Dry
Honey, natural
Milk Whole Dried
Beans, dry
Linseed oil
Cream Fresh
Rapeseed
Sweet potatoes
Soybeans
Linseed
Molasses
Onions
Sesame seed
C
2009
Whey Condensed
Lentils
Grapefruit
Kiwi fruit
Sweet potatoes
Tobacco, unmanufactured
Linseed
Wool, greasy
Nuts, nes
Olive oil, virgin
Beans, dry
Wine
Garlic
Pistachios
Leeks
Honey
Chick peas
Maize oil
Beeswax
Blueberries
Raspberries
Peas, dry
Rye
Rapeseed
Sunflower oil
Currants
Linseed oil
Horse meat
Soybeans
Groundnut oil
Wheat
Milk Whole Evp
Rapeseed oil
Cream Fresh
Sunflower seed
Sesame seed
Cereals, nes
Walnuts, with shell

Figure 1. ER-DP diagram for Italy for years 1961, 1985 and 2009. The indices of dependency DP and export ratio ER are computed over all products available and are visualised on a bubble graph where the horizontal axis shows DP and the vertical axis ER. The production amount is reflected in the size of the bubbles, that also varies among the years. The radius of the bubble is assumed to scale with production amounts according to a logarithmic law, i.e. when the radius doubles, a factor of 10 is applied. Logarithmic scaling allows to appreciate figures for the smallest productions. Everything is then scaled by the numeric amount of the highest production (for years 1961, 1985, 2009 respectively: Grapes 8,5E+06 tonnes, Grapes 9,6E+06 tonnes, Grapes 8,2E+06 tonnes). The colour green indicates primary and processed crops whereas light blue stands for primary and processed livestock. The orange stars indicate the position of beans and tobacco in reference to Figure 2.

only be imported or exported. When both terms are different from zero the smallest of the two can be considered "unnecessary". In other words, if the minimum is zero, there is an understandable dependency of one of the two countries (Italy in the specific case) on the other country. If this minimum is different from zero there is overlapping between I and E and this overlap is the swap amount.

The total (in tonnes) of the swap exchanges for each commodity is calculated by adding up all contributions from each trading country. The translation of these data into quantities of virtual water is subsequently done by multiplying the total swap amounts by the virtual water content per ton of each commodity, assessed for Italy [3], [8], [16]. This turns the product amounts into volumes of virtual water, in m^3, inefficiently exchanged for each commodity; for the assessment of the annual totals it is simply sufficient to add up all products contributions. This calculation is conducted for each year from 1986 to 2009 over all products reported by FAO. Results are presented as a dotted time series in Figure 3. A significant increasing trend is observed for the total volumes of virtual water traded through swap amounts at the nation's level on all trade connection. In order to understand the entity of the swap phenomenon a significant comparison with the total net flux can be done for the Italian case study: in 2009 the average net import of virtual water is 50E+09 m^3 [8] while the swap entails for 3E+09 m^3.

In the supporting information, Figure S1 and Figure S2, shown are synthetic representations of the virtual water carried from unnecessary imports, per commodity, for year 2009. The products with a null result of swap amount were omitted from this already "dense" representation. Of the totality of 397 items in year 2009, 365 undergo the swap phenomenon, with differences in quantity and in the number of countries involved. Of 365 items only for 162 was the virtual water content (m^3/ton) available and therefore the computation of the inefficiency in terms of traded virtual water possible. A logarithmic x axis is used to cover the whole range of trading amounts. For the sake of a clear representation the totality of the 162 products is divided into two categories, called 'crop' and 'livestock'. From this moment onwards the pool of products which will be taken into consideration for further analysis consists in the 162 commodities for which information on water content was obtainable.

While studying the possible explanations behind these vicious exchanges, two main considerations emerge regarding the phenomenon.

Firstly, the existence of swap entails that a country may work as a transit, i.e. products are passing through the country to reach their final destination. A method to pinpoint the manifestations of this phenomenon in the trade of a single country is to detect the products with a higher import than consumption (DP>1). This behaviour, corresponding also to higher export than production (ER>1), indicates a transportation of the product in question through the country, which may just act as a sorter. The values of the ratios ER and DP previously computed for all products and all years were scanned, allowing to identify a few products, such as Cotton lint, Mustard seed, Bananas and Flax fibre & tow, which show such a behaviour in Italy.

Secondly, many of the trade links between countries from different hemispheres are justified by the seasonality of some cultivations and allow to obtain a year-round availability of the commodities in question. The values obtained from FAO refer to the whole year which makes these trade links not immediately detectable. A more sensible analysis is needed in order to avoid considering swap for what is possibly an import or export of seasonal relevance; this could be further attempted by considering only the European countries, with which Italy shares climate and seasonality in crop production. The relevance of this subset finds support in Tamea et al. (8) from which it emerges that 60% of the total import of virtual water for Italy in 2010 comes from Europe and 72% of the export of virtual water from Italy in the same year is destined to Europe. European countries as a whole therefore represent the main trading partners for Italy. In particular, France is the major supplier of virtual water and one of the greatest importer of Italian goods, second only to Germany.

Swap Analysis

The analyses reported so far provide figures that raise questions about the sustainability of swap trade, as reasons behind the anomalies of swap products could be numerous. The attempt made is to investigate each of the previous trade connections which present a swap from an economic point of view and in terms of water resources efficiency. To this purpose, two ratios are introduced which are firstly computed for all trading links and, at a later stage, examined in depth for those products presenting the highest quotes of swap, as they result from the previous section.

The first indicator, referred to as Virtual Water Volume Ratio, is a ratio between the virtual water content (m^3/ton) of an item leaving Italy as export for another country, and the virtual water content (m^3/ton) of that item entering Italy from that same country. The ratio is computed for all items and all countries, generating a matrix of Virtual Water Volume Ratios, with a value for each trade connection. A "spatial" weighted mean is calculated for each product by using the import quantities from each country as weights, resulting into a single average value for each commodity. A Virtual Water Volume Ratio higher than 1 indicates a commodity which averagely requires more water for its production in Italy than in all other countries from which Italy is importing it. Table S1 reports the complete list of items considered in the computation. Of the 162 commodities for which the calculation was conducted, 93 present a Virtual Water Volume Ratio higher than 1.

The second indicator is the ratio between the economic unit value (1000USD/ton) of a certain commodity leaving Italy towards another country and the economic unit value (1000USD/ton) of the same commodity entering Italy from that same country. Information on such figures is obtained from the previously cited FAO Statistics TradeSTAT database. This ratio can justify the swap of products, since it is singularly calculated on each trading link between Italy and every single other nation on products which are being traded both ways. It will be referred to as Value Ratio and allows to pinpoint trade exchanges which are economically convenient in terms of unit value. As was done for

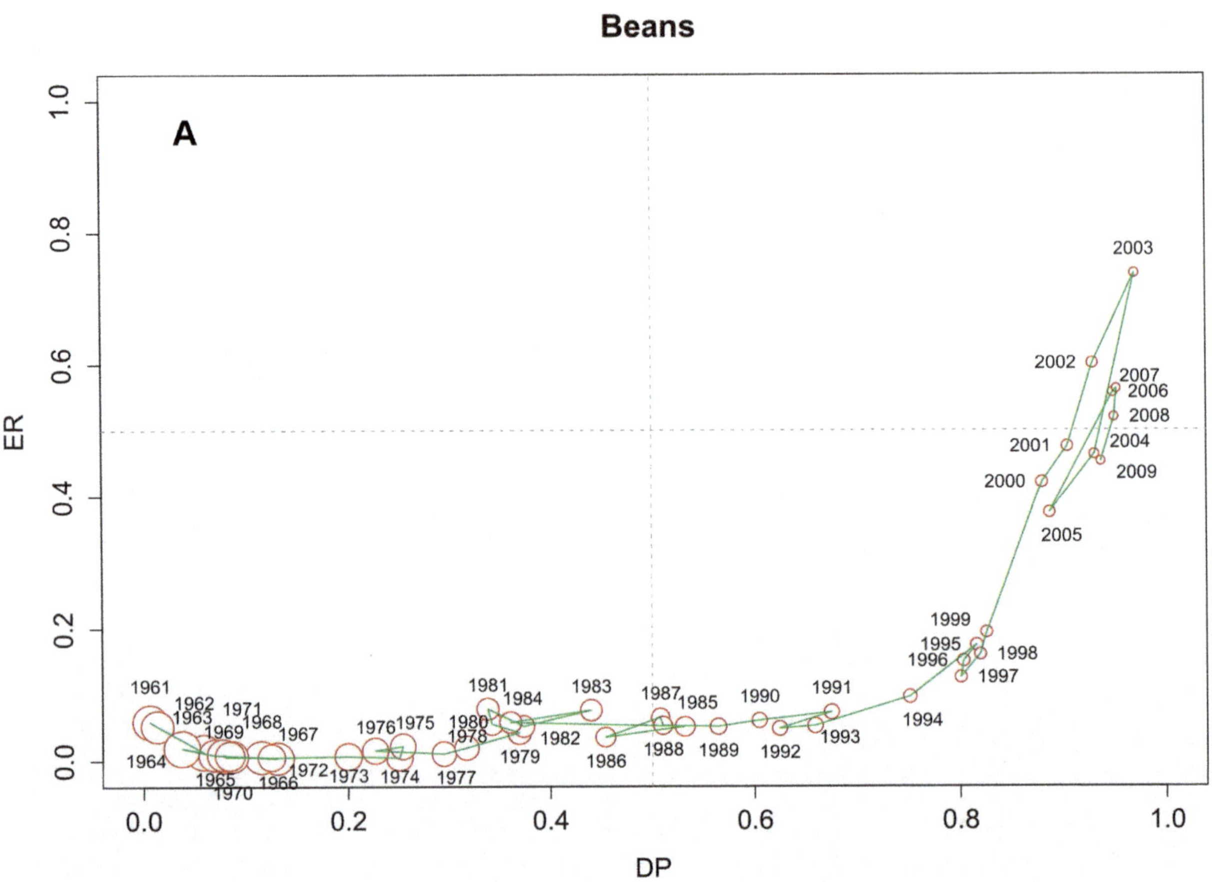
Beans
A
ER
DP
0.0
0.2
0.4
0.6
0.8
1.0
1961 1962 1963 1964 1965 1966 1967 1968 1969 1970 1971 1972 1973 1974 1975 1976 1977 1978 1979 1980 1981 1982 1983 1984 1985 1986 1987 1988 1989 1990 1991 1992 1993 1994 1995 1996 1997 1998 1999 2000 2001 2002 2003 2004 2005 2006 2007 2008 2009

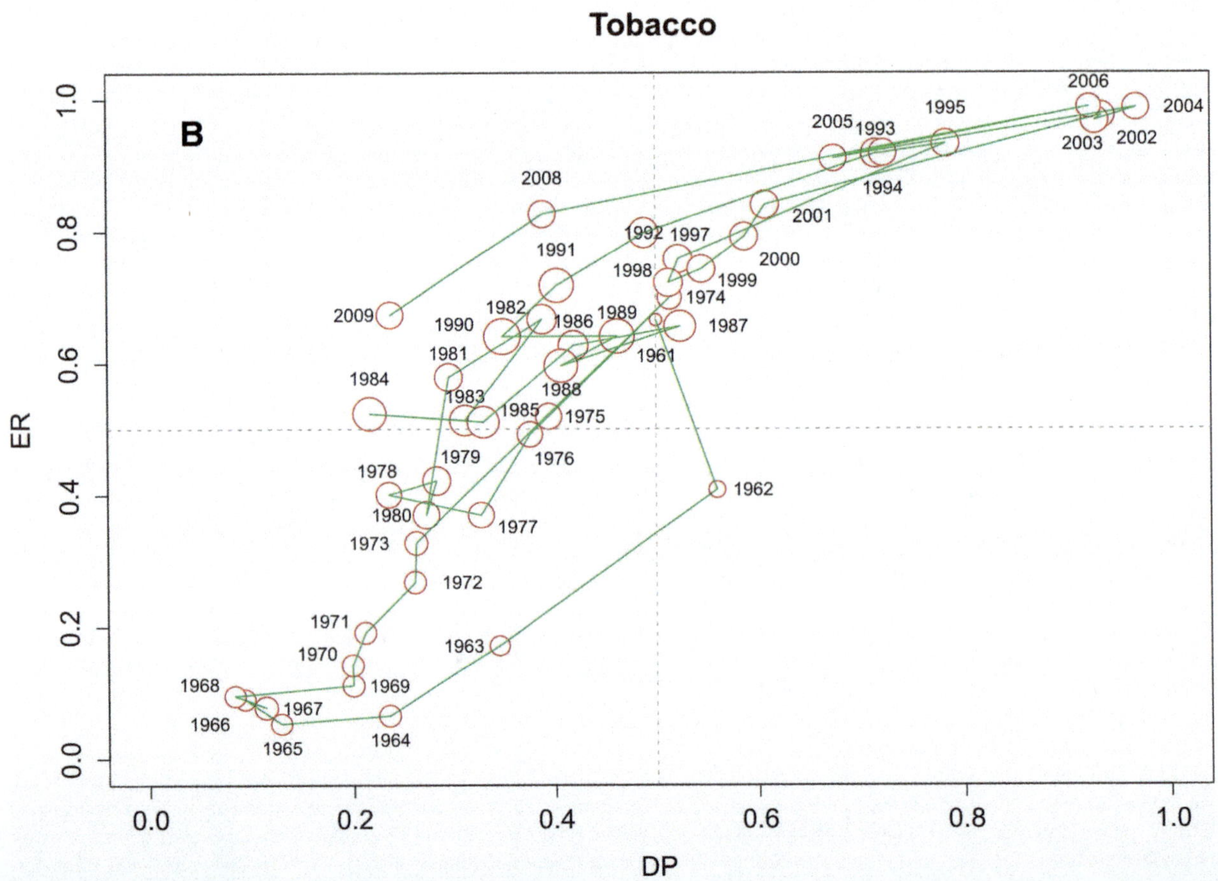
Tobacco
B
ER
DP
0.0
0.2
0.4
0.6
0.8
1.0
1961 1962 1963 1964 1965 1966 1967 1968 1969 1970 1971 1972 1973 1974 1975 1976 1977 1978 1979 1980 1981 1982 1983 1984 1985 1986 1987 1988 1989 1990 1991 1992 1993 1994 1995 1997 1998 1999 2000 2001 2002 2003 2004 2005 2006 2008 2009

Figure 2. ER-DP diagram for Italy of Beans and Tobacco 1961–2009. Visualisation of the time trend of a single product on the ER-DP diagram. The examples chosen for this time-series representation are Beans and Tobacco, whose location in the ER-DP diagram is highlighted by an orange star in Figure 1 (beans in 1961 were below the threshold for labelling). For both of these crops, DP and ER data have been plotted versus time. It is evident that for these commodities there has been an increase in time in both DP and ER. The radius of the bubble is again proportional to the production, but is not scaled in a logarithmic way; the numeric amount of the highest production is 2,1E+05 tonnes in case of Beans in 1964 and 2,1E+05 tonnes for Tobacco in 1990.

the Virtual Water Volume Ratio, a matrix of Value Ratios is obtained when investigating each trading link and the weighted mean for each product is computed by using the swap (in tonnes) of each trade connection as weight. Table S2 presents the complete list of Value Ratio for the pool of commodities considered. The calculated Value Ratio is higher than 1 in 123 cases, meaning that Italian products are on average more valuable than the foreign product with which they are swapped. Goods exported from Italy thus present an added-value that could justify the swap.

Figure 4 presents the totality of commodities taken into consideration (162 items) plotted on a bubble graph where the horizontal axis shows the Value Ratio and the vertical axis the Virtual Water Volume Ratio (both axis are in logarithmic scale). Among those commodities produced in Italy in year 2009, those with a swap higher than 10E+07 m^3 are represented as orange bubbles, while the swap amount is reflected in the size of the bubbles (the radius is scaled according to the swap amount). It can be seen that, with the notable exception of cattle meat, the most important commodities from the point of view of swap fall in the right part of the graph, indicating a profit. The opposite behaviour could be ascribed to other drivers which determine the existence of some trade links, as for instance the ones related to the quality of imported products or to the customs and habits among neighbouring countries. These categories of drivers that influence the global food trade logics are extremely difficult to map and to quantify and are not investigated in this paper.

Data presented in tables 1 and 2 show the 20 most impacting items from the point of view respectively of m^3 of virtual water inefficiently swapped (Table 1) and of what is referred to as net flux (8) (difference between the m^3 of virtual water imported and m^3 of virtual water exported, Table 2). The reported figures are coupled with the value and ranking (in brackets) of the two previous ratios, the Virtual Water Ratio and the Value Ratio. It can be noticed that many of the products which present a high swap, have at the same time a high value of net flux, such as wheat, meat-cattle boneless(beef and veal), maize, horse meat, soybeans. For a product presenting a high value of net flux, the country is strongly relying on import and consequently on the import of virtual water in order to satisfy the national demand of such commodity. A swap in such case would be most unwelcome and unreasonable, even if small in percentage. This is the case of wheat. Wheat, in fact, represents a highly traded crop (net flux of 8.3E+09 m^3 of virtual water) and at the same time a highly swapped crop (swap of 1.3E+08 m^3). By analysing the Virtual Water Volume Ratio one realises that, on average, the virtual water content of imported wheat is higher than the virtual water content of Italian wheat (the Italian contains 0.63 times the water averagely contained by foreign wheat). However, according to the Value Ratio, the wheat imported is economically less valuable than the Italian one (the Italian wheat has an economic unit value which is 1.54 times the average incoming wheat value). One could conclude, also in this case, that Italy swaps wheat in order to profit from it.

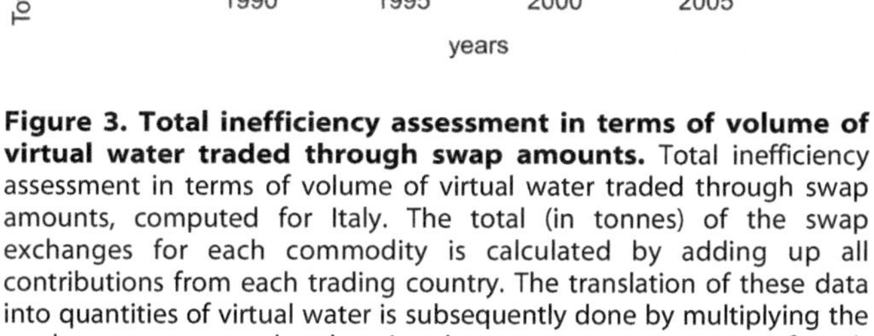

Figure 3. Total inefficiency assessment in terms of volume of virtual water traded through swap amounts. Total inefficiency assessment in terms of volume of virtual water traded through swap amounts, computed for Italy. The total (in tonnes) of the swap exchanges for each commodity is calculated by adding up all contributions from each trading country. The translation of these data into quantities of virtual water is subsequently done by multiplying the total swap amounts by the virtual water content per ton of each commodity. This turns the product amounts into volumes of virtual water inefficiently exchanged for each commodity; for the assessment of the annual totals it is simply sufficient to add up all products contributions. This calculation is conducted for each year from 1986 to 2009 over all products reported by FAO. Results are presented as the dotted time series.

Conclusions

Relations between food production and the related water consumption are investigated by examining the nature and the evolution of bilateral trade between countries, with particular reference to the virtual water efficiency or inefficiency connected to production and trade. The research presented here relates not so much to the marginal production efficiency, which will depend on the characteristics of irrigation techniques or to the variations of productivity according to climate conditions; it rather considers the influence of commercial interests on the global trade of virtual water, by means of the isolation of two-way exchanges of the same products between importers and exporters ("swap").

The Italian case, in this sense, is very representative since Italy is a strong importer of virtual water. By using the concept of swap it is highlighted that, even for products which are abundant in Italy, a substantially inefficient trade from the point of view of the budget of virtual water is practiced with increasing intensity. The swap occurs for the economic benefit of the country in question, but raises doubts about its sustainability from the point of view of global management of water resources, in particular with reference to the growing trend in these exchanges.

The considerations expressed in this paper do not contemplate the shipping costs related to the traded goods nor any aspect related to the organoleptic quality of the imported and exported food.

The extension of this work on a larger scale could allow interesting developments in the assessment of the impact of swap and its economic drivers on a global scale, for example with reference to sudden food shortages, deriving both from climate anomalies or from other human interventions. Future activities in this direction will allow us to evaluate the possible relations

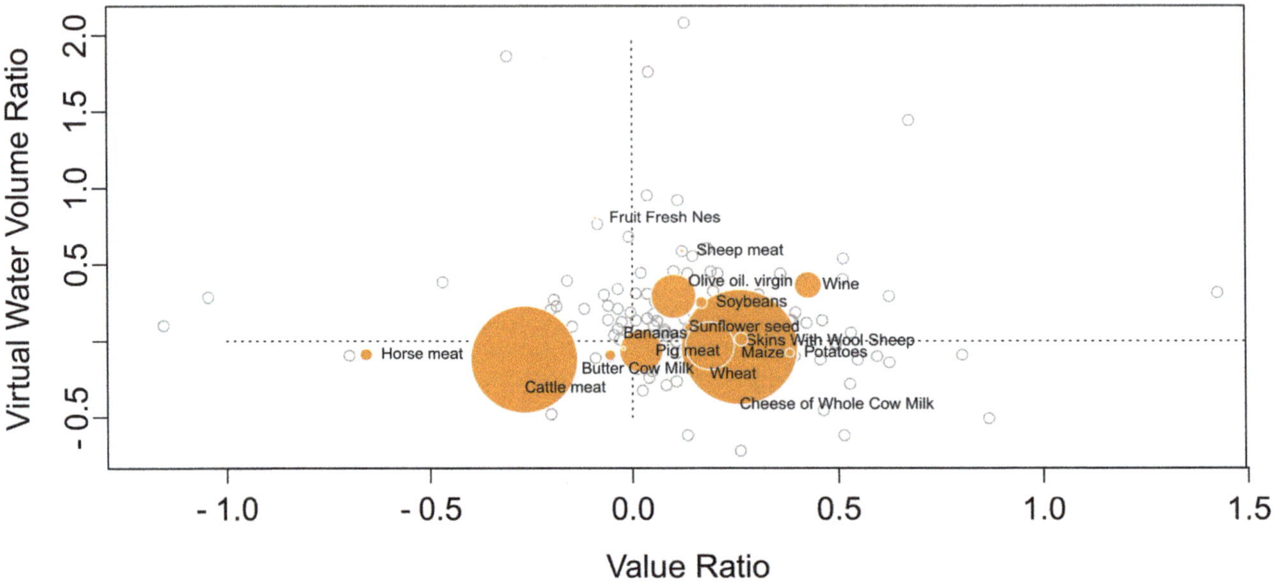

Figure 4. Value Ratio – Virtual Water Volume Ratio diagram for Italy in 2009. The indices of Value Ratio and Virtual Water Volume Ratio are computed over all products available and are presented as a scatter plot. Among the commodities produced in Italy in 2009 those with a swap higher than 10^7 m^3 are represented by bubbles whose dimension is given by the swap amount. The radius of the bubbles is scaled (non logarithmically) from the highest figure which is 3.2E+08 m^3 of Cheese of Whole Cow Milk.

Table 1. The 20 most swapped products.

FAO code	Item	Swap (m3)	Virtual Water Volume Ratio	Value Ratio
870	Meat-CattleBoneless(Beef&Veal)	4.9E+08	0.79 (129)	0.88 (136)
901	Cheese of Whole Cow Milk	3.2E+08	0.90 (106)	1.82 (44)
867	Cattle meat	3.0E+08	0.75 (134)	0.54 (154)
15	Wheat	1.3E+08	0.63 (105)	1.54 (66)
261	Olive oil, virgin	1.2E+08	1.94 (34)	1.26 (93)
1035	Pig meat	1.2E+08	0.85 (112)	1.05 (117)
1039	Bacon and Ham	9.6E+07	0.81 (117)	2.21 (30)
564	Wine	7.4E+07	2.29 (24)	2.67 (19)
920	Hides Wet Salted Cattle	7.4E+07	0.81 (120)	0.95 (128)
921	Hidesdry S.Cattle	7.0E+07	0.77 (132)	0.81 (142)
268	Sunflower oil	5.7E+07	1.35 (64)	1.15 (102)
1232	Food Prep Nes	5.6E+07	2.51 (20)	1.44 (74)
898	Milk Skimmed Dry	4.7E+07	0.84 (113)	0.42 (157)
56	Maize	3.7E+07	1.02 (90)	1.84 (42)
16	Flour of Wheat	3.6E+07	0.63 (142)	1.12 (108)
1097	Horse meat	3.5E+07	0.81 (121)	0.22 (159)
1042	Prep of Pig Meat	3.3E+07	0.82 (116)	2.31 (27)
236	Soybeans	3.2E+07	1.77 (40)	1.47 (72)
886	Butter Cow Milk	3.1E+07	0.80 (124)	0.88 (135)
18	Macaroni	3.0E+07	0.93 (102)	0.97 (126)

The 20 most swapped products in terms of virtual water inefficiently traded, with relative Virtual Water Volume Ratio and Value Ratio.

Table 2. The 20 products with the highest net flux.

FAO code	Item	Net flux (m3)	Virtual Water Volume Ratio	Value Ratio
15	Wheat	8.3E+09	0.63 (105)	1.54 (66)
236	Soybeans	3.1E+09	1.77 (40)	1.47 (72)
238	Cake of Soybeans	2.2E+09	2.02 (29)	1.09 (112)
882	Cow milk, whole, fresh	1.9E+09	4.01 (10)	1.52 (68)
867	Cattle meat	1.7E+09	0.75 (134)	0.54 (154)
56	Maize	1.4E+09	1.02 (90)	1.84 (42)
268	Sunflower oil	1.4E+09	1.35 (64)	1.15 (102)
870	Meat-CattleBoneless(Beef&Veal)	1.3E+09	0.79 (129)	0.88 (136)
271	Rapeseed oil	1.3E+09	0.55 (148)	1.78 (48)
261	Olive oil, virgin	1.1E+09	1.94 (34)	1.26 (93)
1035	Pig meat	1.1E+09	0.85 (112)	1.05 (117)
267	Sunflower seed	9.4E+08	1.22 (75)	1.37 (81)
1097	Horse meat	8.9E+08	0.81 (121)	0.22 (159)
920	Hides Wet Salted Cattle	8.7E+08	0.81 (120)	0.95 (128)
901	Cheese of Whole Cow Milk	8.2E+08	0.90 (106)	1.82 (44)
269	Sunflower Cake	6.0E+08	1.47 (55)	1.59 (61)
44	Barley	5.0E+08	0.88 (109)	1.67 (56)
231	Almonds Shelled	4.2E+08	1.46 (56)	1.45 (73)
237	Soybean oil	3.7E+08	1.86 (36)	1.35 (84)
888	Milk Skm of Cows	3.4E+08	0.81 (122)	6.32 (3)

The highest 20 products from the point of view of the net flux (difference between the virtual water totally imported and exported), with relative Virtual Water Volume Ratio and Value Ratio.

between virtuous (or vicious) trade policies and country (or global) resilience to droughts, famines etc.

Supporting Information

Figure S1 Inefficiency per CROP product in terms of volume of virtual water traded through swap amounts (in 10× units). Synthetic representations of the virtual water carried from unnecessary imports, per commodity per year, for year 2009. A logarithmic x axis is used to cover the whole range of trading amounts. The category represented is 'crop'.

Figure S2 Inefficiency per LIVESTOCK product in terms of volume of virtual water traded through swap amounts (in 10× units). Synthetic representations of the virtual water carried from unnecessary imports, per commodity per year, for year 2009. A logarithmic x axis is used to cover the whole range of trading amounts. The category represented is 'livestock'.

Table S1 Complete list of items considered in the computation of the Virtual Water Volume Ratio. Of the 162 commodities for which the calculation was possible, 93 present a Virtual Water Volume Ratio higher than 1.

Table S2 Complete list of commodities for which Italy is conducting a swap, with the referring Value Ratio. Of the 162 products analysed the calculated Value Ratio is higher than 1 in 123 cases.

Acknowledgments

The authors would like to thank Francesco Laio, Paolo D'Odorico and Hector Malano for the precious contributions and valuable comments. Two anonymous reviewers are gratefully acknowledged. Support from the EU Project WARECALC (2008-230845) to the first author is also acknowledged.

Author Contributions

Conceived and designed the experiments: JW PA PC. Performed the experiments: JW. Analyzed the data: JW PA PC. Wrote the paper: JW.

References

1. Allan JA (1998) Virtual water: A strategic resource global solutions to regional deficits. Ground Water 36: 545–546.
2. Hoekstra AY, Mekonnen MM (2012) The water footprint of humanity. P Natl Acad Sci USA 109: 3232–3237.
3. Chapagain AK, Hoekstra AY (2004) Water Footprints of Nations vol 1 Main Report. (Delft: UNESCO-IHE).
4. Mekonnen MM, Hoeckstra AY (2010) A global and high-resolution assessment of the green, blue and grey water footprint of wheat. Hydrol Earth Syst Sci. : 1259–1276.
5. D'Odorico P, Carr J, Laio F, Ridolfi L (2012) Spatial organization and drivers of the virtual water trade: a community-structure analysis. Environ Res Lett 7: 034007.
6. Van Oel PR, Mekonnen MM, Hoekstra AY (2009) The external water footprint of the Netherlands: Geographically-explicit quantification and impact assessment. Ecological Economics 69: 82–92.
7. Chapagain AK, Hoekstra AY (2007) The water footprints of Morocco and the Netherlands: global water use as a result of domestic consumption of agricultural commodities. Ecol Econ 64: 143–151.

8. Tamea S, Allamano P, Carr JA, Claps P, Laio F et al. (2013) Local and global perspectives on the virtual water trade. HESS, 17: 1205–1215.
9. Mekonnen MM, Hoekstra AY (2010) The green, blue and grey water footprint of crops and derived crop products, Value of Water Research Report Series No. 47. UNESCO-IHE, Delft, the Netherlands.
10. Salmoral G, Aldaya MM, Chico D, Garrido A, Llamas MR (2011) The water footprint of olives and olive oil in Spain. Spanish Journal of Agricultural Research 9: 1089–1104.
11. Bulsink F, Hoekstra AY, Booij MJ (2010) The water footprint of Indonesian provinces related to the consumption of crop products. Hydrol Earth Syst Sci 14: 119–128.
12. Dalin C, Konar M, Hanasaki N, Rinaldo A, Rodiguez-Iturbe I (2012) Evolution of the global virtual water trade network P. Natl Acad Sci USA 109: 5989–5994.
13. Carr J, D'Odorico P, Laio F, Ridolfi L (2012) On the temporal variability of the virtual water network. Geoph Res Letts 39 : L06404.
14. Konar M, Dalin C, Suweis S, Hanasaki N, Rinaldo A et al. (2011) Water for food: the global virtual water trade network. Water Resources Research 47 Issue 5.
15. Hoekstra AY, Chapagain AK (2008) Globalization of water: sharing the planet's freshwater resources. Blackwell, Oxford (UK).
16. Food and Agriculture Organisation, FAOSTAT commodity list. Available: http://www.fao.org/economic/ess/ess-standards/commodity/en/ (verified on February 18, 2014).

15

Where There Is No Toilet: Water and Sanitation Environments of Domestic and Facility Births in Tanzania

Lenka Benova[1]*, Oliver Cumming[2], Bruce A. Gordon[3], Moke Magoma[4], Oona M. R. Campbell[1]

1 London School of Hygiene and Tropical Medicine, Faculty of Epidemiology and Population Health, London, United Kingdom, 2 London School of Hygiene and Tropical Medicine, Faculty of Infectious and Tropical Diseases, London, United Kingdom, 3 World Health Organization, Geneva, Switzerland, 4 Evidence for Action (E4A), Department of Obstetrics and Gynaecology, Bugando Medical Centre and Teaching Hospital, Mwanza, Tanzania

Abstract

Background: Inadequate water and sanitation during childbirth are likely to lead to poor maternal and newborn outcomes. This paper uses existing data sources to assess the water and sanitation (WATSAN) environment surrounding births in Tanzania in order to interrogate whether such estimates could be useful for guiding research, policy and monitoring initiatives.

Methods: We used the most recent Tanzania Demographic and Health Survey (DHS) to characterise the delivery location of births occurring between 2005 and 2010. Births occurring in domestic environments were characterised as WATSAN-safe if the home fulfilled international definitions of improved water and improved sanitation access. We used the 2006 Service Provision Assessment survey to characterise the WATSAN environment of facilities that conduct deliveries. We combined estimates from both surveys to describe the proportion of all births occurring in WATSAN-safe environments and conducted an equity analysis based on DHS wealth quintiles and eight geographic zones.

Results: 42.9% (95% confidence interval: 41.6%–44.2%) of all births occurred in the woman's home. Among these, only 1.5% (95% confidence interval: 1.2%–2.0%) were estimated to have taken place in WATSAN-safe conditions. 74% of all health facilities conducted deliveries. Among these, only 44% of facilities overall and 24% of facility delivery rooms were WATSAN-safe. Combining the estimates, we showed that 30.5% of all births in Tanzania took place in a WATSAN-safe environment (range of uncertainty 25%–42%). Large wealth-based inequalities existed in the proportion of births occurring in domestic environments based on wealth quintile and geographical zone.

Conclusion: Existing data sources can be useful in national monitoring and prioritisation of interventions to improve poor WATSAN environments during childbirth. However, a better conceptual understanding of potentially harmful exposures and better data are needed in order to devise and apply more empirical definitions of WATSAN-safe environments, both at home and in facilities.

Editor: Karol Sestak, Tulane University, United States of America

Funding: This research was made possible with UK Aid from the Department of International Development (DFID) as part of the Sanitation and Hygiene Applied Research for Equity (SHARE) Research Consortium (www.SHAREresearch.org). However, the views expressed do not necessarily reflect the Department's official policies. The funders had no role in study design, data collection and analysis, decision to publish, or preparation of the manuscript.

Competing Interests: The authors have declared that no competing interests exist.

* Email: lenka.benova@lshtm.ac.uk

Introduction

At the end of the 18th century, the causal link between poor-hand hygiene and puerperal sepsis was recognised, eventually enabling reductions in maternal deaths [1–3]. Currently, WHO guidelines for delivery in health facilities advise frequent hand-washing, and clean birth kits have been designed for births in domestic environments [4]. A recent systematic review concluded that a lack of sanitation facilities appears to be associated with maternal mortality, as does lack of water access [5]. This review highlighted the paucity of primary studies assessing the impact of water and sanitation environments on maternal mortality and recommended future assessments of the burden of exposure to poor water and sanitation during pregnancy and delivery.

The United Republic of Tanzania is a sub-Saharan African country with 45 million inhabitants. Despite a 3.5% average annual rate of reduction in maternal mortality between 1990 and 2013, the current maternal mortality ratio of 454 deaths per 100,000 births in 2010 means that Tanzania remains off-track to achieve the Millennium Development Goal 5 target to reduce the maternal mortality ratio by three quarters between 1990 and 2015 [6–8]. Approximately 7,900 women die annually from the largely preventable or treatable complications of pregnancy and childbirth; and sepsis is estimated to account for 9% of these deaths [9].

Globally, an effective intrapartum care strategy, encompassing institutional delivery with referral capacities, has been suggested as a strategy to reduce maternal mortality [10]. Tanzania has seen a modest increase in the proportion of births occurring in health facilities; from 43.5% in 1999 to 50.1% in 2010 [7], but wide socio-economic inequalities in the utilization of skilled birth attendance exist [11]. To reduce maternal mortality, the Tanzanian government proposed scaling-up the availability of basic emergency obstetric and newborn care services at dispensaries and health centres, and improving the ability of rural health centres to perform caesarean sections and blood transfusions [6]. The health service delivery system in Tanzania is characterized as a network of hospitals, health centres and dispensaries (primary care clinics) [12].

In 2010, the proportion of Tanzanian population with access to improved water sources was 53%, a slight decrease from 55% in 1990. Access to improved sanitation was very low at 10% in 2010, a marginal improvement from 7% in 1990 [13]. A survey of 175 public facilities providing maternal care in Southern Tanzania showed only 83% of dispensaries had staff hand-washing facilities. The study did not report on other aspects of water, sanitation and hygiene environment, such as the availability of soap, running water, or hygiene practices among health staff and patients [14]. However, a recent study in Tanzania found that women who rated their local primary care centres as poor quality were more likely to bypass them to deliver in hospitals; upgrading or renovating the clinics reduced bypassing by 60% [15].

The main objective of this paper is to estimate the coverage of water and sanitation (WATSAN) in the various birth environments. We propose using household data to describe the WATSAN environment of home birth settings, and facility surveys to describe the WATSAN environment of facility deliveries. We selected Tanzania for this case study because both types of surveys were available and relatively recent. The secondary objective of this country study is to demonstrate how existing secondary data can create generate useful information for policy initiatives and future primary research. This approach permits an assessment of geographical variability in the coverage of WATSAN in birth environments that may generate useful information for prioritisation and targeting of limited resources.

Methods

Data sources

The Demographic and Health Surveys (DHS) are cross-sectional nationally representative household surveys, conducted in over 90 countries worldwide. The Service Provision Assessments (SPA) are cross-sectional nationally representative facility surveys conducted by the same group, in 15 countries. We used the most recent Tanzania DHS (DHS, 2010), which reported on the number and location of live births occurring between 2005–2010 to women in sampled households [7]. The DHS dataset included a relative socio-economic categorisation of women's households, wealth quintile [16], and information on household water and sanitation.

We used the most recent SPA survey conducted in 2006 to characterise the WATSAN environment of facilities. This survey included a nationally-representative sample of 611 public and non-public facilities [17]. A questionnaire was administered and elements of the delivery room environment were observed during facility visits. The analysis in this paper was limited to those health facilities which reported conducting deliveries. Both DHS and SPA surveys were representative nationally and on the level of eight geographic zones (Central, Western, Lake, Southern Highlands, Southern, Northern, Zanzibar and Eastern).

Definitions

Birth location. We characterised births reported in the DHS by delivery location. Births outside of a health facility were classified as having occurred in the woman's home or in a different location (e.g., parental or traditional birth attendant's home). The duration of residence in the current dwelling was not collected and we were unable to distinguish home births that occurred in the current residence from those in a previous residence. Therefore, all births reported in the woman's home were assumed to have occurred in the current household environment (the dwelling assessed by the household questionnaire). Births which were delivered in health facilities were characterised according to the level of health facility reported (dispensary, health centre or hospital). Births that did not occur in the woman's home or in a health facility were described as having occurred in 'other locations'.

Domestic WATSAN environment. We defined the home birth environment as WATSAN-safe if both the drinking water source and the sanitation facility access could be characterised as 'improved' according to the WHO/UNICEF Joint Monitoring Programme (JMP) definition (Table 1) [18]. A WATSAN-unsafe environment, on the other hand, described homes in which either water or sanitation, or both were classified as 'unimproved'. This construct does not capture many other important components of the environment, such as water quality, consistency of availability, actual use of sanitation facilities or hygienic practices, but it does indicate the existence and location of physical assets required for hygienic behaviour during childbirth and the postpartum period.

Delivery facility WATSAN environment. No uniform definitions of acceptable or 'improved' WATSAN environments of health facilities are currently available for international monitoring. We classified the WATSAN environment in facilities using the limited data collected by the SPA to capture facility environments with different risk profiles and the requisite equipment/supplies for infection control measures. The survey collected information on the WATSAN environment of the facility as a whole and a more detailed description of the delivery room environment. We characterised both environments, defining 'WATSAN-safe' environments as those which fulfilled both the 'improved' water and 'improved' sanitation requirements (Table 2). We reasoned that in hospitals, the delivery room may better describe the environment where the birth occurred, but in smaller facilities, such as dispensaries and health centres, the overall facility environment may be indistinguishable from the delivery room environment. WATSAN profiles of both these environments were therefore used to calculate uncertainty intervals.

Analysis

In analysing both DHS and SPA data we accounted for the complex survey sampling (clustering, stratification and sample weights) by using the *svyset* command in Stata/SEv.13 in order to produce point estimates and their 95% confidence intervals. To assess the WATSAN environment of facility births, we combined the level of health facility where the birth occurred (dispensary, health centre or hospital) with the weighted average of WATSAN-safe facilities of that level in the zone where the birth occurred, from the SPA. No information was available about the WATSAN environment for births occurring in 'other locations'. We combined the estimated number of WATSAN-safe births in the three locations (home, health facility, other) to estimate of the

Table 1. Categorisation of types of domestic drinking water sources and sanitation facilities, Tanzania DHS 2010.

Domain	Drinking water source	Sanitation facility
Improved	Piped into dwelling	Facility which is not shared with other households and is:
	Piped to yard/plot	Flush to piped sewer system
	Public tap or standpipe	Flush to septic tank
	Neighbour's tap	Flush to pit latrine
	Protected well in dwelling	Ventilated improved pit latrine
	Protected well in yard/plot	Pit latrine with slab
	Protected public well	
	Neighbour's borehole	
	Rainwater	
Unimproved	Open well in dwelling	Any facility which is shared with other households, or is:
	Open well in yard/plot	Flush - to somewhere else
	Open public well	Pit latrine - without slab/open pit
	Neighbour's open well	No facility/bush/field
	River, dam, lake, pond, stream, canal or irrigation channel	Other facility
	Spring	
	Tanker truck	
	Cart with small tank	
	Bottled water	

proportion of all births in WATSAN-safe environments, by zone and nationally. The midpoint estimate and the best and worst case scenarios, representing the range of uncertainty, were obtained using the scenarios provided in Table 3.

Ethical procedures and approvals

DHS: Respondents were informed about the purpose of the survey before the start of the interview, informed that their participation was voluntary, and that all information provided was confidential and de-identified. The respondent's verbal consent, if obtained, was noted on the questionnaire with a signature of the enumerator.

SPA: Informed consent was obtained from the facility in-charge and from all respondents for the facility audit questionnaires. Prior to commencing the Delivery and Newborn Care questionnaire module, the enumerator located the manager or most senior health worker and provided them with the details of the survey. The respondent was told the study aims, that the facility was selected randomly, and that no patient names would be recorded or shared. They were informed that participation was voluntary, and that the information collected might be used by the Ministry of Health or other organisations seeking to improve the planning and delivery of health services, and that the name of the facility will be removed from the dataset. Verbal consent of the responding health worker, if obtained, was noted on the questionnaire with a signature of the enumerator.

Both the DHS and the SPA surveys used in this study were implemented by the National Bureau of Statistics and the Office of the Chief Government Statistician - Zanzibar; in collaboration with the Ministry of Health and Social Welfare. ICF Macro provided technical assistance for the survey through the MEASURE DHS programme and The United States Agency for International Development (USAID) funded this technical assistance. The ethical nature of both surveys, including the method of obtaining and recording informed consent received approval from local government authorities. The secondary analysis of the de-

Table 2. Categorisation of WATSAN environment of health facilities conducting deliveries, Tanzania SPA 2006.

Domain	Facility level	Delivery room level
Improved source of water	Piped from protected source (a protected well or a borehole)	Facility level water source is improved
	AND	AND
	Source of water is on site (within 500 m of facility)	Delivery room has running water, either piped or bucket with tap, observed on the day of survey
		AND
		Delivery room has soap for hand-washing, observed on the day of survey
Improved sanitation facility access	Functioning latrine for facility clients	Functioning latrine for facility clients (facility level)

Table 3. Definitions of WATSAN birth environments according to scenario (best, worst, midpoint).

Scenario	Home births	Facility births	Other locations
Best case	Home WATSAN environment	Facility WATSAN environment	All WATSAN safe
Midpoint	Home WATSAN environment	*Hospitals*: Delivery room WATSAN environment	Same as the home WATSAN environment
		Centres and dispensaries: Facility WATSAN environment	
Worst case	Home WATSAN environment	Delivery room WATSAN environment	All WATSAN unsafe

identified datasets was approved by the Observational/Interventions Research Ethics Committee at the London School of Hygiene and Tropical Medicine. Both sources of data are available at www.measuredhs.com.

Results

Home births

The women sampled in the 2010 DHS reported a total of 8,176 live births during the recall period. The overall proportion of these births reported to have occurred at home was 42.9% (95% confidence interval: 41.6%–44.2%), ranging from 19.5% in the Eastern zone to 57.4% in Western zone (Figure 1). Of the remainder, 50.1% (48.8%–51.5%) were reported to be in health facilities and 7.0% (6.3%–7.8%) in other locations. Based on the characteristics of the water and sanitation facilities of the households, the overall proportion of domestic births taking place in a WATSAN-safe environment was 1.5% (1.0%–2.4%), as shown in Figure 2. The proportion of home births that occurred in an environment with both unimproved water and unimproved sanitation was 63.3%, signifying a double-burden. The remaining births occurred in water-safe toilet-unsafe (32.9%) or in toilet-safe water unsafe (2.3%) environments. Zanzibar was the geographic zone with the highest proportion of home births occurring in WATSAN-safe environments (20.9%). The proportion of home births in a double-burden, water-unsafe toilet-unsafe environment was greater than 50% in all zones except for Zanzibar (20.3%).

We used the DHS household wealth quintiles to estimate the proportions of all births occurring at the current home and the proportion of home births occurring in a WATSAN-safe home environment. The results show that births to the poorest quintile of households were more than eight times more likely to have been delivered at home compared to births to the richest quintile (Figure 3). On the other hand, the proportion of home births delivered in a WATSAN-safe home environment was at or below 3% among the poorer four quintiles, increasing to 29% in the richest quintile.

Facility births

Based on women's report of delivery location in the DHS, Figure 4 shows that among the births delivered in facilities, 52.0% (50.2%–53.9%) were in a hospital, 17.7% (16.4%–19.2%) in a health centre and the remaining 30.2% (28.5%–32.0%) in a dispensary. The proportion of facility births occurring in hospitals ranged from 38% in the Western zone to 94% in Zanzibar. The highest proportion of facility births in dispensaries (the lowest facility level) occurred in the Southern Highlands zone - 43%. The proportion of births occurring in health facilities and proportion of facility births occurring in hospitals was higher for wealthier quintiles (Figure 5).

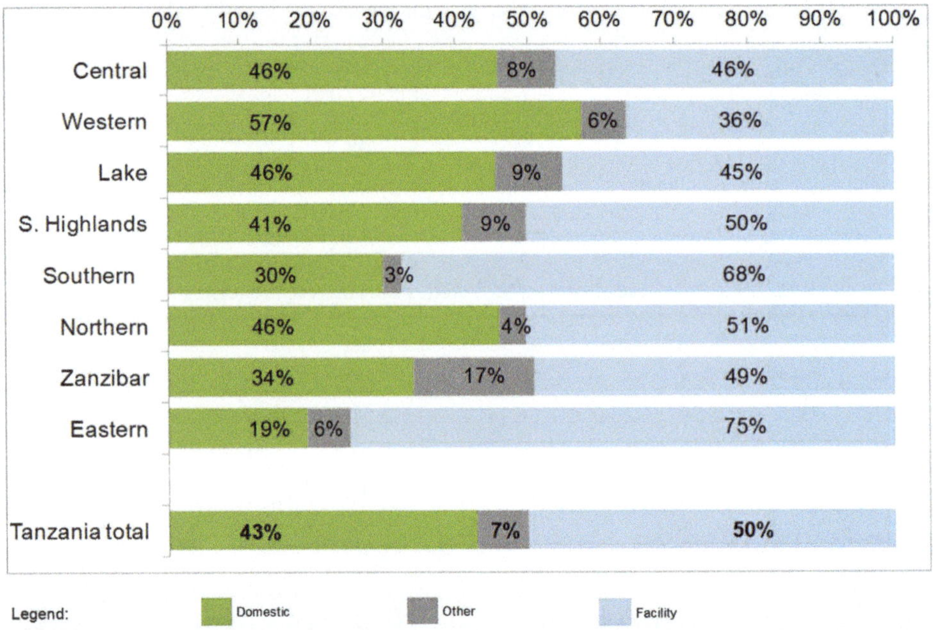

Figure 1. Distribution of all live births in five years prior to survey, by delivery location, by zone, Tanzania DHS 2010 (n = 8176).

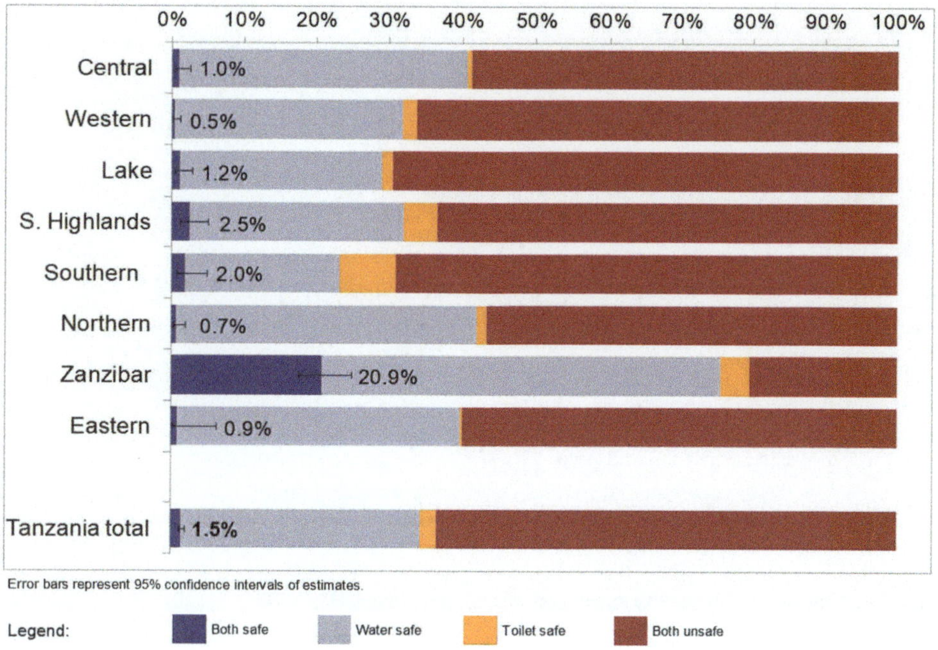

Figure 2. Births delivered in domestic environments by WATSAN environment of the current home, by zone, Tanzania DHS 2010 (n = 3504).

Among the facilities sampled on the SPA, 74.3% (70.2%–77.9%) reported providing delivery care. This ranged from 95.6% of hospitals to 71.8% of dispensaries. Based on our definitions of facility level WATSAN environment, 43.8% (38.6%–49.1%) of all facilities conducting deliveries in Tanzania were classified as WATSAN-safe. However, only 23.6% (19.4%–28.3%) met the more stringent definition of having a WATSAN-safe delivery room environment. The proportion of facilities meeting the facility level WATSAN-safe definition and the delivery room WATSAN-safe definition varied between the three facility levels (Figure 6). The reasons facilities did not meet the definition of WATSAN-safe facility level environment varied. Among hospitals and health centres, more than 90% of facility environments were classified as WATSAN-unsafe as a result of unimproved water sources. Among dispensaries, this was 82%. A substantial proportion of dispensaries (15%) lacked both improved water and improved sanitation. By applying the proportions of WATSAN-safe facilities by facility type within each zone to births reported in the same level of facility and zone on the DHS, we estimated that 69% of all facility births took place in a WATSAN-safe facility level environment and 49% in a WATSAN-safe delivery room.

Estimate of WATSAN-safe environment for all births in Tanzania

Our best estimate for the combined two-survey assessment of birth locations and their WATSAN profile revealed that less than

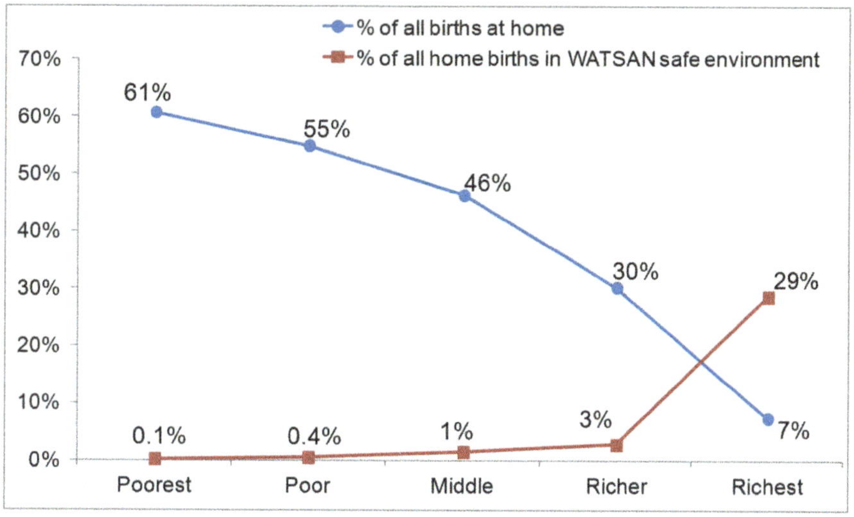

Figure 3. Equity analysis of home births and WATSAN-safe status of home births, by wealth quintile.

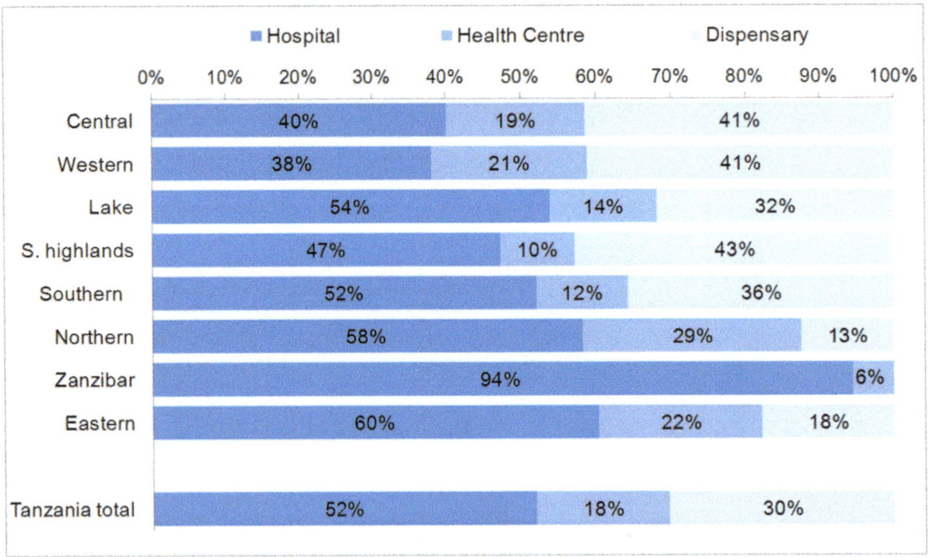

Figure 4. Distribution of facility births by facility type and zone, Tanzania DHS 2010.

one third (30.5%) of all births in Tanzania took place in a WATSAN-safe environment (range of uncertainty 25%–42%, Figure 7). The highest proportion of all births occurring in WATSAN-safe environments occurred in the Eastern zone (63.4%), where Dar es Salaam is the largest city. In a cluster of zones in the North-Western part of the country (Lake, Western and Central) more than 80% of all births took place in WATSAN-unsafe environments (Figure 8). The high levels of WATSAN-unsafe births in these zones stemmed mainly from the low proportion of facilities meeting the WATSAN-safe definitions compared to other zones. For example, fewer than half of the hospital delivery rooms in these three zones met the WATSAN-safe criteria.

Discussion

This study combined the most recent household and facility-level data in Tanzania to assess the proportion of births occurring in an environment with poor water and sanitation. Substantial proportions of births in Tanzania, as in many other countries in sub-Saharan Africa, take place at home. In this delivery context, clean birth is an essential approach. It seems difficult to envisage how clean birth can be ensured in birth environments where even

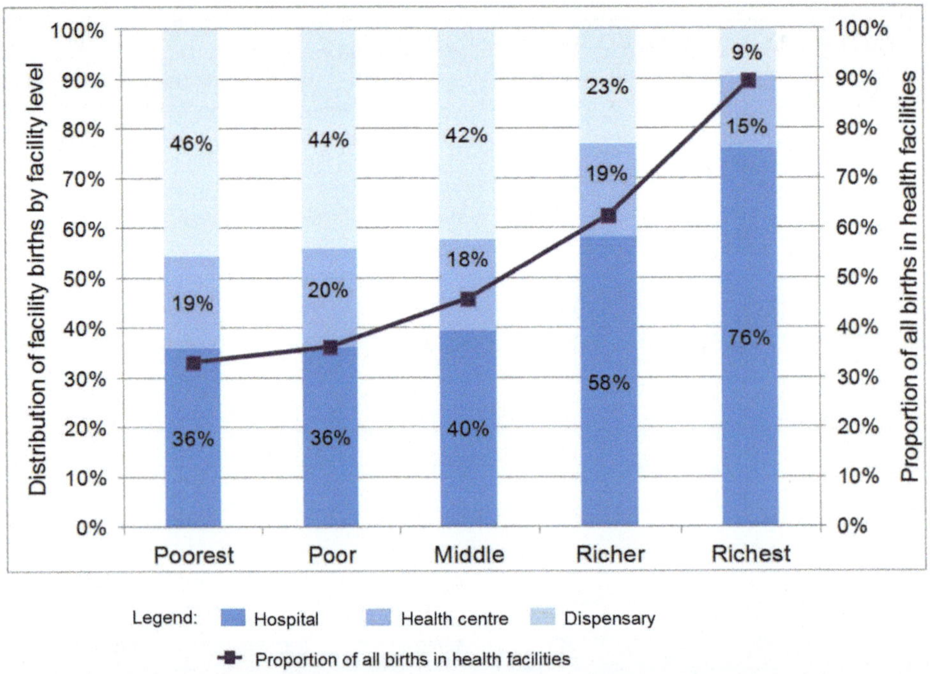

Figure 5. Distribution of facility births by facility type and wealth quintile, Tanzania DHS 2010.

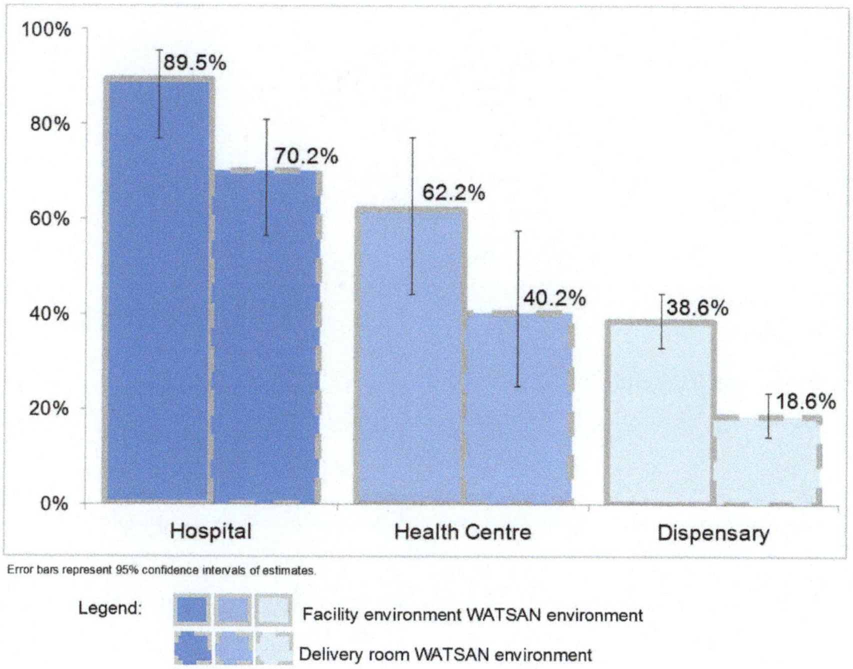

Figure 6. Facilities which conduct deliveries, by WATSAN-safe environments, on facility level and delivery room level, by facility type, Tanzania SPA 2006.

the most basic level of 'improved' water and sanitation access does not exist. Furthermore, even though delivery facilities, were, on average, better than home environments, they were far from universally WATSAN-safe, particularly health centres and dispensaries. This situation has the potential to cause great harm to mothers and newborns. The novel combination of DHS and SPA datasets in Tanzania also revealed large geographic variations in birth locations and in the WATSAN profile of homes and facilities. The large wealth inequities in the proportions of facility delivery and WATSAN-safe home birth environments suggested that any potential associated disease burden falls disproportionately on the poorest.

Limitations

This study's limitations stem primarily from our reliance on secondary data. The births reported in the most recent Tanzanian DHS were from 2005–10, meaning on average two and a half years after the SPA survey, conducted in 2006. The DHS did not collect information about the length of residence in the current household and we were therefore unable to identify home births that may have taken place in a previous dwelling (a dwelling

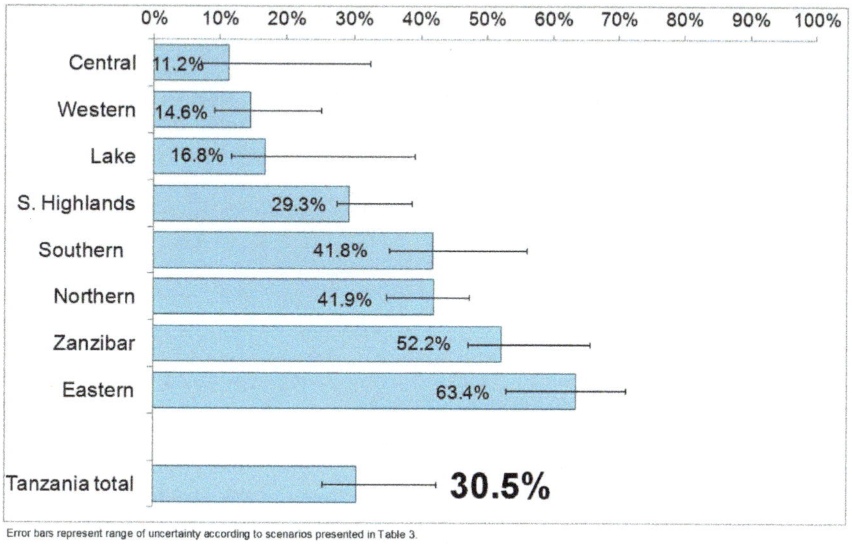

Figure 7. Proportions of all births in five years preceding 2010 DHS that were WATSAN-safe, by zone and overall.

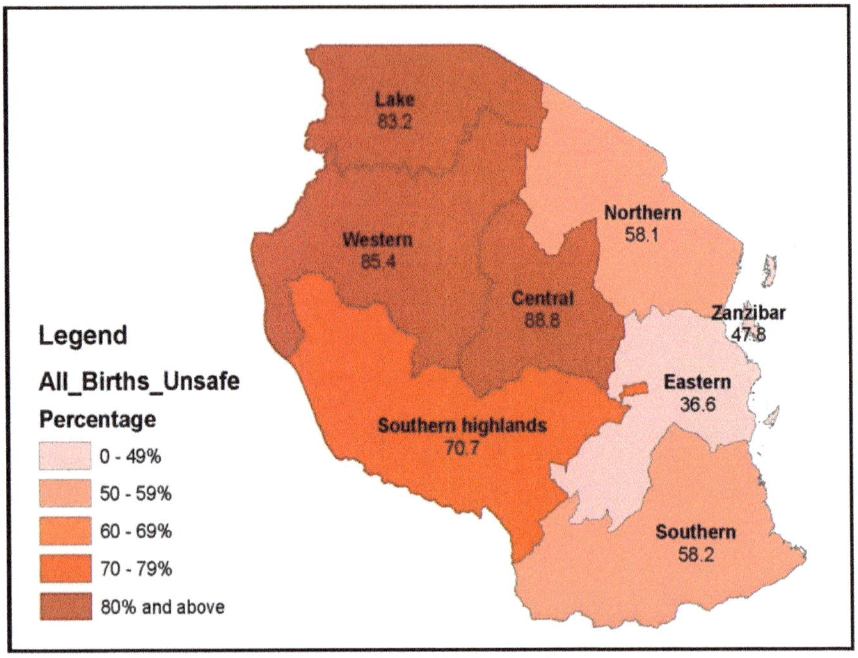

Figure 8. Proportion of all births occurring in WATSAN-unsafe environments, by zone.

different from the one described by the household questionnaire). No information was available about the WATSAN environment for the 7% of births occurring in other locations. All measures of the household WATSAN environment and locations of deliveries were self-reported on the DHS. In combining the DHS and SPA datasets to assess the WATSAN environment of all births, we assumed correct recall of delivery facility level (hospital, health centre or dispensary) by women. We are cognizant of the potential for sub-optimal recall of such data. However, a recent study in Mozambique found that women's self-report of the level of health facility utilised in delivery care was highly sensitive and specific [19]. We assumed that facility births to households living in a certain geographic zone took place in facilities within that zone.

The JMP definitions of 'improved' water sources and sanitation facilities for households omit some important characteristics relating to safe births such as water quality, availability, storage, distance to source and safety of access. Household hygiene environment and hygiene behaviours (i.e., soap availability and use) were not captured by the DHS and were therefore not explored in this analysis. The current JMP definitions require that for a household sanitation facility to be classified as 'improved,' it cannot be shared with other households. However, this may or may not be relevant to describing the sanitation environment as it relates to maternal and neonatal health outcomes. JMP definitions were originally developed to assess the risk of faecal-oral infection and the mechanisms of exposure for maternal health may differ. The exposures relevant for maternal health are poorly understood but are likely complex, cumulative and long-term. As such, the definitions of 'improved' or 'safe' WATSAN are unlikely to adequately capture all the associated risks. Despite these shortcomings, the use of JMP definitions by this study brings the results in line with currently national and international monitoring standards.

In terms of characterising the WATSAN environment of delivery facilities, we relied on data collected in the 2006 SPA survey to construct facility level and delivery room level definitions of WATSAN-safe environments. The lack of data about important elements of a health facility's WATSAN environment meant that these definitions are imperfect. For example, information was not collected about the type of latrine, its state of repair and cleanliness, acceptability to patients and distance from the delivery ward.

Recommendations and Conclusion

Despite these limitations and the fact that our approach resulted in estimates with a range of uncertainty, we are confident that our findings are a useful starting point to identify further research priorities and policy objectives. The main strategy for reducing maternal mortality and improving maternal and neonatal health has focused on ensuring availability of quality intrapartum care, which in most developing countries equates to institutional delivery. Half of deliveries in Tanzania occurred in health facilities and a study in Western Tanzania confirmed that positive perceptions of medical providers' quality of care on the community level were strongly associated with the odds of seeking facility delivery services [20]. A qualitative study among women in south-central Tanzania revealed that concerns over quality of care and issues of shame were major deterrents to seeking facility-based delivery care [21]. The attributes which influenced women to deliver in a health facility are directly observable characteristics of quality of care (i.e., provider attitude, availability of drugs and equipment). Policy simulations suggested that if such attributes were improved, the proportion of women preferring facility delivery would rise to 88% [22]. These authors hypothesized that such observable attributes signal a functioning health system and it is therefore likely that water and sanitation facilities and general perceptions of cleanliness could increase the proportion of births occurring in health facilities.

Deliveries in health facilities should occur in an enabling environment ensuring adequate hygiene [23,24]. We found that half of facility deliveries in Tanzania occurred in WATSAN-unsafe delivery room environments. This may explain the role of poor intrapartum and postnatal infection-control practices in the findings that neonatal infections are 3–20 times higher among hospital-born babies in developing countries compared to developed countries [25]. In both the peripartum and postnatal periods, the lack of essential equipment, such as soap, wash-basin and clean water are critical points in the causation of such infections [25]. Furthermore, in addition to risks to the mother and newborn, poor water, sanitation and hygiene conditions in delivery facilities pose occupational hazards for both medical and non-medical (i.e., porters, cleaners) staff. Improvement in water and sanitation aspects of working conditions may contribute to increased staff retention and uptake of rural postings [26].

A primary study from Tanzania showed a strong independent association between poor water and sanitation access at home and increased odds of maternal mortality [27]. We need a better conceptual and empirical understanding of water, sanitation and hygiene exposures to inform definitions and create better instruments for capturing WATSAN-safe environments in relation to maternal and neonatal health. In both home and facility birth environments, there is a pressing need to understand, measure and ultimately improve far more than just the basic WATSAN assets. Other important factors include the continuous availability and quality of these assets, gender issues related to their use (acceptability of female facilities, the hygienic practices of pregnant and postpartum women, birth attendants and other facility staff), as well as the socio-cultural factors that promote the construction, maintenance and appropriate use of improved WATSAN facilities at the community level. Improving our understanding, definition and measurement of WATSAN safety within health facilities is necessary due to ongoing efforts to encourage women to deliver in facility settings. While more data are needed on the capacities of delivery facilities overall, the development of an international standard for assessing water and sanitation environments of delivery facilities mirroring the JMP home environment definitions would be particularly useful in order to consistently capture the burden of WATSAN exposures, potentially as part of routine quality of care monitoring. This is especially important given the recent proposal that any water and sanitation goals that follow the MDGs should include reference to WASH in health facilities [18].

We demonstrated that secondary data sources, available for many other low and middle-income countries, are useful for assessing the overall burden of poor WATSAN during childbirth. This low-cost exercise in characterisation of both home and facility birth environments in Tanzania showed that great need exists for improvement in this area. There is a need to ensure the implementation of planning, building and maintenance regulations for facilities to obtain and maintain appropriate basic WATSAN infrastructure. Facilities with a basic water supply system should have a simple risk management system in place (i.e., a Water Safety Plan) and provide water at levels consistent with international guidelines, such as the World Health Organization's Essential Environmental Health Standards [28]. Furthermore, due consideration should be given to water quantity in order to enable hygiene, but also to enable maternity rooms to be cleaned, sheets laundered and meals prepared. The quality of drinking water must also be considered and may imply additional point-of-use treatment.

Our findings on the low levels of safe WATSAN in home environments provide extra impetus to existing efforts to improve provision of facility-based delivery care. They can also aid in outlining interim context-specific solutions. In households without an 'improved' water source, household water treatment could be added to a mother's/birth attendant's care pack and antiseptic hand gel might be an effective component of home delivery kits in water-scarce environments [29]. Beyond the delivery event, mothers require high quality drinking water while breastfeeding and water should be treated at home if required. There are challenges associated with changing behaviors on how household water is treated, but we could reasonably expect a mother to achieve sustained behavior change during the postpartum window of time. Scott and colleagues have found that women respond well to behavior change messages around their nurturing role [30] and this may be an important entry point for such interventions. In addition to currently being an important delivery location, homes constitute the primary living environment for mothers, children and their families. Interventions to improve household-level WATSAN may consider prioritizing households with newly married couples and with young children, particularly in geographic areas where health facilities are distant and thus less likely to be used.

Lastly, we propose strengthening national monitoring systems for home and facility delivery environments. With respect to the monitoring of facilities, the last 2012 GLAAS report demonstrated that many countries were unable to report on access to WATSAN in health-care settings, and very likely to not have national monitoring systems in place [31]. At the same time, many countries have national access targets in place for health care facilities, so this provides an entry point for improving monitoring and closing the gap between commitment and capacity. At the international level, the JMP recently convened a consultative process ahead of the recent High Level Panel on post-2015 which called for universal water and sanitation access to include health facility coverage. There is political will at both the national and international levels, but the systems to monitor resulting commitments and progress are not in place. This paper shows how significant progress on monitoring could be made just by using existing publicly available data and may bring greater urgency to ensure women are able to give birth in environments with access to the basic essential water and sanitation facilities.

Author Contributions

Conceived and designed the experiments: LB OMRC. Analyzed the data: LB OMRC OC. Contributed to the writing of the manuscript: LB OMRC OC MM BG.

References

1. Gordon A (1795) A treatise on the epidemic puerperal fever of Aberdeen. London, UK: G.G. and J. Robinson.
2. Semmelweis I (1983) The Etiology, Concept and Prophylaxis of Childbed Fever. Madison, WI: The University of Wisconsin Press.
3. Gould I (2010) Alexander Gordon, puerperal sepsis, and modern theories of infection control–Semmelweis in perspective. Lancet Infect Dis 10: 275–278.
4. World Health Organization (1996) Care in Normal Birth: A Practical Guide. Geneva: World Health Organization.
5. Benova L, Cumming O, Campbell OM (2014) Systematic review and meta-analysis: association between water and sanitation environment and maternal mortality. Trop Med Int Health 19: 368–387.
6. Tanzanian Ministry of Health and Social Welfare (2014) Women and Children First: Countdown to ending preventable maternal, newborn and child deaths. Available at www.countdown2015mnch.org: Published by Countdown to 2015.
7. National Bureau of Statistics (NBS) Tanzania, ICF Macro (2011) Tanzania Demographic and Health Survey 2010. Dar es Salaam, Tanzania: NBS and ICF Macro.
8. WHO UNICEF (2012) Countdown to 2015: Building a Future for women and children - the 2012 Report (Tanzania).
9. World Health Organization (2014) Trends in maternal mortality: 1990 to 2013. Estimates by WHO, UNICEF, UNFPA, The World Bank and the United Nations Population Division. Geneva: WHO.

10. Campbell O, Graham W (2006) Strategies for reducing maternal mortality: getting on with what works. Lancet 368: 1284–1299.
11. Kengia JT, Igarashi I, Kawabuchi K (2013) Effectiveness of health sector reforms in reducing disparities in utilization of skilled birth attendants in Tanzania. Tohoku J Exp Med 230: 241–253.
12. Hanson C, Ronsmans C, Penfold S, Maokola W, Manzi F, et al. (2013) Health system support for childbirth care in Southern Tanzania: results from a health facility census. BMC Res Notes 6: 435.
13. UNICEF WHO (2012) Progress on Drinking Water and Sanitation: 2012 Update. New York, NY: UNICEF and World Health Organization.
14. Penfold S, Shamba D, Hanson C, Jaribu J, Manzi F, et al. (2013) Staff experiences of providing maternity services in rural southern Tanzania - a focus on equipment, drug and supply issues. BMC Health Serv Res 13: 61.
15. Kruk ME, Hermosilla S, Larson E, Mbaruku GM (2014) Bypassing primary care clinics for childbirth: a cross-sectional study in the Pwani region, United Republic of Tanzania. Bull World Health Organ 92: 246–253.
16. Filmer D, Pritchett LH (2001) Estimating wealth effects without expenditure data-or tears: An application to educational enrollments in states of India. Demography 38: 115–132.
17. National Bureau of Statistics Tanzania, Macro International Inc (2007) Tanzania Service Provision assessment Survey 2006. Tanzania: National Bureau of Statistics, Macro International Inc.
18. World Health Organization, UNICEF (2013) Progress on sanitation and drinking-water - 2013 update. Geneva: WHO.
19. Stanton CK, Rawlins B, Drake M, dos Anjos M, Cantor D, et al. (2013) Measuring Coverage in MNCH: Testing the validity of women's self-report of key maternal and newborn health interventions during the peripartum period in Mozambique. PLoS ONE 8: e60694.
20. Kruk ME, Rockers PC, Mbaruku G, Paczkowski MM, Galea S (2010) Community and health system factors associated with facility delivery in rural Tanzania: a multilevel analysis. Health Policy 97: 209–216.
21. Spangler SA, Bloom SS (2010) Use of biomedical obstetric care in rural Tanzania: the role of social and material inequalities. Soc Sci Med 71: 760–768.
22. Kruk ME, Paczkowski M, Mbaruku G, de Pinho H, Galea S (2009) Women's preferences for place of delivery in rural Tanzania: a population-based discrete choice experiment. Am J Public Health 99: 1666–1672.
23. World Health Organization (2012) Hand hygiene in outpatient and home-based care and long-term care facilities: a guide to the application of the WHO multimodal hand hygiene improvement strategy and the "My Five Moments For Hand Hygiene" approach. Geneva: WHO.
24. World Health Organization (2009) WHO Guidelines on Hand Hygiene in Health Care: First Global Patient Safety Challenge Clean Care is Safer Care. Geneva: WHO.
25. Zaidi AK, Huskins WC, Thaver D, Bhutta ZA, Abbas Z, et al. (2005) Hospital-acquired neonatal infections in developing countries. Lancet 365: 1175–1188.
26. Henderson L, Tulloch J (2008) Incentives for retaining and motivating health workers in Pacific and Asian countries. Human Resources for Health 6.
27. Urassa E, Lindmark G, Nystrom L (1995) Maternal mortality in Dar es Salaam, Tanzania: Socio-economic, obstetric history and accessibility of health care factors. African Journal of Health Sciences 2: 242–249.
28. Adams J, Bartram J, Chartier Y (2008) Essential environmental health standards in health care. Geneva: World Health Organization.
29. Hussein J, Fortney JA (2004) Puerperal sepsis and maternal mortality: What role can new technologies play? International Journal of Gynecology and Obstetrics 85: S52–S61.
30. Scott B, Curtis V, Rabie T, Garbrah-Aidoo N (2007) Health in our hands, but not in our heads: understanding hygiene motivation in Ghana. Health Policy and Planning 22: 225–233.
31. World Health Organization (2012) UN-water global annual assessment of sanitation and drinking-water (GLAAS) 2012 report: the challenge of extending and sustaining services. Geneva: WHO.

16

Biological Instability in a Chlorinated Drinking Water Distribution Network

Alina Nescerecka[1,2], Janis Rubulis[1], Marius Vital[2], Talis Juhna[1], Frederik Hammes[2*]

1 Department of Water Engineering and Technology, Riga Technical University, Riga, Latvia, **2** Department of Environmental Microbiology, Eawag, Swiss Federal Institute for Aquatic Science and Technology, Dübendorf, Switzerland

Abstract

The purpose of a drinking water distribution system is to deliver drinking water to the consumer, preferably with the same quality as when it left the treatment plant. In this context, the maintenance of good microbiological quality is often referred to as biological stability, and the addition of sufficient chlorine residuals is regarded as one way to achieve this. The full-scale drinking water distribution system of Riga (Latvia) was investigated with respect to biological stability in chlorinated drinking water. Flow cytometric (FCM) intact cell concentrations, intracellular adenosine tri-phosphate (ATP), heterotrophic plate counts and residual chlorine measurements were performed to evaluate the drinking water quality and stability at 49 sampling points throughout the distribution network. Cell viability methods were compared and the importance of extracellular ATP measurements was examined as well. FCM intact cell concentrations varied from 5×10^3 cells mL^{-1} to 4.66×10^5 cells mL^{-1} in the network. While this parameter did not exceed 2.1×10^4 cells mL^{-1} in the effluent from any water treatment plant, 50% of all the network samples contained more than 1.06×10^5 cells mL^{-1}. This indisputably demonstrates biological instability in this particular drinking water distribution system, which was ascribed to a loss of disinfectant residuals and concomitant bacterial growth. The study highlights the potential of using cultivation-independent methods for the assessment of chlorinated water samples. In addition, it underlines the complexity of full-scale drinking water distribution systems, and the resulting challenges to establish the causes of biological instability.

Editor: Jose Luis Balcazar, Catalan Institute for Water Research (ICRA), Spain

Funding: The authors acknowledge the financial support of the EU project "TECHNEAU" (Nr. 018320) and the NMS-CH project "BioWater: Assessment of biological stability in drinking water distribution networks with chlorine residuals" (Sciex-N-7 12.265). The funders had no role in study design, data collection and analysis, decision to publish, or preparation of the manuscript.

Competing Interests: The authors have declared that no competing interests exist.

* E-mail: frederik.hammes@eawag.ch

Introduction

The goal of public drinking water supply systems is to produce water of acceptable aesthetic and hygienic quality and to maintain that quality throughout distribution until the point of consumption. From a microbiological perspective, the quality of treated water can deteriorate as a result of excessive bacterial growth, which can lead to problems such as a sensory deterioration of water quality (e.g. taste, odor, turbidity, discoloration) as well as pathogen proliferation [1–10]. To avoid this, biological stability during distribution can be achieved by maintaining sufficient residual disinfectants in the water, and/or through nutrient limitations [3,7,11,12]. However, drinking water systems should not be viewed as sterile; complex indigenous bacterial communities have been shown to inhabit both chlorinated and non-chlorinated drinking water distribution systems [5,13–17].

The concept of biological stability and its impact on a system's microbiology has been discussed extensively in the framework of non-chlorinated drinking water distribution systems [3,7,17–20]. However, many treatment plants worldwide employ a final disinfection step to ensure that no viable bacteria enter the distribution system. The latter is often achieved by oxidative disinfection, usually by chlorination [21]. Disinfection has a number of implications for a biological system. During chlorination, one can expect that a considerable fraction of bacteria in the water are killed or damaged, while some residual chlorine may remain in the water (Figure 1). This could be visible through numerous microbial monitoring methods. For example, the number of cultivable bacteria, measured with heterotrophic plate counts, would reduce dramatically [22,23]. Secondly, bacteria cells are likely to display measurable membrane damage irrespective of their cultivability [24], though the rate and extent of damage may differ between different communities. This would be detectable with several staining techniques coupled with epifluorescence microscopy or flow cytometry (FCM). Also, adenosine tri-phosphate (ATP), often used as a cultivation-independent viability method [19,22,25] will be severely affected. Based on data from Hammes and co-workers [4] one may reasonably expect increased levels of extracellular ATP (so-called free ATP) and decreased concentrations of intracellular ATP (bacterial ATP) following oxidative disinfection. Irrespective of the detection method, the overall consequence of disinfection is a considerable decrease in the viable biomass, potentially opening a niche for microorganisms to occupy downstream of the treatment process. Following initial disinfection, residual chlorine might provoke undesirable changes during drinking water distribution. Disinfectants target not only bacteria, but it also react with natural organic matter, pipe surfaces and particles in the network, thus potentially forming/releasing assimilable organic carbon (AOC) [26–30]. AOC can easily be consumed by bacteria, and is therefore seen as a main contributor to biological instability. Moreover, chlorine decay within the network negatively affects its ability to inhibit microbial growth at the far ends of the network [12]. If all factors were considered, the presence of nutrients, a reduction in the number of competing bacteria, and the lack of residual disinfectant would potentially lead to biological instability in the distribution network, manifesting in a subsequent bacterial growth (Figure 1). Besides the importance of nutrients, the extent of bacterial growth will be influenced by a number of factors. For example, increased water temperature can accelerate chlorine decay and favor bacteria growth [19,31], while changes in hydraulic conditions can alter nutrient supply for microorganisms in biofilms and/or bacteria detachment from the pipe surfaces [32,33]. Finally, the quality of materials in contact with drinking water, as well as the presence of sediments and loose deposits, can both affect the general microbial quality of the water [6,34,35].

In the present study we examined some of the above-discussed concepts in a full-scale, chlorinated distribution system in the city of Riga (Latvia) with a number of microbiological methods. The purpose was a detailed investigation of the entire city's distribution network, asking the basic question whether evidence of spatial and/or temporal biological instability exists, and if so, to which degree. Additional goals were to evaluate the use of fluorescent staining coupled with FCM, as well as ATP analysis, for the assessment of chlorinated drinking water in a distribution network with disinfectant residuals.

Materials and Methods

Ethics statement

Permission for sampling at all locations in the present study was obtained from the local water utility (Rīgas Ūdens).

Description of study site

Sampling was performed in the full-scale distribution network of Riga (Latvia) with a total length of about 1400 km. The city is supplied with drinking water from six water treatment plants (WTP) produced from both surface and groundwater (150 000 m^3 d^{-1}). Only the three major WTP, which are continuously operated, were included in the sampling campaign. Average WTP effluent water quality parameters for each treatment plant are shown in Table 1. The distribution network mainly consists of cast iron (80%) and unlined iron (15%) pipes as old as 50 years. The diameters of pipes ranged from 100 to 1200 mm. Three reservoirs are operated in the network to compensate for fluctuations in the daily water demand, while four high-pressure zones are maintained in some distal areas of the network. The high-pressure zones were excluded from the present study. A total of 49 sampling sites were selected across the city to cover the network broadly and to include both proximal and distal zones relative to the treatment plants. The sampling sites were selected according to the approximate water retention times obtained from a validated hydraulic model made in EPANET 2.0 [36,37] based on a total length of 538 km (39% of the total length of the network). Apart from the effluents of the three treatment plants, the sampling sites were in all cases fire hydrants in order to attain some degree of reproducibility between sampling and to avoid localized effects (e.g. household growth). The exact locations of sampled fire hydrants can be obtained from the authors after agreement from the local water utility.

Sampling protocol

A specific sampling protocol was designed and followed in order to avoid artifacts due to water stagnation in unused fire hydrants. Each hydrant was pre-flushed at a high velocity (never exceeding 1.6 m s^{-1}) for no more than 60 s, then immediately adjusted to a low velocity of 0.015–0.25 m s^{-1} and connected to an online system for monitoring pH, temperature, redox potential, electro-conductivity and turbidity. The low sampling velocity was specifically used to ensure a minimal possible impact of cell wall erosion and detachment from biofilms on the samples and measurements. Readings of all parameters were taken at 5–10 minute intervals, and water was only sampled for microbiological analysis once all of the parameters stabilized. The impact of this hydrant flushing is demonstrated in an example in Figure 2 and discussed in detail in the results section. Samples were kept in cold storage (≈5°C) and analyzed within four hours of sampling.

Chemical analysis

Determination of free chlorine was performed according to standard method EN ISO 7393-1, based on the direct reaction with N,N-diethyl-1,4-phenylenediamine (DPD) and subsequent formation of a red compound at pH 6.2–6.5. Afterwards titration by means of a standard solution of ammonium iron (III) sulfate until disappearance of the red color was performed. Determination of total chlorine was performed according to EN ISO 7393-1,

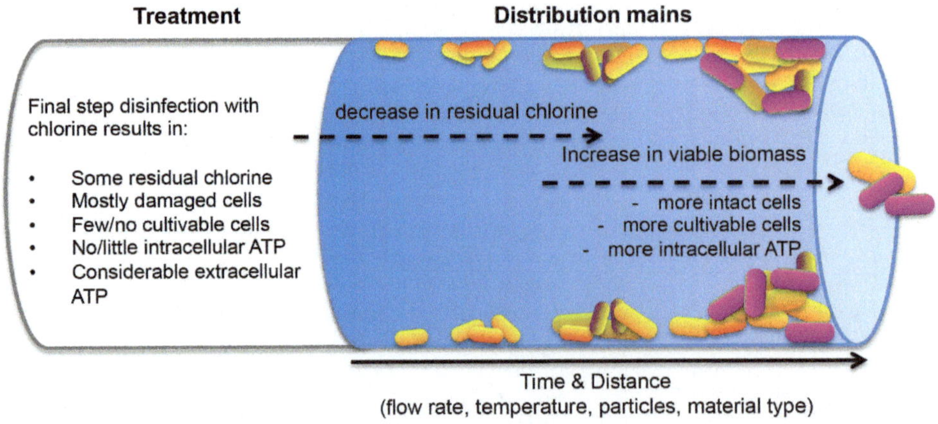

Figure 1. A worst-case-scenario in an unstable, chlorinated distribution network. Prediction of changes in the microbiological state of the water due to the depletion of residual chlorine and the concomitant growth of bacteria, potentially resulting in hygienic and sensory deterioration of the water quality.

Table 1. Average water quality parameters for the final effluents of the the three main treatment plants of Riga (Latvia).

	WTP 1	WTP 2	WTP 3
Source water	surface water	artificially recharged groundwater	groundwater
Final treatment step[a]	Cl_2 (0.5–3 mg L^{-1})	Cl_2 (ca. 1.5 mg L^{-1})	N.A.
Residual chlorine (mg L^{-1})[b]	0.44±0.11	0.51±0.01	0.42±0.26
Total organic carbon (TOC) (mg L^{-1})[a]	6±1	9±3	3
Assimilable organic carbon (AOC) (μg L^{-1})[a]	213±37	209±59	N.A.
Total cell concentration (cells mL^{-1})[b]	$5.31\pm0.97\times10^5$	$5.45\pm0.47\times10^5$	$1.69\pm0.18\times10^5$
Intact cell concentration (cells mL^{-1})[b]	$1.83\pm1.18\times10^4$	$1.4\pm0.86\times10^4$	$1.03\pm0.68\times10^4$
Total ATP (nM)[b]	0.015±0.005	0.029±0.004	0.011±0.002
Intracellular ATP (nM)[b]	0.007±0.003	0.000±0.004	0.001±0.002
HPC 22°C (CFU mL^{-1})[b]	23±24	4±2	4
HPC 36°C (CFU mL^{-1})[b]	16±16	4±2	1
Conductivity (μS cm^{-1}), 25°C[a]	468±101	625±4	272±25
pH[a]	6.63±0.18	7.41±0.04	7.5±0.05

[a]Data supplied by the water utility or measured in previous sampling campaigns.
[b]Data from present study.

based on the reaction with DPD in the presence of an excess of potassium iodide, and then titration as described above.

Fluorescent staining and flow cytometry (FCM) of water samples

Staining and FCM analysis was done as described previously [4,38]. In short, for a working solution, SYBR® Green I (SG) stock (Invitrogen AG, Basel, Switzerland) was diluted 100x in anhydrous dimethylsulfoxide (DMSO) and propidium iodide (PI; 30 mM) was mixed with the SYBR® Green I working solution to a final PI concentration of 0.6 mM. This working solution was stored at −20°C until use. From every water sample, 1 mL was stained with SGPI at 10 μL mL^{-1}. Before analysis, samples were incubated in the dark for 15 minutes. Prior to FCM analysis, the water samples were diluted with 0.22 μm filtered bottled water (Evian, France) to 10% v/v of the initial concentration. FCM was performed using a Partec CyFlow SL instrument (Partec GmbH, Münster, Germany), equipped with a blue 25 mW solid state laser emitting light at a fixed wavelength of 488 nm. Green fluorescence was collected at 520±10 nm, red fluorescence above 630 nm, and high angle sideward scatter (SSC) at 488 nm. The trigger was set on the green fluorescence channel and data were acquired on two-parameter density plots while no compensation was used for any of the measurements. The CyFlow SL instrument is equipped with volumetric counting hardware and has an experimentally determined quantification limit of 1000 cells mL^{-1} [4].

Adenosine tri-phosphate (ATP) analysis

Total ATP was determined using the BacTiter-Glo reagent (Promega Corporation, Madison, WI, USA) and a luminometer

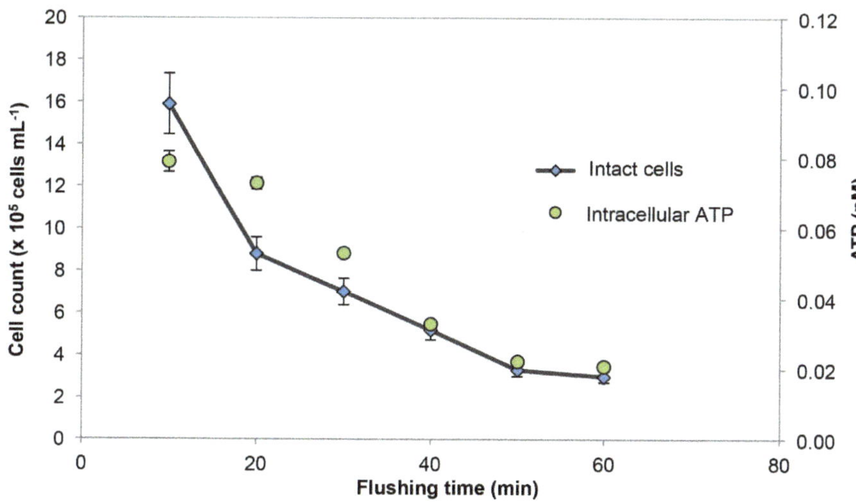

Figure 2. The impact of low velocity flushing on the water quality in a newly-opened fire hydrant. Intracellular adenosine tri-phosphate (ATP) data points were derived from duplicate measurements of extracellular and total ATP concentrations. FCM intact cells (after SYBR Green I and propidium iodide staining) were single measurements, with a relative standard deviation of 9% calculated from all data in the present study.

(Glomax, Turner Biosystems, Sunnyvale, CA, USA) as described elsewhere [25]. A water sample (500 μl) and the ATP reagent (50 μl) were warmed to 38°C simultaneously in separate sterile Eppendorf tubes. The sample and the reagent were then combined and then the luminescence was measured after 20 s reaction time at 38°C. The data were collected as relative light units (RLU) and converted to ATP (nM) by means of a calibration curve made with a known ATP standard (Promega). For extracellular ATP analysis, each sample was filtered through a 0.1 μm sterile syringe filter (Millex-GP, Millipore, Billerica, MA, USA), followed by analysis as described above. The intracellular ATP was calculated by subtracting the extracellular ATP from the total ATP for each individual sample. ATP was measured in duplicate for all samples.

Heterotrophic plate counts

To obtain heterotrophic plate counts (HPC), samples were serially diluted in sterile distilled water and then inoculated onto nutrient yeast agar plates using the spread plate technique. All plates were incubated in dark at 22°C or 36°C for 3 and 7 days, respectively. Results were expressed as colony forming units (CFU) per ml of water sample.

Statistical analysis

Statistical data evaluation was performed with the MS Excel Data Analysis tool (Descriptive statistics, Regression). The reproducibility for indirect/calculated data (e.g., intracellular ATP) was calculated by a propagation-of-uncertainty method. FCM data was not always measured in duplicate, due to practical constraints. In these cases, a 9% error (average coefficient of variation (CV) ($n = 39$)) was applied for representing FCM data. The residual chlorine concentration distribution box plot was created using on-line calculator on http://www.physics.csbsju.edu/stats/.

Results and Discussion

The importance of correct sampling

Sample collection during this study elucidated some of the problems specific for this network and highlighted the broader importance of correct sampling procedures. Fire hydrants were selected as sampling points to enable direct access to the distribution network and avoid potential household effects [39]. We opted for a low velocity water flow in combination with online monitoring to achieve comparable samples. In some cases, the water initially emerging from the fire hydrants were visibly turbid and/or discolored (data not shown). Turbid water is clearly unwanted and serves as a first visual confirmation of some form of system failure. In this regard, a recent study in the Netherlands has established an important link between suspended solids and microbial growth and biological instability [6]. Hence in some instances continuous low velocity flushing of up to 60 minutes was required before stable values for chemical and physical parameters as well as microbiological parameters were obtained (Figure 2; Table S1). The data in Figure 2 demonstrate clearly the need for a carefully planned sampling protocol when assessing full-scale systems. It should be noted that Figure 2 represents an example of some of the worst sampling points in the system. Data from other hydrants often showed less fluctuation during flushing (Figure S1). One potential problem during the sampling procedure is the re-suspension of sediments/particles and sloughing of biofilms from the pipes, causing artifacts in the measurements. In this respect, we specifically employed a low velocity (0.015–0.25 m s^{-1}) pre-sampling flushing procedure. The latter differs from extreme flushing applied for network cleaning, which is operated with high velocities of 1.5–1.8 m s^{-1} [40,41]. According to Antoun and co-workers [40] low-veocity flushing (below 0.3 m s^{-1}) does not cause any scouring actions. However, it should be considered that part of the samples, especially during the first minutes of the flushing, can cointain biofilm bacteria detatched in a result of pre-flushing [35].

The concept of detecting instability: a single point in the distribution network

In the introduction we proposed the straightforward hypothesis that biological parameters would show an increase between the point of treatment and a point during distribution in case of biological instability (Figure 1). Before the relation between different parameters and the impact on the entire network are discussed in detail below, a single sampling point is compared to its source water as an example to illustrate the concept (Figure 3A). The point was selected on the basis of (1) hydraulic data linking it with a specific WTP, (2) its medial distance from WTP (neither too close and nor too far from the WTP) and, (3) the fact that all microbiological parameters (FCM, ATP and HPC) as well as residual chlorine measurements were performed on this sample. For the purpose of clarity, the data was normalized to the values of the treated water and expressed as the relative change (the raw data and standard deviations for the data in Figure 3A are shown in Figure S2). Evidently the data from Figure 3A supports the basic hypothesis. The microbial parameters such as intact cell concentration, ATP and colony forming units all show a considerable increase in their values. Simultaneously, only 12% (0.06 mg L^{-1}) of the initial residual chlorine concentration (0.5 mg L^{-1}) was left in the water sample. The data suggests that the residual chlorine in the network was not sufficient to inhibit microbial growth, concurring with earlier report from Prévost and colleagues [42] showing increased HPC, total direct and direct viable bacteria counts in a distribution network coinciding with chlorine depletion. Other studies also showed the presence of viable bacteria in water with chlorine concentration lower than 0.1 mg L^{-1} [23] and that residual chlorine levels below 0.07 mg L^{-1} allows bacterial growth [12]. Data of residual chlorine concentrations in the drinking water network is summarized in Figure S3. Evidently a considerable fraction of samples (18%) had residual chlorine concentrations below 0.1 mg L^{-1}.

Staining of bacteria with fluorescent dyes was previously suggested as a way to distinguish between viable and damaged bacteria in real water samples [43,44], and the application of this approach has been successfully demonstrated in laboratory scale chlorination studies [24,45]. One focus point of the present study was to determine whether FCM combined with viability staining can be used for a fast and meaningful assessment of viable bacteria in chlorinated drinking water systems. The same samples from Figure 3A, stained with SYBR Green and propidium iodide (SGPI), are shown as density plots obtained with FCM (Figure 3B). The theory behind the staining method and the interpretation of such data are discussed in detail elsewhere [20,24,38,43,46]. In the treatment plant sample, where the water was recently exposed to chlorine, 98% of all cells were measured as membrane compromised, seen by absence of events inside the gated area of the plot (Figure 3B). In the distribution network (DN) sample, a high concentration of intact cells appeared (Figure 3B). Since these intact cells were clearly not present in the influent, the plausible conclusion is that the bacterial growth occurred during distribution.

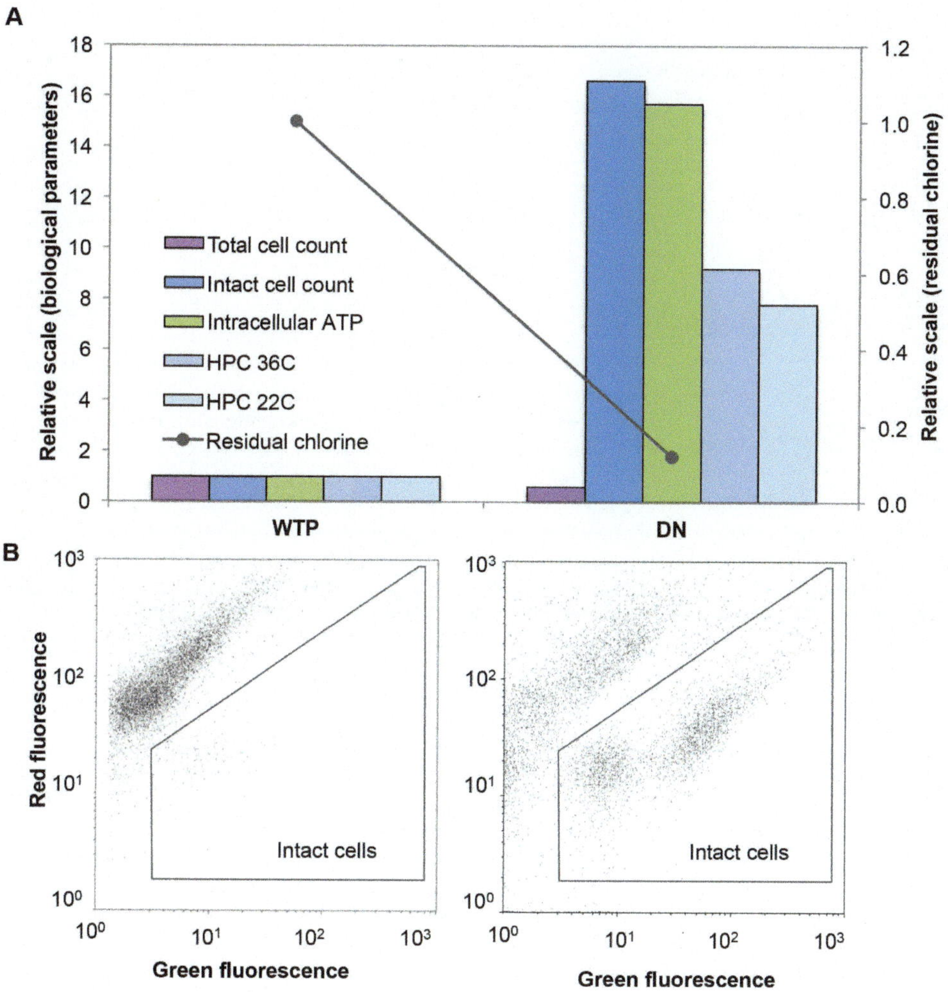

Figure 3. Changes in bacterial parameters between water treatment plant (WTP) and distribution network (DN) sampling points. (A) For comparison, all values at the WTP were set to 1, and values in the DN were expressed relative to their values at the WTP. The original raw data for these samples are shown in Figure S2. Data points are average values for duplicate FCM and ATP measurements and triplicate HPC measurements; (B) Flow cytometric density plots of samples stained with SYBR Green I and propidium iodide, showing the intact cell concentration at the plant and in the specific network point.

Detailed assessment of dynamic changes in a single point

High frequency monitoring of a single sampling point revealed temporal instability in the distribution network. We monitored the effluent of one treatment plant and one point in the network with 1-hour intervals during a day (ca. 21 h). The sampling was arranged in such a way that the network sampling started 15 hours after the treatment plant sampling, which corresponded with the estimated water residence time (WRT) for this location. Figure 4 displays the changes of intracellular ATP and intact cell concentrations in the network and the water treatment plant. Values for both parameters were low in the water samples from the treatment plant ($n = 19$): intracellular ATP varied from 0.0025 nM to 0.0096 nM (mean $= 0.0061 \pm 0.002$ nM) and the intact cell concentration amongst 19 samples varied from 7.5×10^3 to 6.3×10^4 cells mL^{-1} (mean $= 1.6 \times 10^4 \pm 1.2 \times 10^4$ cells mL^{-1} in average). In turn, the values from the distribution network point ($n = 23$) were significantly higher: intact cell concentrations ranged from 1.37×10^5 to 4.66×10^5 cells mL^{-1} (mean $= 2.5 \times 10^5 \pm 9.9 \times 10^4$ cells mL^{-1}), and the ATP concentrations from 0.021 to 0.063 nM (mean $= 0.038 \pm 0.012$ nM). Moreover, a distinct pattern was apparent in the distribution network data, with values peaking at about 05:00–07:00 and again at 12:00–13:00. During both these events, the intracellular ATP data followed a similar pattern as the intact cell concentration data, with a good overall correlation ($R^2 = 0.81$; $p < 0.005$). Although it is not evident exactly why the bacterial concentrations peaked at these specific time periods, a plausible explanation is a change in the flow velocity due to diurnal changes in water consumption by both industrial and domestic consumers. It was previously shown in laboratory scale experiments that increased flow velocity could lead to increased bacterial detachment from biofilms and a re-suspension of loose deposits, thus leading to an increase in suspended cell concentrations [32,33,47]. In addition, it is possible that lower water consumption overnight resulted in considerably reduced flow rates, and consequently a faster decay of chlorine and increased bacterial growth [42,48].

Detailed data sets of diurnal changes in the microbial quality of water mains, such as Figure 4, are particularly scarce in literature. Importantly, this clearly demonstrated temporal instability in the network for which the exact cause remains uncertain. Moreover, it

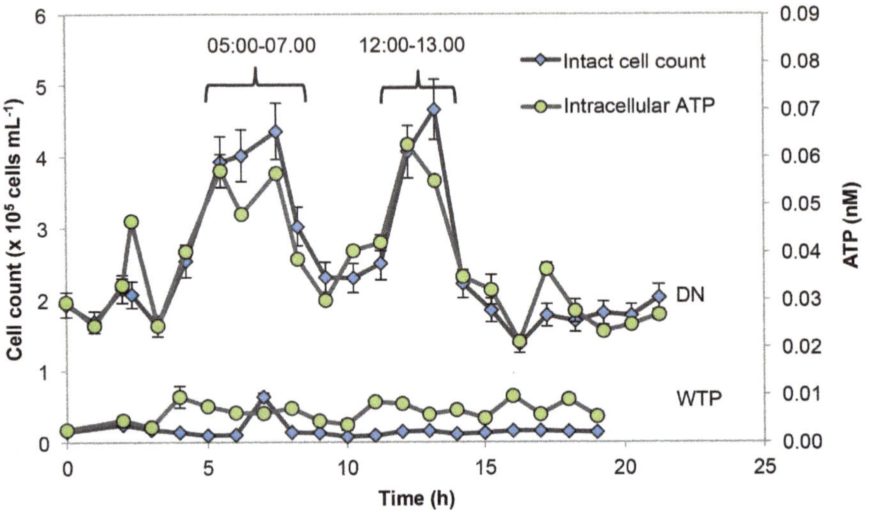

Figure 4. Diurnal changes in bacterial parameters of WTP and DN points. Intensive sampling of one WTP (n = 19) and one point in the DN (n = 23) during 21 hours reveals steady cell concentrations at the treatment plant but clear variations in the distribution network. Intracellular adenosine tri-phosphate (ATP) data points were derived from duplicate measurements of extracellular and total ATP concentrations. FCM intact cells (after SYBR Green I and propidium iodide staining) were single measurements, with a relative standard deviation of 9% calculated from all data in the present study.

shows that the absolute cell concentrations at any sampling point may be influenced by the time of sampling.

Instability data for the entire network

Full-scale distribution networks are complicated systems, not restricted to a single source or a straight distribution line [17]. The Riga distribution network is supplied with drinking water from several separate treatment plants (Table 1). One plant treats surface water from the Daugava River (WTP 1) and the others supply natural groundwater (WTP 3) and artificially recharged groundwater (WTP 2). Chlorination is applied as the final disinfection step at all plants, resulting in low concentrations of intact cells, intracellular ATP and cultivable bacteria in the effluents (Table 1). A large fraction of the active chorine is rapidly consumed due to relatively high levels of organic matter. Despite the fact that the purpose of chlorination and residual chlorine is to limit microbial growth during distribution, a considerable increase in the concentration of intact cells was detected throughout the distribution network. Figure 5A shows the range of intact cell concentrations arranged in ascending order. Treated water contained between 1.84×10^5–5.63×10^5 total cells mL^{-1} and between 9.7×10^3–2.13×10^4 intact cells mL^{-1} (hence 2–5% intact cells) depending on WTP. The data confirms effective final disinfection in all treatment plants. The total cell concentration values of the drinking water samples from the distribution network (n = 49) varied from 1.62×10^5 cells mL^{-1} to 1.07×10^6 cells mL^{-1} and the range of the intact cell concentration was from 5.28×10^3 cells mL^{-1} to 4.66×10^5 cells mL^{-1} (3–59% intact cells). Notably, 50% of all samples contained more than 1.06×10^5 intact cells mL^{-1} corresponding to an increase of at least one order of magnitude in those samples compared to effluent water, which clearly shows that bacterial growth in the distribution network was not an isolated occurrence. The observed increase in intact cell concentration is likely related to the presence of assimilable organic carbon (AOC) in the distributed water. While AOC was not measured in the present study, previous data for two of the treatment plants were high (in the range of 200 µg L^{-1}; Table 1), and nutrient availability in the water is generally regarded as a key factor that promotes microbial growth [29,49]. It cannot be excluded that some variability in the data resulted from bacteria detached from biofilms or re-suspended from sediments during the fire hydrant sampling procedure. However, the potential adverse impact of this was minimized by the low velocity sampling protocol (see above), while the systematic increase in cell concentrations in the network clearly suggests the occurrence of biological instability rather than sampling artifacts. In contrast to these findings, several studies analyzing drinking water distribution systems without any additional residual disinfectants showed no (or only minute) changes in bacterial parameters during distribution [17,19,20]. These distributions systems rely on nutrient limitation to achieve biological stability, and while intact cell concentrations are often relatively high (ca. 1×10^5 cells mL^{-1}) [17,20], changes during distribution tend to be negligible.

To examine the spatial distribution of the growth/instability in the network, the data was divided into four broad categories based on the extent of growth (Figure 5A). These were visualized on the sampling map (Figure 5B). The sampling points with the lowest intact cell concentration (less than 5×10^4 cells mL^{-1}) are marked with green bullets. Yellow and orange colored bullets indicate higher concentrations, while the points with the highest values (over 2×10^5 cells mL^{-1}) are shown as red bullets. As could be expected, the map shows that the points with the lowest cell concentrations are mostly concentrated in areas close to the water treatment plants. Low intact cell concentrations in those areas could be ascribed to (1) disinfection during treatment and (2) growth inhibition from sufficient residual chlorine. Also the flow rate in the outgoing pipes closest to the treatment plants is high, which likely prevents water stagnation, sedimentation and cell adhesion on the pipe surface, and, consequently, biofilm formation and further bacterial growth. A different situation is observed in the distant areas from the water treatment plants and particularly in the so-called mixing zones, where the water from three different water treatment plants potentially mix. The map displays different color points spread in these zones without any visible order. The prevalence of the samples with higher cell concentrations there compared to the areas close to WTPs also corroborates the

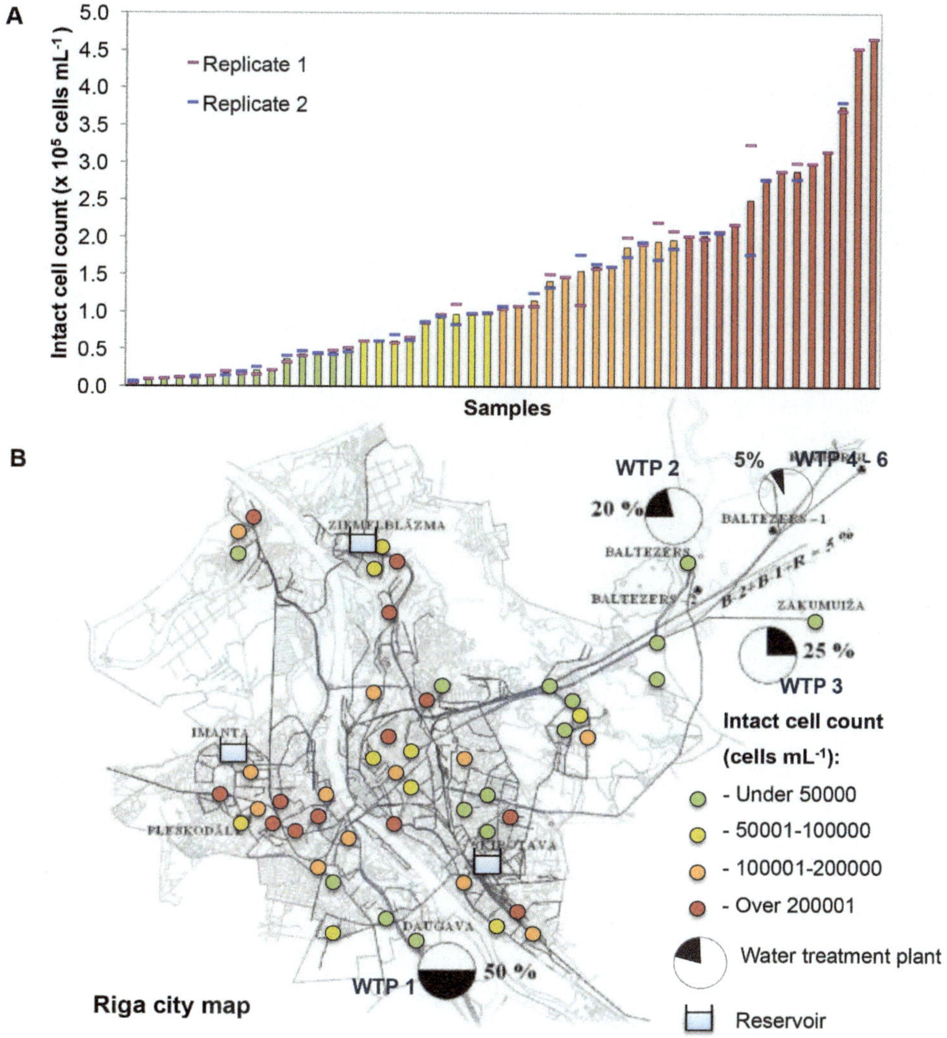

Figure 5. Intact cell concentrations of all samples measured from the distribution network (n = 49). (A) Intact cell concentrations arranged in ascending order and categorized into four main classes (colored bars) according to increasing concentrations. Data points are average values of duplicate measurements. Blue and purple stripes above and below data bars show the measured values. (B) Actual distribution of the classes of intact cells (colored circles) throughout the drinking water distribution network. WTP 1, WTP 2, WTP 3 represent location and productivity of the main water treatment plans supplying the city: WTP 1 operates using surface water, WTP 2 – artificially recharged ground water, WTP 3 – natural groundwater. WTP 4 – 6 indicates on other three pump stations with less significance for the city water supply.

argument that increasing distance and water residence time could lead to chlorine decay with concomitant oxidation of dissolved organic matter; both these events would favor bacterial growth. Moreover, mixing zones are potential hot-spots for bacterial growth, as one water might well contain the nutrients that are growth limiting in the other.

The uneven spatial distribution of the samples with different intact cell concentrations is noteworthy, highlighted for example by the three points in upper-left corner of the map. Based on the long distance from the WTPs, high intact cell concentrations were expected, but the samples taken from the hydrants located in this small area rather show variability (respectively 1.82×10^4, 1.87×10^5, 2.51×10^5 intact cells mL^{-1}). Such different intact cell concentrations could be due to several reasons: the time the samples were taken, which is linked to water consumption and the potential impact of which is shown in Figure 4, the condition of the pipes in this specific area (unknown), the way water flows from the treatment plant, and/or the relative proximity of these sample points to one of the reservoirs (Figure 5B), etc. Other authors showed a decrease in AOC [13] and ATP [19] in the some distal points of the distribution networks. Decrease of AOC concentration was explained by its consumption by bacteria within the network. These authors argued that an insufficient amount of nutrients led to starvation and a decrease in bacterial parameters at the end of the pipelines. However, it is an unlikely reason in the present study, because this phenomenon seems more occasional than systematic.

The combined data demonstrates clearly biological instability throughout the distribution network. However, despite the relative simplicity of the concept (Figure 1; Figure 3A), a complex interplay of chemical, physical and biological parameters and hydraulic conditions should be taken into account for characterization of each particular case of instability.

Comparison of different microbiological parameters

FCM and ATP data showed clear correlations, but these data did not correlate well with conventional HPC data. A total amount of 49 different samples was measured in duplicate with ATP (total and extracellular) and FCM (total and intact cell concentration) analyses, while 38 of those samples were further analyzed with HPC. The significant linear correlation ($R^2 = 0.77$; n = 49) between intracellular ATP and FCM intact cell concentration is shown in Figure 6A. This corroborates previous studies that showed good results comparing total ATP with total cell concentration [19,50] and intracellular ATP with intact cell count as well [20,25]. The strong correlation is encouraging, since FCM and ATP analysis are independent viability parameters – integrity of the cell membrane (FCM) and cellular energy (ATP). A correlation between these parameters during disinfection is not necessarily a given fact. The membrane integrity based PI staining method implies that PI positive cells are damaged and thus considered as inactive, yet extreme examples where living cells became permeable for propidium iodide have been described [46]. In turn, Nocker and co-workers [51] showed that after UV-C exposure cells became inactivated, while their membranes remained essentially intact. Discrepancies between intracellular ATP and intact cell concentration can also result from cell morphology, bacterial species and physiological state, that was discussed in detail previously [25]. The results provided by FCM provide information on single cell level, whereas during ATP analyses the values are evaluated per volume. Hence, intracellular ATP-per-cell was calculated for characterization of biomass activity. In the present study intracellular ATP-per-cell ranges from zero (no cell-bound ATP observed) to 5.92×10^{-10} nM cell^{-1} ($= 3 \times 10^{-17}$ g cell^{-1}) with the average value of 1.68×10^{-10} nM cell^{-1} ($= 8.52 \times 10^{-18}$ g cell^{-1}) (stdev $= 9.58 \times 10^{-11}$ nM cell^{-1}, n = 49). The result is in the same range as ATP-per-cell values obtained from various water sources, which were analyzed with the same methods [20,25]. This suggests that bacterial activity (ATP values) in the intact cells was not affected by any remaining chlorine residuals, and that membrane damage (SGPI values) was in this case reflective of viability in the sample. The good correlation between these two independent parameters is an optimistic prospect for applying these methods for chlorinated water analyses in future studies.

The conventional HPC results were compared with the FCM intact cell concentration values. A weak correlation ($R^2 = 0.18$, n = 38) was observed between HPC (at 22°C) and intact cell concentrations (Figure 6B), similar to reports in previous studies [38,50]. It could be explained by the often described phenomenon, that less than 1% of drinking water bacteria are cultivable on conventional agar plates [25,50,52]. In addition, Mezule and co-workers [53] demonstrated evidence of the presence of so called viable-but-not-cultivable (VNBC) bacterial state, in both drinking water and biofilms for the network investigated here, thus indicating further limitations in the HPC method. Since intracellular ATP showed a good correlation with intact cell concentration, but intact cell count correlated weakly with HPC, it was expected that intracellular ATP and HPC would not correlate well (Figure 6C; $R^2 = 0.11$, n = 38). Various studies were performed to compare ATP and HPC parameters from water samples, but good correlations were never observed e.g., $R^2 = 0.20$ [19], $R^2 = 0.36$ [22] and $R^2 = 0.31$ [50]. Our results combined with those from previous studies cast further doubts on the value of using the HPC method for general microbiological drinking water quality control. In our opinion, the clear correlation between two methodologically independent viability parameters (intracellular ATP and FCM intact cell counts), and the absence of any correlations with

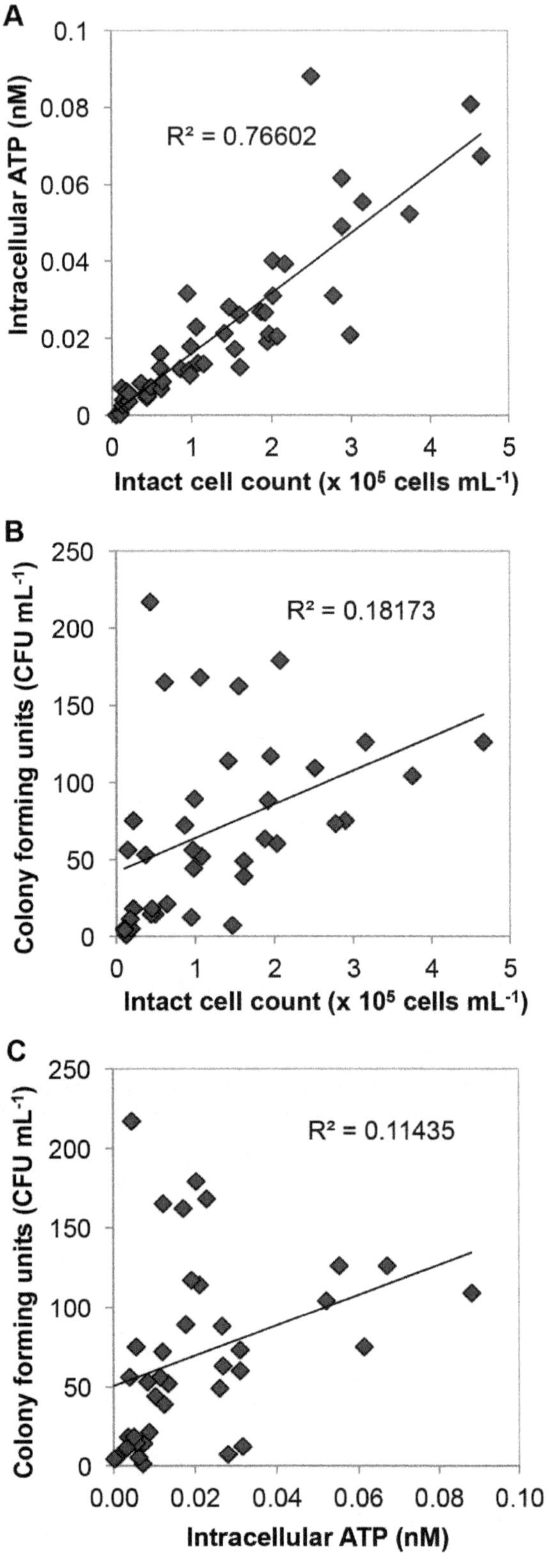

Figure 6. Comparison of the various microbiological parameters. Clear correlations were observed between intact cells and intracellular ATP (n = 49) (A), but no obvious correlations between these two parameters and heterotrophic plate counts at 22°C (n = 38) (B) (C).

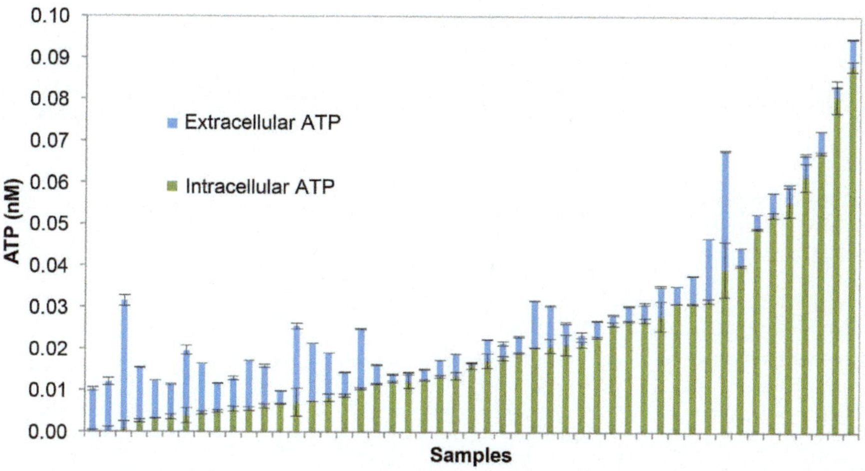

Figure 7. Distribution of intracellular and extracellular ATP in the water samples. In general, higher concentrations and relative percentages of extracellular ATP were measured in samples that exhibited lower intracellular ATP concentrations (n = 49).

two different HPC methods, renders the former methods more meaningful for assessing and understanding biological instability, particularly in chlorinated environments.

Importance of measuring extracellular ATP

Arguments for and against the concept and importance of measuring extracellular ATP have been made [4,19,20,25,54,55]. To understand this better, we arranged our data according to increasing intracellular ATP concentrations, after which the measured extracellular ATP values were added to each corresponding sample (Figure 7). It is evident that extracellular ATP constitutes a considerable fraction of the total ATP amount in some samples – varying from 3% up to 100% – with an average contribution of 36% (n = 49). Moreover, 33% of the samples contain more that 50% of extracellular ATP. This data supports other studies, where analyses showed high extracellular ATP in drinking water samples from the distribution networks [20,25]. Interestingly, the highest extracellular ATP ratio is mostly observed in the samples with relatively low intracellular ATP, in this case samples with close proximity to the treatment plant. In the case of chlorinated water, this could potentially be explained by the oxidative effect of chlorine on bacterial cells. Previous studies have shown extensive damage to bacterial membranes during chlorination [24,45], after which a release of extracellular ATP from the damaged bacteria can occur. This membrane damage was also clearly detected in the present study (e.g., Figure 3B). Although, there is lack of detailed data considering the release of extracellular ATP in water samples affected by chlorination, strong evidence of ATP release during oxidation was presented in previous studies [4,20]. Both these works showed a significant decrease in cell concentrations and intracellular ATP after ozonation, whereas extracellular ATP comprised 83–100% of the total ATP. Moreover, Figure 7 shows that samples with increased intracellular ATP concentrations, which we linked to bacterial growth during distribution, often had considerably less extracellular ATP in relation to total ATP. This could be due to the fact that extracellular ATP can be biodegraded by bacteria or extracellular enzymes in the network [54,56–58]. However, it cannot be excluded that a decrease in extracellular ATP during distribution occurs due to oxidation by residual chlorine present in the network.

Conclusions

- An investigation of a full-scale chlorinated drinking water distribution network with various microbiological methods clearly demonstrated both spatial and temporal biological instability in the network.
- Fluorescent staining with SGPI in combination with ATP measurements provided reliable and descriptive information about bacterial density and viability in chlorinated drinking water samples.
- A good correlation was observed between intracellular ATP and intact cell counts ($R^2 = 0.77$), whereas HPC showed poor correlations with both parameters ($R^2 = 0.18$ with intact cell concentration and $R^2 = 0.11$ with intracellular ATP).
- Extracellular ATP constituted on average 36% of total ATP in the present study, which confirms the necessity of extracellular ATP subtraction from total ATP measurements during chlorinated drinking water analyses.
- Overall the results raise questions with respect to the offset between increased biological safety gained from disinfection opposed to increased risk from instability (uncontrolled bacterial growth). While an improvement of the chlorination procedure could be a solution, the data suggests looking beyond only disinfection for achieving biological stability of drinking water.

Supporting Information

Figure S1 Additional examples of hydrant flushing. Changes in intact cell concentration and intracellular ATP during flushing in 6 newly-opened fire hydrants. Intact cell concentration values are shown as solid lines with blue markers, whereas intracellular ATP results displayed as single green bullets.

Figure S2 Actual data for Figure 3A. Changes in various bacterial parameters between one water treatment plant and a randomly selected point in the distribution network (actual values for Figure 3A).

Figure S3 Residual chlorine concentration in the distri-

bution network. 50% of residual chlorine concentration in the network was between 0.12 (first quartile) and 0.23 (third quartile) mg mL^{-1}, with a mean value of 0.17 mg mL^{-1} (n = 27). The whiskers indicate on minimum and maximum values, whereas bullets show outliers of the population.

Table S1 Physical and chemical parameters of water measured on-line during low velocity flushing of newly-opened fire hydrant. Some measurements were omitted during the first 20 minutes of flushing due to the high fluctuation in measuring tools readings.

Acknowledgments

The authors thank Stefan Kötzsch for critical input, Arturs Briedis, Edgars Grundbergs and Kaspars Neilands for assistance in the sampling campaigns, and sampling/information support from Rigas Udens Ltd.

Author Contributions

Conceived and designed the experiments: FH MV JR TJ. Performed the experiments: FH MV JR. Analyzed the data: FH MV JR TJ AN. Contributed reagents/materials/analysis tools: FH TJ. Wrote the paper: FH MV JR TJ AN.

References

1. Bartram J, Cotruvo J, Exner M, Fricker C, Glasmacher A (2003) Heterotrophic plate count and drinking-water safety: The significance of HPCs for water quality and human health. World Health Organization. 244 p.
2. Boe-Hansen R, Albrechtsen H-J, Arvin E, Jørgensen C (2002) Bulk water phase and biofilm growth in drinking water at low nutrient conditions. Water Res 36: 4477–4486. doi:10.1016/S0043-1354(02)00191-4.
3. Hammes F, Berger C, Köster O, Egli T (2010) Assessing biological stability of drinking water without disinfectant residuals in a full-scale water supply system. J Water Supply Res Technol 59: 31. doi:10.2166/aqua.2010.052.
4. Hammes F, Berney M, Wang Y, Vital M, Köster O, et al. (2008) Flow-cytometric total bacterial cell counts as a descriptive microbiological parameter for drinking water treatment processes. Water Res 42: 269–277. doi:10.1016/j.watres.2007.07.009.
5. Juhna T, Birzniece D, Larsson S, Zulenkovs D, Sharipo A, et al. (2007) Detection of Escherichia coli in biofilms from pipe samples and coupons in drinking water distribution networks. Appl Environ Microbiol 73: 7456–7464. doi:10.1128/AEM.00845-07.
6. Liu G, Lut MC, Verberk JQJC, Van Dijk JC (2013) A comparison of additional treatment processes to limit particle accumulation and microbial growth during drinking water distribution. Water Res 47: 2719–2728. doi:10.1016/j.watres.2013.02.035.
7. Van der Kooij D (2000) Biological stability: A multidimensional quality aspect of treated water. In: Belkin S, editor. Environmental Challenges. Springer Netherlands. pp. 25–34.
8. Vital M, Stucki D, Egli T, Hammes F (2010) Evaluating the growth potential of pathogenic bacteria in water. Appl Environ Microbiol 76: 6477–6484. doi:10.1128/AEM.00794-10.
9. Vital M, Füchslin HP, Hammes F, Egli T (2007) Growth of Vibrio cholerae O1 Ogawa Eltor in freshwater. Microbiology 153: 1993–2001. doi:10.1099/mic.0.2006/005173-0.
10. Vital M, Hammes F, Egli T (2008) Escherichia coli O157 can grow in natural freshwater at low carbon concentrations. Environ Microbiol 10: 2387–2396. doi:10.1111/j.1462-2920.2008.01664.x.
11. LeChevallier MW, Schulz W, Lee RG (1991) Bacterial nutrients in drinking water. Appl Environ Microbiol 57: 857–862.
12. Niquette P, Servais P, Savoir R (2001) Bacterial dynamics in the drinking water distribution system of Brussels. Water Res 35: 675–682.
13. Liu W, Wu H, Wang Z, Ong SL, Hu JY, et al. (2002) Investigation of assimilable organic carbon (AOC) and bacterial regrowth in drinking water distribution system. Water Res 36: 891–898.
14. Eichler S, Christen R, Höltje C, Westphal P, Bötel J, et al. (2006) Composition and dynamics of bacterial communities of a drinking water supply system as assessed by RNA- and DNA-based 16S rRNA gene fingerprinting. Appl Environ Microbiol 72: 1858–1872. doi:10.1128/AEM.72.3.1858-1872.2006.
15. Hong P-Y, Hwang C, Ling F, Andersen GL, LeChevallier MW, et al. (2010) Pyrosequencing analysis of bacterial biofilm communities in water meters of a drinking water distribution system. Appl Environ Microbiol 76: 5631–5635. doi:10.1128/AEM.00281-10.
16. Pinto AJ, Xi C, Raskin L (2012) Bacterial community structure in the drinking water microbiome Is governed by filtration processes. Environ Sci Technol 46: 8851–8859. doi:10.1021/es302042t.
17. Lautenschlager K, Hwang C, Liu W-T, Boon N, Köster O, et al. (2013) A microbiology-based multi-parametric approach towards assessing biological stability in drinking water distribution networks. Water Res 47: 3015–3025. doi:10.1016/j.watres.2013.03.002.
18. Rittmann BE, Snoeyink VL (1984) Achieving biologically stable drinking water. J - Am Water Works Assoc 76: 106–114.
19. Van der Wielen PWJJ, van der Kooij D (2010) Effect of water composition, distance and season on the adenosine triphosphate concentration in unchlorinated drinking water in the Netherlands. Water Res 44: 4860–4867. doi:10.1016/j.watres.2010.07.016.
20. Vital M, Dignum M, Magic-Knezev A, Ross P, Rietveld L, et al. (2012) Flow cytometry and adenosine tri-phosphate analysis: Alternative possibilities to evaluate major bacteriological changes in drinking water treatment and distribution systems. Water Res 46: 4665–4676. doi:10.1016/j.watres.2012.06.010.
21. LeChevallier MW, Au K-K (2004) Water treatment and pathogen control: Process efficiency in achieving safe drinking-water. IWA Publishing. 136 p.
22. Delahaye E, Welté B, Levi Y, Leblon G, Montiel A (2003) An ATP-based method for monitoring the microbiological drinking water quality in a distribution network. Water Res 37: 3689–3696. doi:10.1016/S0043-1354(03)00288-4.
23. Francisque A, Rodriguez MJ, Miranda-Moreno LF, Sadiq R, Proulx F (2009) Modeling of heterotrophic bacteria counts in a water distribution system. Water Res 43: 1075–1087. doi:10.1016/j.watres.2008.11.030.
24. Ramseier MK, von Gunten U, Freihofer P, Hammes F (2011) Kinetics of membrane damage to high (HNA) and low (LNA) nucleic acid bacterial clusters in drinking water by ozone, chlorine, chlorine dioxide, monochloramine, ferrate(VI), and permanganate. Water Res 45: 1490–1500. doi:10.1016/j.watres.2010.11.016.
25. Hammes F, Goldschmidt F, Vital M, Wang Y, Egli T (2010) Measurement and interpretation of microbial adenosine tri-phosphate (ATP) in aquatic environments. Water Res 44: 3915–3923. doi:10.1016/j.watres.2010.04.015.
26. LeChevallier MW, Welch NJ, Smith DB (1996) Full-scale studies of factors related to coliform regrowth in drinking water. Appl Environ Microbiol 62: 2201–2211.
27. Polanska M, Huysman K, van Keer C (2005) Investigation of assimilable organic carbon (AOC) in flemish drinking water. Water Res 39: 2259–2266. doi:10.1016/j.watres.2005.04.015.
28. Ramseier MK, Peter A, Traber J, von Gunten U (2011) Formation of assimilable organic carbon during oxidation of natural waters with ozone, chlorine dioxide, chlorine, permanganate, and ferrate. Water Res 45: 2002–2010. doi:10.1016/j.watres.2010.12.002.
29. Van der Kooij D (1990) Assimilable organic carbon (AOC) in drinking water. In: McFeters GA, editor. Drinking Water Microbiology. New York, NY: Springer New York. pp. 57–87.
30. Weinrich LA, Jjemba PK, Giraldo E, LeChevallier MW (2010) Implications of organic carbon in the deterioration of water quality in reclaimed water distribution systems. Water Res 44: 5367–5375. doi:10.1016/j.watres.2010.06.035.
31. Jjemba P (2010) Guidance document on the microbiological quality and Biostability of reclaimed water following storage and distribution. WateReuse Research Foundation.
32. Lehtola MJ, Laxander M, Miettinen IT, Hirvonen A, Vartiainen T, et al. (2006) The effects of changing water flow velocity on the formation of biofilms and water quality in pilot distribution system consisting of copper or polyethylene pipes. Water Res 40: 2151–2160. doi:10.1016/j.watres.2006.04.010.
33. Manuel CM, Nunes OC, Melo LF (2007) Dynamics of drinking water biofilm in flow/non-flow conditions. Water Res 41: 551–562. doi:10.1016/j.watres.2006.11.007.
34. Bucheli-Witschel M, Kötzsch S, Darr S, Widler R, Egli T (2012) A new method to assess the influence of migration from polymeric materials on the biostability of drinking water. Water Res 46: 4246–4260. doi:10.1016/j.watres.2012.05.008.
35. Douterelo I, Husband S, Boxall JB (2014) The bacteriological composition of biomass recovered by flushing an operational drinking water distribution system. Water Res 54: 100–114. doi:10.1016/j.watres.2014.01.049.
36. Rossman LA (2000) EPANET 2 Users manual. National Risk Management Research Laboratory. U.S. Environmental Protection Agency, Cincinatti, Ohio.
37. Rubulis J, Dejus S, Meksa R (2011) Online measurement usage for predicting water age from tracer tests to validate a hydraulic model American Society of Civil Engineers. pp. 1488–1497. doi:10.1061/41203(425)133.
38. Berney M, Vital M, Hülshoff I, Weilenmann H-U, Egli T, et al. (2008) Rapid, cultivation-independent assessment of microbial viability in drinking water. Water Res 42: 4010–4018. doi:10.1016/j.watres.2008.07.017.
39. Lautenschlager K, Boon N, Wang Y, Egli T, Hammes F (2010) Overnight stagnation of drinking water in household taps induces microbial growth and changes in community composition. Water Res 44: 4868–4877. doi:10.1016/j.watres.2010.07.032.

40. Antoun EN, Dyksen JE, Hiltebrand DJ (1999) Unidirectional flushing: A powerful tool. J - Am Water Works Assoc 91: 62–71.
41. Friedman M, Kirmeyer GJ, Antoun E (2002) Developing and implementing a distribution system flushing program. J - Am Water Works Assoc 94: 48–56.
42. Prévost M, Rompré A, Coallier J, Servais P, Laurent P, et al. (1998) Suspended bacterial biomass and activity in full-scale drinking water distribution systems: Impact of water treatment. Water Res 32: 1393–1406. doi:10.1016/S0043-1354(97)00388-6.
43. Berney M, Hammes F, Bosshard F, Weilenmann H-U, Egli T (2007) Assessment and interpretation of bacterial viability by using the LIVE/DEAD BacLight Kit in combination with flow cytometry. Appl Environ Microbiol 73: 3283–3290. doi:10.1128/AEM.02750-06.
44. Grégori G, Citterio S, Ghiani A, Labra M, Sgorbati S, et al. (2001) Resolution of viable and membrane-compromised bacteria in freshwater and marine waters based on analytical flow cytometry and nucleic acid double staining. Appl Environ Microbiol 67: 4662–4670. doi:10.1128/AEM.67.10.4662-4670.2001.
45. Lisle JT, Pyle BH, McFeters GA (1999) The use of multiple indices of physiological activity to access viability in chlorine disinfected Escherichia coli O157:H7. Lett Appl Microbiol 29: 42–47. doi:10.1046/j.1365-2672.1999.00572.x.
46. Shi L, Günther S, Hübschmann T, Wick LY, Harms H, et al. (2007) Limits of propidium iodide as a cell viability indicator for environmental bacteria. Cytometry A 71A: 592–598. doi:10.1002/cyto.a.20402.
47. Tsai Y-P (2005) Impact of flow velocity on the dynamic behaviour of biofilm bacteria. Biofouling 21: 267–277. doi:10.1080/08927010500398633.
48. Srinivasan S, Harrington GW, Xagoraraki I, Goel R (2008) Factors affecting bulk to total bacteria ratio in drinking water distribution systems. Water Res 42: 3393–3404. doi:10.1016/j.watres.2008.04.025.
49. Van der Kooij D (1992) Assimilable organic carbon as an indicator of bacterial regrowth. J - Am Water Works Assoc 84: 57–65.
50. Siebel E, Wang Y, Egli T, Hammes F (2008) Correlations between total cell concentration, total adenosine tri-phosphate concentration and heterotrophic plate counts during microbial monitoring of drinking water. Drink Water Eng Sci Discuss 1: 71–86. doi:10.5194/dwesd-1-71-2008.
51. Nocker A, Sossa KE, Camper AK (2007) Molecular monitoring of disinfection efficacy using propidium monoazide in combination with quantitative PCR. J Microbiol Methods 70: 252–260. doi:10.1016/j.mimet.2007.04.014.
52. Van der Kooij D, Vrouwenvelder JS, Veenendaal HR (2003) Elucidation and control of biofilm formation processes in water treatment and distribution using the Unified Biofilm Approach. Water Sci Technol J Int Assoc Water Pollut Res 47: 83–90.
53. Mezule L, Larsson S, Juhna T (2013) Application of DVC-FISH method in tracking *Escherichia coli* in drinking water distribution networks. Drink Water Eng Sci 6: 25–31. doi:10.5194/dwes-6-25-2013.
54. Cowan DA, Casanueva A (2007) Stability of ATP in Antarctic mineral soils. Polar Biol 30: 1599–1603. doi:10.1007/s00300-007-0324-9.
55. Venkateswaran K, Hattori N, La Duc MT, Kern R (2003) ATP as a biomarker of viable microorganisms in clean-room facilities. J Microbiol Methods 52: 367–377. doi:10.1016/S0167-7012(02)00192-6.
56. Azam F, Hodson RE (1977) Dissolved ATP in the sea and its utilisation by marine bacteria. Nature 267: 696–698. doi:10.1038/267696a0.
57. Mempin R, Tran H, Chen C, Gong H, Ho KK, et al. (2013) Release of extracellular ATP by bacteria during growth. BMC Microbiol 13: 301. doi:10.1186/1471-2180-13-301.
58. Riemann B (1979) The occurrence and ecological importance of dissolved ATP in fresh water. Freshw Biol 9: 481–490. doi:10.1111/j.1365-2427.1979.tb01532.x.

17

Effects of Favorable Alleles for Water-Soluble Carbohydrates at Grain Filling on Grain Weight under Drought and Heat Stresses in Wheat

Bin Zhang[1]☯, Weiyu Li[1,2]☯, Xiaoping Chang[1], Runzhi Li[2], Ruilian Jing[1]*

1 National Key Facility for Crop Gene Resources and Genetic Improvement/Institute of Crop Science, Chinese Academy of Agricultural Sciences, Beijing, China, **2** College of Agronomy, Shanxi Agricultural University, Taigu, China

Abstract

Drought, heat and other abiotic stresses during grain filling can result in reductions in grain weight. Conserved water-soluble carbohydrates (WSC) at early grain filling play an important role in partial compensation of reduced carbon supply. A diverse population of 262 historical winter wheat accessions was used in the present study. There were significant correlations between 1000-grain weight (TGW) and four types of WSC, viz. (1) total WSC at the mid-grain filling stage (14 days after flowering) produced by leaves and non-leaf organs; (2) WSC contributed by current leaf assimilation during the mid-grain filling; (3) WSC in non-leaf organs at the mid-grain filling, excluding the current leaf assimilation; and (4) WSC used for respiration and remobilization during the mid-grain filling. Association and favorable allele analyses of 209 genome-wide SSR markers and the four types of WSC were conducted using a mixed linear model. Seven novel favorable WSC alleles exhibited positive individual contributions to TGW, which were verified under 16 environments. Dosage effects of pyramided favorable WSC alleles and significantly linear correlations between the number of favorable WSC alleles and TGW were observed. Our results suggested that pyramiding more favorable WSC alleles was effective for improving both WSC and grain weight in future wheat breeding programs.

Editor: Jauhar Ali, International Rice Research Institute, Philippines

Funding: This study was supported by the National High-tech R&D Program (2011AA100501), the National Basic Research Program (2010CB951501), and the National Natural Science Foundation of China (31271720). The funders had no role in study design, data collection and analysis, decision to publish, or preparation of the manuscript.

Competing Interests: The authors have declared that no competing interests exist.

* Email: jingruilian@caas.cn

☯ These authors contributed equally to this work.

Introduction

Wheat (*Triticum aestivum* L.) is one of the most important crops in the world, feeding nearly half the world population [1]. High grain yield is the most important breeding objective in wheat improvement. Drought, heat and other abiotic stresses greatly affect growth and productivity of wheat, especially during grain filling stage. Grain filling in wheat depends on two major sources of carbon: current photosynthate in leaves and non-leaf organs; and carbohydrates stored in the stem and leaf sheath from stem elongation to the early phase of grain filling [2]. The latter can be important in buffering grain yields against unfavorable conditions for photosynthesis during the grain-filling period [3,4].

Water-soluble carbohydrates (WSC) accumulation and utilization depend on growing conditions and genotypes, and there may be differences between internodes [2,5,6]. Among three segments of the main stem (peduncle, penultimate internode and the remainder segments), the remainder segments are the major storage sites and the major source for WSC mobilization during the grain filling period [7]. In general, WSC accumulate until 10–20 days after anthesis, and the reserved WSC can reach more than 40% of total stem dry weight in wheat [8]. The contribution of WSC to final yield and kernel size is 10%–20% of total grain weight under normal condition [9]. Drought stress during grain filling, often involving not only water stress but also heat, inhibits current assimilation and damages photosynthetic organs, especially leaves. When photosynthetic activity is suppressed, the reserved WSC play a more important role in partial compensation of the reduced carbon supply. In addition, drought induced reserved WSC mobilization with higher efficiency, potentially contributing up to 70% of grain dry matter [8,10].

Based on a large group of genotypes with various WSC contents, the ranking of wheat lines for WSC is consistent across diverse environments. Stem WSC content shows high broad-sense heritability ($h^2 = 0.9$). WSC are inversely related to stem number but genotypic ranking persists when compared at similar stem densities [11,12]. In past years, selection for high WSC in stems occurred during development of drought-tolerant wheat varieties in the UK and Australia [13,14]. It has been suggested that the release of representative UK wheat cultivars from 1972 to 1995 was associated with increasing stem WSC content [15]. Therefore, high stem WSC content was suggested as a useful trait for improving grain weight in wheat breeding programs [11,14,15].

Variation in stem WSC among wheat genotypes is an important genetic factor involving grain weight and yield under drought stress conditions [16]. Thus, knowledge of stem WSC is essential for understanding yield-limiting factors and for improving yield potential in wheat. QTL associated with stem WSC have been reported in perennial ryegrass [17], rice [18], maize [19], barley [20], and wheat [21–23]. In wheat, QTL for WSC were mapped on chromosomes 1A, 2D, 4A, 4B, 5D, 6B, 7B and 7D. QTL for drought tolerance also appeared in homologous regions on the group 7 chromosomes [22]. Yang et al. [23] discovered eight, one and two additive QTL for WSC at flowering, grain-filling and maturity, respectively. However, WSC content is a complex quantitative trait controlled by polygenes, and the small effects of many independent QTL limit their direct use for marker-assisted selection in breeding programs [8,24].

Photosynthesis is the all-important metabolic process determining grain yield in wheat. When water deficit occurs during grain filling stage, photosynthetic rates significantly decrease in leaf blades and non-leaf green organs, such as leaf sheath, glume and awn [25–27]. However, non-leaf green organs are relatively more stable than leaf blades [27]. In this study, four types of WSC (Total, Leaf, Non-leaf and Remo) under drought stress (DS) and well-watered (WW) conditions in 262 winter wheat accessions were mainly used, viz. Total, total WSC at the mid-grain filling stage (14 days after flowering) produced by leaves and non-leaf organs; Leaf, WSC contributed by current leaf assimilation during the mid-grain filling; Non-leaf, WSC in non-leaf organs at the mid-grain filling, excluding the current leaf assimilation; Remo, WSC used for respiration and remobilization during the mid-grain filling. The objectives were to (1) detect the relative contributions of leaf and non-leaf organs during grain filling stage to WSC and 1000-grain weight (TGW) under two water regimes; (2) explore genetic resources with high WSC by association analysis; (3) verify the stable favorable WSC alleles with significant effect on TGW under 16 environments (5 drought stress conditions, 3 drought and heat stress (HS) conditions, 5 well-watered conditions, 3 well-watered and heat stress conditions). We observed a significantly positive contribution of WSC to TGW, and the results of favorable WSC allele analysis will be helpful for wheat breeders in selecting genotypes with higher TGW.

Materials and Methods

Ethics statement

Two locations, Changping (116°13′E; 40°13′N) and Shunyi (116°56′E; 40°23′N) in Beijing, are the experiment stations of the Institute of Crop Science, Chinese Academy of Agricultural Sciences. We have obtained the relevant permission for our field studies for growing our plant materials in the field from the corresponding institutions. There was no specific permissions required for these locations/activities. Our field studies did not involve endangered or protected species.

Plant materials and field experiments

A population of 262 common wheat accessions were used in our research [28,29]; 254 were from China, 3 from USA, 2 from Australia, 2 from Italy, and 1 from Romania (Table S1). The Chinese accessions were mainly from the Northern Winter Wheat Zone, and Yellow and Huai River Valleys Facultative Wheat Zone, including landraces and modern cultivars released from the 1940s to 2000s. All were planted at the beginning of October and harvested in the following mid-June. The experimental unit was a 2 m 4-row plot with 0.3 m between rows. There were 40 plants per row. The field was managed under separate rain-fed (drought stress, DS) and well-watered (WW) conditions. The WW plots were watered with 750 m^3/ha (75 mm) at the pre-overwintering, booting, flowering and grain filling stages, respectively.

The materials were grown in Changping, Beijing, the experiment station of the Institute of Crop Science in 2010, for collecting data on WSC at the mid-grain filling (14 days after flowering) and TGW at maturity. The treatments involved cutting spikes at flowering stage, removing leaves at flowering stage and a normal control. The rainfall from sowing to harvest was 131 mm. The field was managed under DS and WW conditions.

The population was planted in Changping and Shunyi, Beijing, the experiment stations of the Institute of Crop Science, over 3 years for measuring TGW at maturity. The planting years were 2009 and 2010 at both Changping and Shunyi, and 2011 in Shunyi. The rainfalls in the growing seasons were 192 mm, 131 mm and 180 mm, respectively. The field was also managed under DS and WW conditions. A greenhouse experiment was conducted at Shunyi; at heading polythene covers were placed over selected heads to increase the temperature and thereby simulate heat stress (HS). Thus there were four treatments (DS, WW, DS+HS and WW+HS), with E1 to E16 indicating the environments of Changping in 2009 under DS and WW; Shunyi in 2009 under DS, DS+HS, WW and WW+HS; Changping in 2010 under DS and WW; Shunyi in 2010 under DS, DS+HS, WW and WW+HS; Shunyi in 2011 under DS, DS+HS, WW and WW+HS, respectively.

Phenotyping of WSC and TGW

The methods of collecting data on WSC were reported earlier [30]. For each genotype, five main stems with the same heading date were selected as samples. The main stem was cut from the soil surface at the mid-grain filling (14 days after flowering). Leaf blades were removed, and stem samples were cut into three parts, the upmost internode (peduncle, Ped), the lower internode (the remainder segments of stem except for peduncle, Low) and the spike. The fresh samples were dehydrated until a constant dry weight. The WSC of the three sections, i.e. peduncle, the lower internode and whole stem (Ste), were determined by different near-infrared reflectance spectroscopy regression models, which were developed for quantitative determination of WSC using samples of 150 doubled haploid lines (Hanxuan 10×Lumai 14) [30]. Briefly, as the first step, partial least square regression models for predicting WSC in the target parts of wheat were developed using selected wavelength regions, spectroscopy pretreatments and latent variables included in each model. The amounts of WSC (mg WSC/g dry weight, mg/g dw) in each modeling sample of 150 doubled haploid lines were also measured by chemical assay (anthrone colorimetric assay), and used for cross validation. WSC were extracted according to the modified procedure described by Wardlaw and Willenbrink [2,30]; the amounts of WSC were measured as fructose equivalents using the anthrone colorimetric assay at 620 nm by 722S spectrophotometer [31]. This showed that the near-infrared reflectance spectroscopy regression models were highly accurate in determining the true values of WSC measured by chemical assay in the wheat organs tested (coefficient of determination $R^2>0.992$ and root mean square error of prediction RMSEP<0.228). In addition, 40 samples per model (i.e. not included in the modeling samples) were used to assess the models. The results confirmed the high quality of the models in evaluating WSC.

We obtained four types of WSC (Total, Leaf, Non-leaf, and Remo), viz. Total, the total WSC at the mid-grain filling produced by leaves and non-leaf organs which was obtained from the treatment of cutting spikes ($WSC_{cutting\ spikes}$); Leaf, WSC contributed by current leaf assimilation during the mid-grain filling, i.e. the reduction in WSC due to cutting leaves which was estimated by comparing WSC between the normal control and the treatment of cutting leaves ($WSC_{untreated} - WSC_{removing\ leaves}$); Non-leaf, the WSC in non-leaf organs at the mid-grain filling (excluding the current leaf assimilation) which was estimated by the treatment of removing leaves ($WSC_{removing\ leaves}$); Remo, WSC used for respiration and remobilization during the mid-grain filling which was obtained by comparing WSC between the normal control and the treatment of cutting spikes ($WSC_{cutting\ spikes} - WSC_{untreated}$).

Spikes corresponding to main stem samples were collected at maturity stage for each accession to obtain TGW. The reduction of TGW due to leaf removal was calculated for each cultivar as: $[(TGW_{untreated} - TGW_{removing\ leaves})/TGW_{untreated}] \times 100\%$.

SSR genotyping and association mapping

Two hundred and nine SSR markers, evenly distributed on the 21 wheat chromosomes, were selected for evaluating population structure, relative kinship, and association mapping. The genetic positions of SSR markers were obtained from the consensus map Ta-SSR-2004 [32] and the Komugi wheat genetic resources database (http://www.shigen.nig.ac.jp/wheat/komugi?/top/top.jsp). Fluorescent primers were synthesized by ABI (Applied Biosystems, Foster City, CA, USA). Amplification products were separated on an ABI3730 DNA Analyzer, and fragment sizes were analyzed by GeneMapper software (Applied Biosystems).

Allele number, allele frequency and polymorphism information content were calculated by PowerMarker V3.25 [33]. Population structure was estimated by STRUCTURE v2.3.2 using data from 209 SSR markers. The number of hypothetical subpopulations was set from $k = 1$ to 10 with a burn-in period of 50,000 iterations and a run of 500,000 replications of Markov Chain Monte Carlo after burn in. The Δk method was applied according to LnP(D) in STRUCTURE [34]. The Q data of five replicate runs were integrated by CLUMPP software [35]. Principal coordinate analysis based on genetic distances was also used to confirm the results of STRUCTURE by NTSYSpc analysis [36]. The relative kinship coefficient (K) was calculated by the SPAGeDi software package [37]. Finally, the Q+K models were performed using mixed linear model in TASSEL V2.1 for association of SWSC [38,39].

Results

Contribution of leaf and non-leaf organs to 1000-grain weight during grain filling

Figure S1 summarizes the relative contributions of leaf and non-leaf organs to final 1000-grain weight during grain filling across the 262 diverse winter wheat genotypes. Reduction in TGW due to cutting leaves (i.e. the contribution of leaves to TGW during grain filling) was 14.79% (6.26 g) at maturity under DS, compared to 19.84% (8.56 g) under WW condition (Figure S1). The lower contribution of leaves to TGW under DS condition reflected the negative effect of water deficit on photosynthetic rates in leaf blades during grain filling. Non-leaf organs contributed 85.21% (36.05 g) to TGW under DS, whereas it was 80.16% (34.60 g) under WW condition (Figure S1).

Variation in WSC of leaves and non-leaf organs in different internodes at the mid-grain filling under two water regimes

The WSC in lower internodes (the remainder segments of stem except for peduncle, Low) were higher than those in peduncles (the upmost internode, Ped; Figure 1) under both water regimes in all types of WSC. The WSC in non-leaf organs at the mid-grain filling ranged from 83.82 to 178.50 mg WSC/g dry weight (mg/g dw), and those contributed by current leaf assimilation were from 41.13 to 68.58 mg/g dw, thus showing the relative importance of stem-reserved WSC for grain filling. The WSC used for Remo at the mid-grain filling ranged from 56.88 to 98.87 mg/g dw (Figure 1). WSC in non-leaf organs were 131.51, 178.50 and 159.37 mg/g dw in peduncles, lower internodes and the whole stem under drought stress, and 83.82, 94.35 and 88.05 mg/g dw under well-watered condition, respectively; the ratios between two water regimes were 156.90%, 189.19% and 181.00%, respectively (Figure 1). This implied that long term drought stress triggered a series of metabolic reactions by increasing fructans for self-protection. At the mid-grain filling, WSC contributed by current leaf assimilation were 57.62 and 55.43 mg/g dw in the lower internode and whole stem under drought stress, compared with 68.58 and 60.05 mg/g dw under well-watered condition (Figure 1). Thus drought during grain filling greatly influences current photosynthesis and dry matter accumulation.

Correlations between WSC at the mid-grain filling and TGW

WSC are recognized as an important source of grain dry matter for grain filling in wheat. There were significant correlations between the four types of WSC (Total, Non-leaf, Leaf and Remo) at the mid-grain filling and TGW (Table 1, Table S2~S3). Moreover, there were higher correlations between the four types of WSC and TGW under drought stress compared to those under well-watered condition (Table 1). Under drought stress, WSC of Total was significantly correlated with TGW ($r = 0.248^{***}$, 0.386^{***} and 0.392^{***}); and correlations between WSC of Non-leaf, Leaf and TGW were $r = 0.140^{*}$ to 0.275^{***}, 0.156^{*} to 0.220^{***}, respectively. Under well-watered condition, there were three instances of significant correlations between WSC of Total and TGW ($r = 0.135^{*}$, 0.146^{*} and 0.176^{**}).

Association analysis for WSC at the mid-grain filling and the search for favorable alleles

Based on the population structure assessment using 209 markers, the 262 wheat accessions were separated into two sub-populations, comprising 126 and 136 accessions (our unpublished data). Association analysis using the 209 SSR markers and four types of WSC at the mid-grain filling was conducted using a mixed linear model, which accounted for population structure (Q) and relative kinship (K matrix). Thirteen, 13, 23 and 14 novel loci were significantly ($P<0.01$) associated with WSC of Total, Leaf, Non-leaf and Remo in 17, 17, 31 and 18 instances, respectively (Table S4~S7). Variances explained by SSR markers (R^2) ranged from 0.11% to 10.51%. Twenty-two loci were identified more than once. *Xcfd17-2D* (associated with WSC of Remo under WW condition; Remo, WW), *Xgwm513-4B* (Non-leaf, DS) and *Xwmc517-7B* (Non-leaf, WW) were detected in all internodes (peduncle, lower internode and the whole stem). *Xbarc228-2D* (Total, DS), *Xgwm169-6A* (Remo, WW) and *Xgwm537-7B* (Leaf, DS) were detected in both peduncle and lower internode. *Xcfd53-2D* (Non-leaf, DS) and *Xcfa2240-7A* (Non-leaf, DS) were identified in both the peduncle and whole stem; *Xgwm630*

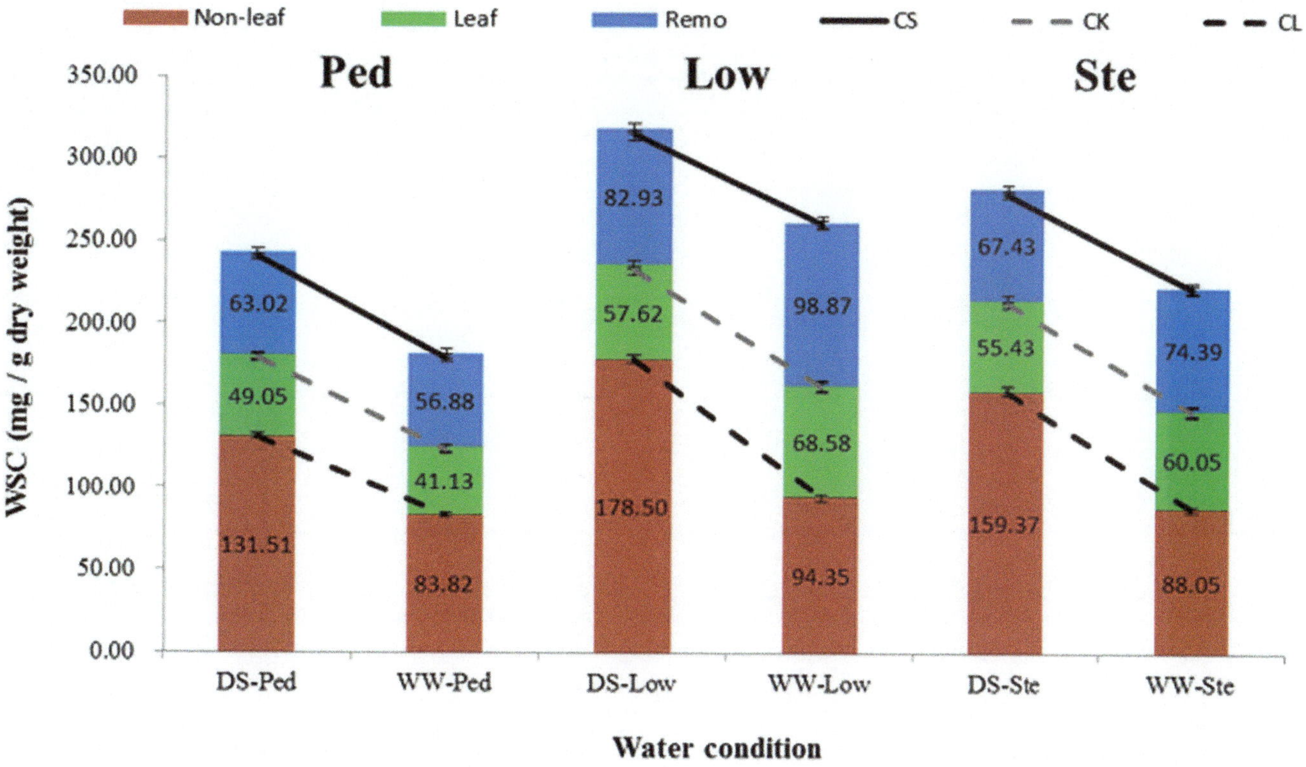

Figure 1. WSC (mg/g dw) of different internodes at the mid-grain filling stage (14 days after flowering) under well-watered and drought stress conditions. Bars indicate 2SE. WSC, water-soluble carbohydrates; DS-Ped, peduncle under drought stress; WW-Ped, peduncle, well-watered; DS-Low, lower internode, drought stress; WW-Low, lower internode, well-watered; DS-Ste, whole stem, drought stress; WW-Ste, whole stem, well-watered; CS, cutting spikes; CL, removing leaves; CK, normal control; Non-leaf, WSC in non-leaf organs at the mid-grain filling, excluding the current leaf assimilation; Leaf, WSC contributed by current leaf assimilation during the mid-grain filling; Remo, WSC used for respiration and remobilization during the mid-grain filling.

(Remo, WW), *Xgwm610-4A* (Leaf, WW), *Xgwm165.1-4D* (Non-leaf, WW) and *Xgwm182-5D* (Total, WW) were similarly identified in both the lower internode and whole stem. *Xbarc125-3D* was associated with WSC of Total in lower internode under both DS and WW conditions. *Xgwm66*, *Xgwm88*, *Xgwm192*, *Xwmc470-2D*, *Xgwm181-3B*, *Xgwm358-5D*, *Xgwm583-5D* and *Xgwm428-7D* were associated with more than one types of WSC (Total, Non-leaf, Leaf and Remo).

For associated loci, we explored favorable WSC alleles by assessing differences in WSC between accessions carrying favorable alleles and those with other alleles using ANOVA (SAS 8.01), i.e. the WSC of the former were significantly ($P<0.05$) higher than those of the latter. There were 7, 10, 12 and 9 novel favorable alleles for WSC of Total, Leaf, Non-leaf and Remo, respectively (Tables S4~S7). *Xcfd17-2D* (Remo, WW) had the same favorable WSC alleles (*Xcfd17-2D*$_{223}$) in peduncle, lower internode and the whole stem estimates, i.e. 74.1 compared with 51.7 mg/g dw ($P<0.001$), 113.5 compared with 94.3 mg/g dw ($P<0.05$), and 89.7 compared with 69.7 mg/g dw ($P<0.01$), respectively. *Xgwm181-3B*$_{131\ and\ 161}$ (Leaf, DS), *Xgwm610-4A*$_{167}$ (Leaf, WW), *Xgwm513-4B*$_{144}$ (Leaf, DS), *Xgwm165.1-4D*$_{199}$ (Non-leaf, WW), *Xwmc517-7B*$_{188}$ (Non-leaf, WW) had positive effects both in lower internode and the whole stem. Higher WSC were associated with *Xgwm169-6A*$_{203}$ (Remo, WW) and *Xgwm537-7B*$_{205}$ (Leaf, DS) in both the peduncle and lower internode. *Xbarc125-3D*$_{147}$ (Total) contributed to higher WSC in lower internodes, not only under well-watered conditions but also under drought stress. Some associated loci, however, had various favorable alleles for different types of WSC; for example, accessions carrying the allele *Xgwm513-4B*$_{144}$ exhibited higher WSC of Leaf, whereas accessions with the 142 bp allele had higher WSC of Non-leaf in lower internodes under drought-stress conditions.

Seven novel favorable WSC alleles individually exhibited positive contributions to TGW under well-watered, drought and heat stress conditions

In order to evaluate the genetic relationship between WSC and TGW, we analyzed the effects of favorable WSC alleles on final TGW by comparing differences in TGW between accessions carrying favorable WSC alleles and those with other alleles. Seven novel favorable WSC alleles exhibited significantly ($P<0.05$) positive contributions to TGW on an individual basis. They were *Xcfd17-2D*$_{223}$, *Xcfd53-2D*$_{263}$, *Xgwm181-3B*$_{140\ and\ 161}$, *Xgwm389-3B*$_{116}$, *Xbarc125-3D*$_{147}$, *Xgwm358-5D*$_{162}$ and *Xgwm537-7B*$_{205}$ (Table 2). For *Xbarc125-3D*$_{147}$, the higher WSC of Total (341.5 compared to 309.9 mg/g dw) in lower internodes led to a higher TGW (44.99 g compared to 41.14 g) under drought stress conditions; likewise, accessions with this allele also produced higher TGW (43.91 g) than accessions with other alleles (41.85 g) under well-watered conditions, with WSC of 284.3 and 255.9 mg/g dw, respectively. In order to verify the positive contributions of these seven favorable alleles to TGW, we used the same population (262 winter wheat accessions) planted in 16 environments (year×site×water and heat regime combinations; 5 drought stress conditions, 3 well-watered and heat stress conditions, 3 drought and heat stress conditions, 5 well-watered conditions) to confirm the above results. The average TGW of

Table 1. Pearson correlation coefficients of WSC at the mid-grain filling and TGW under well-watered and drought stress conditions.

WSC		TGW	
Types	**Internodes**	**DS**	**WW**
Total[a]	Ped	0.248***	0.135*
	Low	0.386***	0.146*
	Ste	0.392***	0.176**
Leaf[a]	Ped	0.218***	0.027
	Low	0.156*	0.100
	Ste	0.220***	0.071
Non-leaf[a]	Ped	0.011	−0.046
	Low	0.177**	−0.000
	Ste	0.140*	−0.011
Non-leaf[b]	Ped	0.207***	0.121
	Low	0.275***	0.011
	Ste	0.274***	−0.001
Remo[a]	Ped	−0.014	0.100
	Low	0.106	0.044
	Ste	0.037	0.105

*Significant at $P=0.05$; **Significant at $P=0.01$; ***Significant at $P=0.001$. Total, total WSC at the mid-grain filling produced by leaves and non-leaf organs; Leaf, WSC contributed by current leaf assimilation during the mid-grain filling; Non-leaf, WSC in non-leaf organs at the mid-grain filling, excluding the current leaf assimilation; Remo, WSC used for respiration and remobilization during the mid-grain filling; Ped, peduncle; Low, lower internode; Ste, whole stem; TGW, 1000-grain weight at maturity.

[a]TGW was measured on the normal control;

[b]TGW was measured with treatment with removing leaves.

accessions carrying favorable WSC alleles were higher than those without the favorable alleles in all environments (Figure 2).

Pyramiding of favorable WSC alleles indicated potential application in wheat breeding

To explore whether the pyramiding of favorable WSC alleles showed additive effects, we also analyzed the mean TGW of accessions with different numbers of favorable WSC alleles in 16 environments (Table 3). The average TGW of genotypes with single favorable WSC allele were 31.75 - 42.82 g; TGW with two favorable alleles were 33.57 - 44.66 g; TGW with three ones were 35.17 - 46.76 g; TGW with more than four favorable alleles were 35.90 - 47.18 g across 16 environments. The average TGW of accessions without favorable WSC allele ranged from 30.30 to 41.01 g. In addition, a significantly linear correlation ($y = 1.579x+ 36.847$, $R^2 = 0.369$) between TGW and number of favorable WSC alleles further confirmed the additive effect (Figure 3A). We also evaluated the distribution of combined favorable WSC alleles in modern varieties from different decades (Figure 3B). The average number of favorable WSC alleles was 0.61 before 1960, and the current average number was 2.59. The increasing number over time reveals a genomic footprint left by breeders, but the relatively low number of 2.59 alleles in current cultivars (post-2000) indicates a potential for pyramiding more favorable alleles [40].

Discussion

Consistency between WSC and TGW under stress conditions

Previous studies suggested that increases in grain yield can mainly be attributed to better partitioning of photosynthetic products [41]. WSC accumulation ability and its remobilization efficiency are much higher in the internodes of drought tolerant cultivars than those of sensitive genotypes under both normal and stress conditions. On the other hand, fructans, the major components of WSC, insert between the head groups of phospholipids, acting as compatible solutes in cells to protect cell membranes and proteins from osmotic damage [42,43]. Stem samples of rainfed wheat have significantly higher average fructan than irrigated samples. In our research four types of WSC (Total, Leaf, Non-leaf and Remo) under drought stress were overall higher than those under well-watered condition (Figure 1). It has been reported that fructan synthesis is induced by drought stress, and that drought tolerant plants can manufacture more fructans. The tolerant cultivars activate their protection mechanisms faster and more efficiently than the sensitive ones to cope with stress conditions [10,44].

Drought stress during grain filling can result in reductions in grain weight, due to lower numbers of endosperm cells and a limited maximum storage capacity of the kernels [45,46]. WSC are recognized as an important source of grain dry matter for grain filling, especially when current photosynthesis is inhibited by drought stress. Water deficit during grain filling stimulates senescence of the whole plant and enhances remobilization of reserved WSC to the grains [47,48]. Thus, the reserved WSC assimilated pre-anthesis and current assimilation are critically important for grain filling. In the present study, we observed that final grain yield mainly depends on pre-anthesis assimilation by green organs and current photosynthesis of non-leaf organs during grain filling, especially under drought stress condition (Figure S1). In addition, compared with those under well-watered condition, higher correlations between the four types of WSC at the mid-grain filling and TGW under drought stress indicate that yield in

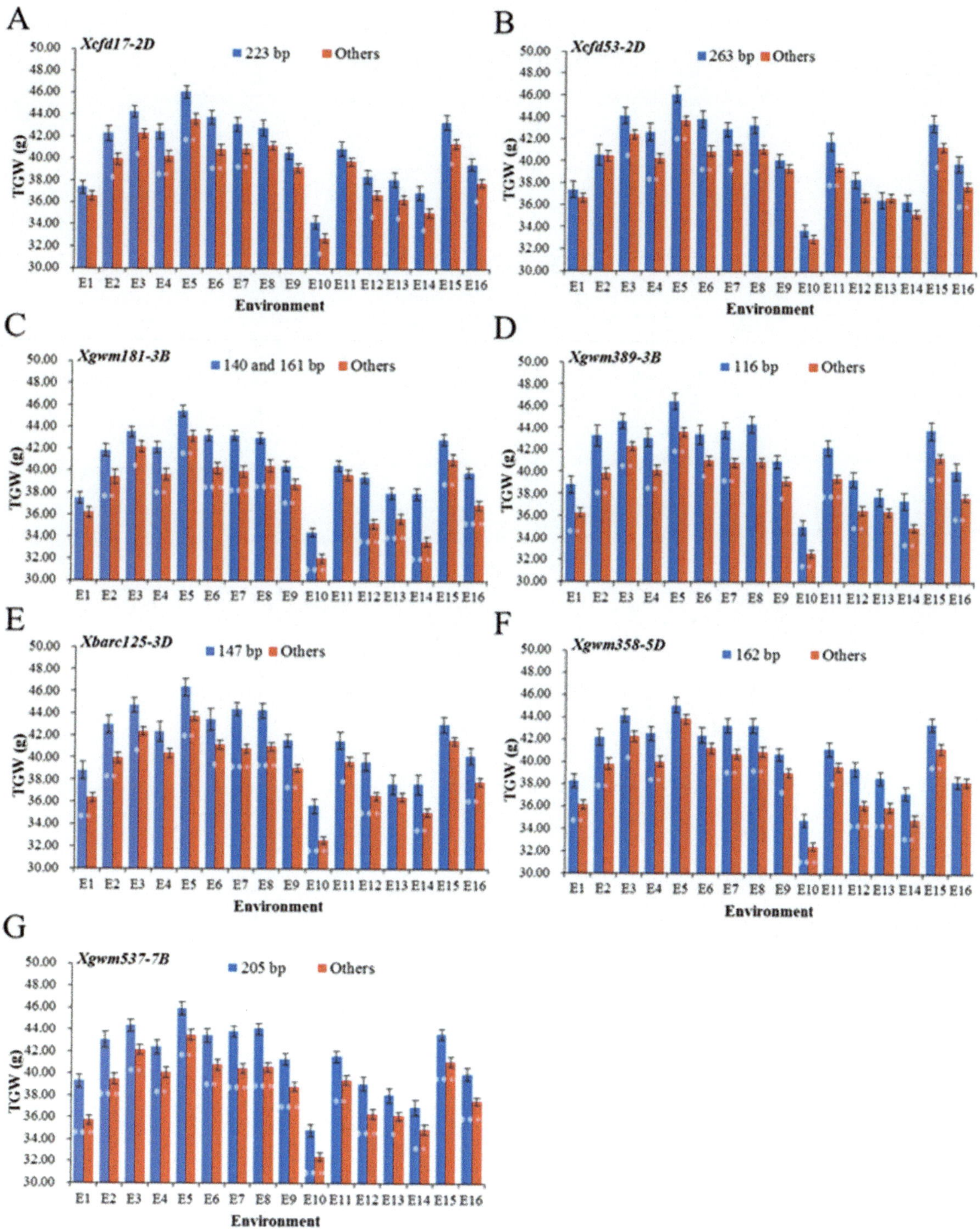

Figure 2. Verification of phenotypic effects of seven novel favorable WSC alleles individually contributing to TGW in sixteen environments. E1, E3, E7, E9 and E13 were drought stress conditions, E6, E12 and E16 were well-watered and heat stress conditions, E4, E10 and E14 were drought and heat stress conditions, E2, E5, E8, E11 and E15 were well-watered conditions. Bars indicate 2SE. *, **, *** Significant at $P = 0.05$, 0.01 and 0.001, respectively.

unfavorable conditions relies more on pre-stored carbohydrates (Table 1). Association and favorable allele analyses were conducted on four types of WSC (Total, Leaf, Non-leaf and Remo) at the mid-grain filling. Seven novel favorable WSC alleles made positive individual contributions to final TGW under well-watered, drought and heat stress conditions (Table 2, Figure 2).

Complex relationship between WSC and TGW

WSC accumulation and remobilization are influenced by many factors, making the relationship between WSC and TGW more complex. For example, WSC remobilization is affected by N fertilizers and water deficit [6,49]. Heavy use of N fertilizers delays plant senescence and reduces the remobilization of prestored

Table 2. Seven favorable WSC alleles individually contributed to significantly higher TGW.

Locus	Trait	Favorable allele (bp)	Freq. (%)	WSC Mean ± SE (mg/g dw)	*P* value	TGW-DS Mean ± SE (g)	*P* value	TGW-WW Mean ± SE (g)	*P* value
Xbarc125-3D	Total-Low-DS	147	16.41	341.5±6.0	0***	44.99±0.61	0.0002***	43.91±0.62	0.0488*
		Others	83.59	309.9±3.3		41.14±0.43		41.85±0.44	
	Total-Low-WW	147		284.3±5.5	0***				
		Others		255.9±2.5					
Xgwm537-7B	Leaf-Ped-DS	205	28.63	58.0±3.5	0.0039**	43.53±0.61	0.0032**	43.81±0.60	0.0067**
		Others	71.37	45.4±2.4		41.06±0.46		41.52±0.48	
	Leaf-Low-DS	205		65.8±4.3	0.0345*				
		Others		54.2±3.0					
Xgwm358-5D	Leaf-Ste-DS	162	30.15	65.4±3.8	0.0064**	43.43±0.72	0.0039**	43.94±0.72	0.0026**
		Others	69.85	51.6±2.7		41.05±0.44		41.43±0.45	
Xcfd53-2D	Non-leaf-Ste-DS	263	20.61	168.0±3.4	0.0163*	43.53±0.75	0.0168*	43.80±0.76	0.0326*
		Others	79.39	157.1±2.1		41.30±0.43		41.77±0.44	
Xgwm181-3B	Non-leaf-Ste-DS	140 and 161	45.42	163.6±2.6	0.0377*	43.45±0.48	0***	42.91±0.52	0.0834
			54.58	155.9±2.5		40.37±0.54		41.58±0.55	
Xgwm389-3B	Remo-Ste-DS	116	19.08	79.4±5.6	0.0152*	44.10±0.74	0.0026**	43.95±0.88	0.0246*
		Others	80.92	64.5±2.7		41.21±0.43		41.76±0.42	
Xcfd17-2D	Remo-Ped-WW	223	24.81	74.1±5.1	0.0002***	43.31±0.63	0.0183*	43.75±0.67	0.0193*
		Others	75.19	51.7±2.8		41.25±0.46		41.67±0.46	
	Remo-Low-WW	223		113.5±7.1	0.0123*				
		Others		94.3±3.6					
	Remo-Ste-WW	223		89.7±6.5	0.0044**				
		Others		69.7±3.3					

*Significant at $P = 0.05$; **Significant at $P = 0.01$; ***Significant at $P = 0.001$. Total, total WSC at the mid-grain filling produced by leaves and non-leaf organs; Leaf, WSC contributed by current leaf assimilation during the mid-grain filling; Non-leaf, WSC in non-leaf organs at the mid-grain filling, excluding the current leaf assimilation; Remo, WSC used for respiration and remobilization during the mid-grain filling; Ped, peduncle; Low, lower internode; Ste, whole stem; DS, drought stress; WW, well-watered; TGW, 1000-grain weight at maturity.

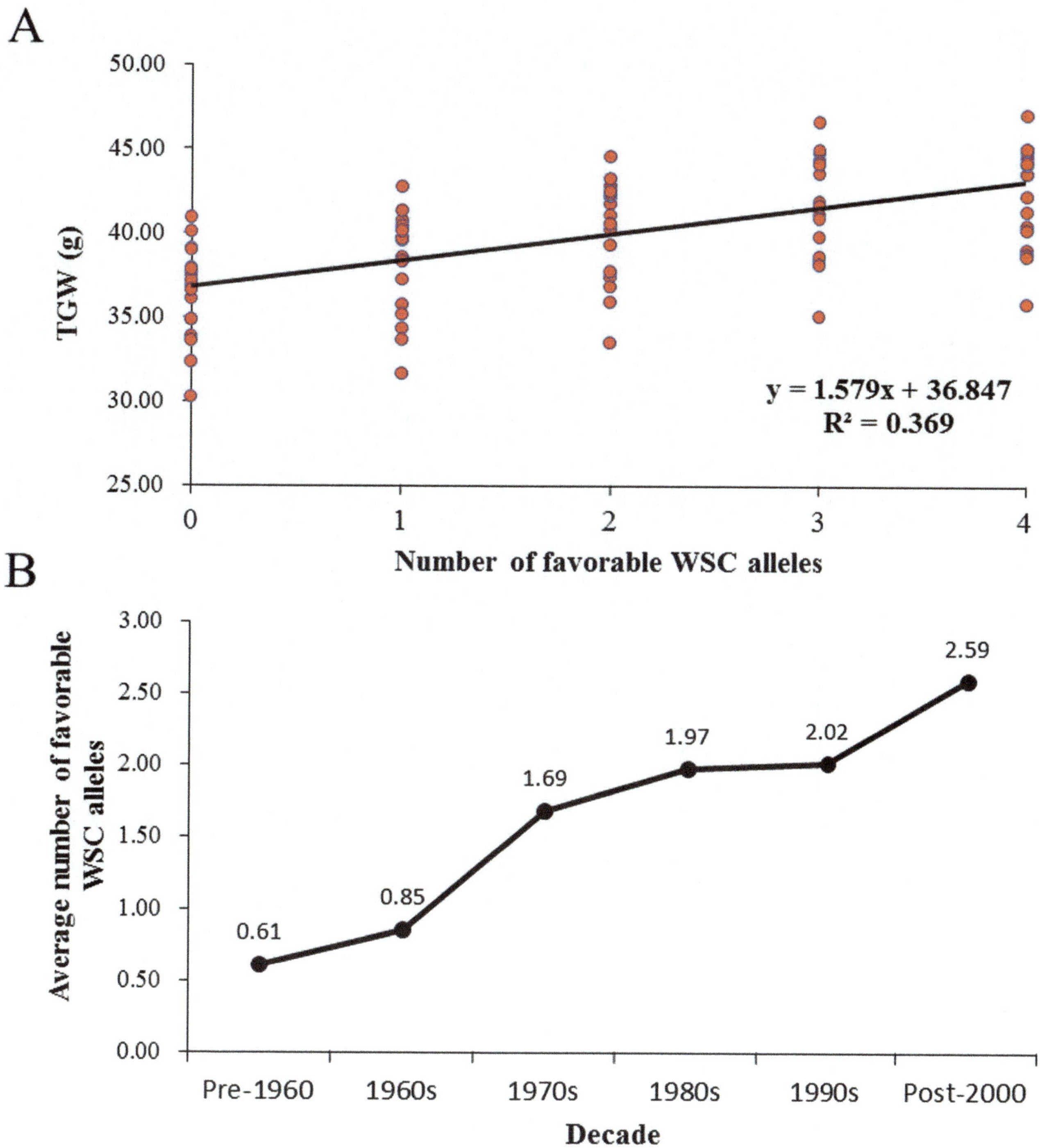

Figure 3. Linear regressions of TGW based on seven favorable WSC alleles in sixteen environments (A), and accumulation of seven favorable WSC alleles in modern varieties released in different decades (B). There were 18, 27, 54, 39, 58 and 51 accessions released in Pre-1960, 1960s, 1970s, 1980s, 1990s and Post-2000, respectively. 15 accessions with unknown released decades were excluded.

assimilates; whereas water deficit performs the reverse function. Thus final TGW is often significantly increased with heavy use of N fertilizers under drought stress, even higher than that under well-watered conditions [49,50]. WSC in the stem are negatively correlated with tiller per unit area, that is, WSC accumulation is dependent on plant density [8,12]. Moreover, WSC accumulation is also affected by stem length (plant height) and stem weight. We calculated Pearson correlation coefficients between the four types of WSC at the mid-grain filling and TGW, and we also evaluated the effects of favorable WSC alleles on TGW to further understand their relationships at the genetic level. The results showed that (1) the correlations between them were significant but not robust, the highest Pearson correlation being only $r = 0.393^{***}$, and there was no relationship between WSC of Remo at the mid-grain filling and TGW; and (2) there were 7, 10, 12 and 9 favorable alleles for WSC of Total, Leaf, Non-leaf and Remo,

Table 3. Pyramiding of favorable WSC alleles contributing to TGW in sixteen environments.

Env.	No. of alleles	TGW[1] Mean ± SE (%)	Fre. (%)	Env.	No of alleles	TGW[1] Mean ± SE (%)	Fre. (%)
E1	≥4	39.17±0.75 (A)	13.73	E9	≥4	41.38±0.65 (A)	13.46
(DS)	3	38.47±0.92 (A)	16.08	(DS)	3	41.25±0.57 (A)	16.92
	2	37.48±0.62 (AB)	27.06		2	40.34±0.54 (AB)	26.54
	1	35.79±0.67 (BC)	23.14		1	38.62±0.72 (B)	23.08
	0	33.94±0.81 (C)	20.00		0	36.69±0.71 (C)	20.00
E2	≥4	43.67±1.09 (A)	12.94	E10	≥4	35.90±0.67 (A)	13.46
(WW)	3	42.01±0.80 (AB)	16.86	(DS+HS)	3	35.17±0.62 (AB)	16.92
	2	41.82±0.70 (AB)	27.06		2	33.57±0.54 (B)	26.54
	1	39.72±0.92 (B)	23.92		1	31.75±0.61 (C)	23.08
	0	36.17±1.04 (C)	19.22		0	30.30±0.67 (C)	20.00
E3	≥4	45.17±0.70 (A)	13.41	E11	≥4	42.35±0.83 (A)	13.51
(DS)	3	45.10±0.61 (A)	16.48	(WW)	3	41.77±0.71 (A)	16.60
	2	43.33±0.62 (AB)	26.82		2	40.68±0.56 (A)	26.64
	1	41.40±0.76 (BC)	23.37		1	38.43±0.79 (B)	23.17
	0	40.16±0.81 (C)	19.92		0	37.91±0.68 (B)	20.08
E4	≥4	43.66±0.89 (A)	13.13	E12	≥4	40.54±0.68 (A)	13.46
(DS+HS)	3	43.67±0.70 (A)	16.99	(WW+HS)	3	40.99±0.64 (A)	16.92
	2	41.15±0.67 (B)	26.25		2	37.81±0.63 (B)	26.54
	1	39.82±0.80 (B)	23.55		1	34.38±0.69 (C)	23.08
	0	36.95±0.77 (C)	20.08		0	33.69±0.61 (C)	20.00
E5	≥4	47.18±0.82 (A)	13.03	E13	≥4	38.93±0.95 (A)	13.62
(WW)	3	46.76±0.63 (AB)	16.86	(DS)	3	38.71±0.71 (A)	16.73
	2	44.66±0.63 (BC)	26.82		2	36.89±0.67 (AB)	26.46
	1	42.82±0.83 (CD)	23.37		1	35.22±0.68 (B)	22.96
	0	41.01±0.82 (D)	19.92		0	34.91±0.73 (B)	20.23
E6	≥4	44.21±1.00 (A)	12.85	E14	≥4	38.77±0.93 (A)	13.62
(WW+HS)	3	44.44±0.71 (A)	16.06	(DS+HS)	3	38.28±0.71 (A)	16.73
	2	42.49±0.70 (AB)	26.91		2	35.98±0.68 (B)	26.46
	1	40.45±0.81 (B)	24.10		1	33.72±0.69 (C)	22.96
	0	37.55±0.85 (C)	20.08		0	32.37±0.62 (C)	20.23
E7	≥4	45.04±0.70 (A)	13.46	E15	≥4	44.32±0.78 (A)	13.67
(DS)	3	44.41±0.61 (A)	16.92	(WW)	3	44.26±0.67 (A)	16.80
	2	42.36±0.59 (B)	26.54		2	42.57±0.63 (A)	26.17
	1	39.68±0.74 (C)	23.08		1	40.21±0.70 (B)	23.05
	0	37.22±0.82 (D)	20.00		0	39.08±0.79 (B)	20.31

Table 3. Cont.

Env.	No. of alleles	TGW[1] Mean ± SE (%)	Fre. (%)	Env.	No of alleles	TGW[1] Mean ± SE (%)	Fre. (%)
E8	≥4	45.15±0.85 (A)	13.46	E16	≥4	40.36±0.84 (A)	13.57
(WW)	3	44.24±0.68 (AB)	16.92	(WW+HS)	3	39.88±0.55 (A)	16.67
	2	42.40±0.57 (B)	26.54		2	39.42±0.61 (A)	26.36
	1	39.72±0.78 (C)	23.08		1	37.36±0.68 (B)	23.26
	0	37.96±0.74 (C)	20.00		0	34.86±0.60 (C)	20.16

[1]Values with different letters are significantly different ($P<0.05$). DS, drought stress; WW+HS, well-watered and heat stress; DS+HS, drought and heat stress; WW, well-watered.

respectively. However, only seven favorable WSC alleles exhibited positive individual contributions to TGW. The complex relationship between WSC and TGW due to many influential factors may help us to understand the reasons for these results. In addition, cutting spikes or removing leaves at flowering change the source-sink relationship during grain filling and therefore the four types of WSC at the mid-grain filling might not fully reflect the situation under normal condition.

Seven favorable WSC alleles will help to improve breeding progress in yield potential

Broad-sense heritability of WSC is relatively high, but shows wide fluctuations under different conditions, i.e. WSC are very sensitive to environments [11,23]. Yang et al. [23] reported that (1) QTL for WSC accumulation and remobilization could have different expression patterns at different growth stages or in different environments; and (2) 7 of 10 significantly additive QTL for WSC interacted with environment. Thus, stable molecular markers for WSC are essential to understand its genetic basis. Moreover, exploration of favorable WSC alleles in germplasm resources could be useful to plant breeders, but the effectiveness of such alleles needs to be verified [29,51]. In this study, seven favorable WSC alleles significantly ($P<0.05$) enhanced TGW on an individual basis (Table 2). An additive QTL for WSC, *QSwscf.cgb-2D.1* (*WMC453.1–WMC18*), was detected in a Hanxuan 10×Lumai 14 doubled haploid population [23]. *Xcfd17-2D* was 1.5 cM from the flanking marker *Xwmc18-2D*. *QReswc.cgb-3B*, controlled WSC and remobilization efficiency, and *Xgwm181-3B* shares one of its flanking markers (*Xgwm547–Xgwm181*). Adjacent chromosome intervals, such as *QAeswc.cgb-3B.1*, *QSwscm.cgb-3B.1* and *QSwscf.cgb-3B*, carry QTL for WSC and its accumulation efficiency [23]. In addition, *Xcfd53-2D*, *Xgwm389-3B* and *Xgwm537-7B* were associated with yield-related traits [49–54]. *Xbarc125-3D* was also associated with TGW (our unpublished data).

The seven favorable WSC alleles for enhancing TGW were verified under 16 environments (5 drought stress conditions, 3 well-watered and heat stress conditions, 3 drought and heat stress conditions, and 5 well-watered conditions) using a population of 262 winter wheat accessions (Table 2 and Figure 2). Many studies show that marker-based strategies of gene pyramiding are effective [27,55,56]. A dosage effect of pyramiding seven favorable WSC alleles (Table 3, Figure 3A) was also demonstrated in the study. The accumulation of favorable WSC alleles over different decades also indicated that they had been individually selected by breeders in the past and that there is potential for further improvement in the future.

Supporting Information

Figure S1 The percentage contributions of leaf and non-leaf organs to 1000-grain weight (TGW) under drought stress (DS) and well-watered (WW) conditions during grain filling. Bars indicate 2SE. The data in the columns were the absolute values of TGW (g).

Table S1 262 common wheat accessions and their origins.

Table S2 Statistic data of WSC (mg/g dw) at the mid-grain filling under well-watered and drought stress conditions.

Table S3 Statistic data of TGW under well-watered and drought stress conditions.

Table S4 Thirteen loci significantly associated with the total WSC at the mid-grain filling produced by leaves and non-leaf organs (Total) and phenotypic values of favorable marker alleles under two water regimes.

Table S5 Thirteen loci significantly associated with the WSC contributed by current leaf assimilation during the mid-grain filling (Leaf) and phenotypic values of favorable marker alleles under two water regimes.

Table S6 Twenty-three loci significantly associated with the WSC in non-leaf organs at the mid-grain filling (excluding the current leaf assimilation, Non-leaf) and phenotypic values of favorable marker alleles under two water regimes.

Table S7 Fourteen loci significantly associated with the WSC used for respiration and remobilization during the mid-grain filling (Remo) and phenotypic values of favorable marker alleles under two water regimes.

Acknowledgments

We thank Professor Robert A. McIntosh (Plant Breeding Institute, University of Sydney, NSW, Australia) for revision and suggestions on the manuscript.

Author Contributions

Conceived and designed the experiments: RJ RL. Performed the experiments: WL BZ XC. Analyzed the data: BZ WL. Contributed to the writing of the manuscript: BZ. Revised the manuscript: BZ RJ.

References

1. Gupta P, Mir R, Mohan A, Kumar J (2008) Wheat genomics: present status and future prospects. Int J Plant Genomics doi:10.1155/2008/896451.
2. Wardlaw IF, Willenbrink J (1994) Carbohydrate storage and mobilisation by the culm of wheat between heading and grain maturity: the relation to sucrose synthase and sucrose-phosphate synthase. Funct Plant Biol 21: 255–271.
3. Gallagher J, Biscoe P, Hunter B (1976) Effects of drought on grain growth. Nature 264: 541–542.
4. Bidinger F, Musgrave R, Fischer R (1977) Contribution of stored pre-anthesis assimilate to grain yield in wheat and barley. Nature 270: 431–433.
5. Spollen WG, Nelson CJ (1994) Response of fructan to water deficit in growing leaves of tall fescue. Plant Physiol 106: 329–336.
6. Blum A (1998) Improving wheat grain filling under stress by stem reserve mobilisation. Euphytica 100: 77–83.
7. Ehdaie B, Madore M, Waines J, Alloush G (2006) Genotypic variation for stem reserves and mobilization in wheat: II. Postanthesis changes in internode water-soluble carbohydrates. Crop Sci 46: 2093–2103.
8. Rebetzke G, Van Herwaarden A, Jenkins C, Weiss M, Lewis D, et al. (2008) Quantitative trait loci for water-soluble carbohydrates and associations with agronomic traits in wheat. Crop and Pasture Sci 59: 891–905.
9. Gebbing T, Schnyder H (1999) Pre-anthesis reserve utilization for protein and carbohydrate synthesis in grains of wheat. Plant Physiol 121: 871–878.
10. Goggin DE, Setter TL (2004) Fructosyltransferase activity and fructan accumulation during development in wheat exposed to terminal drought. Funct Plant Biol 31: 11–21.
11. Ruuska SA, Rebetzke GJ, van Herwaarden AF, Richards RA, Fettell NA, et al. (2006) Genotypic variation in water-soluble carbohydrate accumulation in wheat. Funct Plant Biol 33: 799–809.
12. Dreccer MF, Chapman SC, Rattey AR, Neal J, Song Y, et al. (2013) Developmental and growth controls of tillering and water-soluble carbohydrate accumulation in contrasting wheat (*Triticum aestivum* L.) genotypes: can we dissect them? J Exp Bot 64: 143–160.
13. Foulkes M, Scott R, Sylvester-Bradley R (2002) The ability of wheat cultivars to withstand drought in UK conditions: formation of grain yield. J Agric Sci 138: 153–169.
14. van Herwaarden AF, Richards RA (2002) Water-soluble carbohydrate accumulation in stems is related to breeding progress in Australian wheats. In: Plant Breeding for the 11th Millenium, JAMcComb (ed), Proceedings 12th Australasian Plant Breeding Conference (Perth, 15–20 September, 2002): 878–882.
15. Shearman V, Sylvester-Bradley R, Scott R, Foulkes M (2005) Physiological processes associated with wheat yield progress in the UK. Crop Sci 45: 175–185.
16. Xue GP, McIntyre CL, Jenkins CLD, Glassop D, Van Herwaarden AF, et al. (2008) Molecular dissection of variation in carbohydrate metabolism related to water-soluble carbohydrate accumulation in stems of wheat. Plant Physiol 146: 441–454.
17. Turner LB, Cairns AJ, Armstead IP, Ashton J, Skot K, et al. (2006) Dissecting the regulation of fructan metabolism in perennial ryegrass (*Lolium perenne* L.) with quantitative trait locus mapping. New Phytol 169: 45–57.
18. Nagata K, Shimizu H, Terao T (2002) Quantitative trait loci for nonstructural carbohydrate accumulation in leaf sheaths and culms of rice (*Oryza sativa* L.) and their effects on grain filling. Breed Sci 52: 275–283.
19. Thevenot C, Simond-Cote E, Reyss A, Manicacci D, Trouverie J, et al. (2005) QTLs for enzyme activities and soluble carbohydrates involved in starch accumulation during grain filling in maize. J Exp Bot 56: 945–958.
20. Teulat B, Borries C, This D (2001) New QTLs identified for plant water status, water-soluble carbohydrate and osmotic adjustment in a barley population grown in a growth-chamber under two water regimes. Theor Appl Genet 103: 161–170.
21. Galiba G, Kerepesi I, Snape J, Sutka J (1997) Location of a gene regulating cold-induced carbohydrate production on chromosome 5A of wheat. Theor Appl Genet 95: 265–270.
22. Salem KFM, Roder MS, Borner A (2007) Identification and mapping quantitative trait loci for stem reserve mobilisation in wheat (*Triticum aestivum* L.). Cereal Res Commun 35: 1367–1374.
23. Yang DL, Jing RL, Chang XP, Li W (2007) Identification of quantitative trait loci and environmental interactions for accumulation and remobilization of water-soluble carbohydrates in wheat (*Triticum aestivum* L.) stems. Genetics 176: 571–584.
24. Saint Pierre C, Trethowan R, Reynolds M (2010) Stem solidness and its relationship to water-soluble carbohydrates: association with wheat yield under water deficit. Funct Plant Biol 37: 166–174.
25. Morgan J (1980) Osmotic adjustment in the spikelets and leaves of wheat. J Exp Bot 31: 655–665.
26. Xu X, Wang Z, Zhang J (2000) Effect of heat stress on photosynthetic characteristics of different green organs of winter wheat during grain-filling stage. Acta Bot Sin 43: 571–577.
27. Zhang Y, Wang Z, Wu Y (2006) Stomatal characteristics of different green organs in wheat under different irrigation regimes. Acta Agron Sin 32: 70–75.
28. Li W, Zhang B, Zhang J, Chang X, Li R, et al. (2012) Exploring elite alleles for chlorophyll content of flag leaf in natural population of wheat by association analysis. Acta Agron Sin 38: 962–970.
29. Zhang B, Shi W, Li W, Chang X, Jing R (2013) Efficacy of pyramiding elite alleles for dynamic development of plant height in common wheat. Mol Breed 32: 327–338.
30. Wang Z, Liu X, Li R, Chang X, Jing R (2011) Development of near-infrared reflectance spectroscopy models for quantitative determination of water-soluble carbohydrate content in wheat stem and glume. Anal Lett 44: 2478–2490.
31. Yemm E, Willis A (1954) The estimation of carbohydrates in plant extracts by anthrone. Biochem J 57: 508–514.
32. Somers DJ, Isaac P, Edwards K (2004) A high-density microsatellite consensus map for bread wheat (*Triticum aestivum* L.). Theor Appl Genet 109: 1105–1114.
33. Lui K, Muse S (2005) PowerMarker: integrated analysis environment for genetic marker data. Bioinformatics 21: 2128–2129.
34. Evanno G, Regnaut S, Goudet J (2005) Detecting the number of clusters of individuals using the software STRUCTURE: a simulation study. Mol Ecol 14: 2611–2620.
35. Jakobsson M, Rosenberg NA (2007) CLUMPP: a cluster matching and permutation program for dealing with label switching and multimodality in analysis of population structure. Bioinformatics 23: 1801–1806.
36. Nei M (1972) Genetic distance between populations. Am Nat 106: 283–292.
37. Hardy OJ, Vekemans X (2002) SPAGeDi: a versatile computer program to analyse spatial genetic structure at the individual or population levels. Mol Ecol Notes 2: 618–620.
38. Bradbury PJ, Zhang Z, Kroon DE, Casstevens TM, Ramdoss Y, et al. (2007) TASSEL: software for association mapping of complex traits in diverse samples. Bioinformatics 23: 2633–2635.
39. Zhang Z, Ersoz E, Lai CQ, Todhunter RJ, Tiwari HK, et al. (2010) Mixed linear model approach adapted for genome-wide association studies. Nat Genet 42: 355–360.
40. Barrero RA, Bellgard M, Zhang X (2011) Diverse approaches to achieving grain yield in wheat. Funct Integr Genomics 11: 37–48.
41. Sayre K, Rajaram S, Fischer R (1997) Yield potential progress in short bread wheats in northwest Mexico. Crop Sci 37: 36–42.

42. Rathinasabapathi B (2000) Metabolic engineering for stress tolerance: installing osmoprotectant synthesis pathways. Ann Bot 86: 709–716.
43. Vereyken IJ, Chupin V, Demel RA, Smeekens S, De Kruijff B (2001) Fructans insert between the headgroups of phospholipids. Biochim Biophys Acta Biomembr 1510: 307–320.
44. Aprile A, Mastrangelo A, De Leonardis A, Galiba G, Roncaglia E, et al. (2009) Transcriptional profiling in response to terminal drought stress reveals differential responses along the wheat genome. BMC Genomics 10: 279.
45. Nicolas ME, Gleadow RM, Dalling MJ (1985) Effect of post-anthesis drought on cell division and starch accumulation in developing wheat grains. Ann Bot 55: 433–444.
46. Wardlaw I, Willenbrink J (2000) Mobilization of fructan reserves and changes in enzyme activities in wheat stems correlate with water stress during kernel filling. New Phytol 148: 413–422.
47. Araus J, Slafer G, Reynolds M, Royo C (2002) Plant breeding and drought in C3 cereals: what should we breed for? Ann Bot 89: 925–940.
48. Guoth A, Tari I, Galle A, Csiszar J, Pecsvaradi A, et al. (2009) Comparison of the drought stress responses of tolerant and sensitive wheat cultivars during grain filling: Changes in flag leaf photosynthetic activity, ABA levels, and grain yield. J Plant Growth Regul 28: 167–176.
49. Yang J, Huang Z, Zhu Q, Zhang J, Wang L (2000) Remobilization of carbon reserves is improved by controlled soil-drying during grain filling of wheat. Crop Sci 40: 1645–1655.
50. Yang J, Zhang J, Wang Z, Zhu Q, Liu L (2003) Involvement of abscisic acid and cytokinins in the senescence and remobilization of carbon reserves in wheat subjected to water stress during grain filling. Plant Cell Environ 26: 1621–1631.
51. Famoso AN, Zhao K, Clark RT, Tung C-W, Wright MH, et al. (2011) Genetic architecture of aluminum tolerance in rice (*Oryza sativa*) determined through genome-wide association analysis and QTL mapping. PLoS Genet 7: e1002221.
52. Huang X, Cloutier S, Lycar L, Radovanovic N, Humphreys D, et al. (2006) Molecular detection of QTLs for agronomic and quality traits in a doubled haploid population derived from two Canadian wheats (*Triticum aestivum* L.). Theor Appl Genet 113: 753–766.
53. Marza F, Bai G, Carver B, Zhou W (2006) Quantitative trait loci for yield and related traits in the wheat population Ning7840×Clark. Theor Appl Genet 112: 688–698.
54. Wu X, Cheng R, Xue S, Kong Z, Wan H, et al. (2013) Precise mapping of a quantitative trait locus interval for spike length and grain weight in bread wheat (*Triticum aestivum* L.). Mol Breed 33: 129–138.
55. Werner K, Friedt W, Ordon F (2005) Strategies for pyramiding resistance genes against the barley yellow mosaic virus complex (*BaMMV*, *BaYMV*, *BaYMV-2*). Mol Breed 16: 45–55.
56. Sacco A, Di Matteo A, Lombardi N, Trotta N, Punzo B, et al. (2013) Quantitative trait loci pyramiding for fruit quality traits in tomato. Mol Breed 31: 217–222.

User Perceptions of Shared Sanitation among Rural Households in Indonesia and Bangladesh

Kali B. Nelson[1], Jonathan Karver[2], Craig Kullman[3], Jay P. Graham[4]*

1 Department of Epidemiology and Biostatistics, Milken Institute School of Public Health, The George Washington University, Washington, D.C., United States of America, 2 Washington, D.C., United States of America, 3 World Bank Water and Sanitation Program, Washington, D.C., United States of America, 4 Department of Environmental and Occupational Health and Department of Global Health, Milken Institute School of Public Health, The George Washington University, Washington, D.C., United States of America

Abstract

Background: The practice of sharing sanitation facilities does not meet the current World Health Organization/UNICEF definition for what is considered improved sanitation. Recommendations have been made to categorize shared sanitation as improved sanitation if security, user access, and other conditions can be assured, yet limited data exist on user preferences with respect to shared facilities.

Objective: This study analyzed user perceptions of shared sanitation facilities in rural households in East Java, Indonesia, and Bangladesh.

Methods: Cross-sectional studies of 2,087 households in East Java and 3,000 households in Bangladesh were conducted using questionnaires and observational methods. Relative risks were calculated to analyze associations between sanitation access and user perceptions of satisfaction, cleanliness, and safety.

Results: In East Java, 82.4% of households with private improved sanitation facilities reported feeling satisfied with their place of defecation compared to 68.3% of households with shared improved facilities [RR 1.19, 95% CI 1.09, 1.31]. In Bangladesh, 87.7% of households with private improved facilities reported feeling satisfied compared to 74.5% of households with shared improved facilities [RR 1.15, 95% CI 1.10, 1.20]. In East Java, 79.5% of households who reported a clean latrine also reported feeling satisfied with their place of defecation; only 38.9% of households who reported a dirty latrine also reported feeling satisfied [RR 1.74, 95% CI 1.45, 2.08].

Conclusion: Simple distinctions between improved and unimproved sanitation facilities tend to misrepresent the variability observed among households sharing sanitation facilities. Our results suggest that private improved sanitation is consistently preferred over any other sanitation option. An increased number of users appeared to negatively affect toilet cleanliness, and lower levels of cleanliness were associated with lower levels of satisfaction. However, when sanitation facilities were clean and shared by a limited number of households, users of shared facilities often reported feeling both satisfied and safe.

Editor: David W. Dowdy, Johns Hopkins Bloomberg School of Public Health, United States of America

Funding: The authors have no support or funding to report.

Competing Interests: The authors have declared that no competing interests exist.

* Email: jgraham@gwu.edu

Background

The Millennium Development Goals (MDG) Target 7C calls for halving the proportion of households without sustainable access to basic sanitation by 2015. A projected 2.4 billion people will still lack access to improved sanitation facilities in 2015, and global sanitation improvements towards Target 7C are estimated to fall short by half a billion people [1].

The goal of improved sanitation is to hygienically separate human excreta from human contact and therefore reduce exposure to fecal contamination [2]. UNICEF and the World Health Organization's Joint Monitoring Programme (JMP), the official body in charge of monitoring MDG development in the water and sanitation sector, define improved sanitation by the following types of facilities: toilets connected to sewers or septic systems, water-based toilets that flush into pits, simple pit latrines with slabs, and ventilated improved pit latrines [2]. To be considered improved, these facilities must be privately used by a single household: if any of these improved technologies is shared by more than one household, the household's sanitation access is considered unimproved. Unimproved facilities include any otherwise improved facility that is shared by more than one household as well as infrastructure that does not properly separate human excreta from potential human contact [3]. Unimproved sanitation facilities include, for example, the use of buckets, hanging latrines, or pit latrines without slab coverings. Engaging in open defecation is also considered unimproved sanitation.

Shared Sanitation

Globally, increasing sanitation access has largely focused on shifting households away from engaging in open defecation or using unimproved facilities towards using private household facilities. In some areas, shared sanitation facilities may provide access to sanitation in communities where the installation of private latrines is not practical or sustainable due to issues of cost, space, or other barriers [4]. There is limited research on what specific factors influence households to share sanitation facilities in rural Asia. Research from other regions of the world, however, suggests that driving factors may include the lack of affordable sanitation options and low levels of awareness of the importance of sanitation [5]. Sharing sanitation facilities with extended family members and neighbors has been found to be more acceptable in certain cultures than in others, and shared sanitation may be a feasible alternative in rural and urban communities that lack access to improved sanitation, especially where open defecation persists [6,7]. A recent study based on data from 51 countries found that shared sanitation appeared to be a risk factor for diarrhea, although the authors noted that social and economic context was an important factor for considering and judging shared sanitation [8].

The JMP acknowledges that shared sanitation responds to the need for increased sanitation access. Shared sanitation is being considered along with current proposals to define the Post-2015 MDG goals and indicators for sanitation, which recommend that improved sanitation be shared among no more than 5 households or 30 people, whichever is fewer. Historically, the JMP has hesitated to endorse shared sanitation in part because of potential decreased security and limited access to shared facilities [9]. Furthermore, the JMP has cited concerns about cleanliness, maintenance, lengthy distances from users' homes, long lines, cost barriers, and difficulty of use for elderly or disabled users and children [10]. Therefore, the JMP has recommended further research on the nature of shared facilities, including whether shared access can be tolerated by users [9].

Significant variability exists within the practice of shared sanitation, which suggests that diverse sanitation solutions may be more effective at meeting the needs of unique populations [11]. The location of the sanitation facility, who owns the facility, who is responsible for maintaining and cleaning the facility, and who uses the facility are examples of such divergence [12]. Further, because the division between improved and unimproved facilities is centered on technology instead of function, authorities may be dissuaded from incorporating context-specific solutions that are outside the realm of approved technologies, even if they properly address sanitation issues [13]. In a study of the limitations of the current JMP classification for shared sanitation, Mazeau and colleagues suggested new metrics for judging the adequacy of shared sanitation that would address the number of households that share a facility, the operations and maintenance of the facility, and the cleanliness of the facility [14]. Despite current proposals to include certain types of shared facilities in the JMP classification for improved sanitation, there is limited data on user satisfaction, especially among women and children, with respect to shared facilities in rural Asia.

User Satisfaction with Sanitation

The literature on user satisfaction with shared sanitation is limited partially because little information is collected from users after the initial installation of sanitation infrastructure [15]. Measuring satisfaction is also limited because satisfaction is a complex concept that reflects personal and cultural experiences and expectations. Socio-cultural factors often drive utilization of sanitation facilities; therefore, the potential acceptability of new sanitation facilities must be modeled by socio-cultural acceptance of the technology as well as practical acceptance [12]. Satisfaction is integral to an individual's or community's decision to use available sanitation facilities.

Sanitation in Indonesia and Bangladesh

Indonesia and Bangladesh highlight the tremendous, divergent gaps in access to sanitation and the global need for an intensified focus on addressing household barriers to promote access to improved sanitation. In 2009, half of all households in Indonesia and only one third of rural households had sustainable access to improved sanitation. Indonesia's MDG sanitation target strives for an overall increase to 62% coverage by 2015 [16]. In Bangladesh, almost two thirds of all households had sustainable access to improved sanitation in 2009. The MDG sanitation target for Bangladesh calls for an increase to 70% coverage by 2015 [17]. In 2010, approximately one third of the rural population of Indonesia and an estimated 7% of the rural population of Bangladesh engaged in open defecation [18].

Given the limited information that exists on user perceptions of shared sanitation in rural contexts, this study analyzes the relationship between different types of sanitation and user perceptions in rural areas of East Java, Indonesia, and Bangladesh. The specific aims of the study are to: 1) explore differences in user satisfaction among users who share toilet facilities compared to those who do not share and to those who engage in open defecation; 2) assess user satisfaction based on factors such as: a) facility cleanliness (East Java), b) ownership of the facility (Bangladesh), and c) perceptions of safety among women (East Java); and 3) characterize differences in households' plans to improve their sanitation facilities based on the type of sanitation facility used.

Methods

East Java

In December 2006, the World Bank's Water and Sanitation Program (WSP) began implementation of the Global Scaling Up Rural Sanitation project, a multi-year intervention focused on reducing diarrhea risk and increasing demand for sanitation in India, Indonesia, and Tanzania. In Indonesia, this project is locally known as Sanitasi Total dan Pemasaran Sanitasi (SToPs). In August and September 2008, the project's Global Impact Evaluation Team collected baseline data from households located in eight rural districts in East Java, Indonesia. Selection of the participants consisted of village selection followed by household selection. From the eight selected districts, 20 total villages (160 sub-villages) were included based on criteria such as water and sanitation access and poverty levels. The eight districts include: Probolinggo, Bondowoso, Situbondo, Banyuwangi, Ngawi, Madiun, Jombang, and Blitar. Thirteen households were randomly selected for inclusion in the baseline survey from a household list that included all households with children under the age of two years. Some sub-villages had too few households with children under the age of two; in these cases, households with children under the age of five were substituted. A total of 2,080 households were included in the baseline survey.

Bangladesh

In 2003, the Government of Bangladesh launched a three-year national sanitation campaign focused on achieving total sanitation coverage and eradicating the practice of open defecation in rural populations. To assess the sustainability of sanitation interventions supported by the government and non-governmental organizations (NGO), WSP collected information on sanitation indicators and other household demographics for 3,000 households drawn from 50 of the Union Parishads (UP) that were declared open defecation free (ODF) in 2005. The Union Parishads represented the different geographic regions in Bangladesh and the different intervention approaches used to combat open defecation during the campaign. In places where a particular implementation approach was used more frequently, Union Parishads were selected randomly; however, in regions where a particular implementation approach was used less frequently, all UPs using this less popular approach were included to guarantee a broad representation of the approaches used in the sanitation campaign. Once the Union Parishad was selected, the research team randomly selected three villages or clusters of households with at least 100 households: one village located close to the UP government headquarters, one village located at an intermediate distance from UP headquarters, and a final village located a considerable distance away from UP headquarters (or one that was considered remote). Following this selection process, 100 households from each village or cluster of households were listed using a standard sampling format; from this list, 20 households were identified using a systematic random sampling method for a final sample size of 3,000 households.

Variables

The variables of interest drawn from the East Java and Bangladesh household surveys were compiled from questions directed to the head of household. In East Java, a household includes everyone who lives in the house and shares meals together, including both family and non-family members. In Bangladesh, a household includes everyone who shares the household's food and who usually sleeps in the house at night. Individuals who visit periodically and contribute wages to the household are also included.

This research examines user satisfaction with the household sanitation facility. We recognize that household members may also use other sanitation facilities at work, school, neighbors' houses, and elsewhere. Nonetheless, we seek to understand users' preferences regarding the facilities they use at home given that the household is the basic unit of reference used in the MDG sanitation targets.

Differences exist in the questionnaires used by the two study sites, as noted below. Our study examined the following topics: sharing status; type of sanitation; level of satisfaction; perceived cleanliness (East Java only); improvement plans; open defecation (Bangladesh only); and ownership of the sanitation facility (Bangladesh only). Table S1 provides additional information on our research questions.

In East Java, the household questionnaire asked: "Overall, how satisfied are you with your main defecation facility?" Response options included: *very satisfied*, *somewhat satisfied*, *less than satisfied*, and *completely dissatisfied*. In the analysis for East Java, *very satisfied* and *somewhat satisfied* were combined; similarly, *less than satisfied* and *completely dissatisfied* answers were combined.

In Bangladesh, the household questionnaire asked: "How satisfied are you with your current place of defecation?" The answers included: *satisfied*, *moderately satisfied*, and *unsatisfied*. The answers *satisfied* and *moderately satisfied* were combined in the analysis for Bangladesh. Given the limited sample size in the different categorical responses, the binary satisfaction variables for East Java and Bangladesh provided more statistical power for the analyses.

For this study, we used the improved or unimproved designations based on the sanitation infrastructure only and did not use the facility's status as a private or shared facility to designate the toilet as improved or unimproved. Thus, if a facility met the structural standard for improved, not considering the private or shared status, then for the purposes of this study the facility was considered *improved private* or *improved shared*. We considered a facility to be private if used by only one household; we considered a facility to be shared if used by two or more households. Open defecation was considered a separate category and was not included in either the shared or unimproved categories. Variables S1 provides more information on variable definitions.

Potential confounders controlled for in the relevant analyses include: the age, gender, highest level of education and occupation of the head of household, the household's wealth category, religion, ethnicity (East Java only), ODF approach (Bangladesh only), and the status of the household's water source (improved or unimproved using JMP definitions). Definitions are provided in Covariates S1.

Statistical Analysis

Pearson's chi-squared tests were conducted to test for independence among the covariates. A significant chi-square statistic suggests a dependent relationship between the tested variables. Relative risks (RRs) and 95% confidence intervals (CIs) were calculated using log-binomial regression. In cases where the model experienced problems of convergence, Poisson regression with robust error variance was used to calculate relative risks and 95% CIs. Relative risks were evaluated controlling for variables – selected *a priori* to the analysis – that were found to be significant in bivariate analyses. Relative risks can be interpreted as the likelihood of an outcome of interest occurring in one group of households compared to the likelihood of the same outcome occurring in a different group of households. Analyses were conducted using SAS 9.2 (Cary, NC).

Ethical Consideration

The data used from Bangladesh and East Java are publicly available. The Bangladesh dataset was obtained from Craig Kullman, Water and Sanitation Program, World Bank. The East Java dataset was obtained from Bertha Briceno, Water and Sanitation Program, World Bank. Both datasets were anonymized before data analysis, and the research was determined to be exempt from human subject protection by The George Washington University Institutional Review Board given that it involved the analysis of pre-existing data that are publicly available. Further, ethical oversight was not obtained because the proposed uses and disclosures of protected health information involved no more than minimal risk to the privacy of individuals (45 CFR 46 164.512).

Results

East Java

Approximately 40% of surveyed households reported practicing open defecation. A further 11% used an unimproved sanitation facility and the remaining 49% of households used an improved facility. Among households with improved facilities, 71% of all households had a private improved facility and 29% shared an improved facility. Of households that shared improved or unimproved sanitation facilities, 51% of households shared the facility between two households, and 38% used a facility shared among three to five households. Only 10% shared the facility with

Table 1. East Java Descriptive Statistics.

East Java Descriptive Statistics	
Total number of households surveyed	2,087
Average size of household	8.6 people
Gender of head of household	
Male	1,994 (95.5%)
Female	93 (4.5%)
Age of head of household	
<25	105 (5.0%)
25 to 34	739 (35.4%)
35 to 44	667 (32.0%)
45 to 54	340 (16.3%)
55 to 64	154 (7.4%)
65+	82 (3.9%)
Ethnicity of household (n = 2,073)	
Javanese	1,347 (65.0%)
Maduranese	681 (32.9%)
Other (Osing, Chinese)	45 (2.2%)
Religion of household	
Islam	2,059 (98.7%)
Other (Protestant, Catholic, Hindu)	28 (1.3%)
Highest education level of head of household (n = 1,997)	
Primary or lower	1,127 (56.4%)
Secondary	786 (39.4%)
Tertiary or higher	84 (4.2%)
Primary occupation of head of household (n = 2,016)	
Agriculture, forestry, fishery, hunting, livestock	1,008 (50%)
Mining and exploration	17 (0.8%)
Manufacturing industry	142 (7.0%)
Electricity, gas and water	6 (0.3%)
Construction	168 (8.3%)
Trade, retail, restaurant and hotel	293 (14.5%)
Transportation, warehousing and communication	99 (4.9%)
Finance, insurance, building leasing, land and services	14 (0.7%)
Public service	261 (12.9%)
Other	8 (0.4%)
Household source of drinking water (rainy season) (n = 2,082)	
Improved	*1,757 (84.4%)*
Piped water, into home or yard	130 (6.2%)
Piped water, public tap, tube well, borehole	482 (23.2%)
Protected dug well	757 (36.4%)
Protected spring water	388 (18.6%)
Unimproved	*325 (15.6%)*
Unprotected dug well	210 (10.1%)
Unprotected spring water	47 (2.3%)
Other	68 (3.3%)
Household sanitation facility infrastructure and sharing status (n = 2,031)	
Improved, private	723 (35.6%)
Improved, shared	262 (12.9%)
Unimproved, private	105 (5.2%)
Unimproved, shared	118 (5.8%)
Open defecation	823 (40.5%)

Table 1. Cont.

East Java Descriptive Statistics	
Number of households sharing facility (n = 380)	
2 households	194 (51.1%)
3 households	98 (25.8%)
4 households	26 (6.8%)
5 households	21 (5.5%)
6 or more households	38 (10.0%)
Don't know	3 (0.8%)
User satisfaction with sanitation facility (n = 2,031)	
Very satisfied or somewhat satisfied	*1,375 (67.7%)*
Improved, private	596 (29.3%)
Improved, shared	179 (8.8%)
Unimproved, private	44 (2.2%)
Unimproved, shared	63 (3.1%)
Open defecation	493 (24.3%)
Less than satisfied or completely dissatisfied	*656 (32.3%)*
Improved, private	127 (6.3%)
Improved, shared	83 (4.1%)
Unimproved, private	61 (3.0%)
Unimproved, shared	55 (2.7%)
Open defecation	330 (16.2%)
Perceived cleanliness of sanitation facility by user (n = 1,208)	
Very clean or clean	*1,011 (83.7%)*
Improved, private	665 (55.0%)
Improved, shared	222 (18.4%)
Unimproved, private	59 (4.9%)
Unimproved, shared	65 (5.4%)
Dirty or very dirty	*197 (16.3%)*
Improved, private	58 (4.8%)
Improved, shared	40 (3.3%)
Unimproved, private	46 (3.8%)
Unimproved, shared	53 (4.4%)

six or more households. Approximately 68% of all surveyed households reported feeling satisfied with their sanitation facility. (Table 1).

Sharing status, sanitation facility, and open defecation. We found that 82.4% of households with private improved latrines reported feeling satisfied with their place of defecation, while 70.2% of households who shared an improved facility between two households were satisfied [RR 1.19, 95% CI 1.06, 1.33] and only 68.3% of households with improved facilities shared between two or more households reported satisfaction [RR 1.19, 95% CI 1.09, 1.31]. In contrast, 48.0% of households with unimproved facilities reported feeling satisfied with their place of defecation, which is not significantly different from the 59.9% of households who engaged in open defecation who reported feeling satisfied [RR 1.10, 95% CI 0.94, 1.30]. (Table 2).

Satisfaction levels and the number of households sharing the facility. In our analysis of satisfaction with place of defecation stratified by the number of households sharing the facility, the results were not significant. We found that 67.2% of households who shared their facility between two households reported feeling satisfied; 60.9% of households who shared their facility among three or more households reported feeling satisfied [RR 1.04, 95% CI 0.89, 1.22]. Satisfaction among households sharing between two or three households was reported at 66.0%; satisfaction among households who shared among four or more households was not significantly lower at 57.6% [RR 1.08, 95% CI 0.88, 1.32]. (Table 2). Finally, no significant differences in user satisfaction were observed in a comparison of three sharing levels: 66.0% of households who shared between two or three households were satisfied, 59.6% of households who shared among four or five households were satisfied, and 55.3% of households who shared among six or more households were satisfied [chi-square value 2.17, $p = 0.338$].

Perceived cleanliness, sharing, and satisfaction. Households with clean sanitation facilities reported significantly different levels of satisfaction compared to households with dirty sanitation facilities. We found that 79.5% of households who perceived their sanitation facility to be clean also reported feeling satisfied with their place of defecation; only

Table 2. Relative Risk for Households Reporting Satisfaction with their Main Defecation Facility (East Java).

Analysis		*Crude RR (95% CI)*	*p*	*Adjusted RR (95% CI)**	*p*
Sanitation facility	Improved private facilities	1.38 (1.29, 1.47)	<0.0001	1.41 (1.32, 1.52)	<0.0001
	Improved shared facilities	1.14 (1.03, 1.26)	0.0097	1.25 (1.12, 1.40) $^{\Omega}$	<0.0001
	Unimproved facilities	1.25 (1.08, 1.45)	0.0032	1.10 (0.94, 1.30) $^{\Omega}$	0.2373
	Open defecation	1.00 (ref.)		1.00 (ref)	
Private vs. shared (improved facilities)	Improved private facility	1.21 (1.10, 1.32)	<0.0001	1.19 (1.09, 1.31)	0.0002
	Improved shared facility	1.00 (ref)		1.00 (ref)	
Private vs. limited sharing	Private facility	1.17 (1.05, 1.31)	0.0039	1.19 (1.06, 1.33)	0.0037
	Facility shared between only 2 households	1.00 (ref)		1.00 (ref)	
Stratified limited sharing	Facility shared between only 2 households	1.10 (0.95, 1.28)	0.2020	1.04 (0.89, 1.22) $^{\Omega}$	0.6379
	Facility shared among 3 or more households	1.00 (ref)		1.00 (ref)	
Stratified limited sharing	Facility shared between 2–3 households	1.14 (0.94, 1.40)	0.1844	1.08 (0.88, 1.32) $^{\Omega}$	0.4753
	Facility shared among 4 or more household	1.00 (ref)		1.00 (ref)	
Facility cleanliness (as reported by household)$^{+}$	Clean	2.05 (1.71, 2.44)	<0.0001	1.74 (1.45, 2.08) $^{\Omega}$	<0.0001
	Dirty	1.00 (ref)		1.00 (ref)	

*All adjusted analyses were controlled for gender, age, ethnicity, religion, education level, and occupation of head of household, as well as household's income quartile and JMP status of drinking water source.
$^{+}$Households categorized as "open defecators" were excluded from this analysis.
$^{\Omega}$RRs and 95% CIs were calculated using Poisson regression with robust error variance.

38.9% of households who perceived their facility to be dirty reported feeling satisfied [RR 1.74, 95% CI 1.45, 2.08].

Households with private improved sanitation facilities did not perceive their places of defecation to be significantly cleaner than households that shared improved sanitation facilities between two households. Almost 92% of households with private improved latrines reported cleanliness; 91.4% of households who shared improved latrines between two households reported cleanliness [RR 1.00, 95% CI 0.82, 1.21]. Among households who shared an improved latrine, 91.4% of households who shared between two households perceived their facility to be clean; only 75.7% of households who shared among three or more households perceived their facility to be clean [RR 1.14, 95% CI 1.02, 1.27]. In a separate analysis of households who share an improved latrine, 88.7% of households who shared between 2 to 3 households perceived their facility to be clean; 68% of households who shared among 4 or more households perceived their facility to be clean [RR 1.21, 95% CI 1.00, 1.47]. (Table 3).

Plans to improve sanitation facilities, sharing status, and satisfaction. We found that 42.6% of households with sanitation facilities shared between two or more households reported plans to upgrade their sanitation facility within the next year; in comparison, 44.2% of households with private improved facilities reported plans to upgrade their facility. This difference was not significant [RR 1.09, 95% CI 0.73, 1.63]. Approximately 47% of households with improved sanitation facilities shared between two households reported plans to upgrade their facility, which was not significantly more than the 44.2% of households with private improved facilities who reported plans to upgrade their sanitation facility [RR 1.14, 95% CI 0.74, 1.77]. In our analysis of households with improved facilities shared between two or more households, 46% of households who were satisfied with their place of defecation also reported plans to upgrade their facility within a year; 39.7% of households who were not satisfied with their place of defecation reported plans to upgrade their facility. No significant difference was observed [RR 0.97, 95% CI 0.63, 1.48]. (Table 4).

Perceptions of female users. Additional questions were asked of female respondents regarding their perceptions of safety, privacy, and harassment. Almost 95% of women from households with private improved latrines reported that their facility is safe at night, which was not significantly different than the 91.2% of women from households with shared improved latrines that also reported that their facility is safe at night [RR 1.03, 95% CI 0.99, 1.08]. Approximately 89% of women from households who shared an improved latrine between two households reported that their facility is safe at night, which is significantly greater than the 77% of women from households who shared an improved latrine among three or more households who also reported that their facility is safe at night [RR 1.15, 95% CI 1.08, 1.23]. In a separate analysis, we found that 91.2% of women from households who shared an improved latrine with two or more households reported that their facility is safe at night; just 73.6% of women from households who openly defecate reported that their place of defecation is safe at night [RR 1.24, 95% CI 1.17, 1.31]. (Table 5). Additional results on perceptions of female users are provided in Table S2.

Bangladesh

Just over 2% of surveyed households reported engaging in open defecation, and only 8% used an unimproved facility. Over half of all households had a private improved facility (53%), and the remaining 37% of households had a shared improved facility. Among households that shared, 52% of the facilities were shared

Table 3. Relative Risk for Households Reporting Clean Sanitation Facilities (East Java).

Analysis		*Crude RR (95% CI)*	*p*	*Adjusted RR (95% CI)**	*p*
Private vs. limited sharing (improved facilities)	Improved private facility	1.01 (0.95, 1.06)	0.8144	1.00 (0.82, 1.21) [Ω]	0.9813
	Improved facility shared between only 2 households	1.00 (ref)		1.00 (ref)	
Stratified limited sharing (improved facilities)	Improved facility shared between only 2 households	1.21 (1.08, 1.36)	0.0015	1.14 (1.02, 1.27) [Ω]	0.0215
	Improved facility shared among 3 or more households	1.00 (ref)		1.00 (ref)	
Stratified limited sharing (improved facilities)	Improved facility shared between 2–3 households	1.30 (1.07, 1.59)	0.0080	1.21 (1.00, 1.47) [Ω]	0.0498
	Improved facility shared among 4 or more households	1.00 (ref)		1.00 (ref)	

*All adjusted analyses were controlled for gender, age, ethnicity, religion, education level, and occupation of head of household, as well as household's income quartile and JMP status of drinking water source.
[Ω]RRs and 95% CIs were calculated using Poisson regression with robust error variance.

between two households, 45% of the facilities were shared with three to five households, and 3% of households shared with six or more households. Almost 83% of households with a private or shared improved facility reported feeling satisfied with their sanitation facility. Only 35% of households with a private or shared improved facility reported having plans to improve their latrine or build a new one within a year (Table 6).

Sharing status and sanitation facility. Among households with private improved facilities, 87.7% of households reported feeling satisfied. In comparison, only 74.5% of households with improved facilities shared by two or more households were satisfied [RR 1.15, 95% CI 1.10, 1.20]. Approximately 74% of households with improved facilities who shared between two households reported feeling satisfied with their place of defecation compared to only 64.8% of households who shared an improved facility among three or more households [RR 1.15, 95% CI 1.03, 1.29]. (Table 7).

Satisfaction levels and the number of households sharing the facility. We found that 71.4% of households sharing an improved facility between two to three households felt satisfied, which was not significantly different than the 62.7% of households sharing an improved facility among four or more households who also felt satisfied with their facility [1.15, 95% CI 0.98, 1.34]. No significant differences in user satisfaction were seen in a separate analysis that assessed three levels of sharing: households who shared an improved facility between two or three households, households who shared an improved facility among four or five households, and households who shared an improved facility among six or more households [chi-square value: 5.96, p = 0.051]. (Table 7).

Ownership of the sanitation facility. Significant differences in user satisfaction were observed between households who independently owned a shared latrine compared to households who shared ownership of the shared latrine with other households. Exactly 80% of households who privately owned an improved shared latrine felt satisfied with their place of defecation; 72.7% of households who jointly owned an improved shared latrine felt satisfied [RR 1.10, 95% 1.01, 1.19]. (Table 7).

Plans to improve sanitation facilities, sharing status, and satisfaction. Among households with shared improved facilities, 39.2% reported plans to improve their current latrine or build a new latrine within a short period of time. Among households with private improved facilities, 33.1% reported similar improvement plans [RR 1.14, 95% CI 1.02, 1.27]. In contrast, in an analysis of household plans to improve a current latrine or build a new latrine, 38.4% of households sharing an improved latrine between two households reported plans to improve their latrine; 33.1% of households with a private improved latrine reported improvement

Table 4. Relative Risk for Households Reporting Plans to Improve Sanitation Facility or Build New Sanitation Facility (East Java).

Analysis		*Crude RR (95% CI)*	*p*	*Adjusted RR (95% CI)**	*p*
Private vs. shared (improved facilities)	Improved shared facility	0.96 (0.65, 1.44)	0.8575	1.09 (0.73, 1.63)	0.6696
	Improved private facility	1.00 (ref)		1.00 (ref)	
Private vs. limited sharing (improved facilities)	Improved facility shared between only two households	1.07 (0.70, 1.65)	0.7531	1.14 (0.74, 1.77)	0.5448
	Improved private facility	1.00 (ref)		1.00 (ref)	
Satisfaction level (improved shared facilities)	Satisfied	1.16 (0.75, 1.80)	0.5056	0.97 (0.63, 1.48)	0.8729
	Unsatisfied	1.00 (ref)		1.00 (ref)	

*All adjusted analyses were controlled for gender, age, ethnicity, religion, education level, and occupation of head of household, as well as household's income quartile and JMP status of drinking water source.

Table 5. Relative Risk for Women Reporting Feeling Safe at Night at Defecation Facility (East Java).

Analysis		*Crude RR (95% CI)*	*p*	*Adjusted RR (95% CI)**	*p*
Private vs. shared (improved facilities)	Improved private facility	1.04 (1.00, 1.08)	0.06	1.03 (0.99, 1.08)	0.15
	Improved shared facility	1.00 (ref)		1.00 (ref)	
Stratified limited sharing (improved facilities)	Improved facility shared between only 2 households	1.16 (1.09, 1.23)	<0.0001	1.15 (1.08, 1.23)	<0.0001
	Improved facility shared among 3 or more households	1.00 (ref)		1.00 (ref)	
Sanitation facility	Improved shared facility	1.24 (1.17, 1.31)	<0.0001	–	
	Open defecation	1.00 (ref)		–	

*All adjusted analyses were controlled for gender, age, ethnicity, religion, education level, and occupation of head of household, as well as household's income quartile and JMP status of drinking water source.

plans [RR 1.14, 95% CI 0.96, 1.35]. Among households with an improved shared facility, 63.4% of households who reported feeling unsatisfied with their facility also reported plans to improve their current latrine. In contrast, only 31.1% of households with improved shared facilities who reported feeling satisfied with their facility also reported plans to improve their latrine [RR 2.04, 95% CI 1.76, 2.37]. (Table 8).

Open defecation. Almost 18% of households who shared an improved sanitation facility and who also felt unsatisfied with their sanitation facility continued to openly defecate; in contrast, only 6.3% of households who were satisfied reported continued open defecation [RR 2.68, 95% CI 1.81, 3.96]. In a similar analysis, 11.6% of households that shared their improved facility between two households continued to openly defecate; only 3.9% of households with private improved latrines also reported open defecation [RR 2.18, 95% CI 1.38, 3,45]. (Table 9).

Discussion

East Java

Sharing status, sanitation facility, and open defecation. Our results suggest that households with private improved facilities were more satisfied than households that shared improved facilities, even when sharing the facility was limited to sharing between only two households. Households were much more likely to be satisfied with a private improved facility than were households practicing open defecation. Similarly, households sharing an improved facility were more likely to be satisfied than households practicing open defecation. No significant difference in satisfaction was found between households with unimproved facilities and households practicing open defecation. These results suggest that sharing improved sanitation facilities is preferred over the practice of open defecation and that private improved sanitation is consistently preferred over any other sanitation option.

Satisfaction levels and the number of households sharing the facility. Among sharers, the number of households sharing did not seem to significantly alter satisfaction levels, although we were not able to look at all combinations of the number of households sharing a given facility in our analysis. No significant difference was observed in satisfaction in users of facilities shared between two households compared to users of facilities shared by three or more households. This finding is contrary to our results found in Bangladesh. Other research has indicated that a greater number of households sharing a sanitation facility lead to lower levels of satisfaction. For example, in a study on user satisfaction with sanitation facilities in Kampala, Uganda, the main determinant for reporting dissatisfaction was sharing a facility with too many users [7].

Perceived cleanliness. As mentioned previously, the JMP acknowledges that shared facilities potentially could be regarded as improved facilities but notes concerns about cleanliness and hygiene [10]. User perceptions of latrine cleanliness were not significantly different in users of private latrines compared to users of facilities shared between two households. Our research suggests that sharing sanitation facilities between a limited number of households, namely two, may not alter perceived cleanliness. Similarly, cleanliness was seen as a driver for latrine adoption in a separate study in Benin, West Africa [19]. In Bhopal, India, an assessment of seven communal toilet facilities revealed that 65% of users were satisfied with the condition of the communal toilets, and the facilities were liked mostly because of their convenience and privacy; they were disliked primarily due to their lack of cleanliness and smell [20]. In a study in Kampala, Uganda, the second most frequent reason for reporting dissatisfaction with sanitation facilities was that facilities were not clean and smelled bad [7]. Perceived cleanliness appears to play an important role in satisfaction with sanitation facilities in East Java; similar research in other regions corroborates these results.

Plans to improve sanitation facilities. Household plans to improve a current latrine or build a new latrine within a short period of time were not affected by sharing status or level of user satisfaction. These results indicate that other barriers to procuring a private sanitation facility may exist or that private sanitation is not a main priority relative to other household needs.

Perceptions of female users. In our analysis of surveyed communities in East Java, shared improved facilities were perceived to be safe for use by women during the evening hours. No significant differences were observed in perceived nighttime safety among women from households with private improved latrines compared to women from households with shared improved latrines. However, among sharers of improved latrines, women from households who shared between two households were more likely to feel safe than women from households who shared among three or more households. Further, women from households with shared improved facilities were significantly more likely to report feeling safe compared to women from households that openly defecated. These results suggest that efforts made to reduce the number of households that share sanitation facilities could reduce levels of fear among female users. Other research in rural Benin has shown that increasing privacy and safety,

Table 6. Bangladesh Descriptive Statistics.

Bangladesh Descriptive Statistics	
Total number of households surveyed	3,000
Average size of household	4.9 people
Gender of head of household	
Male	2,771 (92.4%)
Female	229 (7.6%)
Age of head of household (years)	
<25	77 (2.6%)
25 to 34	626 (20.9%)
35 to 44	782 (26.1%)
45 to 54	682 (22.7%)
55 to 64	490 (16.3%)
65+	343 (11.4%)
Religion of household	
Islam	2,553 (85.1%)
Hinduism	386 (12.9%)
Buddhism	61 (2.0%)
Highest education level of head of household	
Primary or lower	2,104 (70.1%)
Secondary	833 (27.8%)
Tertiary or higher	63 (2.1%)
Primary occupation of head of household	
Agriculture, farming, poultry, fish, livestock	1,027 (34.2%)
Business	184 (6.1%)
Housewife	151 (5.0%)
Service	232 (7.7%)
Skilled labor	194 (6.5%)
Unskilled labor	511 (17.0%)
Small business	393 (13.1%)
Other	308 (10.3%)
Household source of drinking water	
Improved	*2,930 (97.7%)*
Piped water, into home or yard	48 (1.6%)
Public tap, standpipe	19 (0.6%)
Tube well, shallow or deep	2,826 (94.2%)
Protected well	17 (0.6%)
Protected spring	20 (0.7%)
Unimproved	*70 (2.3%)*
Unprotected well	3 (0.1%)
Unprotected spring	21 (0.7%)
PSF (Pond Sand Filter - surface water)	46 (1.5%)
Household sanitation facility, sharing status, and ownership status	
Improved facility, private	*1,588 (52.9%)*
Own	1570 (52.3%)
Jointly owned	0 (0%)
Live on rent	18 (0.6%)
Improved facility, shared	*1,098 (36.6%)*
Own	225 (7.5%)
Jointly owned	692 (23.1%)
Owned by others/neighbor	181 (6.0%)
Unimproved facility	*237 (7.9%)*

Table 6. Cont.

Bangladesh Descriptive Statistics	
Open defecation	*77 (2.6%)*
Number of households sharing sanitation facility (n = 515)	
2 households	268 (52.0%)
3 households	145 (28.2%)
4 households	64 (12.4%)
5 households	24 (4.7%)
6 or more households	14 (2.7%)
User satisfaction with sanitation facility (n = 2,487)	
Satisfied or moderately satisfied	*2,060 (82.8%)*
Improved, private	1,377 (55.4%)
Improved, shared	683 (27.5%)
Unsatisfied	*427 (17.2%)*
Improved, private	193 (7.8%)
Improved, shared	234 (9.4%)
Household plans to improve the present latrine or build a new one within twelve months (n = 2,466)	
Yes	*871 (35.3%)*
Improved, private	516 (20.9%)
Improved, shared	355 (14.4%)
No	*1,595 (64.7%)*
Improved, private	1,044 (42.3%)
Improved, shared	551 (22.3%)
Member of household defecates in the open other than in flood time (n = 2,680)	
Yes	*176 (6.6%)*
Improved, private	62 (2.3%)
Improved, shared	114 (4.3%)
No	*2,504 (93.4%)*
Improved, private	1,521 (56.8%)
Improved, shared	983 (36.7%)

especially for women, were documented reasons for wanting a latrine [19]. Likewise, a safe and private sanitation facility was more significant for women in urban Ghana and urban Uganda than for their male counterparts [14]. While more research is needed on the perceptions of women and children to better understand the factors that contribute to safety in different sanitation contexts, safety remains a key matter in the assessment of sanitation facilities.

Bangladesh

Sharing status and sanitation facility. Our results suggest that households were more satisfied with private improved facilities than with shared improved facilities, even if only two households shared the facility. In contrast to East Java, the results in Bangladesh indicate that satisfaction was significantly more likely among households sharing an improved facility between two households compared to households sharing an improved facility among three or more households.

Satisfaction levels and the number of households sharing the facility. Satisfaction was not significantly different between households with improved facilities shared between two to three households compared to households with improved facilities shared among four or more households. These results are similar to our results from East Java. It should be noted that the Government of Bangladesh considers hygienic sanitation facilities shared between two households to be improved sanitation [21]. It is unclear, however, how and why this decision by the government was made and whether it was based on users' perceptions, feasibility issues, or other driving factors.

Ownership of the sanitation facility. Satisfaction was found to be significantly greater among households who privately owned a shared facility compared with those who jointly owned a shared facility. Ownership of a housing structure with a sanitation facility may provide greater satisfaction, as landlords can potentially charge more for housing. In rural Benin, researchers found that households often wanted a latrine because they were embarrassed to ask visitors to openly defecate. Households wanted to gain the respect of visitors, and latrine ownership was seen as a way to raise a household's status among their neighbors [19].

Plans to improve sanitation facilities. Households sharing an improved facility between two or more households were more likely to report plans to improve their current latrine or to build a new latrine compared to households who did not share; however, households who shared an improved latrine between two or more households were not significantly more likely to have these plans compared to households with improved private facilities. These

Table 7. Relative Risk for Households Reporting Satisfaction with their Current Place of Defecation (Bangladesh).

Analysis		*Crude RR (95% CI)*	*p*	*Adjusted RR (95% CI)**	*p*
Private vs. shared (improved facilities)	Improved private facility	1.18 (1.13, 1.23)	<0.0001	1.15 (1.10, 1.20) Ω	<0.0001
	Improved shared facility	1.00 (ref)		1.00 (ref)	
Private vs. limited sharing (improved facilities)	Improved private facility	1.18 (1.10, 1.27)	<0.0001	1.15 (1.07, 1.24) Ω	0.0001
	Improved facility shared between 2 households	1.00 (ref)		1.00 (ref)	
Stratified limited sharing (improved facilities)	Improved facility shared between 2 households	1.15 (1.02, 1.29)	0.0209	1.15 (1.03, 1.29) Ω	0.0122
	Improved facility shared among 3 or more households	1.00 (ref)		1.00 (ref)	
Stratified limited sharing (improved facilities)	Improved facility shared between 2-3 households	1.14 (0.97, 1.34)	0.1157	1.15 (0.98, 1.34) Ω	0.0910
	Improved facility shared among 4 or more households	1.00 (ref)		1.00 (ref)	
Facility ownership (improved shared facilities)	Privately own sanitation facility	1.10 (1.02, 1.19)	0.0184	1.10 (1.01, 1.19) Ω	0.0251
	Jointly own sanitation facility	1.00 (ref)		1.00 (ref)	

* Controlling for head of household's gender, age, religion, education level, and occupation, household income quintile, approach to ODF, and JMP status of drinking water source.

ΩRRs and 95% CIs were calculated using Poisson regression with robust error variance.

results may indicate that households with private improved latrines do not differ in their plans to improve their current latrine or to build a new latrine when compared to households sharing an improved latrine between two households. Further, unsatisfied households were more likely to report improvement plans than were satisfied households. These results suggest that households who share their sanitation facilities with others and who are unsatisfied with their current facility would like to see an improvement in their existing sanitation facility.

Open defecation. The practice of open defecation is wrought with difficulties, discomforts, and dangers, and in many cases, households would like to improve their sanitation situation [19]. It should be noted, however, that households with better sanitation options may still resort to open defecation. In Bangladesh, households with shared improved facilities who reported dissatisfaction with their sanitation facility were more likely to report practicing open defecation compared to households with shared improved facilities who reported satisfaction with their facility. These results suggest that user satisfaction may play a role in helping move households "up" the sanitation ladder and away from the practice of open defecation. Sharing status and sanitation infrastructure are two important factors that must be considered when analyzing sanitation options. Understanding the factors that affect a latrine user's level of satisfaction must be included in planning strategies.

Studies in urban settings have reported that shared sanitation may serve to reduce open defecation in communities where it continues to persist. For example, community sanitation systems were installed in urban Bhopal, India, because they were considered the best option for quickly addressing the sanitation needs of the urban poor. Upon evaluation of the community sanitation systems, the significant determinants of communal latrine use by members of households without a latrine included distance to the latrine, opening hours, facility age, cleanliness, and cost. These shared sanitation facilities were found to reduce open defecation but not eliminate it completely. Interestingly, twice as many males as females used the communal latrines [22]. A study in urban Bangladesh found that public sanitation facilities

Table 8. Relative Risk for Households Reporting Plans to Improve Sanitation Facility or Build New Sanitation Facility (Bangladesh).

Analysis		*Crude RR (95% CI)*	*p*	*Adjusted RR (95% CI)**	*p*
Private vs. shared (improved facilities)	Improved shared facility	1.18 (1.06, 1.32)	0.0020	1.14 (1.02, 1.27)	0.0248
	Improved private facility	1.00 (ref)		1.00 (ref)	
Limited sharing vs. private (improved facilities)	Improved facility shared between 2 households	1.16 (0.98, 1.37)	0.0825	1.14 (0.96, 1.35)	0.1471
	Improved private facility	1.00 (ref)		1.00 (ref)	
Satisfaction level (improved shared facilities)	Unsatisfied	2.04 (1.76, 2.37)	<0.0001	–	–
	Satisfied	1.00 (ref)		1.00 (ref)	

* Controlling for head of household's gender, age, religion, education level, and occupation, household income quintile, approach to ODF, and JMP status of drinking water source.

Table 9. Relative Risk for Households Reporting Open Defecation Plans (Bangladesh).

Analysis		Crude RR (95% CI)	p	Adjusted RR (95% CI)*	p
Satisfaction level (improved shared facilities)	Unsatisfied	2.85 (1.91, 4.25)	<0.0001	2.68 (1.81, 3.96)	<0.0001
	Satisfied	1.00 (ref)		1.00 (ref)	
Private vs. limited sharing (improved facilities)	Improved facilities shared between only 2 households	2.95 (1.96, 4.46)	<0.0001	2.18 (1.38, 3.45) $^{\Omega}$	0.0008
	Improved private facilities	1.00 (ref)		1.00 (ref)	

* Controlling for head of household's gender, age, religion, education level, and occupation, household income quintile, approach to ODF, and JMP status of drinking water source.
$^{\Omega}$RRs and 95% CIs were calculated using Poisson regression with robust error variance.

increased weight-for-height scores in children. The authors hypothesized that the improvements were a result of reductions in environmental contamination by children's feces [23]. Reducing open defecation remains integral to overall improvement in the sanitation sector, and moving people to shared sanitation, under certain conditions, may be beneficial [24].

Limitations

This cross-sectional study provides insight into factors that affect satisfaction and draws inferences about different exposures. Alternative explanations for the results may exist and should be considered. Many of the findings in East Java, Indonesia, and Bangladesh were analogous. However, the variability in our results suggests that the findings may be limited to rural East Java and rural Bangladesh. These results may not be applicable to urban and peri-urban populations within the countries studied or to other countries. Furthermore, this study does not analyze the effects of broader factors such as policies, economic factors, or macro-social influences such as ethnicity or religion on latrine access and use. In certain cases, the limited sample size within different categories of the number of households sharing a sanitation facility hindered the ability to statistically test satisfaction levels.

Additionally, efforts are needed to expand the approaches used to measure satisfaction in different cultural contexts. For example, researchers in India have developed the Pachod Paisa scale, a metric created with the local culture in mind. The numeric Pachod Paisa scale, so named for the one hundred "paisa" that make up the Indian rupee, has been used for measuring patient satisfaction with health care facilities and other related variables concerning personal impressions. The scale was also used in a sanitation study in rural Maharashtra, India, to measure the factors driving latrine use. The Pachod Paisa scale offers insight into satisfaction measurements on the Indian subcontinent and illustrates a workable, culturally appropriate alternative to other scales [25].

Conclusions

From the perspective of users, private sanitation facilities appear to be the preferred form of sanitation. In both rural and urban communities where private sanitation may not be a realistic option, shared sanitation facilities that maintain an acceptable level of cleanliness and limit the numbers of users are feasible alternatives. User satisfaction is one crucial piece in the sanitation sector that may inhibit or encourage acceptance of and commitment to a change or "step up" the sanitation ladder. More research is needed to understand household satisfaction, and future research should integrate metrics for measuring satisfaction into surveys. As seen in rural Bangladesh, when moving communities away from open defecation, satisfaction with the sanitation facility is crucial in order to prevent open defecation practices from recurring. In addition, perceived cleanliness of the facility was positively associated with household satisfaction in East Java. Simple distinctions between improved and unimproved technologies tend to oversimplify the spectrum of sanitation facilities and misrepresent the variability found within the shared sanitation sector. Shared sanitation facilities that are well maintained, safe for women, and shared by a limited number of households may be an acceptable form of sanitation. Efforts to improve sanitation should aim to incorporate these findings in order to more robustly track progress.

Supporting Information

Table S1 Research questions.

Table S2 Female perceptions of household sanitation facilities in East Java.

Variables S1 Variable definitions.

Covariates S1 Covariate definitions for East Java and Bangladesh.

Acknowledgments

The authors acknowledge Dr. Samuel Simmens and Dr. Angelo Elmi for their contributions on the analysis of the data and for providing useful comments and feedback.

Author Contributions

Conceived and designed the experiments: CK JPG. Analyzed the data: KBN JK. Wrote the paper: KBN JPG.

References

1. WHO/UNICEF (2013) Progress on Drinking Water and Sanitation: 2013 Update. Available: http://www.unicef.org/wash/files/JMP2013final_en.pdf. Accessed 15 September 2013.
2. WHO/UNICEF (2012) Proposal for consolidated drinking water, sanitation and hygiene targets, indicators and definitions. Available: http://www.wssinfo.org/fileadmin/user_upload/resources/A-proposal-for-consolidated-WASH-goal-targets-definitions-and-indicators_version7_Nov22_final.pdf. Accessed 05 July 2013.
3. WHO/UNICEF (2010) Improved and unimproved water and sanitation facilities. Available: http://www.wssinfo.org/definitions-methods/watsan-categories/. Accessed 07 October 2012.
4. Norman G (February 2011) When are communal or public toilets an appropriate option? Water and Sanitation for the Urban Poor. Available: http://www.wsscc.org/sites/default/files/publications/wsup_topicbrieficommunalpublictoilets_2011.pdf. Accessed 07 October 2012.
5. Rodgers AF, Ajono LA, Gyapong JO, Hagan M, Emerson PM (2007) Characteristics of latrine promotion participants and non-participants; inspection of latrines; and perceptions of household latrines in Northern Ghana. Trop Med Int Health 12: 772–782. doi: 10.1111/j.1365-3156.2007.01848.x
6. Günther I, Niwagaba BC, Lüthi C, Horst A, Mosler H, et al. (December 2012) When is shared sanitation improved sanitation? Available: http://www.eawag.ch/forschung/ess/gruppen/ehpsy/publications/pdfs/Policy_Brief_2_2012.pdf. Accessed 05 July 2013.
7. Tumwebaze IK, Orach CG, Niwagaba C, Lüthi C, Mosler HJ (2012) Sanitation facilities in Kampala slums, Uganda: users' satisfaction and determinant factors. Int J Environ Health Res 23: 191–204. doi: 10.1080/09603123.2012.713095
8. Fuller JA, Clasen T, Heijnen M, Eisenberg JNS (2014) Shared Sanitation and the Prevalence of Diarrhea in Young Children: Evidence from 51 Countries, 2001–2011. Am J Trop Med Hyg 91:173-80. doi: 10.4269/ajtmh.13-0503
9. WHO/UNICEF (2008) Progress on Drinking Water and Sanitation: Special Focus on Sanitation. Available: http://www.wssinfo.org/fileadmin/user_upload/resources/1251794333-JMP_08_en.pdf. Accessed: 10 October 2012.
10. WHO/UNICEF (2010) JMP Technical Task Force Meeting on Sanitation and Methods for Estimating Progress. Available: http://www.wssinfo.org/fileadmin/user_upload/resources/JMP-Sanitation-Method-Task-Force-Meeting-Report-July-2010-final.pdf. Accessed: 22 February 2013.
11. Mazeau A, Reed B, Sansom K, Scott R (2013) Emerging categories of urban shared sanitation. Water Environ J. doi: 10.1111/wej.12075.
12. Mazeau AP, Reed B (2010) Assessing people's views of infrastructure: methodologies to study urban shared sanitation. Available: http://hal.archives-ouvertes.fr/docs/00/52/14/09/PDF/9-WWW-YES-Mazeau-Paper-2010-04-08.pdf. Accessed 22 February 2013.
13. Kvarnström E, McConville J, Bracken P, Johansson M, Fogde M (2011) The sanitation ladder - a need for a revamp? J Water Sanit Hyg Dev 1: 3–12. doi:10.2166/washdev.2011.014
14. Mazeau A, Tumwebaze IK, Lüthi C, Sansom K (2013) Inclusion of shared sanitation in urban sanitation coverage? Evidence from Ghana and Uganda. Waterlines 32: 334–348. doi: 10.3362/1756-3488.2013.034
15. Roma E, Buckley C, Jefferson B, Jeffrey P (2010) Assessing users' experience of shared sanitation facilities: A case study of community ablution blocks in Durban, South Africa. Water SA 36: 589–594. doi: 10.4314/wsa.v36i5.61992
16. Ministry of National Development Planning/National Development Planning Agency (BAPPENAS) (2010). Report on the Achievement of the Millennium Development Goals Indonesia 2010. Available: http://planipolis.iiep.unesco.org/upload/Indonesia/Indonesia_MDG_2010.pdf. Accessed 05 July 2013.
17. General Economics Division, Planning Commission, Government of the People's Republic of Bangladesh (GED) (February 2012) The Millennium Development Goals: Bangladesh Progress Report 2011. Available: http://www.un-bd.org/pub/MDG%20Progress%20Report%202011.pdf. Accessed 05 July 2013.
18. WHO/UNICEF (April 2014) Estimates on the use of water sources and sanitation facilities: Bangladesh. Available: http://www.wssinfo.org/documents/?tx_displaycontroller%5Btype%5D = country_files. Accessed 12 July 2014.
19. Jenkins MW, Curtis V (2005) Achieving the 'good life': why some people want latrines in rural Benin. Soc Sci Med 61: 2446–59. doi:10.1016/j.socscimed.2005.04.036
20. de Souza K (April 2010) Communal Toilets in Urban Poverty Pockets: Use and user satisfaction associated with seven communal toilet facilities in Bhopal, India. WaterAid. Available: http://www.wateraid.org/~/media/Publications/communal-toilets-user-satisfaction-bhopal-india-briefing-note.pdf. Accessed: 10 October 2012.
21. Zheng Y, Hakim SAI, Nahar Q, Agthoven Av, Flanagan SV (2013) Sanitation coverage in Bangladesh since the millennium: consistency matters. J Water Sanit Hyg Dev 3: 240–251. doi:10.2166/washdev.2013.154
22. Biran A, Jenkins MW, Dabrase P, Bhagwat I (2011) Patterns and determinants of communal latrine usage in urban poverty pockets in Bhopal, India. Trop Med Int Health 16: 854–862. doi: 10.1111/j.1365-3156.2011.02764.x
23. Buttenheim AM (2008) The sanitation environment in urban slums: implications for child health. Popul Environ 30: 26–47. doi: 10.1007/s11111-008-0074-9
24. Montgomery MA, Desai MM, Elimelech M (2010) Comparing the effectiveness of shared versus private latrines in preventing trachoma in rural Tanzania. Am J Trop Med Hyg 82: 693–695. doi: 10.4269/ajtmh.2010.09-0540
25. Kapadia-Kundu N, Dyalchand A (2006) The Pachod Paisa Scale: A Numeric Response Scale for the Health and Social Sciences. Available: http://www.ihmp.org/pp_scale_demography_india_dec19_2006.pdf. Accessed 07 July 2012..

19

Preliminary Studies on Membrane Filtration for the Production of Potable Water: A Case of Tshaanda Rural Village in South Africa

Gomotsegang F. Molelekwa[1,2]*, Murembiwa S. Mukhola[2], Bart Van der Bruggen[1], Patricia Luis[3]

1 ProcESS, Department of Chemical Engineering, KU Leuven, Leuven, Belgium, **2** Department of Environmental Health, Tshwane University of Technology, Pretoria, South Africa, **3** Materials and Process Engineering (iMMC-IMAP), Université Catholique de Louvain, Louvain-la-Neuve, Belgium

Abstract

Ultrafiltration (UF) systems have been used globally for treating water from resources including rivers, reservoirs, and lakes for the production of potable water in the past decade. UF membranes with a pore size of between 0.1 and 0.01 micrometres provide an effective barrier for bacteria, viruses, suspended particles, and colloids. The use of UF membrane technology in treating groundwater for the supply of potable water in the impoverished and rural village, Tshaanda (i.e., the study area) is demonstrated. The technical and administrative processes that are critical for the successful installation of the pilot plant were developed. Given the rural nature of Tshaanda, the cultural and traditional protocols were observed. Preliminary results of the water quality of untreated water and the permeate are presented. *Escherichia coli* in the untreated water during the dry season (i.e., June and July) was 2 cfu/100 ml and was <1 cfu/100 ml (undetected) following UF, which complied with the WHO and South African National Standards and Guidelines of <1 cfu/100 ml. During the wet/rainy season (February) total coliform was unacceptably high (>2419.2 cfu/100 ml) before UF. Following UF, it dramatically reduced to acceptable level (7 cfu/100 ml) which is within the WHO recommended level of <10 cfu/100 ml. Additionally, during the wet/rainy season *E. coli* and enterococci were unacceptably high (40.4 cfu/100 ml and 73.3 cfu/100 ml, respectively) before UF but were completely removed following UF, which are within the WHO and SANS recommended limit. The values for electrical conductivity (EC) and turbidity were constantly within the WHO recommended limits of 300 μS/cm corrected at 25°C and <5 NTU, respectively, before and after UF, during dry season and wet season. This suggests that there is no need for pre-treatment of the water for suspended particles and colloids. Considering these data, it can be concluded that the water is suitable for human consumption, following UF.

Editor: Vishal Shah, Dowling College, United States of America

Funding: The funding was received from VLIR-UOS, Belgium (Project No.: ZEIN2011PR386, URL:http://www.vliruos.be/). The funders had no role in study design, data collection and analysis, decision to publish, or preparation of the manuscript.

Competing Interests: The authors have declared that no competing interests exist.

* Email: gomotsegangfred.molelekwa@cit.kuleuven.be

Introduction

Water is a precious and valuable resource that plays a critical role in the economic and social growth of any population. Most countries around the world are still struggling to meet the demand for potable water. This is evident by the uneven distribution of and accessibility to potable water between developed and developing countries as well as between urban and rural communities [1].

About 780 million people do not have access to improved drinking water sources and 2.5 billion people lack improved sanitation [2]. Most of these people reside in rural settings where piped and other improved water supplies have failed to reach the dwellers [3], thus forcing rural communities to collect drinking water from untreated sources such as rivers and springs [4]. Unsafe drinking water, with poor sanitation and hygiene, accounts for nearly 10% of the total burden of disease worldwide. This includes an estimated 4 billion cases of diarrhea annually, causing 1.8 million deaths, mostly among children under 5 years of age [5].

South Africa is home to 51.8 million people [6] and approximately 40% of this population lives in small rural villages and scattered settlements [7]. Universal access to potable water and decent sanitation is a governmental objective in South Africa [8]; thus, the government has set a target of 100% coverage of potable water supply by 2014 [9]. Since 1994, access to potable water supply has improved significantly, from 59% of the population [10] to 95.2% in 2011 [11]. Still, around five million South Africans, mainly residing in rural areas, do not have a reliable source of safe drinking water [6,11]. This situation exposes the rural communities to risks of acquiring water-borne diseases with resultant negative financial implications on the government when providing medical treatment to sick people, as these people predominantly use government clinics or hospitals for treatment [12]. For example, the South African government loses about $340 millionannually, due to approximately three million diarrhoeal cases and 50 000 related mortalities [13].

A systematic review by Wright et al. [14] showed that water contamination occurs between source and point-of-use. This pattern has been confirmed by subsequent studies of water contamination in rural Sierra Leone [15] and rural Honduras [16]. The extent of water contamination varies across the water supply chain (i.e., from source to cup used in a household), whereby more contamination is at the source than in a household setting [17].

Studies by Mpenyana-Monyani et al. [18] and Zamxaka et al. [19] reported concentrations of total coliform and *E. coli* which were higher than the limits set by WHO [20], SANS [21] and DWAF [22] for drinking water quality. As a result of the potential disease-causing characteristics of certain *E. coli*, the removal of *E. coli* from raw water is a major step in all drinking water purification systems [23].

Several studies have been done on the bacteriological water quality in some of the villages next to Tshaanda [17,24–30]. The results of these studies showed that the levels of bacteriological concentration, which included *E. coli*, were above the recommended WHO and South African National Standards for drinking water quality and thus suggesting that the water was not fit for human consumption unless treated. Subsequently, ceramic filters [24,31] were applied in an attempt to address water pollution at household level in the affected villages. Although this intervention significantly improved the quality of water and also reduced the incidence of diarrhoea, there was no follow up study to determine sustained use of filters [24]. The back drop of household water treatment options is that the responsibility of treating the water rests on the individual household. This could present variation in the quality of drinking water in the same village. For ceramic filters, tasks such as daily backwashing and cleaning of the filters could be considered tedious by some households and they may decide not to follow the instructions resulting in compromised quality of drinking water. Furthermore, some households may not have time to filter the water and would therefore drink untreated water. Monitoring of household water quality may be difficult to conduct due to restricted access to some households, for instance when there is no one in the house, thus increasing the duration of monitoring. On the other hand, population-based or source treatment is easy to monitor, can be accessed at any time and it saves time because the focus is on a single source. Furthermore, it provides uniform water quality at a given time thus becoming easy to determine which intervention should be implemented to improve water quality. Membrane filtration technology, particularly ultrafiltration (UF) system, can be used for source water treatment to provide potable water. The application of membrane filtration technology for potable water supply is well documented [32–40]. However, fouling is the most important factor that has limited the widespread use of membrane technology for the removal of microorganisms from drinking water. Feed pretreatment and membrane cleaning [41] are common strategies to minimize the fouling effect. Ultrafiltration systems have been used globally for treating various water resources including rivers, reservoirs, and lakes [42–44] for the production of potable water in the past decade. Ultrafiltration membranes with a pore size of between 0.1 and 0.01 micrometres provide an effective barrier for bacteria, viruses, suspended particles and colloids [45–47]. Ultrafiltration can be used on its own for treating drinking water where the feed water concentration is not too high in terms of organic contents [48–50]; for instance, if water turbidity is <100 NTU [51]. Accordingly, it has been used as pre-treatment for nanofiltration to reduce its fouling [50,52] and for reverse osmosis (RO) to provide an excellent treated water quality [46,47]. This has led to UF membranes replacing conventional pre-treatment procedures that consist of acid addition, coagulant/flocculant addition, chlorination, media filtration, and cartridge filtration [47].

Conventional large-scale ultrafiltration systems are operated at a transmembrane pressure of between 0.5–1.0 bar and require pumps for operation and backflushing. However, if operated by gravity, pump costs are avoided, and this can be an attractive option for decentralized, small-scale applications [45]. Presently, only few gravity-driven ultrafiltration systems for decentralized application exist (SkyJuice for community water supply and LifeStraw Family for household water treatment [51,53]. These systems can be operated at ultra-low pressure (100–150 mbar) and require little maintenance compared to the conventionally operated UF [48]. Boulestreau et al. [51] successfully conducted trials at Ogunjini village in Kwa-Zulu Natal, South Africa with a decentralized, gravity-driven membrane system to treat river water and to produce 5 m^3/d of drinking water. However, unlike our study, they used flat-sheet UF membranes. Furthermore, their system was not incorporated into any water supply infrastructure and treated water was not used for human consumption, instead it was conveyed back into the river.

Population-level potable water supply interventions such as UF, that are based on infrastructure network received less attention compared to point-of-use water treatment interventions which are being applied to improve water quality within homes and personal hygiene practices within households in rural areas of developing countries [3,4,14–17,24,54]. Furthermore, treatment of groundwater using decentralized gravity-fed ultrafiltration membrane technology for village-wide supply of potable water in South Africa's impoverished rural areas has not been tested. The aim of this study is to ascertain the use of ultrafiltration membrane for the provision of potable water in Tshaanda. The purpose of this article is to present the preliminary results of the ultrafiltration of groundwater for potable water supply in Tshaanda. The objectives of this study are to:

1. Demonstrate the feasibility of using membrane technology to provide potable water in a rural area;
2. Determine the bacteriological quality of raw water during low and high rainfall periods;
3. Determine the quality of the permeate against the WHO and South African national standards for drinking water quality.

Methods

Contextualization of Study Area

Tshaanda village is situated in Vhembe District Municipality, in the Limpopo province, South Africa. The location of Tshaanda is shown in Figure 1. Tshaanda has approximately 800 people and 150 households. The profile of Tshaanda is presented in Table 1. The community gets water from an unprotected spring, which is the only source of water supply in the area. Tshaanda was chosen because it is one of the small villages that do not have safe drinking water infrastructure in Vhembe District Municipality jurisdiction. Furthermore, the village is impoverished and excluded from the mainstream water network in Vhembe and it is unlikely that the village will be connected to the mainstream water infrastructure in the future because the mountains that are surrounding it make Tshaanda difficult to access. The spring is situated deep in the mountain valleys and it can only be accessed by foot, thus posing a challenge, especially to older persons, people with disability, women, children and AIDS sufferers. The community shares this source with domestic animals. Furthermore, the community uses this source for bathing and washing clothes. Water supply infrastructure in Tshaanda is constituted of a diesel-driven engine, 60 000 L concrete water reservoir and five street taps.

No research study was done on water quality before in Tshaanda, therefore this study presented an opportunity to maximize the use of the water source and to also show case,

Figure 1. Map of South Africa showing Tshaanda.

based on the success of the project, the possibility of using membrane technology to provide safe drinking water in a rural area, especially since drinking water sources in rural areas are not monitored.

Preliminary water sampling and analysis

On 18th June 2012, three water samples were taken in Tshaanda in order to create baseline data of the quality of drinking water before treatment in order to compare it with the quality of water after treatment with UF membrane.

The quality of drinking water in Tshaanda was analysed in relation to the WHO [20] and South African [22] guidelines for drinking water quality. The water samples were analyzed for turbidity, pH, electrical conductivity, total coliform, enterococci and *E. coli* [55]. One water sample (500 ml) was taken at each of the three source points (*i.e.*, *Spring*, *Tap 1* and *Tap 2*). Distilled water (500 ml) which was obtained from the laboratory of the Department of Agriculture, Veterinary section in Makhado, Limpopo province, was used as a control.

Defined Substrate Technology was used for the bacteriological analysis of the water samples. The IDEXX Colilert was used to process the Total coliform and *E. coli* water samples, whereas IDEXX Enterolert was used to process enterococci samples in the laboratory. Water samples of IDEXX Colilert and IDEXX Enterolert were incubated for 22 hours at 37°C and 41°C, respectively. After incubation, total coliform was determined visually by identifying all test wells that turned yellow and

Table 1. Tshaanda community profile [39].

Number of households	153
Population	835
Number of households with government subsidized houses	44
Number of households with Ventilated Improved Pit Toilets	63
Number of households without toilets	21
Source of water supply (spring)	1
Number of street water taps	8
Number of People with Matric (Grade 12)	50
Number of People with Tertiary qualification	12
Primary School	1
Clinic	0

subsequently, the Most Probable Number (MPN) was determined by counting the number of test wells that showed yellow colour. After determining the MPN, all the IDEXX Colilert Quanti-Tray plates were put individually under long wave (366 nm) ultraviolet light in order to detect any wells that had fluorescence [56]. The yellow wells that fluoresced were counted to determine the MPN for *E. coli*. A multi-meter (HACH) was used to measure the pH and electrical conductivity of the water samples. A portable turbidity meter (HACH) was used to measure the turbidity of the samples.

Social and administrative processes (non-technical)

Figure 2 highlights critical and fundamental issues that need to be adhered to in order to efficiently and successfully plan, implement and wind-up a water supply and management project in rural villages which, in most cases, are administered by traditional authorities in conjunction with local and/or district municipalities. Maintaining constant communication with the municipality, traditional authority and the community is very critical towards ensuring seamless implementation of the project and to promptly address concerns of all the key stakeholders. Furthermore, the water group will play a critical role of ensuring that the membrane pilot plant operates properly and that water related issues are resolved promptly and amicably. The water group will also ensure that the community takes ownership of the pilot plant, use it properly and protect it from vandalism and abuse. The group will educate the community about prevention of water pollution, responsible use of water and promotion of public health and personal hygiene.

Ethics statement

The process started with seeking and obtaining approval to conduct the research in Tshaanda from Vhembe District Municipality, Mutale Local Municipality and Tshitandani Traditional Council, all of which have authority over the Tshaanda jurisdiction. The first meeting was held with the local chief of Tshitandani Traditional Authority in order to explain the purpose of the project to him and to obtain approval from the traditional authority to conduct the study in Tshaanda. Subsequent meetings were held with other key stakeholders (e.g., Vhembe District Municipality, Mutale Local Municipality and Tshaanda community) in order to present the research background, obtain approval and support from them and to discuss their roles during the entire duration of the research.

Written ethics approval for this research was obtained from Vhembe District Municipality, which has overall jurisdiction over Mutale Local Municipality, where Tshaanda is situated. Tshitandani Traditional Authority also gave a written ethics approval for the study to be conducted in Tshaanda, as it has sole traditional authority over the Tshaanda jurisdiction. The participants (whom were adults) from the community of Tshaanda gave their verbal consent to participate in this study during a community meeting, which was organized by the the local chief of Tshitandani Traditional Authority. They did so in the presence of the members of Tshitandani Traditional Authority. The community gave consent for instance, to participate in the discussions during community meetings and when establishing a local water group. Their consent was then recorded as part of the minutes of that meeting. It was sensible to ask for consent from the entire community (excluding any minor) in a meeting, because the study focuses on solving a community problem rather than an individual problem. Furthermore, the engagements with the participants were mainly through community meetings.

This study also received ethical approval from the institution review board committees at Tshwane University of Technology (TUT) and Katholieke Universiteit Leuven (KU Leuven).

Public participation

A public participation process for the installation of a membrane filtration unit in Tshaanda was conducted by with the community of Tshaanda, Vhembe District Municipality, Mutale Local Municipality, Tshitandani Traditional Authority and Tshitandani Primary School Governing Body. Some of the key aspects that needed community involvement were the identification of a suitable location where the membrane filtration system should be installed, matters related to the security, operation and maintenance of the filtration system and the need for the establishment of local 'Water Group'.

Technical implementation

Raw water supply. Raw water was drawn from the local spring and pumped into a 2500 L raw water storage tank, which was placed at the top of the membrane filtration system. An existing diesel driven engine, which belongs to Vhembe District Municipality, was used to pump water from the spring to the raw water storage tank.

Membrane filtration system design, operation and maintenance. Gravity-driven ultrafiltration pilot plant was purchased from Ikusasa Water (A Division of Ikusasa Chemicals (Pty) Ltd and Chris Swartz Eng. (Water Utilization Engineers)) and it was installed in April 2013 at Tshaanda. The membrane filtration system was housed inside a 6 m maritime steel container. A hollow fibre UF membrane (Multibore membrane housed in a dizzer XL 1.5 MB 40 module), supplied by Inge watertechnologies AG, Germany), was used to drive the experiment in Tshaanda. The design of the pilot plant is shown in Figure 3 and Figure 4.

The plant was set to operate automatically even though it can be operated manually. The water flow rate was maintained between 300 and 400 L/h at a pressure of 2 kPa. The plant can produce 5 m^3/day of potable water.

Due to the operational simplicity of the pilot plant, the operation and maintenance of the pilot plant was carried out by a member of the community, who was recommended by the community. He was then trained (1-day training) on how to operate and maintain the pilot plant by a technician from Ikusasa Water. Subsequent training of the plant operator was coupled with the main maintenance schedules. The plant maintenance included backflushing of the membrane module, cleaning of bagfilter,

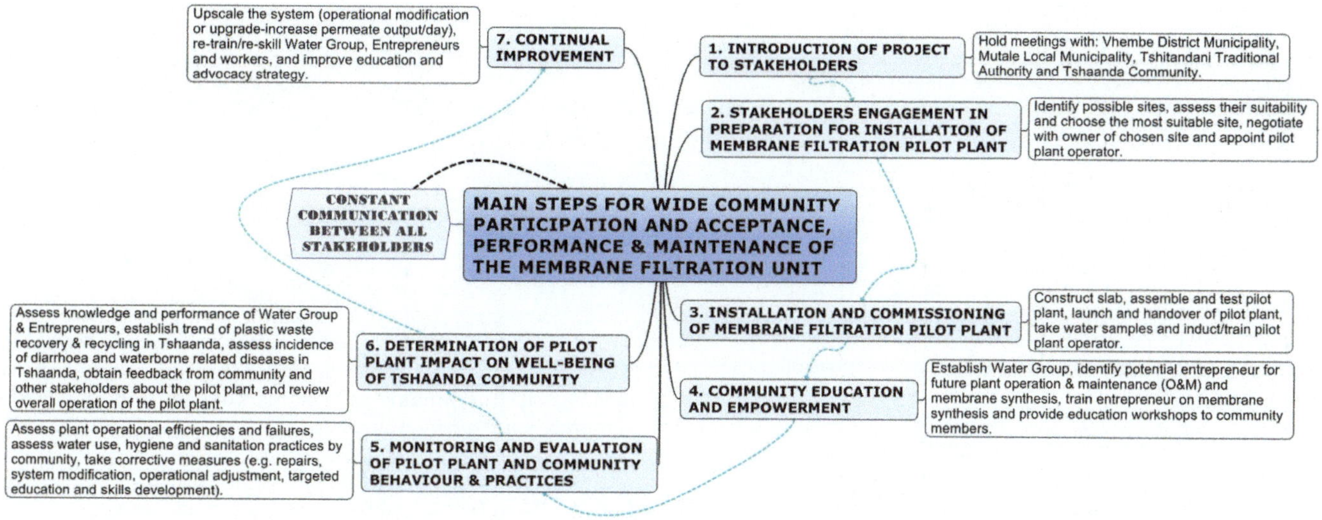

Figure 2. Main steps for wide community participation and acceptance, performance and maintenance of membrane filtration pilot plant.

calibration of the control panel and correcting the flow rate. The main maintenance of the pilot plant was done once, every two months between April 2013 and February 2014 by Ikusasa Water, however; cleaning of the bagfilter was done once every week by the pilot plant operator. The first main maintenance was done in June 2013.

Water sampling and analysis

Three months after the pilot plant was installed, water samples of raw (untreated) and permeate (treated) were taken for analysis on the 25th July 2013 in order to determine whether the pilot plant was able to produce water that met the requirements of the WHO and South African water quality guidelines [20,22] and standards [21]. Two sets of raw water samples (one from the spring, labelled as ***Spring*** and another sample at the inlet of the membrane filtration plant just before the bagfilter, labelled as ***Plant inlet before filtration***) and four samples of treated water were taken (one from the storage tank inside the filtration plant, labelled as ***Permeate*** and three from the three taps outside the plant, labelled as ***Tap 1***, ***Tap 2*** and ***Tap 3***, respectively). Furthermore, other samples were taken on the 25th February 2014 in order to determine the quality of water during wet season. The water samples were taken aseptically and sterile zip-lock water bags were used. The water samples were put inside a cooler bag with ice to maintain the temperature at ~4°C, until they were processed for analysis in the laboratory. The distance between the area where the samples were taken and the laboratory where they were analysed was 160 km. Accordingly, the samples were processed within 6 hours after collection. The samples were processed according to the same procedure that was used during the preliminary investigation of the water quality.

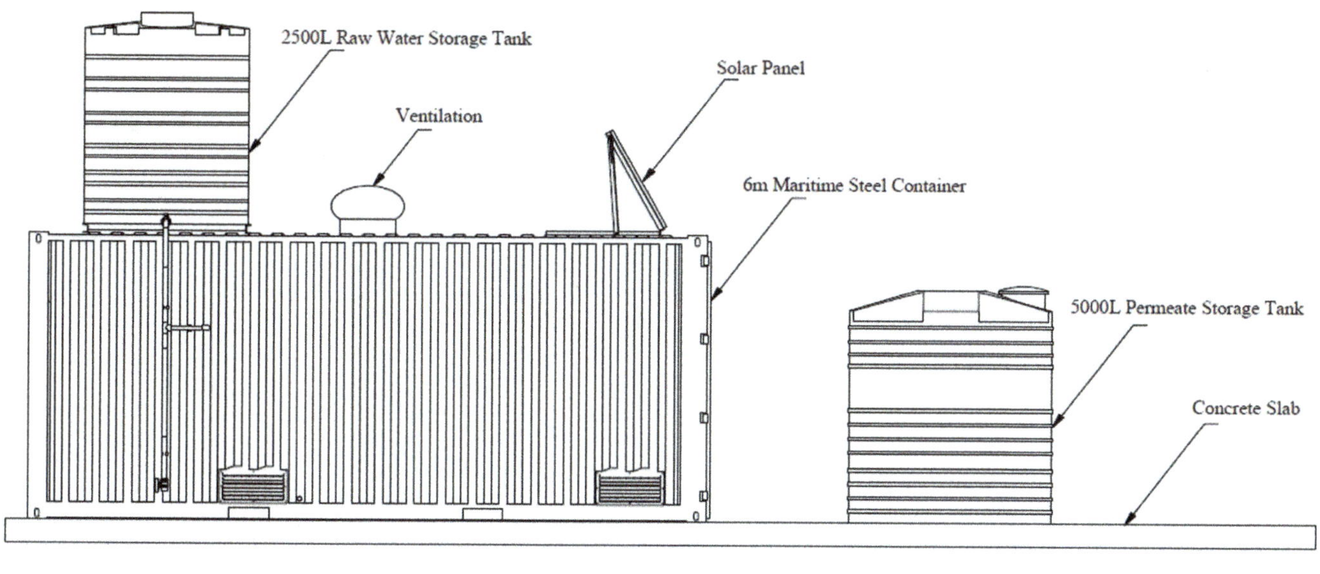

Figure 3. External design of the pilot plant in Tshaanda.

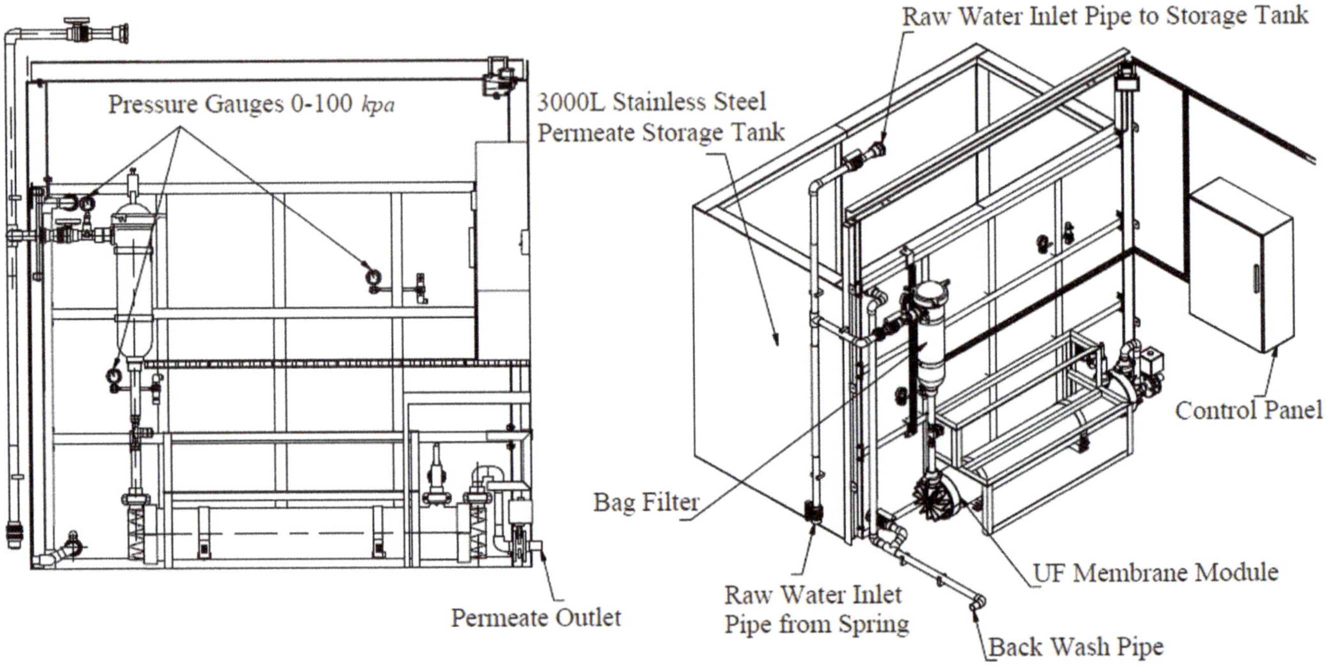

Figure 4. Internal design of the pilot plant in Tshaanda.

Results and Discussion

Installation and operation of pilot plant

The pilot membrane filtration plant was installed on the premises of the local primary school following an agreement between the community and the school governing body. The pilot plant was installed approximately 600 m away from the water source (*i.e.*, spring), construction and commissioning phases were completed within one month. The system operated through gravity for 24 h/day and produced 5000 L of potable water per day for the community of Tshaanda.

Water quality analysis

The water samples were processed and analyzed for microbiological and physico-chemical parameters. Microbiological analysis was carried out in order to determine the presence of *E. coli*, enterococci and total coliform in the water samples. Physico-chemical analysis was carried out to determine the levels of pH, turbidity and electrical conductivity, respectively.

The results of the preliminary water quality investigation are presented in Table 2.

Critical analysis of Table 2 shows that turbidity values of all the samples were within the WHO and South African standards limits for drinking water quality. The results also show the presence of *E. coli*, enterococci and total coliform in concentrations above the recommended limits.

The results of the raw water and treated water samples are shown in Table 3 and Table 4, respectively.

Total Coliform

Detection of total coliform in water indicates potential faecal pollution and thus provide information on treatment efficiency and after growth [21]. After a period of more than 20 months (June 2012–February 2014), total coliform counts of raw water was above the acceptable SANS limit. Furthermore, there was a significant increase in the total coliform count in February 2014 (>2419.2 cfu/100 ml) due to heavy rains that were experienced in the area two weeks before sampling. In July 2013, total coliform of the permeate was about 4 times (200.5 cfu/100 ml) more than that of the untreated water source (48 cfu/100 ml). This was because the permeate tank was not covered on top and therefore it presented an easy passage of dust into the permeate. The tank was subsequently cleaned and covered with a canvas. In February 2014, the coliform count of the permeate was 7.4 cfu/100 ml, following the cleaning of the permeate tank, thus being within the acceptable limit.

Escherichia coli and enterococci

Escherichia coli is the best indicator microorganism of faecal pollution [57]. The overall quality of the water at the spring was found to be consistently unsafe for drinking purposes due to the presence of *E.coli* in the water samples taken. After a period of more than 20 months (June 2012–February 2014), *E. coli* in raw water increased by more than 2 times (17.4 cfu/100 ml in June 2012 as against 40.4 cfu/100 ml in February 2014). The number of enterococci at the water source increased ten folds between June 2012 and February 2014 (7 cfu/100 ml in June 2012 and 73.3 cfu/100 ml) due to rain. This was expected as increased levels of microbial pollution are usually experienced during rainy seasons, since large numbers of microorganisms are washed from various point- and non-point pollution sites [58]. However, the concentration of enterococci was significantly lower at the filtration unit (Plant inlet before filtration). This could probably be attributed to die-off as bacteria compete for limited oxygen and nutrients in the water [59].

This water poses a public health risk to the community of Tshaanda if consumed untreated and could lead to waterborne diseases, such as: hepatitis A and E, cholera, typhoid and poliomyelitis [3]. Consuming water contaminated with *E. coli* produce a 15% chance of of having diarrhea.

Following ultrafiltration, *E. coli* and enterococci were undetected (<1 cfu/100 ml) in the permeate samples taken in July 2013 and February 2014 respectively. Maintaining this water

Table 2. Results of the preliminary water samples taken in Tshaanda on 18 June 2012.

SOURCE	CHEMICAL	GUIDELINES *(DWAF 1996 & WHO 2011)*	BACTERIOLOGICAL *(Indicator of fecal pollution)*	MPN	GUIDELINES *(DWAF 1996 & WHO 2011)*
SPRING	pH = 7.78	5 to 8	Total Coliform = (29/4)	35	<10 cfu/100 ml
	Conductivity = 46.7 µs/cm		*E. Coli* = (15/2)	17.4	0 cfu/100 ml
	Turbidity = 0.52	1 to 5 NTU	Enterococci = (7)	7	0 cfu/100 ml
TAP 1 *(street tap)*	pH = 7.38	5 to 8	Total Coliform = (26/2)	29	<10 cfu/100 ml
	Conductivity = 43.1 µs/cm		*E. Coli* = (16)	16.1	0 cfu/100 ml
	Turbidity = 0.51	1 to 5 NTU	Enterococci = (0/0)	<1	0 cfu/100 ml
TAP 2 *(street tap)*	pH = 7.39	5 to 8	Total Coliform = (15)	15.1	<10 cfu/100 ml
	Conductivity = 40.8 µs/cm		*E. Coli* = (3)	3	0 cfu/100 ml
	Turbidity = 4.31	1 to 5 NTU	Enterococci = (0)	<1	0 cfu/100 ml
CONTROL			Total Coliform = (0)	<1	0 cfu/100 ml
			E. Coli = (0)	<1	0 cfu/100 ml
			Enterococci = (0)	<1	0 cfu/100 ml

Note: **DWAF**-Formerly, Department of Water Affairs & Forestry, now Department of Water Affairs (DWA). **MPN**-Most Probable Number.

quality and ensuring that potable water is available in sufficient quantity will help to reduce the diarrhea episodes [12,60–62].

Turbidity

The turbidity values recorded for all the water samples were within the WHO [20] and SANS [21] acceptable limits for drinking water (<5 NTU) during dry and rainy seasons (June 2012, July 2013 and February 2014). Interestingly, the turbidity of the permeate was almost two times lower than that of the spring between July 2013 and February 2014. These observed lower turbidity values suggest that there is no need for pre-treatment of the water for suspended particles and colloids [51].

Table 3. Results of the water samples taken in Tshaanda on 25 July 2013.

SOURCE	CHEMICAL	GUIDELINES *(DWAF 1996 & WHO 2011)*	BACTERIOLOGICAL *(Indicator of fecal pollution)*	MPN	GUIDELINES *(DWAF 1996 & WHO 2011)*
SPRING	pH = 6.45	5 to 8	Total Coliform = (46)	118.4	<10 cfu/100 ml
	Conductivity = 49.3 µs/cm		*E. Coli* = (6)	6.4	0 cfu/100 ml
	Turbidity = 1.01	1 to 5 NTU	Enterococci = (3)	3.1	0 cfu/100 ml
PLANT INLET (Before filtration)	pH = 6.31	5 to 8	Total Coliform = (21)	27.1	<10 cfu/100 ml
	Conductivity = 47.1 µs/cm		*E. Coli* = (0)	<1.0	0 cfu/100 ml
	Turbidity = 1.69	1 to 5 NTU	Enterococci = (1)	1.0	0 cfu/100 ml
PERMEATE	pH = 6.14	5 to 8	Total Coliform = (50)	200.5	<10 cfu/100 ml
	Conductivity = 53.6 µs/cm		*E. Coli* = (0)	<1	0 cfu/100 ml
	Turbidity = 0.61	1 to 5 NTU	Enterococci = (0)	<1	0 cfu/100 ml
TAP 1	pH = 6.10	5 to 8	Total Coliform = (50)	200.5	<10 cfu/100 ml
	Conductivity = 48.5 µs/cm		*E. Coli* = (0)	<1	0 cfu/100 ml
	Turbidity = 2.09	1 to 5 NTU	Enterococci = (0)	<1	0 cfu/100 ml
TAP 2	pH = 6.11	5 to 8	Total Coliform = (50)	200.5	<10 cfu/100 ml
	Conductivity = 48.1 µs/cm		E. Coli = (0)	<1	0 cfu/100 ml
	Turbidity = 1.34	1 to 5 NTU	Enterococci = (0)	<1	0 cfu/100 ml
TAP 3	pH = 6.08	5 to 8	Total Coliform = (50)	200.5	<10 cfu/100 ml
	Conductivity = 48.2 µs/cm		E. Coli = (0)	<1	0 cfu/100 ml
	Turbidity = 0.88	1 to 5 NTU	Enterococci = (0)	<1	0 cfu/100 ml
CONTROL			Total Coliform = (0)	<1	0 cfu/100 ml
			E. Coli = (0)	<1	0 cfu/100 ml
			Enterococci = (0)	<1	0 cfu/100 ml

Note: **DWAF**-Formerly, Department of Water Affairs & Forestry, now Department of Water Affairs (DWA). **MPN**-Most Probable Number.

Table 4. Results of the water samples taken in Tshaanda on 25 February 2014.

SOURCE	CHEMICAL	GUIDELINES *(DWAF 1996 & WHO 2011)*	BACTERIOLOGICAL *(Indicator of fecal pollution)*	MPN	GUIDELINES *(DWAF 1996 & WHO 2011)*
SPRING	pH = 7.04	5 to 8	Total Coliform = (49/48)	>2419.2	<10 cfu/100 ml
	Conductivity = 45.4 µs/cm		*E. Coli* = (27/2)	40.4	0 cfu/100 ml
	Turbidity = 0.82	1 to 5 NTU	Enterococci = (37/5)	73.3	0 cfu/100 ml
PLANT INLET (Before filtration)	pH = 7.17	5 to 8	Total Coliform = (49/10)	204.6	<10 cfu/100 ml
	Conductivity = 46.8 µs/cm		*E. Coli* = (24/0)	31.7	0 cfu/100 ml
	Turbidity = 0.72	1 to 5 NTU	Enterococci = (7/1)	8.5	0 cfu/100 ml
PERMEATE	pH = 6.93	5 to 8	Total Coliform = (15)	15.1	<10 cfu/100 ml
	Conductivity = 51.9 µs/cm		*E. Coli* = (0)	<1	0 cfu/100 ml
	Turbidity = 0.46	1 to 5 NTU	Enterococci = (0)	<1	0 cfu/100 ml
CONTROL			Total Coliform = (0)	<1	0 cfu/100 ml
			E. Coli = (0)	<1	0 cfu/100 ml
			Enterococci = (0)	<1	0 cfu/100 ml

Note: **DWAF**-Formerly, Department of Water Affairs & Forestry, now Department of Water Affairs (DWA). **MPN**-Most Probable Number.

pH and electrical conductivity

The pH and electrical conductivity levels of all the water samples were within the recommended limits of WHO [20] and SANS [21] for drinking water quality.

Challenges of the project

The main challenge encountered was that there was no formal procedure for ordering and supplying diesel which at times delayed the supply of the diesel to Tshaanda. This affected pumping of raw water to the pilot plant, consequently leading to operational stoppage of the pilot plant. In order to address these challenges, a plan was developed, whereby Vhembe District Municipality supplied 220 L of diesel on a monthly basis. Furthermore, the plant operator would notify the local councillor when it was time to order diesel, the local councillor would request the responsible manager at Vhembe District Municipality to deliver the diesel to Tshaanda. The plan also included an institutional arrangement whereby other parties (i.e., KU Leuven or Tshwane University of Technology) would provide assistance to repair the engine or the water pump in case Vhembe District Municipality needed support.

Conclusions

A gravity driven ultrafiltration membrane pilot plant was used for ten (10) months (from April 2013 to February 2014) to produce potable water to the rural community of Tshaanda. The preliminary results showed that the microbiological quality of the permeate was within recommended and acceptable limits of the WHO and South African standards for drinking water quality.

Following the successful removal of *E. coli*, it can be inferred that the membrane system in Tshaanda was able to produce safe drinking water. This further confirms that the integrity of the filtration membrane was still maintained and that it was still highly effective in removing indicator microorganisms, suspended solids and colloidal matter [63] after ten (10) months of operation.

Though the UF pilot plant in Tshaanda could produce safe drinking water, there is still much work to be done before this technology could be rolled-out country wide. Subsequent research will be expanded to include aspects of evidence of sustained use, positive health impact, water quality over extended periods of use, operation and maintenance, cost-effectiveness, system performance and optimization (considering seasonal variations), and water quality monitoring, whereby the frequency of water testing will be increased. Furthermore, future research will concentrate on the water distribution network to determine bacterial regrowth and water quality deterioration, public perception, household health behaviors, acceptability of the technology, community outreach and awareness raising on water, sanitation and hygiene. Acquiring this information would help in ensuring that the system continues to deliver safe water in the long run.

Acknowledgments

We thank Professor Sadiku R, Professor Maree J and Professor Mathipa E.R for providing invaluable comments on the drafts of this paper. We would also like to thank Ikusasa Water (A Division of Ikusasa Chemicals (Pty) Ltd and Chris Swartz Eng. - Water Utilization Engineers) for their contribution in conceptualizing the design of the pilot plant and technical assistance during the operation and maintenance of the pilot plant. Thanks also go to Mr. David Nemakonde and Mrs. Maria Mufamadi from Vhembe District Municipality and Councillor Esther Mudzielwana from Mutale Local Municipality for providing logistical support in arranging meetings with the Tshitandani Traditional Authority and the community of Tshaanda. We also thank Tshitandani Traditional Authority for arranging the initial community meeting with the people of Tshaanda and for providing logistical support during the preparation for the official launch of the pilot plant. Our gratitude also goes to the community of Tshaanda for allowing us to conduct this study in their village and also for agreeing to participate in the study. We thank Mr. Mike Mokwena for his assistance during the collection and interpretation of the results of the water samples. The findings and conclusions of this paper are those of the authors and do not necessarily represent the views of the organizations they individually represent.

Author Contributions

Conceived and designed the experiments: GFM MSM BVB PL. Performed the experiments: GFM. Analyzed the data: GFM MSM BVB PL. Contributed reagents/materials/analysis tools: MSM. Wrote the paper: GFM. Reviewed and contributed to manuscript: MSM BVB PL.

References

1. World Health Organization and UNICEF (2013) Progress on Drinking Water and Sanitation: 2013 Update. Available: http://www.wssinfo.org/fileadmin/user_upload/resources/JMPreport2013.pdf. Accessed: 2013 September 07.
2. World Health Organization, UNICEF (2012) Progress on Drinking Water and Sanitation: 2012 Update. Available: http://www.unicef.org/media/files/JMPreport2012.pdf. Accessed: 2013 May 24.
3. Rosa G, Clasen T (2010) Estimating the scope of household water treatment in low-and medium-income countries. The American Journal of Tropical Medicine and Hygiene 82: 289–300.
4. Sobsey MD, Water S (2002) Managing water in the home: accelerated health gains from improved water supply. Geneva: World Health Organization.
5. World Health Organization (2005) The World Health Report 2005: Make every mother and child count. Geneva: World Health Organization.
6. Statistics South Africa (2012) Census 2011 Statistical release – P0301.4/Statistics South Africa. Pretoria. Available: http://www.statssa.gov.za/Publications/P03014/P030142011.pdf. Accessed: 2013 May 24.
7. de la Harpe J (2011) South Africa: Lessons for Rural Water Supply; Assessing progress towards sustainable service delivery. The Hague: IRC International Water and Sanitation Centre. Available: http://www.washinschools.info/docsearch/title/179989. Accessed: 2013 April 30.
8. Republic of South Africa (1996) Constitution of the Republic of South Africa Act, 1996 (Act No. 108 of 1996). Pretoria: Government Printing Works.
9. Department of Water Affairs (DWA) (2012) Briefing of the Portfolio Committee: Water & Environmental Affairs: Progress and Perspectives on Water Supply Millennium Development Goals. Available: http://www.pmg.org.za/report/20121018-department-water-affairs-attainment-millennium-development-goals. Accessed: 2013 May 24.
10. Wall K (2011) SAICE infrastructure report card for South Africa 2011. SAICE. Available: http://www.csir.co.za/enews/2011_jun/download/infrastructure_report_card_sa_2011.pdf. Accessed: 2013 May 24.
11. Department of Water Affairs (2013) Speech by the Minister of Water and Environmental Affairs, Mrs BEE Molewa, on the occasion of the presentation of Budget Vote Number 37 on behalf of the Department of Water Affairs at the Old Assembly Chamber, Parliament, Cape Town, 21 May 2013. Available: http://www.dwa.gov.za/Communications/PressReleases/2013/STATEMENT%20BY%20MINISTER%20EDNA%20MOLEWA%20-BUDGET%20VOTE%20%20MEDIA%20BRIEFING%202013%20L%20PA.pdf. Accessed: 2013 May 24.
12. Cameron J, Jagals P, Hunter PR, Pedley S, Pond K (2011) Economic assessments of small-scale drinking-water interventions in pursuit of MDG target 7C. Science of the Total Environment 410: 8–15.
13. Momba M, Obi C, Thompson P (2009) Survey of disinfection efficiency of small drinking water treatment plants: Challenges facing small water treatment plants in South Africa. Water SA 35: 485–493.
14. Wright J, Gundry S, Conroy R (2004) Household drinking water in developing countries: a systematic review of microbiological contamination between source and point-of-use. Tropical Medicine & International Health 9: 106–117.
15. Clasen T, Bastable A (2003) Faecal contamination of drinking water during collection and household storage: the need to extend protection to the point of use. J Water Health 1: 109–115.
16. Trevett A, Carter R, Tyrrel S (2005) The importance of domestic water quality management in the context of faecal-oral disease transmission. J Water Health 3: 259–270.
17. Gundry S, Wright J, Conroy R, Du Preez M, Genthe B (2013) Contamination of drinking water between source and point-of-use in rural households of South Africa and Zimbabwe: implications for monitoring the Millennium Development Goal for water. Water Practice and Technology 1: 1–9.
18. Mpenyana-Monyatsi L, Onyango MS, Benteke Momba MN (2012) Groundwater Quality in a South African Rural Community: A Possible Threat to Public Health. Polish Journal of Environmental Studies 21: 1349–1358.
19. Zamxaka M, Pironcheva G, Muyima N (2004) Microbiological and physico-chemical assessment of the quality of domestic water sources in selected rural communities of the Eastern Cape Province, South Africa. Water SA 30: 333–340.
20. World Health Organization (2011) Guidelines for Drinking Water Quality. Available: http://whqlibdoc.who.int/publications/2011/9789241548151_eng.pdf. Accessed 2013 October 15.
21. South African Bureau of Standards (2011) Drinking Water - SANS 241-1: 2011. Part 1: Microbiological, physical, aesthetic and chemical determinands. Pretoria: SABS Standards Division.
22. Department of Water Affairs and Forestry (DWAF) (1996) South African Water Quality Guidelines Volume 1: Domestic Water Use Second Edition, Pretoria. Available: http://www.dwaf.gov.za/IWQS/wq_guide/domestic.pdf. Accessed: 2013 April 22.
23. Varkey A, Dlamini M (2012) Point-of-use water purification using clay pot water filters and copper mesh. Water SA 38: 721–726.
24. du Preez M, Conroy RM, Wright JA, Moyo S, Potgieter N, et al. (2008) Short report: use of ceramic water filtration in the prevention of diarrheal disease: a randomized controlled trial in rural South Africa and Zimbabwe. American Journal of Tropical Medicine and Hygiene 79: 696–701.
25. Potgieter N, Obi CL, Bessong PO, Igumbor EO, Samie A, et al. (2005) Bacteriological contamination of vhuswa, a local weaning food, and stored drinking water in impoverished households in the Venda region of South Africa. J Health, Pop Nutr 23: 150–155.
26. Potgieter N, Mudau LS, Maluleke FRS (2006) The microbiological quality of ground water sources used by rural communities in the Limpopo Province, South Africa. Water, Science and Technology 54 (11): 371–377.
27. Wright J, Gundry SW, Conroy R, Du Preez M, Genthe B, et al. (2009) Child dysentery in the Limpopo Valley: a cohort study of water, sanitation, and hygiene risk factors. Journal of Water and Health 7: 259–266.
28. Obi CL, Potgieter N, Bessong PO, Matsaung G (2002) Assessment of the microbiological quality of river water sources in rural Venda communities in South Africa. Water SA 28: 287–291.
29. Obi CL, Potgieter N, Bessong PO, Matsaung G (2003) Scope of potential bacterial agents of diarrhoea and microbial assessment of quality of river water sources in rural Venda communities in South Africa. Wat Sci Tech 47: 59–64.
30. Du Preez M, le Roux W, Potgieter N, Venter SN (2008) The genetic relatedness of E. coli associated with post-collection drinking water contamination in rural households. Water SA 34: 107–111.
31. Potgieter N, Becker PJ, Ehlers MM (2009) Evaluation of the CDC safe water-storage intervention to improve the microbiological quality of point-of-use drinking water in rural communities in South Africa. Water SA 35: 505–516.
32. Watson BM, Hornburg CD (1989) Low-Energy Membrane Nanofiltration for Removal of Color, Organics and Hardness from Drinking-Water Supplies. Desalination 72: 11–22.
33. Cartwright P (1991) Zero discharge/water reuse—The opportunities for membrane technologies in pollution control. Desalination 83: 225–241.
34. Rachwal A, Khow J, Colbourne J, O'Donnell J (1994) Water treatment for public supply in the 1990's—A role for membrane technology? Desalination 97: 427–436.
35. Kunikane S, Magara Y, Itoh M, Tanaka O (1995) A comparative study on the application of membrane technology to the public water supply. Journal of membrane science 102: 149–154.
36. Ventresque C, Bablon G (1997) The integrated nanofiltration system of the Mery-sur-Oise surface water treatment plant (37 mgd). Desalination 113: 263–266.
37. Arnal JMA, Fernandez MS, Martin GV, Garcia JL, Zafrilla JMG, et al. (2002) Design and construction of a water potabilization membrane facility and its application to the third world countries. Preliminary tests. Desalination 145: 305–308.
38. Van der Bruggen B, Vandecasteele C (2003) Removal of pollutants from surface water and groundwater by nanofiltration: overview of possible applications in the drinking water industry. Environmental Pollution 122: 435–445.
39. Muhammad N, Sinha R, Krishnan ER, Patterson CL (2009) Ceramic filter for small system drinking water treatment: Evaluation of membrane pore Size and importance of integrity monitoring. Journal of Environmental Engineering 135: 1181–1191.
40. Fane A, Tang C, Wang R (2011) Membrane technology for water: microfiltration, ultrafiltration, nanofiltration, and reverse osmosis. Treatise on Water Science, Academic Press, Oxford 4: 301–335.
41. Pangarkar BL, Sane MG, Guddad M (2011) Reverse osmosis and membrane distillation for desalination of groundwater: A review. ISRN Materials Science. http://dx.doi.org/10.5402/2011/523124.
42. Laîné JM, Vial D, Moulart P (2000) Status after 10 years of operation—overview of UF technology today. Desalination 131: 17–25.
43. Murthy Z, Chaudhari L (2009) Treatment of distillery spent wash by combined UF and RO processes. Global NEST Journal 11: 235–240.
44. Manjunath D, Madhumala M, Prasad R, Kalyani S, Sridhar S (2013) Processing of surface and ground water by hydrostatic pressure-driven membrane techniques: design and economic aspects. Desalination and Water Treatment 51: 5873–5885.
45. Peter-Varbanets M, Johnston R, Meierhofer R, Kage F, Müller S, et al. (2011) Gravity Driven Membrane Disinfection for household drinking water treatment. 35th WEDC International Conference, Loughborough, UK.
46. Pearce G (2007) The case for UF/MF pretreatment to RO in seawater applications. Desalination 203: 286–295.
47. Fernández G, Plaza F, Garralón G, Garralón A, Pérez JI, et al. (2012) A comparative study of ultrafiltration and physicochemical process as pretreatment of seawater reverse osmosis. Desalination and Water Treatment 42: 73–79.
48. Peter-Varbanets M, Zurbrügg C, Swartz C, Pronk W (2009) Decentralized systems for potable water and the potential of membrane technology. Water Research 43: 245–265.
49. Yu Z, Chu H, Cao D, Ma Y, Dong B, et al. (2012) Pilot-scale hybrid bio-diatomite/dynamic membrane reactor for slightly polluted raw water purification. Desalination 285: 73–82.
50. Zheng X, Yu M, Liang H, Qi L, Zheng H, et al. (2012) Membrane technology for municipal drinking water plants in China: progress and prospect. Desalination and Water Treatment 49: 281–295.
51. Boulestreau M, Hoa E, Peter-Verbanets M, Pronk W, Rajagopaul R, et al. (2012) Operation of gravity-driven ultrafiltration prototype for decentralised water supply. Desalination and water treatment 42: 125–130.

52. Ferella F, De Michelis I, Zerbini C, Veglio F (2013) Advanced treatment of industrial wastewater by membrane filtration and ozonization. Desalination 313: 1–11.
53. Clasen T, Naranjo J, Frauchiger D, Gerba C (2009) Laboratory assessment of a gravity-fed ultrafiltration water treatment device designed for household use in low-income settings. The American Journal of Tropical Medicine and Hygiene 80: 819–823.
54. Brown J, Proum S, Sobsey M (2008) Escherichia coli in household drinking water and diarrheal disease risk: evidence from Cambodia. Water Science & Technology 58: 757–763.
55. World Health Organization (2008) Guidelines for Drinking-water Quality. Available: http://www.who.int/water_sanitation_health/dwq/fulltext.pdf. Accessed: 2013 September 23.
56. Landre J, Gavriel A, Lamb A (1998) False-positive coliform reaction mediated by Aeromonas in the Colilert defined substrate technology system. Letters in applied microbiology 26: 352–354.
57. Edberg S, Rice E, Karlin R, Allen M (2000) Escherichia coli: the best biological drinking water indicator for public health protection. Journal of Applied Microbiology 88: 106S–116S.
58. Dobrowksy P, van Deventer A, De Kwaadsteniet M, Ndlovu T, Khan S, et al. (2013) Prevalence of virulence genes associated with pathogenic Escherichia coli strains isolated from domestically harvested rainwater during low and high rainfall periods. Appl Environ Microbiol. doi:10.1128/AEM.03061-13.
59. Momba MN, Notshe T (2003) The microbiological quality of groundwater-derived drinking water after long storage in household containers in a rural community of South Africa. Journal of Water Supply: Research & Technology-AQUA 52: 67–77.
60. Cameron J, Hunter P, Jagals P, Pond K (2011) Valuing water, valuing livelihoods. London: IWA Publishing.
61. Esrey SA, Habicht J (1985) The impact of improved water supplies and excreta disposal facilities on diarrheal morbidity, growth, and mortality among children. Cornell International Nutrition Monograph Series 15: 1–85.
62. Esrey SA, Feachem RG, Hughes JM (1985) Interventions for the control of diarrhoeal diseases among young children: improving water supplies and excreta disposal facilities. Bulletin of the World Health organization 63: 757–772.
63. Hofman J, Beumer M, Baars E, Van der Hoek J, Koppers H (1998) Enhanced surface water treatment by ultrafiltration. Desalination 119: 113–125.

20

Characterization of Gene Expression Associated with Drought Avoidance and Tolerance Traits in a Perennial Grass Species

Peng Zhou[1,2], Yuan An[1], Zhaolong Wang[1], Hongmei Du[1], Bingru Huang[2]*

1 School of Agriculture and Biology, Shanghai Jiao Tong University, Shanghai, P. R. China, **2** Department of Plant Biology and Pathology, Rutgers, the State University of New Jersey, New Brunswick, New Jersey, United States of America

Abstract

To understand molecular mechanisms of perennial grass adaptation to drought stress, genes associated with drought avoidance or tolerance traits were identified and their expression patterns were characterized in C4 hybrid bermudagrass [Cynodon dactylon (L.) Pers.×C. transvaalensis Burtt Davy, cv. Tifway] and common bermudagrass (C. dactylon, cv. C299). Plants of drought-tolerant 'Tifway' and drought-sensitive 'C299' were exposed to drought for 5 d (mild stress) and 10 d (severe stress) by withholding irrigation in a growth chamber. 'Tifway' maintained significantly lower electrolyte leakage and higher relative water content than 'C299' at both 5 and 10 d of drought stress. Four cDNA libraries via suppression subtractive hybridization analysis were constructed and identified 277 drought-responsive genes in the two genotypes at 5 and 10 d of drought stress, which were mainly classified into the functional categories of stress defense, metabolism, osmoregulation, membrane system, signal and regulator, structural protein, protein synthesis and degradation, and energy metabolism. Quantitative-PCR analysis confirmed the expression of 36 drought up-regulated genes that were more highly expressed in drought-tolerant 'Tifway' than drought-sensitive 'C299', including those for drought avoidance traits, such as cuticle wax formation (CER1 and sterol desaturase), for drought tolerance traits, such as dehydration-protective proteins (dehydrins, HVA-22-like protein) and oxidative stress defense (superoxide dismutase, dehydroascorbate reductase, 2-Cys peroxiredoxins), and for stress signaling (EREBP-4 like protein and WRKY transcription factor). The results suggest that the expression of genes for stress signaling, cuticle wax accumulation, antioxidant defense, and dehydration-protective protein accumulation could be critically important for warm-season perennial grass adaptation to long-term drought stress.

Editor: Ji-Hong Liu, Key Laboratory of Horticultural Plant Biology (MOE), China

Funding: The authors wish to thank the Joint Research Fund for the Chinese Natural Science Foundation (Grant No. 31128016) and Chinese Natural Science Foundation Projects (Grant No. 31072071 and No. 31272198). The funders had no role in study design, data collection and analysis, decision to publish, or preparation of the manuscript.

Competing Interests: The authors have declared that no competing interests exist.

* Email: huang@aesop.rutgers.edu

Introduction

With the decline in water resources and the increase in human demand for water, water for irrigation is becoming increasingly limited. Drought stress is becoming a significant abiotic stress limiting plant growth and production in many areas. Plants develop various stress resistance mechanisms involving avoidance and tolerance strategies, which vary with plant species, duration and severity of the stress [1]. Dehydration avoidance of leaves is characterized by reducing water loss through mechanisms, such as stomatal closure and accumulation of wax on leaf surfaces while dehydration or desiccation tolerance has been associated with traits, such as osmotic adjustment, sugar accumulation, and maintenance of the integrity of membranes and proteins from dehydration damage [1–3].

At the molecular level, various studies identified drought-regulated genes in different plant species, mostly in annual crop plants and the model plant *Arabidopsis* [4,5]. For example, using cDNA microarray analysis in barley (*Hordeum vulgare* L. cv. Tokak), Ozturk et al. [6] identified significant up-regulation of jasmonate-responsive, metallothionein-like, late-embryogenesis-abundant (LEA) and ABA-responsive proteins and down-regulation of genes for photosynthesis under short-term (6 and 10 h) drought stress. Seki et al. [7] reported that 44 full-length cDNA clones were modulated by 2 fold or greater at the mRNA level in *Arabidopsis* exposed to 2 h of drought stress, including LEA 76 type 1 protein, a nonspecific lipid transfer protein, a putative water channel protein, and HVA22 homolog. In alfalfa (*Medicago sativa*) exposed to 3- and 8-h drought stress, Chen et al. [8] identified some drought-responsive genes, such as two heat shock-related protein genes, dehydrin and xyloglucan endotransglycosylase. Xue et al. [9]reported by using quantitative RT-PCR and cDNA microarray, genes encoding chloroplast enzymes involved in carbon fixation were down-regulated in wheat (*Triticum aestivum*) leaves exposed to a prolonged period (7 d) of drought stress, while those encoding cytoplasmic and vacuolar enzymes in the pathways leading to glucose, fructose and fructan production were up-regulated. Previous studies have demonstrated that drought-regulated genes and their expression patterns varied with plant species and stress duration or severity. However, knowledge of drought-regulated gene expression patterns associated with intraspecific genetic variation and temporal regulation is generally limited in perennial grass species, which often endure long-term drought stress in natural environments. Such information is important for developing effective engineering strategies to improve stress tolerance.

Dehydration avoidance or tolerance to mild or moderate drought level is important for maintaining grain yield production for annual crops, whereas survival mechanisms through long-term or severe drought is critical for perennial grass species due to their perennial nature. In a resurrection grass species, *Sporobolus stapfianus*, Blomstedt et al. [10] identified some unique genes encoding abundant drought-induced proteins or low-abundance transcripts that were not previously found in other species by differential screening. Using comparative analysis between desiccation-tolerant *S. stapfianus* and desiccation-sensitive grass *S. pyramidalis*, Gaff et al. [11] found 12 novel proteins associated with desiccation tolerance in the tolerant grass species. Perennial forage grasses and turfgrass species are cultivated in many areas with limited irrigation. Many cultivated perennial grass species, such as bermudagrass (*Cynodon* spp.), exhibit a wide range of genetic variation, which is a valuable germplasm for studying drought tolerance mechanisms in warm-season perennial grasses [12–15]. Triploid hybrid bermudagrass (*Cynodon dactylon*×*C. transvaalensis*) has been developed to produce highly desirable turf quality with limited irrigation [13], which exhibited better drought tolerance than common bermudagrass [*C. dactylon* (L.) Pers.] due to greater maintenance of photosynthetic processes, water status, and antioxidant defenses [16–18]. The identification of genes associated with genetic variation in drought tolerance in bermudagrass in response to short-term, mild drought and long-term, severe drought will provide further insight into the molecular basis for drought tolerance in perennial grass species.

The objectives of this study were to identify up-regulated genes in bermudagrass in response to short-term or mild drought stress (withholding irrigation for 5 d) and long-term, severe drought stress (withholding irrigation for 10 d) and to compare differentially-expressed genes between a drought-tolerant hybrid bermudagrass genotype ('Tifway') and a drought-sensitive common bermudagrass genotype ('C299') using the subtraction suppression hybridization (SSH) technique. SSH has successfully been used to identify genes responsive to various biotic and abiotic stresses in various plant species [19–21]. Expression of selected genes from SSH libraries were confirmed using quantitative RT-PCR, and major metabolic processes and pathways regulating bermudagrass adaptation to mild and severe drought were discussed using the functional analysis of known protein sequences through BLAST searches.

Materials and Methods

Plant materials, growing conditions, and treatments

Plants of a drought-tolerant genotype ('Tifway' or Tifway) of hybrid bermudagrass (*C. dactylon* L.×*C. transvaalensis* L.) and a drought-sensitive genotype ('C299') of common bermudagrass (*C. dactylon*) were collected from 3-year-old field plots at the research farm in Shanghai Jiao Tong University, Shanghai, China. Plants (approximately 50) were vegetatively propagated in each plastic pot (15-cm diameter and 14 cm in depth) filled with a mixture (1:3, v/v) of sand and sandy loamy soil (fine-loamy, mixed mesic Typic Hapludult). Plants were maintained in a growth chamber with a temperature regime of 30/25°C (day/night), a 14-h photoperiod, 75% relative humidity, and a photosynthetically active radiation of 480 μmol m^{-2} s^{-1} at the canopy level. Plants were irrigated three times per week until soil water reached field capacity or drainage occurred from the bottom of the pot and fertilized biweekly with controlled-release fertilizers (15 N-15 P_2O_5-10 K_2O) at a total amount of 57 kg N ha^{-1}. Plants were maintained in the above conditions for 40 d to allow the establishment of shoot canopy and root systems.

The experiment consisted of three treatments. In the drought stress treatment, plants for each genotype were not watered for 5 d or 10 d. The control plants of both genotypes were watered thoroughly every other day until soil water content reached field capacity (drainage observed from the pot). Each treatment for each grass genotype had three replicates (pots). Two treatments and two genotypes were arranged randomly inside the growth chamber.

Physiological analysis

Differences in drought tolerance between the two genotypes were characterized by whole plant physiological characteristics, including cell membrane stability and relative water content (RWC) of leaves. Cell membrane stability was determined as electrolyte leakage (EL) on 10 second and third fully-expanded leaves in each container. For EL analysis, 0.1 g fresh leaf segments (approximately 0.5 cm long) from each sample were incubated in 15 mL distilled deionized water on a shaker for 24 h. The conductance of the incubation solution was measured as the initial level of electrolyte leakage (Ci) using a conductance meter (YSI Model 32, YSI Incorporated, Yellow Springs, OH, USA). Leaf tissue in the incubation solution was killed in an autoclave at 120°C for 30 min. The conductance of the incubation solution with killed tissues (Cmax) was determined following 24-h incubation on a shaker. Relative EL was calculated as (Ci/Cmax)×100 [22].

Leaf RWC was determined using 10–15 second and third fully-expanded leaves in each container according to Barrs and Weatherley (1962). Leaf samples were detached from the plants and immediately weighed to determine fresh weight (FW). Samples were placed into covered petri dishes filled with water for leaves to reach full hydration. After approximately 24 h at 4°C, leaf samples were blotted dry with paper towels, and weighed to determine turgid weight (TW). Leaf tissue was then dried in an oven at 80°C for 48 h to determine dry weight (DW). Leaf RWC was calculated as: (FW-DW)/(TW-DW)×100.

Construction of suppression subtractive hybridization (SSH) library

Fully-expanded leaves from each genotype exposed to the well-watered control, and at drought stress for 5 d and 10 d were collected for the extraction of total RNA and construction of the SSH library to identify responsive genes differentially expressed in the two genotypes. Total RNA was extracted using the Trizol Reagent (Invitrogen, Carlsbad, CA, USA). Poly (A_+) RNA was isolated from total RNA using mRNA Oligotex isolation midi kit (Qiagen, Valencia, CA, USA). The subtraction library was prepared using PCR-Select cDNA subtraction kit (Takara BIO, Inc., Otsu, Shiga, Japan) according to the manufacturer's instructions. In order to identify drought up-regulated genes, the cDNAs generated from plants under drought stress for 5 or 10 d served as the testers, while the cDNAs generated from the well-watered control plants were used as the drivers. The tester and the driver cDNAs were synthesized separately from 2 μg poly (A_+) RNA isolated from plants grown under normal or drought stress conditions. Both the tester and driver cDNAs were digested with

Table 1. Primers sequences for RT-PCR.

Primers	Accession number	Forward sequence 5′-3′	Reverse sequence 5′-3′
18S	AB536694	GTGACGGGTGACGGAGAATT	GACACTAATGCGCCCGGTAT
CdL1	JK340473	ACTGGTTCAGAACTCTGCAA	ACCACATCGAAATGCATCAC
CdL2	JK340477	ACACTGTCATGCATGTTGGC	ACCTCCTGCACAGGCTTCTT
CdL4	JK340476	TAAAGACAGGCAGAGCCAGC	ACCCTTAAGCCCTACACGAA
CdL9	JK340478	ACTTTGAACCTTCACGGAGGCT	ACCGTGCATGTCAATGGTTGAT
CdL8	JK340518	ACACATACATACGTGATACGTCAA	TCCTGGACTACTGCGAATCT
CdL19	JK340483	ACTTCTCATAAATTCGTTGAAGGC	ACAGCCAGCATTTTATGAAGCTTT
CdL20	JK340484	ACCTCTGGTTGCACTCGTTG	TGTATGCATAGAGCCACACG
CdL35	JK340626	ACAAGCCTGGTGACACGATA	TACCCTACCACATTGACGGC
CdL36	JK340585	ACCTGTAGCCTTTATCTGTC	ATTGGCTCTGGTTTACCTTC
CdL37	JK340586	ACATTCCGACTGTGATGTCG	AGGTGATTGAGGCTATGTCC
CdL40	JK340589	GCTCACGGAAGCCCAGATTA	CCACGCTCATCCTTCTCCAT
CdL42	JK340615	ACGAACTGCCATAGAGGATC	ACTGGCTTTCCAGTCAAACG
CdL44	JK340578	TCCACGTCGTGCTTGTCCTT	CGTGAAGAGGAAGAAGAAGAAGGG
CdL49	JK340642	GGGAGAAGTAACAACGGATG	GCCGTTATCTCATCTCCAAT
CdL52	JK340628	CAAGTGACGGGCTATGGGAC	CGACCACGCAGGTTATGTTG
CdL56	JK340591	ACCAACCAAACTAAACCCAAGC	GCCTCTACGCTTCAGTTCAC
CdL61	JK340592	CGCCCAGAACCTTCTTCATATA	GGAGATCGTGAGCACATACT
CdL62	JK340580	TGCACAAGGAACACAGATCG	GTTGCAACTTGCCTAGGTTCA
CdL67	JK340607	GGTGTGGCTCCACGTCTTCT	ACGAGAACATCGACGAGCTG
CdL68	JK340608	GCAGAGAGAGTGAGAGGGCA	CGGAGCAGGAGTTCTACAGG
CdL69	JK340599	CTGAAGGAAGCCCTATGGTT	GTTGACGGGCTACTAACTAC
CdL74	JK340625	ACACGTATGGGTTTAGGTGGTG	TTCAAGAAGGAGCTGGACGG
CdL79	JK340618	CATCGTTGCGTTGCACCGTC	ACGGGCAGAAGCAAATTGAGG
CdL80	JK340610	TGCCAATACACAACACAAGAGC	CATCCACAAGGACGCAGAAG
CdL82	JK340582	ACCCCTTGTATGCGTCGGTG	CAGATCGGGATCCAGTAGATGAGG
CdL85	JK340595	CGACTAAGGAAGAACGGTCC	CAGGGATAAGACAGGAGCGA
CdL96	JK340598	ATCCTGTGCTTGCTCCTGGC	AGGTTCTTGACGAGCGGGAG
CdL111	JK340584	GAATCCGAGCACTCTCAGGA	CTCAAGACAACAGTCAGACACA
CdL113	JK340639	ATAGTTGAGATTCCAGGCTC	TCAACTGAGAAGGCTAAGAC
CdL10-27	JK340525	AGTTGTGGAAGGATGGAGAA	CCTGGGAAGGACTGATGAAG
CdL10-39	JK340529	AGGCTGCCCTGATGTGATTT	ATTGGAATGACTCGTGTCGC
CdL10-52	JK340708	AGCGAGCGACTTACAATGGC	AACACTGAAGGCGTGGCTGA
CdL10-99	JK340714	TTTGATCCCTGATCAGGGCA	GCCTGAAGAGTCCTCATGGT
CdL10-127	JK340747	GTTGGTGAGGCTTTCTGGGA	GTGGTGATGGCAAGTGGATTG
CdL10-97	JK340737	AGTCCACTGGTGTCTTCACT	GACACCAACAACGAACATGG
CdL10-47	JK340720	TCACGATGGAGGAGAACCAC	AGCATTTCCTTGTGGACGGG
Tifway-10-46	JK340856	ACCATACAGACCCTTCAGAC	GCGAACAGTCCGTGATAACT

Rsa I and then the tester cDNAs were divided into two portions, each ligated with two different cDNA adaptors (Adaptor 1: 5′-CTAATACGACTCACTATAGGGCTCGAGCGGCCGCCCG-GGCAGGT -3′; Adaptor 2: 5′-CTAATACGACTCACTA-TAGGGCAGCGTGGTCGCGGCCGAGGT-3′). Two rounds of hybridization and PCR amplification were performed to enrich for the differentially-regulated gene sequences. The subtracted cDNA population was then subjected to primary PCR amplification of differentially-expressed genes using the adaptor specific primer (5′-CTAATACGACTCACTATAGGGC-3′). Subsequently, a second PCR amplification was performed with nested primers (Primer 1: 5′- TCGAGCGGCCGCCCGGGCAGGT-3′; Primer 2:5′-AGCGTGGTCGCGGCCGAGGT-3′) to reduce the background and to further enrich for differentially-expressed genes. After the second PCR, the products were directly cloned into the pMD18-T vector using TA Cloning kit (Takara BIO, Inc., Otsu, Shiga, JapanThe ligated product was transformed into *E. coli* competent cells (DH5α) which were then plated onto LB agar medium containing 100 μg ml^{-1} ampicillin and incubated at 37°C overnight. A total of 1302 positive clones were obtained from 'C299' and 'Tifway'. Using PCR with nested primers, 803 clones were confirmed to have inserted cDNA fragment in these four SSH

Table 2. Leaf relative water content (RWC) and electrolyte leakage (EL) of two bermugdagrass genotypes ('Tifway' and 'C299') at 5 and 10 d of drought stress and well-watered conditions.

Treatments	Genotypes	EL (%)	RWC (%)
Well watered	Tifway	23.7aA*	93.5aA
	C299	25.3cA	93.1aA
5-d drought	Tifway	27.2aB	88.0bA
	C299	48.6bA	69.4bB
10-d drought	Tifway	41.9bB	68.8cA
	C299	73.1aA	29.4cB

*Means within a column for either EL or RWC followed by the same letters were not significantly different based on LSD test at $P=0.05$. Lower case letters for the comparison between treatments and uppercase letters for the comparison between two genotypes at a given treatment.

libraries. Each clone was picked individually and grown overnight in 1.0 mL liquid LB medium with 100 µg mL^{-1} ampicillin. The clones were stored as glycerol stocks at −80°C for sequencing.

DNA sequencing and analysis

All 803 clones were sequenced by Invitrogen sequencing company (Shanghai China) and 757 sequences were obtained ranging in length from 88 to 823 bp. After removing contaminated vectors, mitochondrial, ribosomal, and *E.coli* sequences, 444 high quality ESTs were obtained and analyzed using the BlastN and BlastX algorithm (http://www.ncbi.nlm.nih.gov/BLAST) to identify their putative functions. All ESTs were submitted to genebank. Their accession numbers are from JK340473 to JK340916.

Quantitative (Q)-PCR analysis

Total RNA was extracted as described above. The first strand cDNA was generated with a RETROscript Kit (Applied Biosystems/Ambion Inc.,Austin Tx) using oligo (dT) as primer. Thirty seven genes were detected by Quantitative (Q)-PCR. Q-PCR analysis was performed with specific primers as shown in Table 1 to detect the transcripts selected from the SSH library in both genotypes under the well-watered control and drought stress conditions. Q-PCR amplification mixtures (25 µl) contained 25 ng template cDNA, 2*SYBR Green I Master Mix buffer (12.5 ul)(Applied Biosystems) and 300 nM forward and reverse primer. Reactions were run on an ABI PRISM 5700 Sequence Detector (Applied Biosystems). The cycling conditions comprised 4 min polymerase activation at 94°C and 40 cycles at 94°C for 30 sec, 58°C for 30 sec and 72°C for 30 sec. 18S ribosomal RNA gene (primers in table 1, accession number: AB536694) was selected as endogenous control. Each gene was detected three times. All PCR efficiencies were above 95%. Sequence Detection Software (version 1.3) (Applied Biosystems) results were exported as tab-delimited text files and imported into Microsoft Excel for further analysis. The median coefficient of variation (based on calculated quantities) of duplicated samples was 6%.

Hierarchical clustering analysis

Thirty seven drought-responsive genes confirmed with Q-PCR were subjected to clustering analysis. For comparison among treatments, let G_i equal the primary data for gene G in condition *i*. The transformation used was: $g_i=(G_i-G_{average})/G_{average}$, with $G_{average}$ referring to the average expression quantity of gene G relative to the total expression. After transformation, g_i values were used for cluster analysis. Significantly expressed genes were hierarchically clustered with average linkage and Euclidean distance as a measurement of similarity using Genesis version 1.7.5 (Graz University of Technology, http://www.genome.tugraz.at). To display the clustering results, each gene in a certain condition was assigned a cell. Red, green, and black boxes represent genes that increased, decreased, and had equal expression levels at time points after withholding water, respectively. This allows for ease of visualization of the relationship of each gene and the expression patterns.

Results

Physiological responses of two bermudagrass genotypes to drought stress

Leaf relative water content (RWC) was maintained over 90% in both genotypes under well-watered conditions, but declined to 88% for 'Tifway' and 69% for 'C299' by 5 d of drought stress, and to 69% for 'Tifway' and 29% for 'C299' by 10 d of drought stress (Table 2). Leaf cell membrane stability was evaluated by measuring electrolyte leakage (EL). Leaf EL increased with drought stress, to 27% for 'Tifway' and 49% for 'C299' by 5 d of drought stress, and to 42% for 'Tifway' and 73% for 'C299' by 10 d of drought stress (Table 2). The differences in both RWC and EL between the drought-tolerant 'Tifway' and the drought-sensitive 'C299' and between treatments were statistically significant.

Identification of genes up-regulated by drought stress in either genotype from the SSH libraries

Four SSH libraries were constructed to identify drought-up regulated genes in 'C299' and 'Tifway' in response to short-term (5 d) or mild drought and long-term (10 d) or severe of drought stress, as indicated by the levels of physiological responses described above.

Using PCR with nested primers, 803 clones were confirmed to have inserted cDNA fragments from a total of 1272 individual clones in these SSH libraries. High throughput sequencing was conducted for all clones and with a 94% success rate. A total of 757 ESTs were generated with the length from 88 to 823 bp. After data assembly, 368 ESTs were found to be singlets, while the other 389 ESTs clustered into 76 contigs and each contig had 2–53 ESTs (Table 3). The ESTs generated from each library were also assembled and analyzed separately. With all contigs and singlets, 210 transcripts were identified in 'Tifway' at 10 d of drought stress, 145 in 'Tifway' at 5 d of drought stress, 66 in 'C299' at 10 d of drought stress, and 52 in 'C299' at 5 d of drought stress. In total, 421 genes were differentially expressed in at least one

Table 3. Summary of ESTs from bermudagrass drought stress SSH libraries.

Drought (days)	SSH clones	contigs	singlets	gene transcripts	contigs (refer to gene transcripts)
Tifway 10 d	230	24 (10.4%)	205 (89.6%)	210	5
5 d	196	74 (37.7%)	124 (62.3%)	145	21
C299 10 d	172	130 (75.6%)	42 (24.4%)	66	24
5 d	157	124 (79.0%)	33 (21.0%)	52	19
total	757	389(51.4%)	368(48.6%)	444	76

genotype at least one treatment duration period. The number of genes expressed in the two genotypes at the two different treatment periods is depicted with a four-way Venn diagram in Fig. 1.

BLAST analysis of all the cDNA sequences was carried out to identify their putative functions (http://www.ncbi.nlm.nih.gov/BLAST). Sequences without significant homology with known proteins or genes were further analyzed with BLASTn using the EST others database (Supporting Information in File S1). About 84% ESTs had hit similarities in gene bank. Among them, 277 ESTs showed high homology to reported genes or proteins with E-value less than 1E-5. Of the 167 ESTs with no significant homology to reported genes or proteins, 105 were similar to reported ESTs, including one EST "Tifway-10–46" (Accession Number: JK340856) from 'Tifway'-10-d-drought SSH library that had similarity to DHN4 gene encoding dehydrin protein. In the 'C299' SSH libraries, 25 clones expressed at 5-d drought and 41 clones expressed at 10-d drought had high homology to reported genes or proteins. In the 'Tifway' SSH libraries, 76 clones expressed at 5-d drought and 143 clones at 10-d drought had high homology to reported genes or proteins. Among these clones, many with known functions have also been demonstrated to be enhanced or induced by abiotic stress in other plant species, including *Arabidopsis thaliana*, maize (*Zea mays*) and rice (*Oryza sativa*).

Functional classification of up-regulated genes in both bermudagrass genotypes under mild drought and severe drought stress

In order to understand the molecular mechanisms underlying the adaptation of two bermudagrass genotypes differing in drought tolerance, up-regulated candidate genes isolated by SSH in response to short-term (5 d) and long-term (10 d) drought stress were categorized into various functional groups according to their main putative functions indicated by BLASTx and BLASTn analysis. Although some genes were involved in multiple metabolic processes, they were categorized according to their main functions in plant metabolism. A total of 277 gene transcripts with match to the GenBank database were grouped into nine functional categories: 1) stress defense and aging, 2) metabolism, 3) osmoregulation, 4) membrane system, 5) signal and regulator, 6) structure protein, 7) protein synthesis and degradation, 8) energy and 9) others which cannot be classified nor with unknown functions (Fig. 2). About 23% of gene transcripts annotated with unknown function were included in the category of unclassified proteins. The largest group of genes with known functions was metabolism-related genes and approximately 27.2% of the transcripts were included in this category, which was followed by signal and regulator (16.5%), protein synthesis and degradation (10.4%), stress defense and aging (9.6%), membrane system (6.5%), osmoregulation (6.2%), energy (2.8%), and structure protein (2%).

The percentages of drought-responsive genes were compared between the two genotypes. Most of the up-regulated genes in 'C299' with known functions were involved in metabolism (30.4% at 5-d drought, 29.0% at 10-d drought), stress defense (13.0% at 5-d drought, 6.5% at 10-d drought), and signal and regulator (13.0% at 5-d drought, 22.6% at 10-d drought) (Fig. 3A). For 'Tifway', most of the up-regulated genes were classified in metabolism (24.7% at 5-d drought, 24.5% at 10-d drought), stress defense (10.4% in drought 5-d, 8.4% at 10-d drought), and signal and regulators (14.3% at 5-d drought, 16.1% at 10-d drought) (Fig. 2B). In addition, a large proportion of up-regulated genes in the protein synthesis and degradation category (15.6% at 5-d drought, 19.6% at 10-d drought) were detected in 'Tifway', but a small fraction of those genes were detected in 'C299' (6.5% at 10-d drought, 0% at 5-d drought) (Fig. 3B).

The percentages of up-regulated gene transcripts in both genotypes changed with drought stress duration or degree of

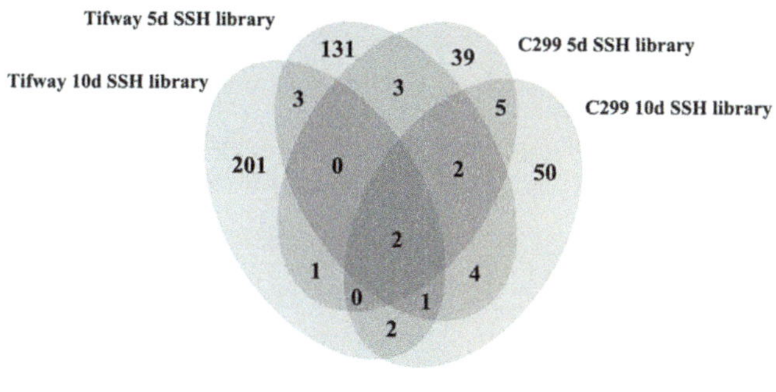

Figure 1. Four-way Venn diagram showing the number of genes differentially-expressed in 'Tifway' and 'C299' at 5 and 10 d of drought.

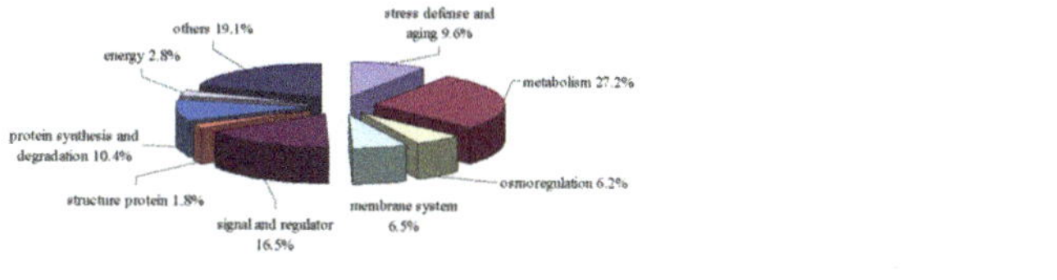

Figure 2. Classification of differentially-expressed genes between two genotypes at 5 and 10 d of drought. The differentially-expressed genes in 'C299' and 'Tifway' SSH libraries at 5 and 10 d of drought stress were grouped according to their putative functions generated by BlastX and BlastN analysis The percentage of genes in each functional group was listed.

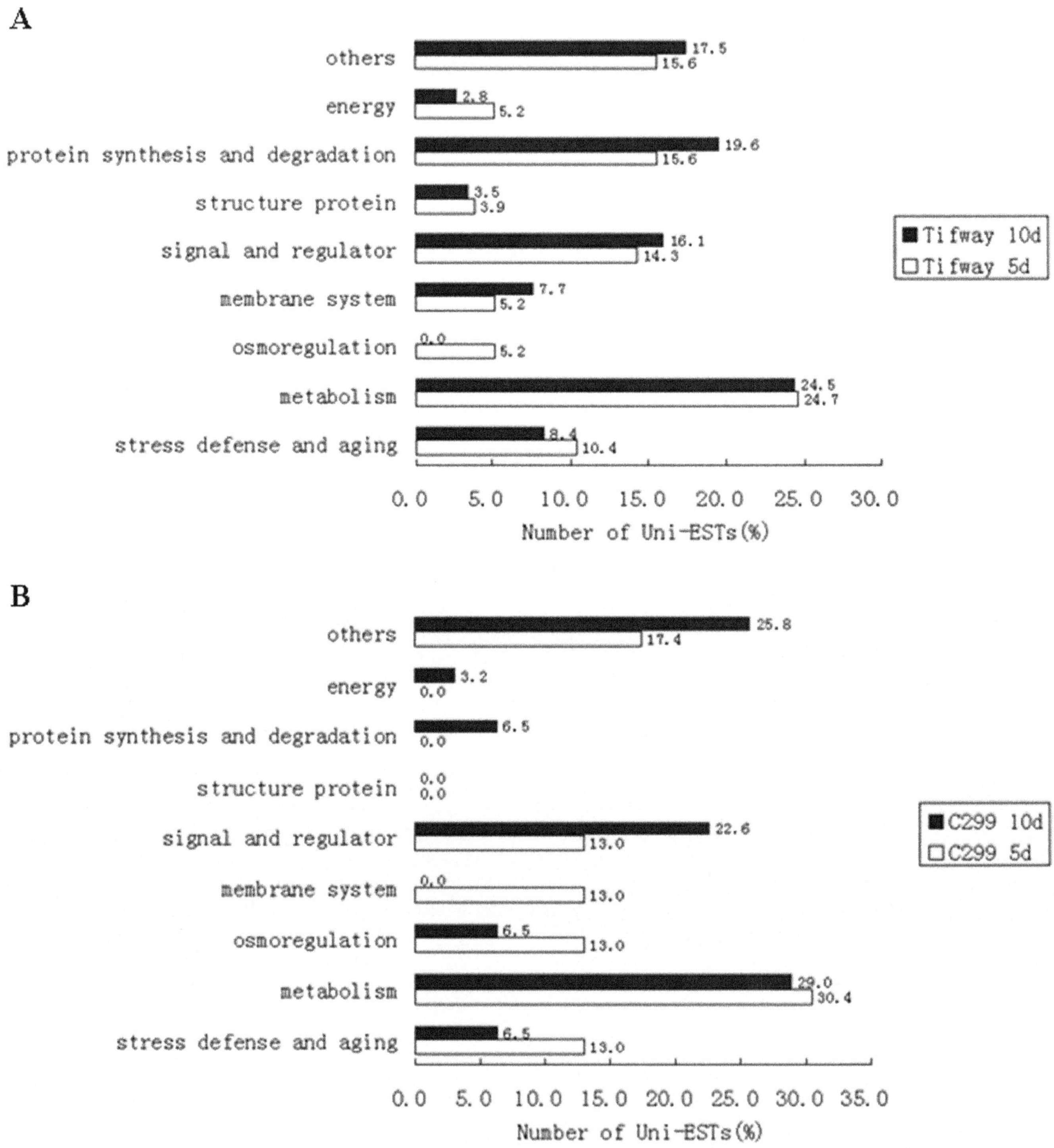

Figure 3. Comparison of the percentages of gene transcripts in each functional classification between 5-d drought and 10-d drought SSH libraries of 'Tifway' and 'C299'. A: Comparison for 'Tifway'; B: Comparison for 'C299'. The bars represent the percentages of gene transcripts for both genotypes in each functional category.

severity. A higher percentage of up-regulated genes were found in the category protein synthesis and degradation in both 'Tifway' and 'C299' plants exposed to 10-d drought than those at 5-d drought. However, the percentage of genes in osmoregulation was higher in both 'Tifway' and 'C299' plants exposed to 5-d drought than those at 10-d drought. The percentage of up-regulated genes in membrane system was also higher in 'C299' exposed to 5-d drought than those at 10-d drought, while for 'Tifway', the percentage of genes in membrane system was lower at 5 d of drought than those at 10 d of drought.

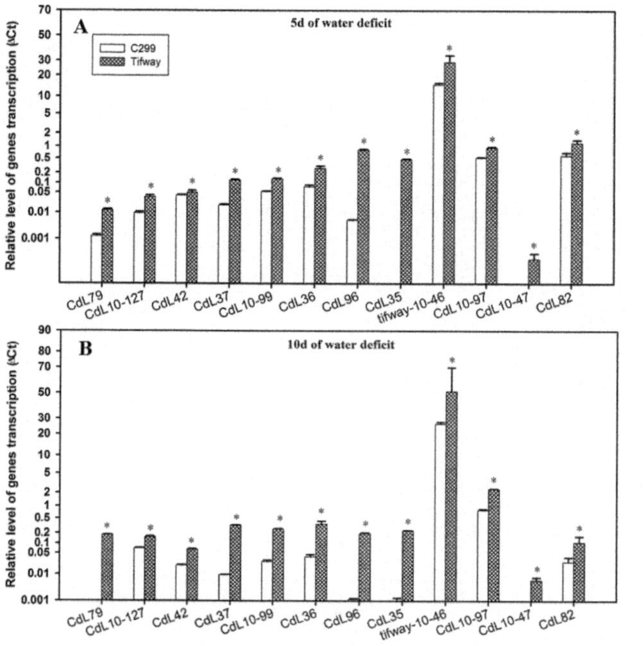

Figure 4. Q-PCR analysis of expression of 12 genes from four SSH libraries, exhibiting higher levels in 'Tifway' and 'C299' at 5 and 10 d of drought. A; Comparison of the relative gene expression levels of 'Tifway' and 'C299' at 5 d of drought; B: Comparison of the relative gene expression levels of 'Tifway' and 'C299' at 10 d of drought. The error bars represent standard deviations of the means. For each gene, the significant differences based on LSD test at P = 0.05 between 'Tifway' and 'C299' were shown by asterisk on the top of the error bars.

Q-PCR confirmation and comparative analysis of transcript levels for drought up-regulated genes between 'C299' and 'Tifway'

Thirty seven genes up-regulated by drought in either or both genotypes and differentially expressed at different transcript levels between the two genotypes at 5 and 10 d of drought that were identified from the SSH analysis were further examined using Real time-PCR. The Real time-PCR analysis confirmed that 36 genes displayed differential expression levels between the two bermudagrass genotypes under drought conditions (Figure S1). One gene, CdL10–39 (JK340529, putative starch branching enzyme 4), did not exhibit differences in its expression level between the two genotypes at 0, 5, or 10 d of drought.

Twelve genes had significantly higher levels of expression in 'Tifway' than 'C299' at 5 and 10 d of drought (Fig. 4). These genes included CdL79 (JK340618, putative WRKY transcription factor 50), CdL10–127 (JK340747, putative DHAR), CdL42 (JK340615, putative EREBP-4 like protein), CdL37 (JK340586, putative cytosolic NADP malic enzyme), CdL10–99 (JK340714, putative 2-cys peroxiredoxin), CdL36 (JK340585, putative sterol desaturase family protein), CdL96 (JK340598, putative CER1), CdL35 (JK340626, putative ATP-dependent Clp protease ATP-binding subunit precursor), Tifway-10–46 (JK340856, putative dehydrin(DHN4)), CdL10–97 (JK340737, putative GAPDH - glyceraldehyde-3-phosphate dehydrogenase), CdL10–47 (JK340720, putative aldehyde oxidase), and CdL82 (JK340582, putative HVA22-like protein).

Seven genes had significantly higher level of expression in 'Tifway' than in'C299' at 10 d of drought, but no significant differences were detected between two genotypes at 5 of drought (Fig. 5). The genes were CdL80 (JK340610, putative Remorin), CdL4 (JK340476, putative DNA repair ATPase-related), CdL62 (JK340580, putative leaf senescence related protein-like), CdL49 (JK340642, putative LEC14B protein), CdL20 (JK340484, putative permease I), CdL10–52 (JK340708, putative superoxide dismutase [Cu-Zn]), and CdL44 (JK340578, putative SK3-type dehydrin).

Nine genes showed significantly lower expression levels in 'Tifway' than in 'C299' at both 5 and 10 d of drought (Fig. 6). These genes were CdL40 (JK340589, putative 4-hydroxyphenylpyruvate dioxygenase), CdL85 (JK340595, putative pyruvate dehydrogenase E1 alpha subunit), CdL111 (JK340584, maturation-associated protein), CdL8(JK340518, putative PR17c precursor), CdL1 (JK340473, putative Cytokine induced apoptosis inhibitor), CdL2(JK340477, putative alcohol dehydrogenase), CdL69 (JK340599, putative delta 1-pyrroline-5-carboxylate synthetase), CdL68 (JK340608, putative stachyose synthase), and CdL52(JK340628, putative protein phosphatase 2C).

Eight genes exhibited significantly lower levels of expression in 'Tifway' than in 'C299' only at 5 d of drought (Fig. 7A).They included CdL56(JK340591, putative beta-galactosidase), CdL61 (JK340592, gibberellin 20 oxidase 2), CdL19 (JK340483, putative sucrose synthase), CdL9 (JK340478, putative carbonic anhydrase), CdL74 (JK340625, putative actin depolymerization factor-like protein), CdL113 (JK340639, putative CTP synthase), CdL67 (JK340607, putative galactinol synthase 1), CdL10–27 (JK340525, putative plastid starch synthase I precursor). At 10 d of drought, CdL56, CdL61, CdL67, CdL113, and CdL10–27 had significantly higher expression level in 'Tifway' than 'C299' and other three did not exhibit genotypic differences (Fig. 7B).

Hierarchical clustering analysis of up-regulated genes under drought

Hierarchical clustering analysis was performed to analyze the similarity of the expression patterns for 37 up-regulated genes based on average linkage and Euclidean distance as a measurement of similarity by using Genesis, which provides information on co-expressed genes [23]. The 37 up-regulated genes by drought in either or both genotypes were clustered into three groups with similar expression patterns for genes within each group (Fig. 8).

Group A included three genes, CdL82 (JK340582, putative HVA22-like protein, CdL96 (JK340598, putative CER1), and CdL35 (JK340626, putative ATP-dependent Clp protease ATP-binding subunit precursor). The expression level of genes in Group A increased at 5 d of drought, to a greater extent in 'Tifway' than in 'C299', but decreased as drought was prolonged to 10 d in both genotypes.

Group B had 17 genes, including five genes involved in osmotic regulation - CdL19 (JK340483, putative sucrose synthase), CdL68 (JK340608, putative stachyose synthase), CdL56 (JK340591, putative beta-galactosidase), and CdL67 (JK340607, putative galactinol synthase 1), and CdL69 (JK340599, putative delta 1-pyrroline-5-carboxylate synthetase 1). In Group B, the expression level of most genes except CdL8 (JK340518, putative PR17c precursor), CdL2 (JK340477, putative alcohol dehydrogenase), and CdL85 (JK340595, putative pyruvate dehydrogenase E1 alpha subunit) decreased to a lower level at 10 d of drought compared to that at 5 d in 'C299', but the gene expression level was significantly higher in 'C299' than in 'Tifway' at 5 and 10 d of drought.

Group C had 17 genes, including four genes associated with ROS scavenging - CdL10–52 (JK340708, putative superoxide dismutase [Cu-Zn]), CdL10–99 (JK340714, putative 2-Cys

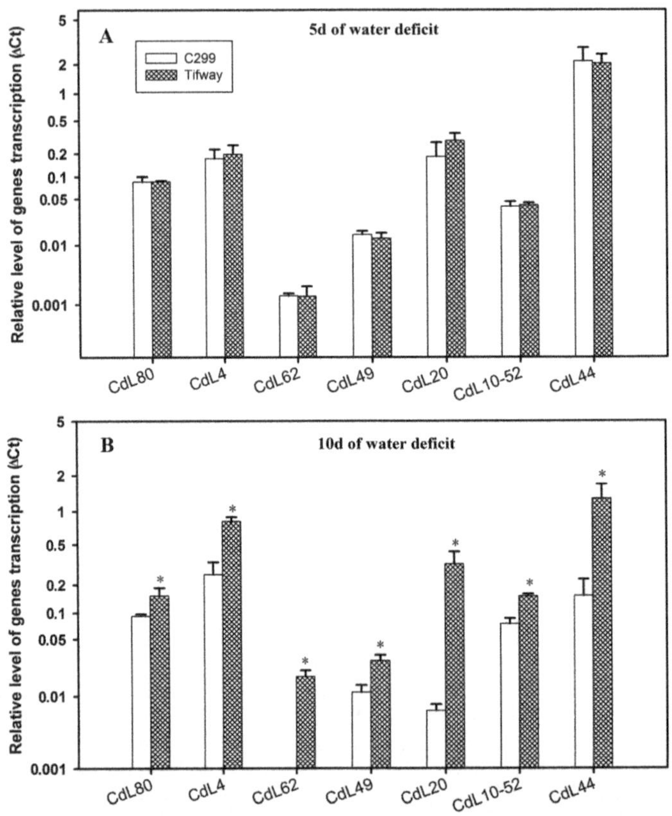

Figure 5. Q-PCR analysis of expression levels of seven genes from four SSH libraries, exhibiting higher expression in 'Tifway' than in 'C299' only 10 d of drought. Comparison of the relative gene expression levels of 'Tifway' and 'C299' at 5 d of drought; B: Comparison of the relative gene expression levels of 'Tifway' and 'C299' at 10 d of drought. The error bars represent standard deviations of the means. For each gene, the significant differences based on LSD test at P = 0.05 between 'Tifway' and 'C299' were shown by asterisk on the top of the error bars.

peroxiredoxins), CdL10–47 (JK340720, putative aldehyde oxidase), and CdL10–127(JK340747, putative dehydroascorbate reductase). Group C also included CdL10–97 (JK340737, putative GAPDH) with diverse functions and a gene associated with dehydration protection (Tifway-10–46, JK340856, putative dehydrin (DHN4). Genes in Group C exhibited increased level of expression with drought at both 5 and 10 d, and had significantly higher expression level in 'Tifway' than 'C299'.

Discussion

Genotypic variations in differential gene expression in response to drought stress are also reflected at the physiological levels. Physiological analysis with 'Tifway' and 'C299' exposed to drought stress demonstrated that 'Tifway' was able to maintain higher cell membrane stability and water status (Table 2), as well as greater photosynthetic rate, photochemical efficiency, and antioxidant defenses [16–18]. The physiological data suggested that 'Tifway' exhibited superior drought resistance to 'C299'. The gene expression analysis in this study provided further insights on molecular factors associated with superior drought resistance in 'Tifway' bermudagrass, as manifested by the physiological traits.

Drought avoidance is a key strategy in drought resistance of plants, which ischaracterized by minimizing water loss through various mechanisms such as formation of wax on leaf surfaces, which provides primary protection against leaf dehydration (Post-Beittenmiller, 1996; Islam et al., 2009) [24,25]. A CER1 gene (CdL96, JK340598) which encodes a protein involved in a key step in wax biosynthesis [26] was detected at 5 and 10 d of drought stress, with the expression level being higher in drought-tolerant 'Tifway' than in drought-sensitive 'C299'. Another gene, sterol desaturase family protein (CdL36, JK340585), putatively encoding a sterol desaturase that has been reported to be involved in epicuticular wax biosynthesis [25], was also much more highly expressed in 'Tifway' than in 'C299' at 5 and 10 d of drought stress. In a previous study, water loss rate from leaves was significantly slower in 'Tifway' than 'C299' under drought stress [17], which was corresponded to the enhanced expression of genes controlling wax synthesis in 'Tifway'. The up-regulation of CER1 and sterol desaturase may enhance cuticle wax synthesis, allowing plants to avoid dehydration. 'Tifway' indeed exhibited higher RWC in leaves compared to 'C299' under drought stress.

Drought tolerance mechanisms have been associated with cellular protection from dehydration by accumulating protective proteins, most notably dehydrins and other late-embryogenesis abundant (LEA) proteins [27]. Although the function of many dehydrins and LEA proteins is not fully understood, at least part of their function is to act as chaperones that protect protein and membrane structure [28,29]. Compatible solutes can also protect protein and membrane structure under dehydration [30]. In 'Tifway' plants exposed to 10 d of drought, one EST (Tifway-10–46, JK340856) showed high homology to DHN4 gene [*Hordeum vulgare*] with E-value 2.00E-98, which is putative encoding dehydrin protein. Q-PCR result confirmed that its expression level increased with progressive drought and was significantly higher in 'Tifway' than 'C299'. The expression pattern of DHN4 gene in

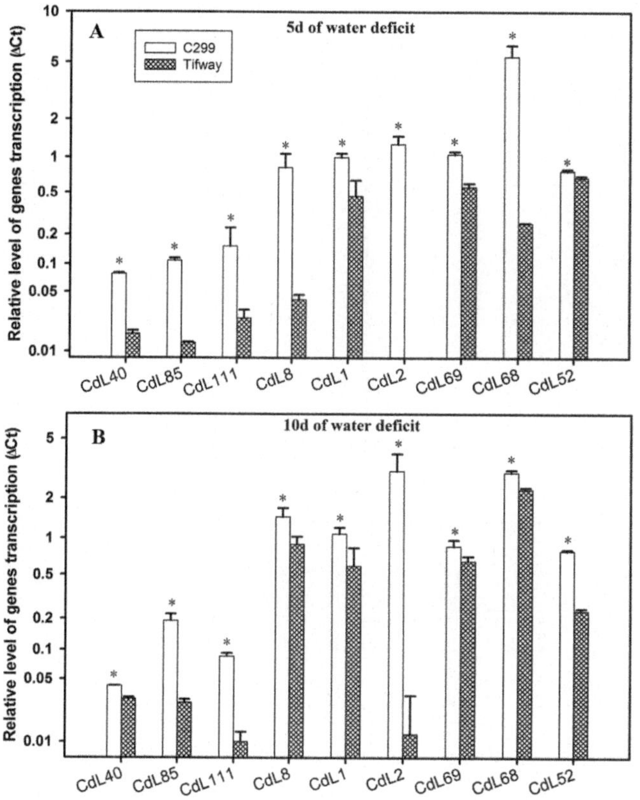

Figure 6. Q-PCR analysis of expression of nine genes from four SSH libraries, exhibiting higher levels in 'C299' than in 'Tifway' at 5 and 10 d of drought. Comparison of the relative gene expression levels of 'Tifway' and 'C299' at 5 d of drought; B: Comparison of the relative gene expression levels of 'Tifway' and 'C299' at 10 d of drought. The error bars represent standard deviations of the means. For each gene, the significant differences based on LSD test at P = 0.05 between 'Tifway' and 'C299' were shown by asterisk on the top of the error bars.

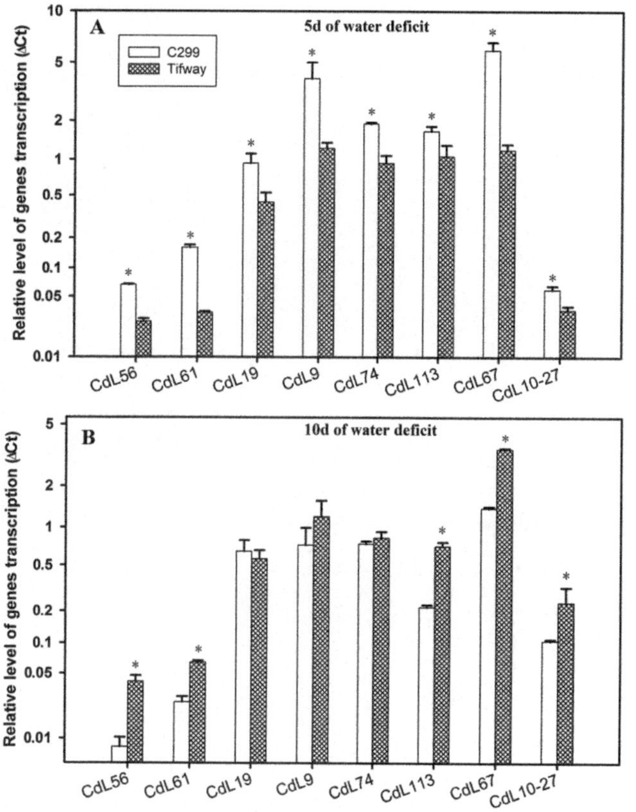

Figure 7. Q-PCR analysis of expression of eight genes from four SSH libraries, exhibiting higher levels in 'C299' than in 'Tifway' only at 5 d of drought. Comparison of the relative gene expression levels of 'Tifway' and 'C299' at 5 d of drought; B: Comparison of the relative gene expression levels of 'Tifway' and 'C299' at 10 d of drought. The error bars represent standard deviations of the means. For each gene, the significant differences based on LSD test at P = 0.05 between 'Tifway' and 'C299' were shown by asterisk on the top of the error bars.

the current study was similar to the expression of 40-kDa dehydrins in the report by Hu et al. [17] for the same cultivar 'Tifway' exposed to 10 d of drought stress. Hu et al. [17] detected multiple dehydrin expression patterns in 'Tifway' and 'C299' under drought stress by immunoblotting analysis, and the level of expression of 40-kDa dehydrin was positively associated with drought tolerance in bermudagrass. Our results at the transcript level in combination with that at the protein level [17] suggested that dehydrins may play an important role in protecting bermudagrass leaves from dehydration. Another gene, *HAV22* (CdL82, JK340582), which was reported to be involved in ABA-induced gene expression and have some homolog with LEA protein [31], was expressed significantly higher in 'Tifway' than in 'C299' at 5 and 10 d of drought stress. A GC-rich element from barley HVA22 gene was reported to be essential for ABA or stress-inducible gene expression [32]. But the expression of *HAV22* decreased significantly at 10 d of stress in 'Tifway', revealing *HAV22* expression may have short-term effects on stress defense.

Another aspect of drought tolerance is the control of the level of ROS or minimizing the damage caused by ROS. In our experiments, some genes involved in ROSs scavenging were detected in 'Tifway' plants exposed to 10 d of drought stress, including superoxide dismutase (SOD; CdL10–52, JK340708), aldehyde oxidase (CdL10–47, JK340720), dehydroascorbate reductase (DHAR; CdL10–127, JK340747), and 2-cys peroxiredoxin bas1 (CdL10–99, JK340714). SOD catalyzes the dismutation of superoxide into oxygen and hydrogen peroxide [33]. Aldehyde oxidase (AO) produces hydrogen peroxide and under certain conditions can catalyze the formation of superoxide [34]. DHAR is responsible for regenerating ascorbic acid (Asc) from an oxidized state, regulates the cellular Asc redox state, which in turn affects cell responsiveness and tolerance to environmental ROS [35]. Plant 2-Cys peroxiredoxins are post-translationally targeted to chloroplasts, protecting the photosynthetic membrane against photooxidative damage [36]. Q-PCR analysis confirmed that these genes had increased expression level in the drought-tolerant genotype. Under drought stress, their expression level was significantly higher in 'Tifway' than in 'C299', especially at 10 d. Zhao et al. [18] also reported high abundance of SOD protein in 'Tifway' than in 'C299' at 10 d of drought stress. Chen et al. [37] found bermdagrass mutants with superior drought tolerance had than the wild type. The results at both transcript and protein levels indicated that the maintenance of active antioxidant genes and proteins could contribute to the superior drought tolerance in 'Tifway', as manifested by greater cell membrane stability (Table 2) The hierarchical clustering analysis revealed that glyceraldehyde-3-phosphate dehydrogenase (GAPDH) (CdL10–97, JK340737) was closely related to the four genes involved in ROSs scavenging. By correlation coefficient analysis,

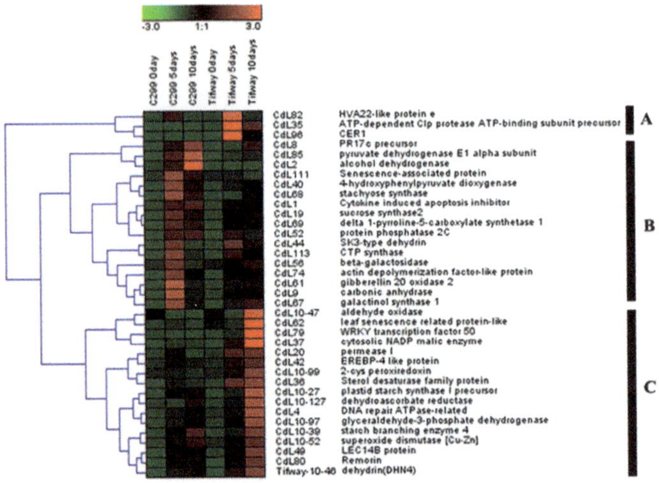

Figure 8. Cluster analysis of 37 genes differentially-expressed between well-watered control and drought in 'Tifway' and 'C299'. Cluster analysis for each group of genes was performed using hierarchical clustering of Genesis 1.7.5 with average linkage and Euclidian distance measurement. Rows represent differentially-expressed genes, while columns represent the genotypes with 5 d or 10 d of drought treatment. Each gene expressed in different conditions is presented as a cell. Red, green, and black boxes represent genes that increased, decreased, and had equal expression levels at a given duration of drought treatment, respectively. The gene name and annotation of each gene are listed on the right of the figure, and the cluster numbers are listed on the left. The color scale shown at the top illustrates the relative expression ratios of gene across all samples.

the expression of CdL10–97 had high positive correlation coefficient with CdL10–52 and CdL10–127 which were related to ROSs scavenging ($r>0.8$, $p<0.05$), revealing GAPDH may had other functions related to ROSs scavenging. GAPDH is an enzyme that catalyzes the sixth step of glycolysis. But increasing evidence suggests that GAPDH is multifunctional, it displays a number of diverse activities unrelated to its glycolytic function [38] such as a signaling role in mediating ROS responses [39–42].

Osmotic adjustment or regulation has also been considered an important drought tolerance mechanism, which is accomplished through the accumulation of solutes in plants, contributing to cell turgor maintenance during drought [43]. Genes classified in the osmoregulation category were detected at 5 d of drought stress in both genotypes, but the percentage of genes in this category decreased with stress duration up to 10 d in both genotypes. These data suggested that osmotic regulation could be involved in early response of bermudagrass to drought stress, but not long-term drought tolerance in this species. Drought up-regulated genes with putative functions as solute synthase were detected in both genotypes of bermudagrass exposed to drought stress, including sucrose synthase2 (CdL19, JK340483), galactinol synthase 1(CdL67, JK340607), stachyose synthase (CdL68, JK340608) and delta 1-pyrroline-5-carboxylate synthetase 1(CdL69, JK340599), but the expression level was lower in the drought-tolerant than in the drought-sensitive genotype. Sucrose synthase, galactinol synthase, and stachyose synthase are related to the synthesis of oligosaccharides, which are involved in osmotic adjustment [44]. Delta 1-pyrroline-5-carboxylate synthetase catalyzes proline synthesis [45]. Previous studies have shown that proline accumulate was responsive to drought stress and serves as a protective solute to maintain cell turgor against dehydration in various plant species [46], oxidative protection [47], and function as molecular chaperone stabilizing the structure of proteins [48]. The up-regulation of those genes associated with solute accumulation under drought stress, particularly in the drought-sensitive genotype reflected that sugar and proline accumulation was sensitive to mild or short-term drought stress in bermudagrass, but may not contribute to superior drought tolerance in this species under long-term stress.

Several genes involved in other functional categories were also found differentially expressed between the two genotypes under drought conditions. CdL35 (JK340626) was confirmed up-regulated in 'Tifway' at 5 and 10 d of drought stress, and presented a higher expression level in 'Tifway' than in 'C299'. This gene is encoding a subunit of ATP-dependent Clp protease. ATP-dependent Clp protease is involved in a variety of processes, ranging from developmental changes to stress tolerance [49,50]. CdL37 (JK340586) is encoding cytosolic NADP malic enzyme, were expressed at a significantly higher level in 'Tifway' than in 'C299' at both 5 and 10 d of drought stress. Cytosolic NADP malic enzyme is related to plant defense to various abiotic stresses [51]. Other genes that were differentially expressed between 'C299' and 'Tifway' included transcription factors and genes regulated by those transcription factors. Transcription Factors (TFs) EREBP-4 like protein (CdL42, JK340615) and WRKY transcription factor 50 (CdL79, JK340618) were expressed at a significantly higher level in 'Tifway' than in 'C299' at 5 and 10 d of drought. EREBP-4 like protein (CdL42, JK340615) belongs to AP2/EREBP family members, which bind DRE "TACCGACAT" motifs and regulate a series of drought responsive genes [52]. Many genes activated by AP2/EREBP TFs, such as *rd29A*, were reported to be associated with better drought tolerance [53,54]. The WRKY genes encode a large group of transcription factors. They bind specifically to W box motif "(T)(T)TGAC(C/T)" which was presented in many co-regulated defense gene promoters [55]. Over-expression of OsWRKY11 in rice (*Oryza sativa*) enhanced drought tolerance [56]. The differential expression of these transcription factors between the two genotypes suggest that the downstream responses to drought stress varied with genotypes differing in drought tolerance. Leaf senescence related protein-like (CdL62, JK340580) regulated by WRKY gene was more highly expressed in 'Tifway' than 'C299'. It may take part in initiating the process of leaf senescence induced by drought [57,58], which has been associated with plant survival of drought stress by reducing leaf area for transpiration to limit water loss from the plant canopy and diverting carbon partitioning.

Summary

The SSH analysis identified 277 drought-responsive genes in two genotypes of bermudagrass in response to mild or short-term drought (5 d) and severe or long-term (10 d) drought stress. Those genes were mainly classified in the functional categories of stress defense and aging, metabolism, osmoregulation, membrane system, signal and regulator, structure protein, protein synthesis and degradation, and energy metabolism. Q-PCR analysis confirmed the expression of 36 drought up-regulated genes that were more highly expressed in drought-tolerant 'Tifway' than drought-sensitive 'C299' at 5 and 10 d of drought stress, such as those for leaf cuticular wax accumulation (CER1, sterol desaturase), stress signaling (EREBP-4 like protein, WRKY transcription factor), dehydration protective proteins (dehydrins and HVA22-like protein), and oxidative stress defense (SOD, DHAR, 2-Cys peroxiredoxins, GAPDH) for bermudagrass tolerance to drought. Molecular markers for the above-mentioned genes could be utilized for marker-assisted selection of drought-tolerant warm-season perennial grasses. This approach is particularly relevant for grasses grown in shallow soil profiles.

Supporting Information

Figure S1 Q-PCR analysis of all the 37 genes expression in both genotypes under 0-d, 5-d, and 10-d water deficit. For each gene, the significant differences based on LSD test at P = 0.05 between 'Tifway' and 'C299' were shown by different letters on the top of the error bars.

File S1 Supporting tables from Table S1 to Table S8 of four SSH libraries clones analyzed by BLAST.

Author Contributions

Conceived and designed the experiments: PZ BH. Performed the experiments: PZ HD BH. Analyzed the data: PZ BH. Contributed reagents/materials/analysis tools: YA ZW. Contributed to the writing of the manuscript: PZ BH.

References

1. Nilsen ET, Orcutt DM (1996) The physiology of plants under stress. John Wiley and Sons, New York, New York, USA.
2. Oliver M, Tuba Z, Mishler BD (2000) The evolution of vegetative desiccation tolerance in land plants. Plant Ecol 151:85–100, 2000.
3. Davies WJ, Wilkinson S, Loveys B (2002) Stomatal control by chemical signalling and the exploitation of this mechanism to increase water use efficiency in agriculture. New Phytol 153:449–460.
4. Cushman JC, Bohnert HJ (2000) Genomic approaches to plant stress tolerance. Curr Opin Plant Biol 3:117–124.
5. Umezawa T, Fujita M, Fujita Y, Yamaguchi-Shinozaki, Shinozaki K (2006) Engineering drought tolerance in plants: discovering and tailoring genes to unlock the future. Curr Opin Biotech 17:113–122.
6. Ozturk ZN, Talamé V, Deyholos M, Michalowski CB, Galbraith DW, et al. (2002) Monitoring large-scale changes in transcript abundance in drought- and salt-stressed barley. Plant Mol Biol 48:551–573.
7. Seki M, Narusaka M, Abe H, Kasuga M, Yamaguchi-Shinozaki K, et al. (2001) Monitoring the expression pattern of Arabidopsis genes under drought and cold stresses by using a full-length cDNA microarray. Plant Cell 13:61–72.
8. Chen D, Liang M, Dewald D, Weimer B, Peel MD, et al. (2008) Identification of dehydration responsive genes from two non-nodulated alfafa cultivars using Medicago truncatula microarray. Acta Physiol Plant 30:183–199.
9. Xue GP, McIntyre LC, Glassop D, Shorter R (2008) Use of expression analysis to dissect alterations in carbohydrate metabolism in wheat leaves during drought stress. Plant Mol Biol 67:197–214.
10. Blomstedt CK, Gianello RD, Gaff DF, Hamill JD, Neale AD (1998) Differential gene expression in desiccation-tolerant and desiccation-sensitive tissue of the resurrection grass, Sporobolus stapfianus. Aust J Plant Physiol 25:937–946.
11. Gaff DF, Bartels D, Gaff JL (1997) Changes in gene expression during drying in a desiccation-tolerant grass Sporobolus stapfianus and a desiccation-sensitive grass Sporobolus pyramidalis. Aust J Plant Physiol 24:617–622.
12. Taliaferro CM (1995) Genetic diversity and vulnerability of bermuda turfgrass species. Crop Sci 35:327–332.
13. Hanna W (1998) The future of bermudagrass. Golf Course Manag 66:49–52.
14. Wu Y, Taliaferro CM, Martin DL, Anderson JA, Anderson MP (2007) Genetic variability and relations for adaptive, morphological, and biomass traits in Chinese bermudagrass accessions. Crop Sci 47:1985–1994.
15. Jewell MC, Zhou Y, Loch DS, Godwin ID, Lambrides CJ (2012) Maximizing genetic, morphological, and geographic diversity in a core collection of Australian bermudagrass. Crop Sci 52:879–889.
16. Hu L, Wang Z, Huang B (2009). Photosynthetic responses of bermudagrass to drought stress associated with stomatal and metabolic limitations. Crop Sci 49:1902–1909.
17. Hu L, Wang Z, Du H, Huang B (2010) Differential accumulation of dehydrins in response to water stress for hybrid and common bermudagrass genotypes differing in drought tolerance. J Plant Physiol 167:103–109.
18. Zhao Y, Du H, Wang Z, Huang B (2011) Identification of proteins associated with drought tolerance in C4 perennial grass species [Cynodon dactylon (L.) Pers.×C. Transvaalensis Burtt Davy] and (C. dactylon). Physiol Plant 141:40–55.
19. Watt DA (2003) Aluminium-responsive genes in sugarcane: identification and analysis of expression under oxidative stress. J Exp Bot 54:1163–74.
20. Zhang Y, Mian MAR, Chekhovskiy K, So S, Kupfer D, et al. (2005) Differential gene expression in Festuca under heat stress conditions. J Exp Bot 56:897–907.
21. Gorantla M, Babu PR, Reddy Lachagari VB, Reddy AMM, Wusirika R, et al. (2007) Identification of stress-responsive genes in an indica rice (Oryzea sativa L.) using ESTs generated from drought-stressed seedlings. J Exp Bot 58:253–65.
22. Blum A, Ebercon A (1981) Cell membrane stability as a measure of drought and heat tolerance in wheat. Crop Sci 21:43–47.
23. Eisen MB, Spellman PT, Brown PO, Botstein D (1998) Cluster analysis and display of genome-wide expression patterns. Proc Natl Acad Sci USA 95:14863–14868.
24. Post-Beittenmiller D (1996) Biochemistry and molecular biology of wax production in plants. Ann Rev Plant Physiol 47:405–430.
25. Islam MA, Du H, Ning J, Ye H, Xiong L (2009). Characterization of Glossy1-homologous genes in rice involved in leaf wax accumulation and drought resistance. Plant Mol Biol 70:443–456.
26. Aarts GMM, Keijzer JC, Stiekema JW, Pereira A (1995) Molecular characterization of the CER1 gene of Arabidopsis involved in epicuticular wax biosynthesis and pollen fertility. The Plant Cell 7: 2115–2127.
27. Close TJ (1997) Dehydrins: a commonalty in the response of plants to dehydration and low temperature. Physiologia Plantarum 100, 291–296.
28. Hara M, Terashima S, Kuboi T (2001) Characterization and cryoprotective activity of cold-responsive dehydrin from Citrus unshiu. J Plant Physiol 158:1333–1339.
29. Bravo LA, Gallardo J, Navarrete A, Olave N, Martinez J, et al. (2003) Cryoprotective activity of a cold-induced dehydrin purified from barley. Physiol Plant 118:262–269.
30. Hincha DK, Hagemann M (2004) Stabilization of model membranes during drying by compatible solutes involved in the stress tolerance of plants and microorganisms. Biochem J 383:277–283.
31. Shen Q, Chen C, Brands A, Pan S, Ho DT (2001) The stress- and abscisic acid-induced barley gene HVA22: developmental regulation and homologues in diverse organisms. Plant Mol Biol 45:327–340.
32. RoyChoudhury A, Gupta B, Sengupta DN (2008) Trans-acting factor designated OSBZ8 interacts with both typical abscisic acid responsive elements as well as abscisic acid responsive element-like sequences in the vegetative tissues of indica rice cultivars. Plant Cell Report 27:779–794.
33. McCord JM, Fridovich I (1969) Superoxide dismutase, an enzymatic function for erythrocuprein. J Biol Chem 244:6049–6055.
34. Badwey JA, Robinson JM, Karnovsky MJ, Karnovsky ML (1981) Superoxide production by an unusual aldehyde oxidase in guinea pig granulocytes. Characterization and cytochemical localization. J Biol Chem 256:3479–3486.
35. Chen Z, Gallie DR (2006) Dehydroascorbate reductase affects leaf growth, development, and function. Plant Physiol 142:775–787.
36. Baier M, Dietz KJ (1999) Protective function of chloroplast 2-Cysteine peroxiredoxin in photosynthesis. Evidence from transgenic Arabidopsis. Plant Physiol 119:1407–1414.
37. Chen C, Lu S, Chen Y, Wang Z, Niu Y, et al. (2009) A gamma-ray–induced dwarf mutant from seeded bermudagrass and its physiological responses to drought stress. J Amer Soc Hort Sci 134:22–30.
38. Sirover MA (1999) New insights into an old protein: the functional diversity of mammalian glyceraldehyde-3-phosphatedehydrogenase. Biochem Biophysics Acta 1432: 159–184.
39. Dastoor Z, Dreyer JC (2001) Potential role of nuclear translocation of glyceraldehyde 3-phosphate dehydrogenase in apoptosis and oxidative stress. J Cell Sci 114:1643–1653.

40. Sweetlove LJ, Heazlewood JL, Herald V, Holtzapffel R, Day DA, et al. (2002) The impact of oxidative stress on Arabidopsis mitochondria. Plant J 32:891–904.
41. Hancock TJ, Henson D, Nyirenda M, Desikan R, Harrison J, et al. (2005) Proteomic identification of glyceraldehyde 3-phosphate dehydrogenase as an inhibitory target of hydrogen peroxide in Arabidopsis. Plant Physiol Biochem 43:828–835.
42. Ralser M, Wamelink MM, Kowald A, Gerisch B, Heeren G, et al. (2007) Dynamic rerouting of the carbohydrate flux is key to counteracting oxidative stress. J Biol 6:10.
43. Mundree SG, Baker B, Mowl S, Peters S, Marais S, et al. (2002) Physiological and molecular insights into drought tolerance. Afric J Biotech 1:28–38.
44. Taji T, Ohsumi C, Iuchi S, Seki M, Kasuga M, et al. (2002) Important roles of drought- and cold-inducible genes for galactinol synthase in stress tolerance in Arabidopsis thaliana. Plant J 29:417–426.
45. Kishor PBK, Hong Z, Miao GH, Hu CAA, Verma DPS (1995) Overexpression of delta1-pyrroline-5-carboxylate synthetase increases proline production and confers osmotolerance in transgenic plants. Plant Physiol 108:1387–1394.
46. Hare PD, Cress WA (1997) Metabolic implications of stress-induced proline accumulation in plants. Plant Growth Reg 21:79–102.
47. Smirnoff N, Cumbes QJ (1989) Hydroxyl radical scavenging activity of compatible solutes. Phytochemistry 28:1057–1060.
48. Verbruggen N, Hermans C (2008) Proline accumulation in plants: a review. Amino Acids 35:753–759.
49. Porankiewicz J, Wang J, Clarke AK (1999) New insights into the ATPdependent ClpP protease: E. coli and beyond. Mol Microbiol 32:449–58.
50. Clarke A (1999) ATP-dependent Clp proteases in photosynthetic organisms-A cut above the rest! Ann Bot 83:593–599.
51. Lai LB, Tausta SL, Nelson TM (2002) Differential regulation of transcripts encoding cytosolic NADP-malic enzyme in C3 and C4 Flaveria species. Plant Physiol 128:140–149.
52. Kizis D, Lumbreras V, Pages M (2001) Role of AP2/EREBP transcription factors in gene regulation during abiotic stress. FEBS Letters 498:187–189.
53. Liu Q, Kasuga M, Sakuma Y, Abe H, Miuraa S, et al. (1998) Two transcription factors, DREB1 and DREB2, with an EREBP/AP2 DNA binding domain separate two cellular signal transduction pathways in drought- and low-temperature-responsive gene expression, respectively, in Arabidopsis. Plant Cell 10: 1391–1406.
54. Kasuga M, Liu Q, Miura S, Yamaguchi-Shinozaki K, Shinozaki K (1999) Improving plant drought, salt, and freezing tolerance by gene transfer of a single stress-inducible transcription factor. Nature Biotech 17:287–291.
55. Somssich IE (2006) Networks of transcriptional regulation underlying plant defense responses towards phytopathogens. In Grasser KD. eds. Regulation of Transcription in Plants. Blackwell Publishing, pp 266–284.
56. Wu X, Shiroto Y, Kishitani S, Ito Y, Toriyama K (2009) Enhanced heat and drought tolerance in transgenic rice seedlings overexpressing OsWRKY11 under the control of HSP101 promoter. Plant Cell Report 28:21–30.
57. Yang J, Zhang J, Wang Z, Zhu Q, Liu L (2001). Water deficit–induced senescence and its relationship to the remobilization of pre-stored carbon in wheat during grain filling. Agron J 93:196–206.
58. Munné-Bosch S, Alegre L (2004) Die and let live: leaf senescence contributes to plant survival under drought stress. Func Plant Biol 31:203–216

Permissions

All chapters in this book were first published in PLOS ONE, by The Public Library of Science; hereby published with permission under the Creative Commons Attribution License or equivalent. Every chapter published in this book has been scrutinized by our experts. Their significance has been extensively debated. The topics covered herein carry significant findings which will fuel the growth of the discipline. They may even be implemented as practical applications or may be referred to as a beginning point for another development.

The contributors of this book come from diverse backgrounds, making this book a truly international effort. This book will bring forth new frontiers with its revolutionizing research information and detailed analysis of the nascent developments around the world.

We would like to thank all the contributing authors for lending their expertise to make the book truly unique. They have played a crucial role in the development of this book. Without their invaluable contributions this book wouldn't have been possible. They have made vital efforts to compile up to date information on the varied aspects of this subject to make this book a valuable addition to the collection of many professionals and students.

This book was conceptualized with the vision of imparting up-to-date information and advanced data in this field. To ensure the same, a matchless editorial board was set up. Every individual on the board went through rigorous rounds of assessment to prove their worth. After which they invested a large part of their time researching and compiling the most relevant data for our readers.

The editorial board has been involved in producing this book since its inception. They have spent rigorous hours researching and exploring the diverse topics which have resulted in the successful publishing of this book. They have passed on their knowledge of decades through this book. To expedite this challenging task, the publisher supported the team at every step. A small team of assistant editors was also appointed to further simplify the editing procedure and attain best results for the readers.

Apart from the editorial board, the designing team has also invested a significant amount of their time in understanding the subject and creating the most relevant covers. They scrutinized every image to scout for the most suitable representation of the subject and create an appropriate cover for the book.

The publishing team has been an ardent support to the editorial, designing and production team. Their endless efforts to recruit the best for this project, has resulted in the accomplishment of this book. They are a veteran in the field of academics and their pool of knowledge is as vast as their experience in printing. Their expertise and guidance has proved useful at every step. Their uncompromising quality standards have made this book an exceptional effort. Their encouragement from time to time has been an inspiration for everyone.

The publisher and the editorial board hope that this book will prove to be a valuable piece of knowledge for researchers, students, practitioners and scholars across the globe.

List of Contributors

Salme Timmusk, Islam A. Abd El-Daim, Elna Stenström and Lawrence Behers
Dept. of Forest Mycology and Plant Pathology, Uppsala BioCenter, SLU, Uppsala, Sweden

Lucian Copolovici, Triin Tanilas and Astrid Kännaste
Dept. of Plant Physiology, Estonian University of Life Sciences, Tartu, Estonia

Eviatar Nevo
Institute of Evolution, The University of Haifa, Haifa, Israel

Gulaim Seisenbaeva
Dept. of Chemistry and Biotechnology, Uppsala BioCenter, SLU, Uppsala, Sweden

Ülo Niinemets
Dept. of Plant Physiology, Estonian University of Life Sciences, Tartu, Estonia
Estonian Academy of Sciences, Tallinn, Estonia

Marisa Mazari-Hiriart
Laboratorio Nacional de Ciencias de la Sostenibilidad, Departamento de Ecología de la Biodiversidad, Instituto de Ecología, Universidad Nacional Autónoma de México, México, D.F., México

Gustavo Pérez-Ortiz
Laboratorio Nacional de Ciencias de la Sostenibilidad, Departamento de Ecología de la Biodiversidad, Instituto de Ecología, Universidad Nacional Autónoma de México, México, D.F., México
Departamento de Biología, Facultad de Ciencias, Universidad Nacional Autónoma de México, México, D.F., México

María Teresa Orta-Ledesma and Isaura Yañez-Noguez
Coordinación de Ingeniería Ambiental, Instituto de Ingeniería, Universidad Nacional Autónoma de México, México, D.F., México

Felipe Armas-Vargas
Posgrado en Ciencias de la Tierra, Instituto de Geología, Universidad Nacional Autónoma de México, México, D.F., México

Rosa Solano-Ortiz
Posgrado en Ciencias Bioló gicas, Facultad de Ciencias-Instituto de Ecología, Universidad Nacional Autónoma de México, México, D.F., México

Marco A. Tapia
Departamento de Biología, Facultad de Ciencias, Universidad Nacional Autónoma de México, México, D.F., México
Posgrado en Ciencias Bioló gicas, Facultad de Ciencias-Instituto de Ecología, Universidad Nacional Autónoma de México, México, D.F., México

Miguel A. Silva
Posgrado de Ciencias Bioquímicas, Departamento de Ingeniería Celular y Biocatálisis, Instituto de Biotecnología, Universidad Nacional Autónoma de México, Cuernavaca, Morelos, México

Yolanda López-Vidal
Programa de Inmunología Molecular Microbiana, Departamento de Microbiología y Parasitología, Facultad de Medicina, Universidad Nacional Autónoma de México, México, D.F., México

Carlos Díaz-Ávalos
Departamento de Probabilidad y Estadística, Instituto de Investigaciones en Matemáticas Aplicadas y en Sistemas, Universidad Nacional Autónoma de México, México, D.F., México

Zhuoyu Wang and Fenny Dane
Department of Horticulture, Auburn University, Alabama, United States of America

Hongtao Hu and Leslie R. Goertzen
Department of Biological Sciences, Auburn University, Alabama, United States of America

J. Scott McElroy
Department of Crop, Soil and Environmental Sciences, Auburn University, Alabama, United States of America

Nora Honsdorf
Chair of Plant Breeding, Institute of Agricultural and Nutritional Sciences, Martin-Luther University Halle-Wittenberg, Halle (Saale), Germany
Interdisciplinary Center for Crop Plant Research (IZN), Halle (Saale), Germany

Timothy John March
School of Agriculture, Food and Wine, University of Adelaide, Waite Campus, Adelaide, Australia

Bettina Berger
The Plant Accelerator, University of Adelaide, Waite Campus, Adelaide, Australia

Mark Tester
Center for Desert Agriculture, King Abdullah University of Science and Technology, Thuwal, Saudi Arabia

Klaus Pillen
Chair of Plant Breeding, Institute of Agricultural and Nutritional Sciences, Martin-Luther University Halle-Wittenberg, Halle (Saale), Germany

Fuhui Xu, Zhixue Liu, Hongyan Xie and Jian Zhu
School of Life Sciences and Technology, Tongji University, Shanghai, China

Juren Zhang
School of Life Science, Shandong University, Shandong, China

Josef Kraus, Tasja Blaschnig and Reinhard Nehls
KWS SAAT AG, Einbeck, Germany

Hong Wang
School of Life Sciences and Technology, Tongji University, Shanghai, China
KWS SAAT AG, Einbeck, Germany

Pengsen Sun and Shirong Liu
Institute of Forest Ecology, Environment and Protection, Chinese Academy of Forestry, Beijing, P. R. China

Said Wahbi
Laboratoire de Biotechnologie et Physiologie Végétale, Faculté des Sciences Semlalia, Université Cadi Ayyad, Marrakech, Morocco

Tsonko Tsonev
Institute of Plant Physiology and Genetics, Bulgarian Academy of Sciences, Sofia, Bulgaria

Matthew Haworth and Mauro Centritto
Trees and Timber Institute, National Research Council, Sesto Fiorentino, Florence, Italy

Agnes Agunos and David Léger
Laboratory for Foodborne Zoonoses, Public Health Agency of Canada, Guelph, Ontario, Canada

Lisa Waddell
Laboratory for Foodborne Zoonoses, Public Health Agency of Canada, Guelph, Ontario, Canada
Department of Population Medicine, University of Guelph, Guelph, Ontario, Canada

Eduardo Taboada
Laboratory for Foodborne Zoonoses, Public Health Agency of Canada, Lethbridge, Alberta, Canada

Hanna Rintala, Sirpa Hakkarainen, Juhani Merentie and Martti Pärnänen
Siilinjärvi municipality, Finland

Leena Maunula, Joana Revez and Marja-Liisa Hänninen
Department of Food Hygiene and Environmental Health, University of Helsinki, Helskinki, Finland

Vicente Gomez-Alvarez, Hodon Ryu and Jorge W. Santo Domingo
Office of Research and Development, United States Environmental Protection Agency, Cincinnati, Ohio, United States of America

Pia Räsänen, Sallamaari Siponen, Outi Nyholm, Aino Kyyhkynen, Markku Kuusi, Anja Siitonen, Marja Palander, Jukka Ollgren and Katri Jalava
Department of Infectious Disease Surveillance and Control, National Institute for Health and Welfare, Helsinki, Finland

Raisa Loginov and Jenni Antikainen
Laboratory HUSLAB, Helsinki University Hospital, Helsinki, Finland

Ilkka Miettinen, Tarja Pitkänen and Ari Kauppinen
Department of Environmental Health, National Institute for Health and Welfare, Kuopio, Finland

Roberto O. Chávez, Jan G. P. W. Clevers, Jan Verbesselt and Martin Herold
Laboratory of Geo-Information Science and Remote Sensing, Wageningen University, Wageningen, The Netherlands

Paulette I. Naulin
Laboratorio de Biología de Plantas, Departamento Silvicultura y Conservación de la Naturaleza, Universidad de Chile, Santiago, Chile

Hui Xia, Xiaoguo Zheng, Liang Chen, Huan Gao, Hua Yang, Ping Long, Jiajia Li and Lijun Luo
Shanghai Agrobiological Gene Center, Shanghai, China

Jun Rong
Center for Watershed Ecology, Institute of Life Science and Key Laboratory of Poyang Lake Environment and Resource Utilization, Ministry of Education, Nanchang University, Nanchang, China

Baorong Lu
Ministry of Education Key Laboratory for Biodiversity and Ecological Engineering, Fudan University, Shanghai, China

Ji-Yeon Kim
Department of Pediatric Dentistry, Dental Research Institute, Pusan National University, Pusan, Korea

Yan Zhang, Yukiko Nakano, Li Zhu and Pamela K. Den Besten
Department of Orofacial Sciences, University of California San Francisco, San Francisco, California, United States of America

Orapin Horst and Sunita Ho
Department of Preventive and Restorative Dental Sciences, University of California San Francisco, San Francisco, California, United States of America

Ralf J. Radlanski
Charité- Campus Benjamin Franklin at Freie Universität Berlin, Center for Dental and Craniofacial Sciences, Department of Craniofacial Developmental Biology, Berlin, Germany

Esther Kohler, Jörg Villiger, Thomas Posch, Tanja Shabarova, Jakob Pernthaler and Judith F. Blom
Limnological Station, Institute of Plant Biology, University of Zurich, Kilchberg, Switzerland

Nicolas Derlon
Eawag: Swiss Federal Institute of Aquatic Science and Technology, Dübendorf, Switzerland

Eberhard Morgenroth
Eawag: Swiss Federal Institute of Aquatic Science and Technology, Dübendorf, Switzerland
Institute of Environmental Engineering, ETH Zurich, Zurich, Switzerland

Eva L. van Donkelaar and Harry W. M. Steinbusch
School for Mental Health and Neuroscience, Faculty of Health, Medicine and Life Sciences, Maastricht University, Maastricht, The Netherlands

Koen R. D. Vaessen
School for Mental Health and Neuroscience, Faculty of Health, Medicine and Life Sciences, Maastricht University, Maastricht, The Netherlands
Central Laboratory Animal Research Facility, Faculty of Veterinary Medicine, University of Utrecht, Utrecht, The Netherlands

Jodi L. Pawluski
School for Mental Health and Neuroscience, Faculty of Health, Medicine and Life Sciences, Maastricht University, Maastricht, The Netherlands
University of Liege, GIGA-Neurosciences, Liège, Belgium
Department of Biological Sciences, Irvine Hall, Ohio University, Athens, Ohio, United States of America

Arjan Blokland
Department of Neuropsychology and Psychopharmacology, Faculty of Psychology and Neuroscience, Maastricht University, Maastricht, The Netherlands

Annerieke S. Sierksma
School for Mental Health and Neuroscience, Faculty of Health, Medicine and Life Sciences, Maastricht University, Maastricht, The Netherlands
VIB Center for the Biology of Disease, Leuven, Belgium
Center for Human Genetics, KU Leuven, Leuven, Belgium

Ramón Cañete
Unidad de Endocrinología Pediátrica del Hospital Universitario Reina Sofía, Instituto Maimónides de Investigación Biomédica de Córdoba (IMIBIC), Córdoba, Spain

Aline Lima Leite
Department of Biological Sciences, Bauru Dental School, University of São Paulo, Bauru, São Paulo, Brazil

Janete Gualiume Vaz Madureira Lobo, Marília Afonso Rabelo Buzalaf, Tatiani Martini and Fernanda Zucki
Department of Biological Sciences, Bauru Dental School, University of São Paulo, Bauru, São Paulo, Brazil

Heloísa Aparecida Barbosa da Silva Pereira and Mileni Silva Fernandes
Department of Genetics and Evolution, Center of Biological Sciences and the Health, Federal University of São Carlos, São Carlos, São Paulo, Brazil

Dóris Hissako Sumida
Department of Basic Sciences, Araçatuba Dental School, UNESP - Universidade Estadual Paulista, Araçatuba, São Paulo, Brazil

Alfredo Rigalli
Bone Biology and Mineral Metabolism Laboratory, National Scientific and Technical Research Council (CONICET) School of Medicine, Rosario National University, Rosario, Santa Fe, Argentina

Lenka Benova and Oona M. R. Campbell
London School of Hygiene and Tropical Medicine, Faculty of Epidemiology and Population Health, London, United Kingdom

Oliver Cumming
London School of Hygiene and Tropical Medicine, Faculty of Infectious and Tropical Diseases, London, United Kingdom

Bruce A. Gordon
World Health Organization, Geneva, Switzerland

Janis Rubulis and Talis Juhna
Department of Water Engineering and Technology, Riga Technical University, Riga, Latvia

Moke Magoma
Evidence for Action (E4A), Department of Obstetrics and Gynaecology, Bugando Medical Centre and Teaching Hospital, Mwanza, Tanzania

Alina Nescerecka
Department of Water Engineering and Technology, Riga Technical University, Riga, Latvia,
Department of Environmental Microbiology, Eawag, Swiss Federal Institute for Aquatic Science and Technology, Dübendorf, Switzerland

Marius Vital and Frederik Hammes
Department of Environmental Microbiology, Eawag, Swiss Federal Institute for Aquatic Science and Technology, Dübendorf, Switzerland

Bin Zhang, Xiaoping Chang and Ruilian Jing
National Key Facility for Crop Gene Resources and Genetic Improvement/Institute of Crop Science, Chinese Academy of Agricultural Sciences, Beijing, China

Weiyu Li
National Key Facility for Crop Gene Resources and Genetic Improvement/Institute of Crop Science, Chinese Academy of Agricultural Sciences, Beijing, China
College of Agronomy, Shanxi Agricultural University, Taigu, China

Runzhi Li
College of Agronomy, Shanxi Agricultural University, Taigu, China

Kali B. Nelson
Department of Epidemiology and Biostatistics, Milken Institute School of Public Health, The George Washington University, Washington, D.C., United States of America

Jonathan Karver
Washington, D.C., United States of America

Craig Kullman
World Bank Water and Sanitation Program, Washington, D.C., United States of America

Jay P. Graham
Department of Environmental and Occupational Health and Department of Global Health, Milken Institute School of Public Health, The George Washington University, Washington, D.C., United States of America

Bart Van der Bruggen
ProcESS, Department of Chemical Engineering, KU Leuven, Leuven, Belgium

Gomotsegang F. Molelekwa
ProcESS, Department of Chemical Engineering, KU Leuven, Leuven, Belgium
Department of Environmental Health, Tshwane University of Technology, Pretoria, South Africa

Murembiwa S. Mukhola
Department of Environmental Health, Tshwane University of Technology, Pretoria, South Africa

Patricia Luis
Materials and Process Engineering (iMMC-IMAP), Université Catholique de Louvain, Louvain-la-Neuve, Belgium

Peng Zhou
School of Agriculture and Biology, Shanghai Jiao Tong University, Shanghai, P. R. China
Department of Plant Biology and Pathology, Rutgers, the State University of New Jersey, New Brunswick, New Jersey, United States of America

Yuan An, Zhaolong Wang and Hongmei Du
School of Agriculture and Biology, Shanghai Jiao Tong University, Shanghai, P. R. China

Bingru Huang
Department of Plant Biology and Pathology, Rutgers, the State University of New Jersey, New Brunswick, New Jersey, United States of America

Index